**Wireless Power Transfer for
Electric Vehicles and Mobile Devices**

Wireless Power Transfer for Electric Vehicles and Mobile Devices

Chun T. Rim
Gwangju Institute of Science and Technology, South Korea

Chris Mi
San Diego State University, California, USA

IEEE PRESS

WILEY

Registered Offices
John Wiley & Sons, Inc., 111 River Street, Hoboken, NJ 07030, USA
John Wiley & Sons Ltd, The Atrium, Southern Gate, Chichester, West Sussex, PO19 8SQ, UK

Editorial Office
The Atrium, Southern Gate, Chichester, West Sussex, PO19 8SQ, UK

For details of our global editorial offices, customer services, and more information about Wiley products visit us at www.wiley.com.

Wiley also publishes its books in a variety of electronic formats and by print-on-demand. Some content that appears in standard print versions of this book may not be available in other formats.

Library of Congress Cataloging-in-Publication data is available for this book.

ISBN 9781119329053 (hardback)

Cover Design: Wiley
Cover Images: (electric car) © Chesky_W/Gettyimages;(smartphone and smart watch) © lcs813/Gettyimages

Set in 10/12pt WarnockPro by Aptara Inc., New Delhi, India

10 9 8 7 6 5 4 3 2 1

Contents

Preface

We cannot talk about wireless power transfer (WPT) without mentioning Nikola Tesla, who carried out historic experiments on WPT and wireless communications a hundred years ago. Even though he failed in sending wireless power to another continent, he stimulated the idea of sending electric power wirelessly. Nearly every trial and effort on WPT since then is within the scope or originated from Nikola Tesla's idea. One of his inventions, the three-phase induction motor, is a kind of WPT because the stator and rotor of an induction motor constitute a transmitter (Tx) and receiver (Rx) set, forming a transformer. A transformer is also a kind of WPT where the primary and secondary windings are the Tx and Rx of a WPT system. It is amazing that a loosely coupled transformer appears in the equivalent circuit of an inductive power transfer (IPT) system, which is currently the most widely used WPT technology. Note that AC transformers made it possible for Nikola Tesla to electrify human society – in the light of this, we can say that WPT has been one of the roots of modern society.

The authors of this book have been interested in WPT since 2008, which is just one year after the renowned experiment by Soljacic at MIT in sending 60 W power wirelessly over a 2.1 m distance.

Dr Chun Rim started with IPT for road-powered electric vehicles (RPEVs), when the Partners for Advance Transit and Highways of California (PATH) team had already shown their RPEV buses in the 1990s. As a power electronic engineer, he was not fully convinced of the success of RPEVs at first after reviewing a final report on the development of the PATH project. The low power efficiency of 60% and small airgap of 7.6 cm of the PATH team looked impractical for commercialization. He cannot but mention Dr Nam P. Suh, the former President of KAIST, who insisted that RPEV is one of the most competitive solutions for the commercialization of electric vehicles (EVs), compared to purely battery-powered EVs. With his strong support and guidance, the development and commercialization of the Korean version of RPEV, called an on-line electric vehicle (OLEV), had been accelerated. Dr Rim became, fortunately, one of the core members of this nationwide project valued at more than $40 million US dollars. He was one of the key team members responsible for developing IPT systems for OLEVs and developed four generations of IPT systems for OLEV until 2013, at which time he embarked on independently developing his own version of the fifth generation IPT system. He has developed U-, W-, I-, and S-type power rails for OLEV buses, sports utility vehicles, and trains. The OLEV bus and train were deployed and commercialized in 2014 and 2010, respectively. OLEV was selected as one of 'The 50 Best Inventions of 2010' by TIME and 'The first of 10 Emerging Technologies in 2013' by the World Economic Forum.

After succeeding in the development of OLEVs, Dr Rim expanded his research areas in WPT to stationary EV charging, mobile device charging, robot power, and drone power. In 2013, he showed for the first time that Soljacic's coupled magnetic resonance of four coils is not new and can be equivalently degenerated to the conventional IPT of two coils. Moreover, he demonstrated that a 5 m-long-distance WPT system of 209 W can be achieved by a conventional IPT with innovative dipole coils. Recently, he attained 10 W using IPT at 12 m distance using the same technology. By adopting cross-dipole coils to IPT, six-degree-of-freedom (6 DoF) mobile power was achieved for the first time for both plane-type Tx and Rx in 2015, enabling the provision of continuous wireless power in free space. A series of dipole-coil-based IPT that he has developed shows the possibility of replacing canonical loop-coil-based IPT, because the dipole coil reduces the coil dimension of the loop coil from two to one. In this manner, two-dimensional dipole coils can generate the same magnetic flux distribution as three-dimensional loop coils. This dimensionless characteristic is crucial for compact and long-range applications such as the Internet of Things (IoT) and wearables. Dr Rim is currently developing special purpose WPT technologies such as a wireless slip ring, wireless nuclear instrumentation, and synthesized magnetic field focusing (SMF), which can improve the resolution of a magnetic field by synthesizing individually controlled currents of arrayed coils that he invented in 2013.

Besides new developments in the aforementioned WPT technologies, Dr Rim has also envisioned new theories for WPT, as he found that there are few theories available for WPT. Maxwell equations cannot be directly used for coil design and the magnetic field canceling problem of WPT. This is why he has developed a magnetic mirror model for coil design and a gyrator circuit model for resonant circuit design in WPT. Phasor transformation that he developed a long time ago has been adopted for the analysis and design of static and dynamic WPT.

In parallel, Dr Chris Mi started his independent research in WPT in 2008. In the past 8 years, his team has developed a number of unique topologies of WPT systems, including the double-sided LCC-compensated topology, the large-power capacitive wireless power transfer (CPT) technology, low-ripple dynamic WPT, etc. The LCC topology reached 8 kW at a 150 mm distance and efficiency of more than 96%. The dynamic charging based on the IPT principle had an out-power ripple of less than 5%. The CPT system reached a power level of 2.5 kW at a distance of 150 mm, with an efficiency of more than 92%. A combined IPT and CPT further improved the system efficiency to 96%.

WPT is one of the hottest topics being actively studied and is being widely commercialized. In particular, there has been a rapid expansion of WPT in mobile phone chargers and stationary EV chargers, and RPEV has drawn worldwide attention recently. It is broadly expected that the WPT industry will grow persistently in the coming decades.

Unfortunately, there are not many books on WPT that comprehensively explain the subject from fundamental theories to industry applications. One of the reasons is the rapid change of technologies in WPT. Even though the authors' expertise cannot cover all areas of WPT, they finally decided to write a WPT book in 2016 as a textbook for the biannual lecture of Wireless Power Electronics at KAIST. Dr Rim is the main author who wrote most of the book except a chapter that was written by Dr Mi who also reviewed and edited the whole book. Hence, 'I' in the book means Dr Rim if not specified otherwise.

Considering the rapid evolution of WPT, this book should be revised in a few years, but we hope that the first edition of this book will provide fruitful state-of-the-art technical insights with a strong basis of theories on WPT. This book is intended to be used as a textbook for graduate students or senior undergraduate students, but it can also be used as a design and analysis guide book for industry engineers.

The authors hope that the readers of this book learn that engineering requires philosophical thoughts and fundamental theories. Innovative ideas and century-lasting inventions come from profound understanding of design principles and experiments. In particular, engineers should neither make too much of experience nor underestimate theories. Actually, people still do not have sufficiently good theories in WPT for magnetic and electric field estimation when metal or core is involved.

In Part I, concepts of mobile power electronics, wireless power, and EVs are briefly introduced.

In Part II, theories for IPT such as the coupled inductor model, gyrator circuit model, magnetic mirror model, and general unified dynamic phasor are explained, where the last three theories are developed by Dr Rim. This part could be difficult for beginners of WPT and can be skipped until a theoretical background is strongly needed.

In Part III, IPTs for RPEVs are widely covered, making this part the most important portion of the book. From RPEV history to OLEV-IPT, technologies on I-type and S-type as well as specific design issues, such as the controller, compensation circuit, electromagnetic field cancel, large tolerance, and power rail segmentation, are extensively explained. Introduction to dynamic charging is primarily written by Nam P. Suh, former President of KAIST.

In Part IV, IPTs for static charging for EVs are explained. Large tolerance and capacitive charging, issues are addressed, written by Chris Mi. Foreign object detection techniques are introduced in this part, which can be used in dynamic charging. Considering the wide body of literature on EV charging, this part is relatively short compared to the treatment of dynamic charging because of the focus on mobile power transfer of this book. Numerous references are provided for the reader interested in more details on static charging than we have had space to include here. However, many design issues in static charging are similar to those in dynamic charging; therefore, the design principles and issues in Part III can be applicable to this part. Note that the reverse is not true, because static charging is a special form of dynamic charging whereas dynamic charging is not just an extension of static charging.

In Part V, IPT mobile applications for phones and robots are covered, starting from a review of a coupled magnetic resonance system. Mid-range and long-range IPT, as well as free space omnidirectional IPT by dipole coils, are explained and two-dimension omnidirectional IPT for robots is addressed.

In Part VI, special applications of WPT such as SMF and wireless nuclear instrumentation are explained and the future of WPT is discussed.

A few Questions and Problems are included in each chapter, which are intended to be potential research topics.

In the near future, almost every valuable object including human beings will be connected to worldwide networks, which is why the IoT has drawn so much attention in transportation, logistics, securities, public services, home appliances, factory automation, military, health care, robots, drones, and many other fields. Because the IoT comprises sensors, communication devices, and power sources, WPT will play significantly important roles in the future IoT. Considering relatively mature technologies in sensors and communication, it can be said that

"Wireless power will be crucial to the future of IoT."

Since this is the first edition of the book, the authors welcome any input and comments from readers (http://tesla.gist.ac.kr) and will ensure that any corrections or amendments as needed are incorporated into future editions.

The authors are grateful to all those who helped to complete the book. In particular, a large portion of the material presented is the result of many years of work by the authors as well as other members of their research groups. Thanks are due to the many dedicated staff and graduate students who made enormous contributions and provided supporting material to this book.

The authors also owe debts of gratitude to their families, who gave tremendous support and made sacrifices during the process of writing this book.

Sincere acknowledgment is made to various sources that granted permission to use certain materials or pictures in this book. Acknowledgments are included where those materials appear. The authors used their best efforts to get approval to use those materials that are in the public domain and on open Internet web sites. If any of these sources were missed, the authors apologize for that oversight and will rectify this in future editions of the book if brought to the attention of the publisher. The names of any product or supplier referred to in this book are provided for information only and are not in any way to be construed as an endorsement (or lack thereof) of such a product or supplier by the publisher or the authors.

Finally, the authors are extremely grateful to John Wiley & Sons, Ltd and its editorial staff for giving them the opportunity to publish this book and helping in all possible ways, and to Mr Ji H. Kim at KAIST who helped to prepare the manuscript of this book.

January 1, 2017

Professor Chun T. Rim, GIST, South Korea
Professor Chris C. Mi, SDSU, USA

Part I

Introduction

This part introduces the concept of mobile power electronics and the very basic knowledge related to wireless power transfer. The most fundamental principles and philosophy of mobile power electronics and wireless power transfer will be explained.

1

Introduction to Mobile Power Electronics

1.1 General Overview of Mobile Power Electronics

The methods of power transfer for various sources and loads have evolved since the advent of electricity in the nineteenth century. As shown in Figure 1.1, more and more loads are movable now and it has become important to provide seamless power to moving things such as electric transportation, robots, and electric airplanes. Currently, we mainly rely on electric cords and batteries to provide power to movables. As we notice daily, smartphones, tablets, and desktop computers should operate continuously even in the event of disconnection of utility power. The electric cord, however, has a limited range of powering and the battery has a limited time of powering; hence, they inevitably accompany anxiety of range and time. It is important to overcome this range and time limitation for movable things. This was the motivation for "mobile power electronics," a term the author (Dr Rim) coined in 2010. In this light, the motto of mobile power electronics can be said to be "to supply electric energy to all movable things freely."

In general, power transfer (PT) can be classified as stationary and mobile depending on the movement of power receiving (Rx) loads, as shown in Figure 1.2. Stationary PT (SPT) traditionally has been used in the major form of electricity use, which includes fixed SPT of a firmly unchanged configuration of power systems and detachable SPT of a variable configuration of power systems. A majority of power use is still fixed STP such as high-voltage power lines, street lights, and home appliances. Nowadays, detachable STP is more widely used to charge movable things such as cable-type electric vehicles (EVs) and electric shavers, where an electric cord with a naked contact is used. These types of plugged-in chargers have an inconvenient user interface and bring exposure to potential danger of electric shock and fire.

To cope with the strong demand for mobility of Rx loads, various mobile PT (MPT) technologies have been studied; they can be further classified as close MPT and remote MPT depending on the range between the power transmitting (Tx) source and the Rx loads. For the closed MPT, the WPT range is usually from a few cm to a few m. It is remarkable that the inductive, capacitive, and conductive PT correspond to L, C, and R circuit components, respectively. Each close PT uses inductive coupling, capacitive coupling, and conductive coupling between the Tx and Rx. Among the close MPTs, inductive PT (IPT) has been used widely due to its high power transfer capability at relatively low frequency, whereas capacitive PT (CPT) is not as commonly used due to its high operating frequency and small power transfer distance [1, 2]. Note that conductive PT was widely used for a century as a practical means for mobile PT until the advent of IPT.

Among remote MPT strategies, radio frequency (RF) PT and optical PT have been researched to extend the range limit of other PT techniques [3, 4]. RF PT uses electromagnetic waves of frequency ranging from MHz to GHz in practice and is quite different from IPT. For example,

Wireless Power Transfer for Electric Vehicles and Mobile Devices, First Edition. Chun T. Rim and Chris Mi.
© 2017 John Wiley & Sons Ltd. Published 2017 by John Wiley & Sons Ltd.

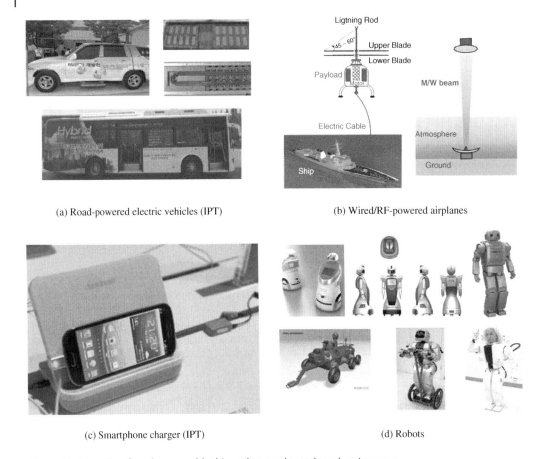

(a) Road-powered electric vehicles (IPT)

(b) Wired/RF-powered airplanes

(c) Smartphone charger (IPT)

(d) Robots

Figure 1.1 Examples of modern movable things that need seamless electric power.

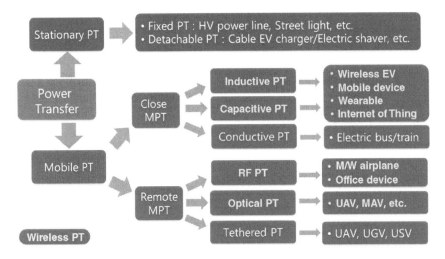

Figure 1.2 A general classification of power transfer in terms of mobility, distance, and means of powering.

the Rx power density of RF PT is usually proportional to the inverse of the square of distance, but that of IPT is typically proportional to the inverse of the sixth power of distance because the Rx magnetic flux density of IPT is typically proportional to the third power of distance. Furthermore, there is no magnetic coupling between the Tx and Rx devices of the RF PT. On the other hand, the tethered PT can provide power over a flexibly long distance if properly designed [5, 6]. As depicted in Figure 1.2, wireless PT (WPT) is not only limited to close MPT such as IPT and CPT but also remote MPT such as RF PT and optical PT. Furthermore, WPT is not only electrical but also optical or even acoustic.

In the era of the ubiquitous, IPT is the most widely used [8, 58]. More mobile devices, home appliances, industry sensors, and EV chargers are becoming wireless due to their convenience, safety against electric shock, cleanness, and competitive power efficiency and price. Eventually, most devices including wearable devices, ubiquitous sensors, and smart cars will merge to the Internet of Things (IoT) and WPT will play a significantly important role in the realization of IoT, which includes compact communication devices, sensors, and power sources.

Question 1 (1) How can you classify electric shavers and vacuum cleaners that have a stretchable cable? (2) Are the items of (1) SPT or MPT? (3) What are the benefits of the classification of SPT and MPT? (4) Is there any fundamental distinction between SPT and MPT as well as IPT, CPT, RF PT, and Optical PT?

1.2 Brief History of Mobile Power Electronics

We cannot discuss mobile power or wireless power without talking about Nikola Tesla, who had carried out many experiments on WPT, as shown in Figure 1.3, and invented a "world system" for "the transmission of electrical energy without wires." Even though he did not succeed in transmitting wireless power over the continent as he desired, he has inspired numerous engineers and scientists with respect to transferring power remotely without wires. Nikola Tesla was the pioneer of wireless telegraphy, which is known to be competitively invented by Guglielmo Marconi in 1895. There were many competitors at that time in utilizing electromagnetic waves, which were discovered by Heinrich Hertz in 1886, for telecommunications. Including his work

Figure 1.3 Nikola Tesla (1856–1943) and his experiment on WPT.

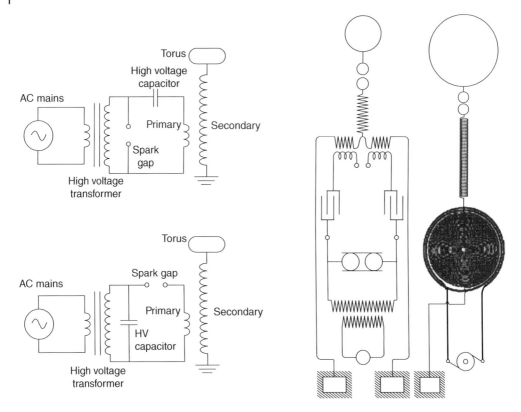

Figure 1.4 Examples of the Tesla coil, which is a type of resonant transformer (25 kHz–2 MHz), invented in 1891.

on three-phase AC power systems and induction motors, Nikola Tesla made the most significant contributions to the era of electricity of the twentieth century.

Nikola Tesla was very interested in AC magnetic fields, which were the basis of many of his
inventions, such as wireless communications, induction motors, and WPT systems. As easily
recognized by experienced engineers, magnetic fields are very difficult to deal with compared
to electric fields and electric circuits. The design of magnets and coils is regarded as one of the
most challenging tasks in electrical engineering. A Tesla coil, as shown in Figure 1.4, is one of
the examples that is difficult to design and to understand in terms of its behavior. He invented
this coil as a means of generating a high-frequency and high-voltage power source and it is
still being used to generate electric sparks. At that time, there were no semiconductor switches
that could withstand high voltage, so he used a mechanical switch to ignite resonant ringing of
LC circuits. Utilizing the parasitic capacitances and inductances of the secondary transformer,
he boosted the resonant voltage to several tens of kV. This "resonant transformer" is a unique
characteristic of the Tesla coil and quite different from conventional transformers, which are
close to an ideal transformer. For example, the output voltage of a resonant transformer is not
simply proportional to the turn ratio of primary and secondary windings of the transformer
and it usually becomes much higher than the turn ratio when tuned. This strange phenomenon
has been a mysterious issue in the design of the Tesla coil and can be explained by theories such
as the coupled inductor model and the gyrator circuit model in Part II of this book.

Understanding the Tesla coil is a good start in understanding WPT and gives us many useful
design tips for WPT because the Tesla coil design involves transformer design, coil design, stray
inductance and capacitance modeling, compensation circuit design, insulation issues, ground

Figure 1.5 A conductively powered tram (left) and its pantograph (right), which is one of the oldest types of MPT.

issues, switching, and snubber circuit design. Nikola Tesla used the Tesla coil in Figure 1.4 for his wireless power experiments. For graduate students who wish to understand a "resonant transformer," I would like to recommend that they make a Tesla coil kit and conduct experiments with special caution regarding high-voltage electric shock when operating. Because of safety reasons, it is not recommended that young scientists deal with the Tesla coil even if the source voltage of the experiment kit is only a few volts. For beginners, it will be much easier and safer to carry out a WPT experiment with a two-coil IPT experiment kit.

As discussed above, one of the oldest types of MPT is conductive power for trams and trains, which fetch AC or DC utility power through a detachable pantograph, as shown in Figure 1.5. This conductively powered tram has been used for a century but is being replaced with battery-powered or wireless-powered trams. Because of cumbersome power lines in the air and the problem of pantograph wear, the conductively powered tram is no longer widely used in urban areas, but conductive power is still widely used in subway trains and high-speed trains in many countries due to the absence of an available candidate solution.

The last case that I would like to talk about in the history of mobile power electronics is the tethered electric helicopter, first built and flown by Gustave Trouvé in 1887 and developed for possible military use by the Nazis during World War II. This tethered drone is useful for continuous surveillance and watch missions without landing for several hours to several weeks if desired. Considering the rapid growth of drone markets, this tethered powering is noteworthy even though it has no relationship to WPT.

Note that MPT is not necessarily WPT but could take many different forms, which will be discussed in the following section. If we extend our concern to mobile energy transfer, batteries as well as petroleum, natural gas, coal, and hydrogen can be the means of energy transfer. In particular, the battery is a good means of electrical energy transfer and an excellent power source. Thus far, as we are dealing with MPT issues, we need to consider the battery as a means of delivering power continuously to remote locations as we now daily rely on it.

Question 2 (1) What is energy harvesting? (2) What are the merits and limitations of energy harvesting compared to MPT? (3) Discuss the potential applications of energy harvesting to IoT when combined with MPT.

1.3 Remote Mobile Power Transfer (MPT)

Because the close MPT of Figure 1.2 will be explained in Chapter 2, remote MPT is explained in this section briefly. One of the purposes of this section is to familiarize readers with other MPT techniques aside from conventional WPTs.

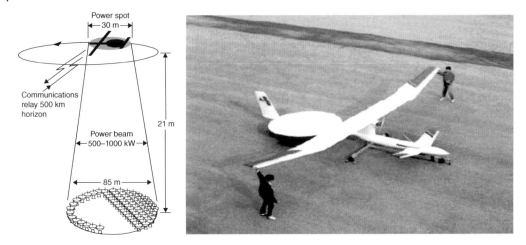

Figure 1.6 An RF-powered airplane by the Canada SHARP (Stationary High Altitude Relay Platform) project.

1.3.1 RF Power Transfer (RF PT)

RF power or energy has been widely used in radars, microwave ovens, electromagnetic pulse (EMP) weapons, and WPT. One of the potential applications of RF PT is wireless-powered airplanes, as shown in Figure 1.6.

According to the Canada SHARP (Stationary High Altitude Relay Platform) project, an electrically propelled airplane is under study at a frequency of 2.45 or 5.8 GHz, using a rectenna array whose RF-to-DC power conversion efficiency is 80%. The rectenna is a device to convert RF receiving power to DC power. The total power efficiency is known to be 10% at 150 m altitude from ground for 10 kW transmission. The airplane is targeted to operate at stratosphere altitude of about 20–30 km, which is a promising altitude where there are almost always no strong wind flows and provides long-distance monitoring comparable to an Earth orbiting low-altitude satellite.

As shown in Figure 1.7, I and Prof. Chul Park with the Department of Aerospace Engineering at KAIST have studied the feasibility of the stratosphere RF-powered airplane, which has a tandem wing antenna structure to receive RF power from a ground Tx antenna and to obtain lift force with a propeller operating by electricity [59]. The mass and height of the airplane are m_s and h_s, respectively.

The required velocity and power of the airplane to obtain lifting force against gravity are ideally determined as follows:

$$F_g = 0.5 C_L \rho V^2 A = m_s g \Rightarrow V = \sqrt{\frac{m_s g}{0.5 C_L \rho A}} \tag{1.1a}$$

$$P \equiv F_D V = 0.5 C_D \rho V^3 A = 0.5 C_D \rho \left[\frac{m_s g}{0.5 C_L \rho A} \right]^{\frac{3}{2}} A, \tag{1.1b}$$

where C_L, C_D, ρ, A, and g are lift coefficient, drag coefficient, air density, wing area, and gravity constant, respectively.

As shown in Figure 1.8, the required velocity and power are found to be 22 m/s and 8.5 kW, respectively, for a wing span of 30 m and weight of 200 kg, which is a reasonable airplane system.

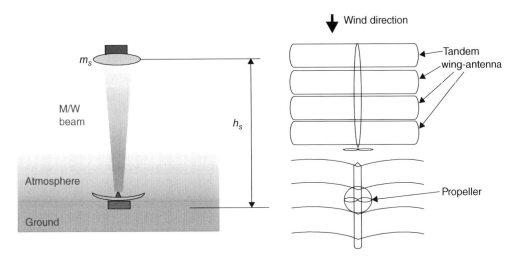

Figure 1.7 An RF-powered airplane designed by KAIST (Chun T. Rim and Chul Park).

As shown in Figure 1.9, the diameter of a ground station to transmit RF power can be calculated for a wing span of 30 m, altitude of 30 km, and RF frequencies of L-band (2.45 GHz) and X-band (10.0 GHz) as follows:

$$L_{WS} \cong \frac{\lambda}{D} h_s = \frac{c h_s}{fD} \quad \Rightarrow D \cong \frac{c h_s}{fL_{WS}}, \tag{1.2a}$$

$$D_{2.45\,\text{GHz}} = \frac{c h_s}{fL_{WS}} = \frac{3 \times 10^8 \cdot 30\,\text{k}}{2.45\,\text{G} \cdot 30} = 122\,\text{m} \tag{1.2b}$$

$$D_{10\,\text{GHz}} = \frac{c h_s}{fL_{WS}} = \frac{3 \times 10^8 \cdot 30\,\text{k}}{10\,\text{G} \cdot 30} = 30\,\text{m} \tag{1.2c}$$

As identified from (1.2b) and (1.2c), the diameters of a ground station for L-band and C-band are 122 m and 30 m, respectively, which are quite reasonable to build. Frequency can be selected considering the ground station size and total power efficiency, where the L-band has higher power efficiency than the C-band. Considering currently available RF components and propagation loss, the power requirement of a ground station is found to be roughly 200 kW [59].

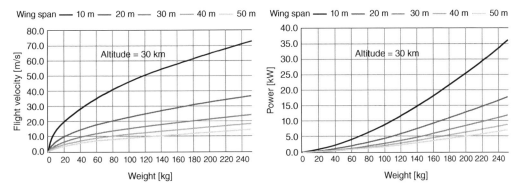

Figure 1.8 Required flight velocity (top) and power (bottom) for the given weight of the KAIST RF-powered airplane.

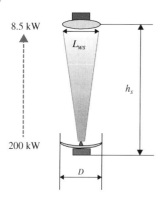

8.5 kW

200 kW

Figure 1.9 Required ground power and diameter of a ground station of the KAIST RF-powered airplane.

It has been outlined that the stratosphere RF airplane can be built if appropriate RF components are used and the RF airplane is properly designed.

The NASA plan of sending solar power generated at a geostationary orbit satellite through microwaves started in 1978. It was designed to have 1 km and 10 km diameter Tx and Rx antennas at 2.45 GHz to achieve 750 MW receiving power on Earth ground. Experiments at a few kW power level with reduced size Tx and Rx antennas were conducted at Goldstone in California in 1975 and Grand Bassin on Reunion Island in 1997, respectively. This NASA plan was finally cancelled due to several issues such as low Tx efficiency, potentially harmful Rx power density, and extremely poor cost-effectiveness compared to ground solar power generation.

Recently, low-power applications of RF PT have been widely explored as the energy source of distributed sensor networks and IoT. RF energy harvesting [60] is currently a hot issue, where very low power of less than 1 mW or a very small amount of energy lower than 1 mJ is pursued. RF power delivery to mobile devices in an office or room is also an interesting application, where dynamic directing of the Tx antenna and the narrow receiving angle of the Rx antenna of arbitrarily positioned mobile devices are important problems. Avoiding harmful RF power exposure to the human body and adjacent electronic equipment is also a challenging issue together with expensive Tx and Rx devices and strong RF interference regulations.

Question 3 (1) What happens to a stratosphere drone with RF PT if the power transfer is abruptly stopped due to breakdown of power systems or bad weather? (2) What are the remedies for (1)?

Question 4 (1) Estimate the cost of launching and maintaining a geostationary satellite for solar power generation and transmitting. (2) Compare (1) to the cost of a conventional ground solar power generator.

1.3.2 Optical Power Transfer (Optical PT)

Optical power is a good candidate for wireless power if good clearance between Tx and Rx is maintained. As shown in Figure 1.10(a), the NASA Marshall Space Flight Center has developed a laser-powered drone whose total power efficiency from the input power of a laser Tx to the output power of solar cell is 6.8%. The rationale for this efficiency is as follows:

– Current laser efficiencies of 25% (it can be improved up to 50% in the near future)
– Solar cell conversion efficiencies of 50%
– Power conditioning efficiency of 80%
– Receiver efficiency of 75%
– Atmospheric transmission efficiency of 90%.

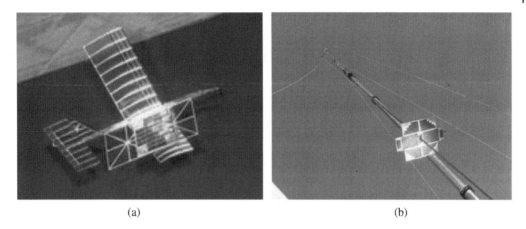

(a)   (b)

Figure 1.10 (a) NASA laser-powered drone and (b) a climber at the NASA Beam Power Challenge test.

Even though the power efficiency of 6.8% is much lower than that of a modern IPT device, the optical drone shows the possibility of low-power WPT to indoor mobile devices without any electromagnetic interference (EMI) problems. If the wavelength of the laser is infrared (IR), it is quite safe to the human body unless the power level is very high. Like RF PT, this optical PT has the problem of dynamic directing of the Tx and a narrow receiving angle of Rx of arbitrarily positioned mobile devices. Furthermore, it is very difficult for the optical PT to have electronic beam steering, which is available for RF PT, although it is expensive and has a limited steering angle.

As shown in Figure 1.10(b), a climber built by the University of Saskatchewan Space Design Team reached 40 feet up a 200-ft climbing ribbon in the NASA Beam Power Challenge test in 2005. Sunlight is free and abundant but is not available in cloudy weather and at night; therefore, an artificial LED or laser light is crucial for an optical PT in order to provide a reliable light power source. One of the fundamental drawbacks of optical PT is that power can be delivered only in the line of sight and cannot be delivered through obstacles or opaque materials, which can be easily overcome by IPT for instance.

Question 5 (1) What would happen to a drone with an optical PT when the power transfer is abruptly stopped due to a breakdown of power systems or bad weather? (2) What are the remedies for (1)? Discuss whether an on-board battery is a good solution. (3) What if the angle of incidence to the drone is variable and sometimes extremely large?

1.3.3 Tethered Power Transfer (Tethered PT)

As discussed, tethered PT is suitable for a stationary drone for persistent missions such as surveillance, environment monitoring, fire and crime monitoring, traffic control, communication relay, broadcasting, search and rescue, and video capture. There had not been many studies when I started to research a tethered unmanned helicopter (UTH) in 2007, as shown in Figure 1.11 [61]. As an example of tethered drones, a summary of this study [61] is provided in the following.

The target altitude, total mass, and total power of the UTH are 1 km, 200 kg, and 25 kW, respectively, as listed in Table 1.1. A power cable of 1 kV, 25 A ratings is used, where the mass and resistance for 1 km of power cable are 95 kg and 12.1 ohms, respectively. Considering a

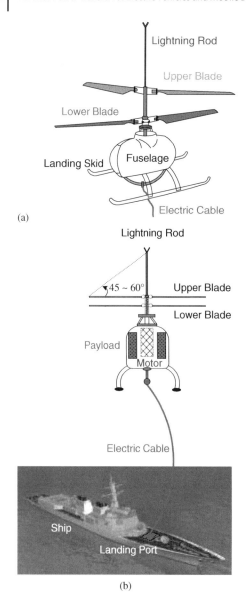

(a)

(b)

Figure 1.11 A tethered helicopter designed by KAIST (Chun T. Rim): (a) configuration and (b) and operation concept.

9.5 kW total power loss including the power cable loss, the delivered power to the UTH is 15.5 kW, which is enough power to lift the UTH and to provide for mission payloads.

As shown in Figure 1.11, a lightning rod is installed to the UTH, which resulted from the lesson learnt by Nazis when the tethered helicopter failed due to lightning. It is assumed that the cable is wrapped on the ground and a brush contact is installed to have an electric connection of the cable to the ground power source.

Another tethered PT is for ground vehicles, as shown in the example in Figure 1.12(a). This tethered ground vehicle (TGV) has been developed by my team since 2011. For the TGV, cable is wound on the vehicle and a constant tension is always provided to the cable so that it can travel around corners without fear of getting stuck. The conventional cabletype ground vehicle carries the cable but easily becomes stuck at corners.

Table 1.1 The mass and power budgets of KAIST UTH designed by Chun T. Rim

Items	Mass (kg)	Power (kW)
TUH platform (lifting)	25	12.0
TUH platform (others)	40	0.5
TUH payload (radar, IR)	30	1.0
Electric cable	95	9.5
Design margin	10	2.0
Total	200	25.0

Recently, my team has developed tethered drones, where the cable is wound either on the ground or on the drone, as shown in Figure 1.12(b) [62]. When the tension control system is on the drone, it is appropriate for roaming missions; in contrast, when the tension control system is on the ground, it is good for stationary missions. By designing a novel cable wrapping mechanism, there is no brush contact for the ground tension control case [62].

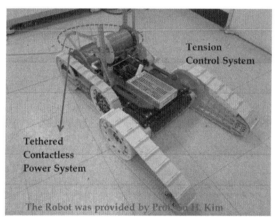

(a) Tethered ground vehicle for surveillance

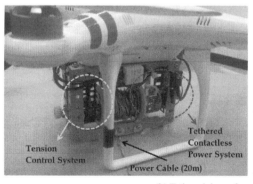 

(b) Tethered drone for environment monitoring

Figure 1.12 Tethered ground vehicle and small drone designed by KAIST (Chun T. Rim).

Question 6 Discuss detail methods for protecting a tethered drone from lightning. (1) For example, what about using a current fuse that is blown up when large lightning current flows? (2) How can the tethered electric cable be grounded for lightning current bypass? Remember that the wound electric cable under lightning is exposed to an extremely high voltage (~MV) and may not withstand the electric shock.

1.4 Conclusion

An overview of MPT has been provided in this chapter. The most important competitors in mobile power electronics will be the battery versus WPT. If a very light, small size, cheap, long lasting, and quick chargeable battery is available, then batteries will dominate over WPT. However, WPT is becoming more important in MPT because of the convenience and inherent safety and batteries need to be recharged. WPT is thus not only a competitor but also an ally of the battery. Moreover, WPT may substitute or dominate the battery, as identified by RPEV. Tethered PT is also a good candidate for MPT and may provide a replacement for the battery. Most WPT and tethered PT require batteries for their emergency power to provide a reliable power source to the system. Therefore, all parts of MPT should be enhanced together to achieve progress in mobile power electronics.

References

1 M. Kline, I. Izyumin, B. Boser, and S. Sanders, "Capacitive power transfer for contactless charging," in *2011 ECCE Conference*, pp. 1398–1404.

2 B. Choi, D. Nguyen, S. Yoo, J. Kim, and C. Rim, "A novel source-side monitored capacitive power transfer system for contactless mobile charger using class-E converter," in *2014 VTC Conference*, pp. 1–5.

3 E.Y. Chow, "Wireless powering and the study of RF propagation through ocular tissue for development of implantable sensors," *IEEE Trans. Antennas Propag.*, vol. 59, no. 6, pp. 2379–2387, June 2011.

4 N. Wang *et al.*, "One-to-multipoint laser remote power supply system for wireless sensor networks," *IEEE Sensors J.*, vol. 12, no. 2, pp. 389–396, February 2012.

5 I. Shnaps and E. Rimon, "Online coverage by a tethered autonomous mobile robot in planar unknown environments," *IEEE Trans. Robot.*, vol. 30, no. 4, pp. 966–974, August 2014.

6 S. Choi *et al.*, "Tethered aerial robots using contactless power systems for extended mission time and range," in *2014 ECCE Conference*, pp. 912–916.

7 O.C. Onar, J. Kobayashi, and A. Khaligh, "A fully directional universal power electronic interface for EV, HEV, and PHEV applications," *IEEE Trans. Power Electron.*, vol. 28, no. 12, pp. 5489–5498, December 2013.

8 E. Waffenschmidt, "Free positioning for inductive wireless power system," in *2011 ECCE Conference*, pp. 3481–3487.

9 W. Zhong, X. Liu, and S. Hui, "A novel single-layer winding array and receiver coil structure for contactless battery charging systems with free-positioning and localized charging features," *IEEE Trans. Ind. Electron.*, vol. 58, no. 9, pp. 4136-4143, September 2011.

10 C. Park, S. Lee, G. Cho, S. Choi, and Chun T. Rim, "Omni-directional inductive power transfer system for mobile robots using evenly displaced multiple pick-ups," in *2012 ECCE Conference*, pp. 2492–2497.

11 C. Park, S. Lee, G. Cho, S. Choi, and Chun T. Rim, "Two-dimensional inductive power transfer system for mobile robots using evenly displaced multiple pickups," *IEEE Trans. Ind. Appl.*, vol. 50, no. 1, pp. 538–565, June 2013.

12 B. Che *et al.*, "Omnidirectional non-radiative wireless power transfer with rotating magnetic field and efficiency improvement by metamaterial," *Appl. Phys. A*, vol. 116, no. 4, pp. 1579–1586, April 2014.

13 W. Ng, C. Zhang, D. Lin, and S. Hui, "Two- and three-dimensional omnidirectional wireless power transfer," *IEEE Trans. Power Electron.*, vol. 29, no. 9, pp. 4470–4474, January 2014.

14 H. Li, G. Li, X. Xie, Y. Huang, and Z. Wang, "Omnidirectional wireless power combination harvest for wireless endoscopy," in *2014 BioCAS Conference*, pp. 420–423.

15 X. Li *et al.*, "A new omnidirectional wireless power transmission solution for the wireless endoscopic micro-ball," in *2011 ISCAS Conference*, pp. 2609–2612.

16 R. Carta *et al.*, "Wireless powering for a self-propelled and steerable endoscopic capsule for stomach inspection," *Biosens. Bioelectron.*, vol. 25, no. 4, pp. 845–851, December 2009.

17 T. Sun *et al.*, "Integrated omnidirectional wireless power receiving circuit for wireless endoscopy," *Electron. Lett.*, vol. 48, no. 15, pp. 907–908, July 2012.

18 B. Lenaerts and R. Puers, "An inductive power link for a wireless endoscope," *Biosens. Bioelectron.*, vol. 22, no. 7, pp. 1390–1395, February 2007.

19 B. Choi, E. Lee, J. Kim, and Chun T. Rim, "7m-off-long-distance extremely loosely coupled inductive power transfer system using dipole coils," in *2014 ECCE Conference*, pp. 858–863.

20 C. Park, S. Lee, G. Cho, and Chun T. Rim, "Innovative 5-m-off-distance inductive power transfer systems with optimally shaped dipole coils," *IEEE Trans. Power Electron.*, vol. 30, no. 2, pp. 817–827, November 2014.

21 Chun T. Rim and G. Cho, "New approach to analysis of quantum rectifier-inverter," *Electron. Lett.*, vol. 25, no. 25, pp. 1744–1745, December 1989.

22 Chun T. Rim, "Unified general phasor transformation for AC converters," *IEEE Trans. Power Electron.*, vol. 26, no. 9, pp. 2465–2475, September 2011.

23 J. Huh, W. Lee, S. Choi, G. Cho, and Chun T. Rim, "Frequency-domain circuit model and analysis of coupled magnetic resonance systems," *J. Power Electron.*, vol. 13, no. 2, pp. 275–286, March 2013.

24 A. Kurs, A. Karalis, R. Moffatt, J.D. Joannopoulos, P. Fisher, and M. Soljacic, "Wireless power transfer via strongly coupled magnetic resonance," *Science*, vol. 317, no. 5834, pp. 83–86, June 2007.

25 A.P. Sample, D.A. Meyer, and J.R. Smith, "Analysis, experimental results, and range adaption of magnetically coupled resonators for wireless power transfer," *IEEE Trans. Ind. Electron.*, vol. 58, no. 2, pp. 544–554, February 2011.

26 T. Imura and Y. Hori, "Maximizing air gap and efficiency of magnetic resonant coupling for wireless power transfer using equivalent circuit and Neumann formula," *IEEE Trans. Ind. Electron.*, vol. 58, no. 10, pp. 4746–4752, October 2011.

27 T.C. Beh, T. Imura, and Y. Hori, "Basic study of improving efficiency of wireless power transfer via magnetic resonance coupling based on impedance matching," in *2010 ISIE Conference*, pp. 2011–2016.

28 J. Park, Y. Tak, Y. Kim, Y. Kim, and S. Nam, "Investigation of adaptive matching methods for near-field wireless power transfer," *IEEE Trans. Antennas Propag.*, vol. 59, no. 5, pp. 1769–1773, May 2011.

29 J. Huh, W.Y. Lee, S.Y. Choi, G.H. Cho, and Chun T. Rim, "Explicit static circuit model of coupled magnetic resonance system," in *2011 ECCE-Asia Conference*, pp. 2233–2240.

30 E. Lee, J. Huh, X.V. Thai, S. Choi, and Chun T. Rim, "Impedance transformers for compact and robust coupled magnetic resonance systems," in *2013 ECCE Conference*, pp. 2239–2244.

31 R. Hui, W. Zhong, and C. Lee, "A critical review of recent progress in mid-range wireless power transfer," *IEEE Trans. Power Electron.*, vol. 29, no. 9, pp. 4500–4511, September 2014.

32 G. Covic, M. Kissin, D. Kacprzak, N. Clausen, and H. Hao, "A bipolar primary pad topology for EV stationary charging and highway power by inductive coupling," in *2011 ECCE Conference*, pp. 1832–1838.

33 S. Li and C. Mi, "Wireless power transfer for electric vehicle applications," *IEEE Trans. Emerg. Sel. Topics Power Electron.*, vol. 3, no. 1, pp. 4–17, March 2015.

34 S. Choi, J. Huh, W. Lee, and Chun T. Rim, "Asymmetric coil sets for wireless stationary EV chargers with large lateral tolerance by dominant field analysis," *IEEE Trans. Power Electron.*, vol. 29, no. 12, pp. 6406–6420, December 2014.

35 M. Budhia, G. Covic, and J. Boys, "Design and optimization of circular magnetic structures for lumped inductive power transfer systems," *IEEE Trans. Power Electron.*, vol. 26, no. 11, pp. 3096–3108, November 2011.

36 M. Budhia, J. Boys, G. Covic, and C. Huang, "Development of a single-sided flux magnetic coupler for electric vehicle IPT charging systems," *IEEE Trans. Ind. Electron.*, vol. 60, no. 1, pp. 318–328, January 2013.

37 T. Nguyen, S. Li, W. Li, and C. Mi, "Feasibility study on bipolar pads for efficient wireless power chargers," in *2014 APEC Conference*, pp. 1676–1682.

38 P. Meyer, P. Germano, M. Markovic, and Y. Perriard, Design of a contactless energy-transfer system for desktop peripherals," *IEEE Trans. Ind. Applic.*, vol. 47, no. 4, pp. 1643–1651, July 2011.

39 J. Shin *et al.*, "Design and implementation of shaped magnetic-resonance-based wireless power transfer system for roadway-powered moving electric vehicles," *IEEE Trans. Power Electron.*, vol. 61, no. 3, pp. 1179–1192, March 2014.

40 G. Elliott, J. Boys, and G. Covic, "A design methodology for flat pick-up ICPT systems," in *2006 ICIEA Conference*, pp. 1–7.

41 S. Lee *et al.*, "On-line electric vehicle using inductive power transfer system," in *2010 ECCE Conference*, pp. 1598–1601.

42 J. Huh, S. Lee, C. Park, G. Cho, and Chun T. Rim, "High performance inductive power transfer system with narrow rail width for on-line electric vehicles," in *2010 ECCE Conference*, pp. 647–651.

43 J. Huh, W. Lee, B. Lee, G. Cho, and Chun T. Rim, "Characterization of novel inductive power transfer systems for on-line electric vehicles," in *2011 APEC Conference*, pp. 1975–1979.

44 J. Huh, S. Lee, W. Lee, G. Cho, and Chun T. Rim, "Narrow-width inductive power transfer system for on-line electrical vehicles," *IEEE Trans. Power Electron.*, vol. 26, no. 12, pp. 3666–3679, December 2011.

45 S. Lee *et al.*, "Active EMF cancellation method for I-type pickup of on-line electric vehicles," in *2011 APEC Conference*, pp. 1980–1983.

46 W. Lee *et al.*, "Finite-width magnetic mirror models of mono and dual coils for wireless electric vehicles," *IEEE Trans. Power Electron.*, vol. 28, no. 3, pp. 1413–1428, March 2013.

47 S. Choi, J. Huh, W. Lee, S. Lee, and Chun T. Rim, "New cross-segmented power supply rails for road powered electric vehicles," *IEEE Trans. Power Electron.*, vol. 28, no. 12, pp. 5832–5841, December 2013.

48 S. Lee, B. Choi, and Chun T. Rim, "Dynamic characterization of the inductive power transfer system for online electric vehicles by Laplace phasor transform," *IEEE Trans. Power Electron.*, vol. 28, no. 12, pp. 5902–5909, December 2013.

49 S. Choi, B. Gu, S. Jeong, and Chun T. Rim, "Ultra-slim S-type inductive power transfer system for road powered electric vehicles," in 2014 *EVTeC Conference*, pp. 1–7.

50 S. Choi *et al.*, "Generalized active EMF cancel methods for wireless electric vehicles," *IEEE Trans. Power Electron.*, vol. 29, no. 11, pp. 5770–5783, November 2014.

51 C. Wang, O. Stielau, and G. Covic, "Design considerations for a contactless electric vehicle battery charger," *IEEE Trans. Ind. Electron.*, vol. 52, no. 5, pp. 1308–1314, October 2005.

52 C. Wang, G. Covic, and O. Stielau, "Power transfer capability and bifurcation phenomena of loosely coupled inductive power transfer systems," *IEEE Trans. Ind. Electron.*, vol. 51, no. 1, pp. 148–157, February 2004.

53 G. Covic and J. Boys, "Modern trends in inductive power transfer for transportation applications," *IEEE Trans. Emerg. Sel. Topics Power Electron.*, vol. 1, no. 1, pp. 28–41, March 2013.

54 O. Onar *et al.*, "A novel wireless power transfer for in-motion EV/PHEV charging," in *2013 APEC Conference*, pp. 2073–3080.

55 S. Choi, B. Gu, S. Jeong, and Chun T. Rim, "Advances in wireless power transfer systems for roadway-powered electric vehicles," *IEEE Trans. Emerg. Sel. Topics Power Electron.*, vol. 3, no. 1, pp. 18–35, March 2015.

56 B. Lee, H. Kim, S. Lee, C. Park, and Chun T. Rim, "Resonant power shoes for humanoid robots," in *2011 ECCE Conference*, pp. 1791–1794.

57 B. Choi, E. Lee, J. Huh, and Chun T. Rim, "Lumped impedance transformers for compact and robust coupled magnetic resonance systems,"*IEEE Trans. Power Electron.*, vol. PP, no. 99, pp. 1, January 2015 (Early access article).

58 J. Kim *et al.*, "Coil design and shielding methods for a magnetic resonant wireless power transfer system," *Proc. IEEE*, vol. 101, no. 6, pp. 1332–1342, June 2013.

59 Chun T. Rim, "Feasibility study on pseudo anti-gravity spaceship and flying saucer," *Korea Aerospace Spring Conference*, April 2008, pp. 809–812.

60 U. Olgun *et al.*, "Investigation of rectenna array configurations for enhanced RF power harvesting," *IEEE Antennas and Wireless Propagation Letters*, vol. 10, pp. 262–265, April 2011.

61 Chun T. Rim, J S. Lee, and B.M. Min, "A tethered unmanned helicopter for aerial inspection: design issues and practical considerations," *International Forum on Rotorcraft Multidisciplinary Technology*, October 2007.

62 B.W. Gu, S.Y. Choi, Y.S. Choi, C. Cai, L. Seneviratne, and Chun T. Rim, "Novel roaming and stationary tethered aerial robots for continuous mobile missions in nuclear power plants," *Nuclear Engineering and Technology*, vol. 48, no. 4, pp. 982–996, August 2016.

2

Introduction to Wireless Power Transfer (WPT)

2.1 General Principle of WPT

2.1.1 General Configuration of a WPT System

An experiment kit for WPT is shown in Figure 2.1, where two air coils are used to transfer power wireless and a single switch inverter, called a Class-E resonant inverter, is used for higher switching frequency. A load transformer is used to adapt the received Rx coil voltage to the load voltage level.

Now let us introduce the WPT in general. A WPT is composed of transmitting (Tx) and receiving (Rx) parts, as shown in Figure 2.2. The Tx part is composed of an AC or DC power source, primary converter, primary compensator, primary communication link, primary controller, and a Tx device such as a coil, metal plate, antenna, and light source. The Rx part is composed of an Rx device (coil, metal plate, rectenna, and solar cell), secondary compensator, secondary controller, secondary converter, secondary communication link, and DC load or AC load.

For IPT and CPT, the primary converter consists of an AC–DC rectifier (if an AC source is used) and a high-frequency (HF) inverter (for DC–AC power conversion), whereas the secondary converter consists of an HF AC–DC rectifier (not required for an HF AC load) and a DC–DC regulator (if the DC load needs voltage or current regulation). The HF inverter and rectifier (passive or active switching) are often used because wireless power through open space tends to be easier for high frequency in most cases. This is, however, not always true and a utility power source of 50 Hz, 60 Hz, or 400 Hz is sometimes used for IPT without the HF inverter. Therefore, the reason for using the HF inverter should always be asked when starting to design an IPT. The HF inverter and HF rectifier are usually soft switched where either the voltage or current of each switch becomes nearly zero when turned on and off. For RF PT and optical PT, the primary converter includes a regulated DC power and RF generator (for the RF PT case), whereas the secondary converter is used for regulating DC power from the Rx device.

The power source and load type could be AC or DC depending on different applications. For example, the power source ranges from AC utility power to a battery, which is a typical DC power source. Induction heating is an AC load case, where the inverter usually operates at a frequency of less than a few kHz and the power level goes up to several kW to several MW. Another AC power load is an electric motor, which is crucial for electric vehicle (EV) applications.

The primary and secondary compensators are usually lossless LC circuits in IPT and CPT for higher power efficiency, and they play the roles of increasing power factors of the source and load sides and decreasing harmonic currents and voltages so that a large power transfer and low electromagnetic interference (EMI) can be achieved. For RF PT, the compensators could

Wireless Power Transfer for Electric Vehicles and Mobile Devices, First Edition. Chun T. Rim and Chris Mi.
© 2017 John Wiley & Sons Ltd. Published 2017 by John Wiley & Sons Ltd.

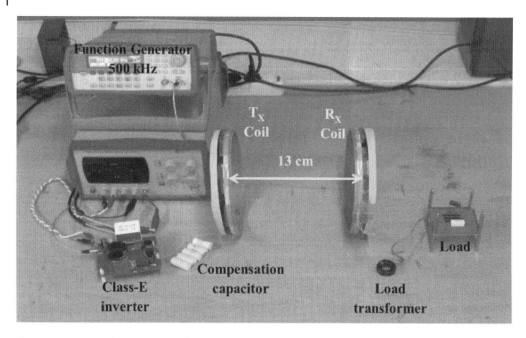

Figure 2.1 An example WPT system of two coils, operating at 500 kHz.

be RF resonant circuits or filters. For optical PT, the compensators could be DC filters or current balancing circuits.

The primary and secondary communication links exchange information such as output voltage, load current, operating status, and temperatures of components for the control of converters. Wireless communication instead of wired communications is inevitable due to physical separation of the Tx and Rx parts. Examples for the communication links are RF communication such as ZigBee and Bluetooth, infrared (IR) communication, and magnetic or electric coupling communication, which is mainly used for IPT and CPT, respectively, by modulating the output or input voltage and current to transmit communication signals.

In order to manage input and output power such as turn-on and turn-off of converters and devices, primary and/or secondary controller(s) are crucial. Each Tx and Rx controller gets the

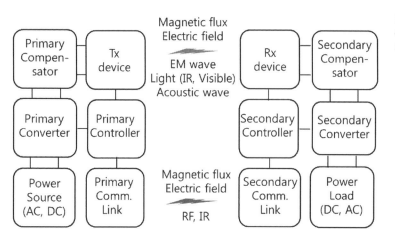

Figure 2.2 General configuration of a WPT system.

status, voltage, and current information from the source, load, converters, compensators, and devices by self-monitoring of components in its own part or through the communication links of components in another part. Different from ordinary controllers, which are typically single in a system, the controllers in WPT are separated from each other but they must be in very good harmony for successful wireless power delivery. Because the Tx and Rx parts could be arbitrarily apart and could be multiple with unknown states, control of the whole WPT system requires highly reliable communication links with good flexibility for managing various situations such as no load, open circuit, short circuit, over voltage, under voltage, transient overshoot, in-rush current, and failure of communication. The controllers are usually composed of microprocessors, but sometimes simple traditional proportional–integral–derivative (PID) controllers are employed. Despite difficulties and inefficiencies in removing the communication links, it would be ideal if Tx and Rx parts could be controlled individually or even independently. This has been explored by innovative engineers in efforts to eliminate cumbersome communication links and to avoid complicated control.

The Tx and Rx devices are designed so that they can transfer large power efficiently regardless of distance and displacement between the Tx and Rx devices. For the design of coils (for IPT) and metal plates (for CPT), mutual couplings of the magnetic flux and electric field, respectively, should be maximized for given dimensions. For the design of antennas (for RF PT) and light sources or solar cells (for optical PT), considerations of power efficiency and directivity of the Tx and Rx devices are of paramount importance, because they usually have a very narrow beam width. There should be reliable RF power links (for RF PT) and optical power links (for optical PT), which are a type of RF and optical couplings, respectively.

The design of a WPT system highly depends upon the media of power transfer. For example, the design principle on the coils of IPT for increasing magnetic flux linkage cannot be used in the design of the antennas of RF PT to increase the beam directivity of electromagnetic (EM) waves. Among all the modules of WPT in Figure 2.2, the Tx and Rx devices may be the most challenging to design in many cases.

Note that the Tx and Rx parts such as Tx and Rx devices, power source, power load, converters, and compensators in Figure 2.2 are not necessarily single but also multiple depending on applications.

Question 1 Should the wireless power be transferred from Tx to Rx? If not, when the wireless power is transferred back to Tx from Rx, how can it be? You will get some clues in the subsequent Section 2.2 on IPT.

2.1.2 General Requirements of WPT

What do you think that an ideal WPT system should be in general? It would be impossible to provide such a general answer that would be valid for any WPT. We can start, however, with a generally acceptable ideal feature of WPT such as power efficiency. A summary of general requirements is presented in Figure 2.3. Those readers who wish to see the specific requirements for road-powered electric vehicles may go to Chapters 8 and 9. Now let me talk about overall requirements of a WPT system.

1) Power efficiency. The higher the power efficiency of a WPT system is, the better the performance will be (is this really right?). The definition of power efficiency, often referred to as "efficiency", should be made carefully. In general, efficiency does not necessarily mean power efficiency but it may mean energy efficiency, even economic efficiency, and managing efficiency. It would thus be wise to use the term "power efficiency" when non-technical

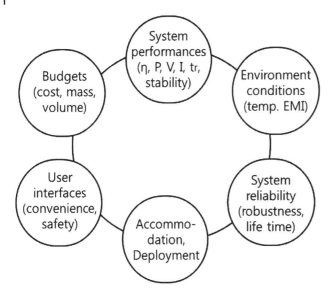

Figure 2.3 General requirements of a WPT system.

issues are involved in discussion. Another viewpoint is the points at which the efficiency is measured. The term "system efficiency" is often used in WPT, and is usually defined as the ratio of the input power measured at the power source and the output power measured at the load of Figure 2.2. Therefore, the system efficiency is the most conservative value because it is the lowest among all the efficiencies defined in a WPT system. Another viewpoint is the time at which the efficiency is measured. If not specifically notified, the efficiency is measured over a very long period of time in the steady state of a WPT system. However, the instantaneous efficiency or quasi-steady state efficiency is sometimes used in WPT to reflect the dynamic behavior of loads such as in the case of electric vehicles (EVs) and movable electronic devices. Moreover, the efficiency is frequently measured over different locations in WPT.

Now let me talk about "good efficiency" instead of "high efficiency." In order to obtain high efficiency of a WPT system, large cables and electronic components of high current ratings are usually used. Of course, professional engineers strive for higher efficiency of WPT for a given cost, mass, volume, and performance. In conclusion, the efficiency inevitably involves a trade-off with other system parameters to obtain good efficiency, which is not necessarily too high.

2) Cost, mass, and volume. As discussed above, the cost, mass, and volume of a WPT system are, in many cases, more important than the efficiency. The cost of a WPT system does not always mean the development cost: the cost in mass production is often more important. As we try to reduce the cost, mass, and volume of a WPT system, not only the efficiency but also accommodation of components, heat dissipation, system reliability, and EMI deteriorates. A compact size WPT system has the problems of difficult accommodation of components and high temperature of components due to a reduced heat dissipation capability. For many applications of smart phones, smart pads, and EVs, a flat shape of the Tx and Rx is preferred to a volumetric shape. Light material and small use of material in a WPT system tends to increase EMI if not properly designed. Therefore, again the cost, mass, and volume should not be minimized alone but should be traded off with accommodation, device temperature, system reliability, and EMI. These design issues can be divided into two categories, as explained in the following.

3) System reliability. Reliable power transfer and reliable operation of a WPT system are of prime importance for commercialization. As discussed above, the reliability of electronic components such as semiconductor switches degrades as the temperature of components increases. The magnetic characteristics of the core, such as permeability and hysteresis loss, are drastically changed by temperature, which may deteriorate the system performance of the WPT system. Mechanical robustness of the WPT system against external forces and shocks is also important. In order to mitigate this reliability decrease, sufficient design margin as well as efficient heat dissipation is imposed to realize a highly reliable WPT system.

In many WPT systems, the distance, lateral displacement, and orientation of Tx and Rx are varied, which results in changes of the wireless power level and resonant frequency due to the changed mutual coupling between the Tx and Rx. If not appropriately handled, these problems may deteriorate reliable operation of the WPT system. For a reliable WPT system, regulation of power or current/voltage by controlling the operating frequency or duty cycle of the HF inverter or HF rectifier is preferred. Sometimes, the primary and secondary compensation circuits are externally varied to cope with the variation of operating conditions.

For high system reliability, multiple Tx and Rx can be used to eliminate dead zones of specific Tx and Rx. Because a high-quality factor (Q) of an LCR resonant circuit, often referred to as a "resonant tank," is subject to sensitive frequency change, a low Q is often selected for increased frequency agility of IPT and CPT. This important design principle had not been well addressed in the WPT society until I emphasized this point, and an unnecessarily too large Q, for example, 2000, is often chosen by researchers in order to increase the distance of wireless power.

A long lifetime is desirable for commercialization; however, the lifetime needs to be moderate in practice to avoid over design and an undesirably robust system for disposal.

4) Environment conditions: ambient temperature, humidity, EMI, etc. Operating conditions for a WPT system such as ambient temperature, humidity, contamination, weather (rain, snow, ice, frost, salt water, wind, etc.), mechanical vibrations, and EMI noise from utility power and adjacent equipment are of paramount important in the design and use of the WPT system. For outdoor applications of WPT such as EVs and military devices, exposed parts of a WPT system (usually Tx and Rx) must withstand the large variation of ambient temperature. The temperature inside the WPT system is usually higher than the ambient temperature due to internal heat generation.

EMI regulations on WPT must also be met and should be considered at a very early design phase in order to avoid significant design changes in the long run. Because the magnetic field (B-field), electric field (E-field), and radiated electromagnetic waves (EM waves) must be regulated over a large range of frequency, passive and active shields and canceling techniques as well as grounding techniques are widely used in WPT systems. The International Commission on Non-Ionizing Radiation Protection (ICNIRP) guidelines are widely used in WPT, where specific values of B-field, E-field, and EM waves are given for each frequency. For example, the B-field should be lower than 27 μT for 3 kHz to 10 MHz according to the ICNIRP 2010 guidelines. Note that the ICNIRP guidelines do not actually regulate WPT or guarantee safety, but are widely adopted by most countries except for a few European countries. In order to commercialize a WPT system, it should also meet conventional EMI regulations or worldwide standards such as Qi for IPT.

5) System performance: power, voltage, current, response time, stability, etc. Needless to say, the WPT should meet the desired system performance if the above-mentioned basic requirements are fulfilled. There are many design considerations of a WPT system such as input and output powers and power factors, rated voltages and currents of each component, harmonic currents and voltages, transient response time, and frequency stability of the WPT system

for different operating conditions and load changes. Nonlinear characteristics of magnetic material, switching converters, controllers, and relative movements of Tx and Rx, which are expected to remain challenging issues in the future WPT, make it difficult to preserve good system performance. In this sense, the development of WPT is a never-ending story.

6) Accommodation and deployment. As discussed above, accommodation of components in a WPT system is important in terms of EMI, thermal design, and overall configuration of the system. Accommodation of the WPT system to a mission system is also an important issue. For example, accommodation of a wireless charger, which is a WPT system, to an electric vehicle (EV) is a formidable system task that requires accurate mechanical and electrical interfaces.

   Another important consideration for the commercialization of WPT systems is the deployment of the system. Without sufficient infrastructure of the Tx part, a movable Rx part cannot be used where desired. Deployment of numerous Tx parts of the system requires a lot of cost, time, and effort. For example, deployment of millions of EV chargers is needed in Japan to recharge a similar number of EVs. Also, to deploy the Tx part of a WPT system on the road is the most challenging problem in road-powered EVs due to road construction and installation of equipment such as inverters and power rails. Like Wi-Fi, numerous Tx parts are required for IoT applications in order to provide a ubiquitous environment of wireless power.

7) User interfaces: convenient, safe, and beautiful design. Even though the technical performance is excellent and cost is competitive, the WPT system may not be widely used if users do not feel convenience and aesthetic satisfaction from the products to which WPT is applied. Safety is also of great concern to the public, which is suspicious about the potential danger of electromagnetic fields (EMFs). Mechanical and electrical safety requirements are crucial for the WPT system.

8) Summary. The above-mentioned aspects on the requirements of a WPT system are not necessarily in the order of importance and are strongly correlated with each other. Like other systems, the WPT system is a complex system of mechanical and electric compartments, and the design of a WPT system requires comprehensive trade-offs among the requirements. Note that the system efficiency has been merged into the system performance in Figure 2.3. It can be said that the development of a WPT system is not a systematic science but an art of system engineering.

2.2 Introduction to Inductive Power Transfer (IPT)

In general, the fundamental principles and configuration of IPT are explained in this section. Specific discussions on IPT for different applications and issues will follow in the successive parts of this book.

2.2.1 Fundamental Principles of IPT

As shown in Figure 2.4, IPT is based on multiple inductors whose magnetic fluxes are interchanged with each other, where the Tx coil of N_1 number of turns is driven by a voltage (or current) source and the Rx coil of N_2 number of turns is connected to a load (or sometimes active loads such as a battery and a current source).

   Let us consider the no-load case first, as shown in Figure 2.4(a). As the source current flows, magnetic flux is generated from the Tx coil, where part of the magnetic flux circulates (ϕ_{11}) and the rest of it intersects the Rx coil (ϕ_{12}), which incurs an induced voltage. Because there is no

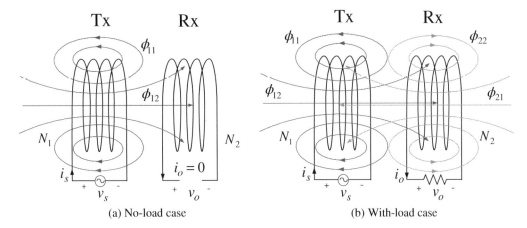

Figure 2.4 Magnetic fluxes of two coupled inductors in the instantaneous time domain.

load current, there is no magnetic flux generated from the Rx due to its current; however, there is an induced open voltage.

For the with-load case, as shown in Figure 2.4(b), the Rx coil also generates magnetic flux in a similar way to the Tx. How about the induced voltage at the Tx coil, which is incurred by the intersected magnetic flux to the Tx coil (ϕ_{21})? There should be no discrimination between the Tx and Rx coils, and this induced voltage at the Tx coil may affect the source current, which again results in a change of the Tx magnetic flux (ϕ_{11} and ϕ_{12}). The two inductors are not only magnetically coupled but also electrically coupled through source and load side electric circuits in most cases.

Question 2 (1) Is the superposition theorem applicable to magnetic flux of Tx and Rx in Figure 2.4? (2) What if the core is used in the coils? (3) Does the linearity of the core affect the applicability of the superposition theorem?

In general, it can be said that the IPT is governed by Ampere's law and Faraday's law among four Maxwell equations, as shown in Figure 2.5. The governing equations of IPT in the steady state for a sinusoidal magnetic field, electric field, and current density at low frequency, which means that the EM wavelength of operating frequency is much longer than the size of the IPT system, are approximated by letting $D = 0$ as follows:

$$\nabla \times \mathbf{H} = \mathbf{J} \qquad \text{(Ampere's law)} \tag{2.1a}$$

$$\nabla \times \mathbf{E} = -j\omega\mathbf{B} \quad \text{(Faraday's law)} \tag{2.1b}$$

The power delivery principle can be explained as follows:

1) The time-varying magnetic flux is generated from the Tx coil according to Ampere's law.
2) Induced voltage is generated at the Rx coil according to Faraday's law, where magnetic flux comes from the Tx coil.
3) Induced voltage, called back-emf (counter electromotive force), is also generated at the Tx coil by the Rx coil current.
4) Power is provided to the Tx coil and fetched from the Rx coil; thus, wireless power is delivered through the coils.

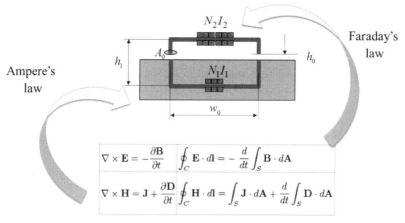

(a) Governing equations of IPT in general

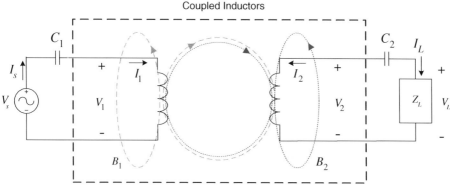

(b) An equivalent circuit of an IPT system in phasor domain, including compensation circuits in the steady state

Figure 2.5 The fundamental principle of IPT.

As identified from the above four steps, the IPT is essentially a transformer. A major difference from an ideal transformer, however, is relatively large leakage inductances at both the primary and secondary sides because of magnetically uncoupled inductors. This is why we need the capacitors, as shown in Figure 2.5(b), which are mostly used to nullify the reactance of leakage inductances. Depending on the source and load types, the compensating capacitors can be connected in series or parallel; Figure 2.5(b) shows a series–series compensation example. The characteristics of compensation circuits are analyzed in detail in Chapter 14.

For a better understanding, the conceptual magnetic circuit of Figure 2.5(a) is approximately analyzed, assuming an equal cross-section area A_0 and thickness w_1 of the magnetic core with permeability μ_c, air gap h_0, width w_0, height h_1 of coils, and number of turns of primary (Tx) and secondary (Rx) coils N_1, N_2. The sinusoidal current of frequency f_s is assumed and all variables are phasors.

The magnetic intensity due to the source current I_s ($= I_1$) can be calculated by Ampere's law as follows:

$$\oint \vec{H} \cdot d\vec{l} = H_{core}\{2w_0 + 2(h_1 - h_0)\} + H_{air}2h_0 = N_1 I_1 \tag{2.2}$$

From the magnetic flux continuity principle, we obtain

$$\phi_{core} = B_{core}A_0 \cong B_{air}A_0 = \phi_{air} \quad \Rightarrow B_{core} = \mu_c\mu_0 H_{core} \cong B_{air} = \mu_0 H_{air} \tag{2.3}$$

where a uniform magnetic flux in the air is assumed for $h_0 \ll w_1$, which is of course not true for $h_0 > w_1$ due to the fringe effect and there is no way to apply a simple calculation.

Applying (2.3) to (2.2), we can determine the magnetic flux density as follows:

$$H_{core}\{2w_0 + 2(h_1 - h_0)\} + H_{air}2h_0 \cong \frac{B_{air}}{\mu_c\mu_0}2(w_0 + h_1 - h_0) + \frac{B_{air}}{\mu_0}2h_0 = N_1 I_1$$

$$\Rightarrow B_{air} = \frac{\mu_0 N_1 I_1}{\dfrac{2(w_0 + h_1 - h_0)}{\mu_c} + 2h_0} \cong \frac{\mu_0 N_1 I_1}{2h_0} \quad \text{for} \quad \mu_c \gg \frac{w_0 + h_1}{h_0} \tag{2.4}$$

Note that the simplified equation of (2.4) is valid for $\mu_c > 100$, where the air gap h_0 is larger than 1 centimeter and the total circular core path is less than 1 meter.

Finally, the induced voltage at the secondary coil can be obtained as follows:

$$V_2 = j\omega_s N_2 \phi_{core} = j\omega_s N_2 B_{air} A_0 \cong \frac{j\omega_s \mu_0 A_0 N_1 N_2}{2h_0} I_1$$

$$\because \phi_{core} = B_{air} A_0 \cong \frac{\mu_0 N_1 I_1 A_0}{2h_0}, \quad \omega_s \equiv 2\pi f_s \tag{2.5}$$

It is identified from (2.5) that the induced voltage of the IPS is proportional to the operating frequency and the number of turns of Tx and Rx, but it is inversely proportional to the airgap. This explains why a higher frequency is preferred in IPS: it provides a higher induced voltage, which may result in higher wireless power. This higher frequency, however, results in increased switching loss of converters, increased core loss, and conduction loss of wire due to decreased skin depth as follows:

$$\delta = \sqrt{\frac{2}{\omega_s \mu_0 \mu_c \sigma}} \tag{2.6}$$

where σ is the conductivity of the wire of coils.

Note from (2.5) that IPT is insensitive to permeability if the air gap is larger than just a few centimeters. This is of course not true for higher frequency when the permeability of the core tends to decrease. It is also noticeable from (2.5) that the phase of induced voltage V_2 is 90 degrees ahead of that of the source current I_1, which is completely different from the characteristics of an ideal transformer, as given in the following:

$$V_2 = \frac{N_2}{N_1}V_1 \quad \text{and} \quad I_1 = \frac{N_2}{N_1}I_2 \tag{2.7}$$

This explains why the behavior of a Tesla coil, a form of IPT, is so strange to engineers accustomed to an ideal transformer. As shown in Figure 2.4, (2.5) is a gyrator with an imaginary trans-resistance gain of R_m in the steady state as follows:

$$R_m \cong \frac{\omega_s \mu_0 A_0 N_1 N_2}{2h_0} \tag{2.8}$$

The gyrator behavior of IPT will be extensively dealt with in Chapter 5, where not only coupled inductors but also any resonating LC circuits are analyzed.

Note that the equivalent circuit of Figure 2.6 is valid only when the primary side is driven by a current source, where the secondary induced voltage is solely determined by the primary

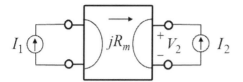

Figure 2.6 An equivalent circuit of IPT (the Tesla coil) for a current source input in the static phasor domain.

current regardless of the load current. Hence, the output voltage of the coupled inductors driven by a current source resembles an ideal voltage source.

Question 3 Note from (2.8) that the induced voltage at Rx is proportional not only to the operating frequency but also to the number of turns N_1 and N_2. Is it therefore possible to increase delivered power by just increasing N_1 or N_2? If it is true, what is the penalty for that?

2.2.2 Configuration of IPT

As shown in Figure 2.7, the configuration of IPT can be drawn by a slight change from the general WPT of Figure 2.2.

The primary and secondary converters are usually an inverter and rectifier, operating normally at several kHz to several MHz. Instead of switching converters, sometimes analog amplifiers or RF amplifiers are used for low-power high-frequency applications, where power efficiency is not of concern or switching loss increases substantially.

The Tx and Rx coils typically constitute a weak magnetic coupling depending on the distance and orientation of the coils. The Tx and Rx coils may or may not have a core, depending on the application purpose and operating frequency. When a core is used, magnetic coupling may be increased and EMF leakage can be reduced; however, core loss and core mass are the penalties of using a core. Note that the magnetic coupling does not depend on the operating frequency as long as circuit parameters are changed, which will be addressed in Chapter 4. Of course, the magnetic coupling changes in practice because the permeability of the core usually drops and parasitic resistance and capacitance become severe at higher frequencies. It is important, however, to differentiate indirect reasons from profound reasons of observed phenomena, which cannot be thoroughly explored by experiments but can be clearly understood by appropriate theories or models.

The compensation circuits constitute mostly LC resonant circuits, where the Tx and Rx coils provide inductance L. Note that the coil inductance of the IPT system can be used as a smoothing inductor of the switching converter.

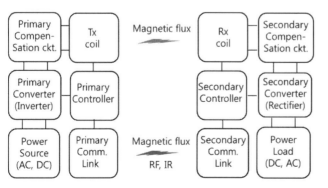

Figure 2.7 Configuration of an IPT system.

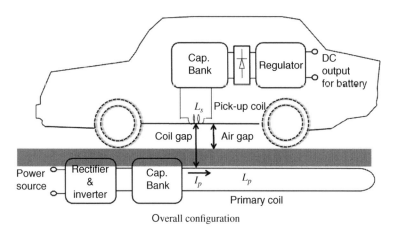

Overall configuration

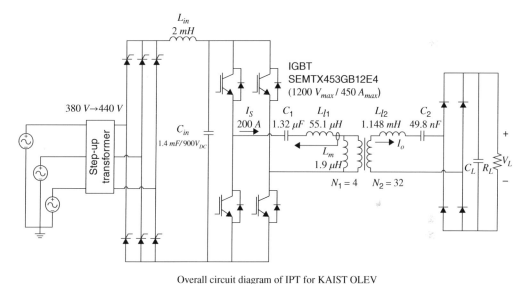

Overall circuit diagram of IPT for KAIST OLEV

Figure 2.8 An example IPT system for EVs.

Depending on applications, the converters, compensation circuits, and coils of an IPT system are different from each other. For instance, the IPT can be classified into (1) resonant versus non-resonant, (2) contact versus contactless, (3) static versus dynamic charging, (4) unidirectional versus bidirectional power flow, (5) current source versus voltage source, (6) degree of freedom (omnidirectional), etc. An example IPT configuration for EVs and its circuit diagram are illustrated in Figure 2.8, where the IPT is resonant, contactless, static or dynamic charging, unidirectional power flow, current source type, and single degree of freedom. Detailed analysis and design of this system will follow in Part III of this book.

Question 4 What is the physical meaning of the power factor of an inverter of IPT? If it is not unity, what is the penalty? How about the power factor of a rectifier of IPT? How can the power factor be calculated? Note that the output of an inverter should be inductive, that is, a non-unity power factor, for a zero voltage switching operation.

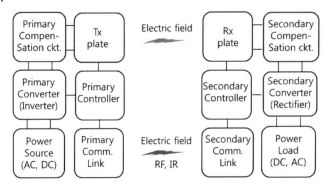

Figure 2.9 Configuration of a CPT system.

2.3 Introduction to Capacitive Power Transfer (CPT)

As discussed, CPT has not been widely used in WPT compared to IPT due to its low power delivery capacity and high operating frequency. CPT has advantages for compact and thin layer applications with low power.

The configuration of a CPT system is shown in Figure 2.9, where other modules are very similar with IPT except for Tx and Rx plates and electric field-based communication links.

As shown in Figure 2.10(a), two pairs of parallel metal plates, which also can be used for the communication link, are crucial for the CPT system. Ideally, there is no cross-linked electric field between two parallel plates and the design of compensation circuits would be very simple;

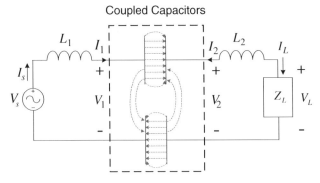

Figure 2.10 An equivalent circuit of a CPT system in the phasor domain, including compensation circuits in the steady state.

(a) Distributed capacitance model with electric field distribution

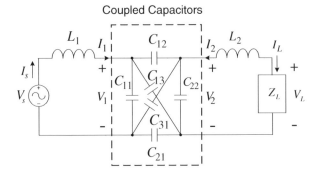

(b) Lumped capacitance model with six capacitors

however, strong leakage electric flux is usually generated in the coupled capacitors. There are six capacitances for the two-port circuit consisting of two pairs of coupled capacitors, as shown in Figure 2.10(b). These capacitances vary according to the air gap and displacement of Tx and Rx plates. Therefore, compensation circuits should cope with these capacitance changes together with load changes so that power delivery can be maximized with high efficiency and high power factor.

Fortunately, the complex capacitance circuits can be degenerated to four capacitances without the loss of generality, which will be shown in Chapter 21 together with a detailed design of CPT for stationary EV chargers.

Like all other WPTs, the CPT has the capability of galvanic isolation between the Tx and Rx plates if their surfaces are insulated, but high voltage protection, grounding, and shielding of the Tx and Rx plates are essential for commercialization.

Question 5 Compared to IPT, why is CPT said to have the lower level of WPT? Note that the operating frequency and voltage ratings of components are typically limiting the WPT power level.

2.4 Introduction to Resonant Circuits

As discussed above, the resonant circuits, which are typically composed of LC components, are mostly used in IPT and CPT as compensation circuits. It is quite often misunderstood, however, that the purpose of resonating LC circuits in the IPT and CPT is to amplify power or energy by the quality (Q) factor. Actually, resonance is not mandatory if the required power delivery is small such that the voltage drop of leakage inductance of the coil may not be large due to a small current. The delivered power or efficiency could be maximized by appropriate resonating circuits; of course, the efficiency is always less than unity.

2.4.1 Non-Resonant IPT

To make the rationale for the resonant circuit clear, a non-resonant IPT circuit is examined, as shown in Figure 2.11. As introduced in previous sections, the IPT is a sort of coupled inductor, that is, a transformer that has a turn ratio of $N_1 : N_2$, leakage inductances L_{l1}, L_{l2}, and magnetizing inductance L_m, assuming no internal resistances. If the source voltage is sinusoidal, the DC gain of the system, which corresponds to the output voltage, can be determined as follows:

$$G_V \equiv \frac{V_o}{V_s} = \frac{nL_m}{L_{l1} + L_m} \cdot \left| \frac{R_o}{j\omega_s \{(L_m//L_{l1})n^2 + L_{l2}\} + R_o} \right| \ll \frac{nL_m}{L_{l1} + L_m} \; for \; small \; R_o, \quad \because n \equiv \frac{N_2}{N_1}$$

$$(2.9)$$

Figure 2.11 A non-resonant IPT circuit in the steady state.

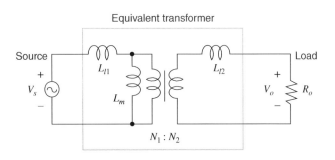

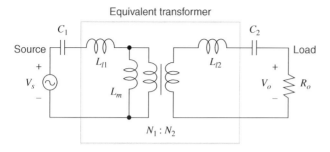

Figure 2.12 A series–series resonant IPT circuit in the steady state.

Note from (2.9) that the output voltage drops rapidly for a small load resistance compared to reactance, which makes large power delivery impossible. For an ordinary transformer, the leakage inductances are relatively small and not a problem, but the leakage inductances of Tx and Rx coils of IPT tend to be large because of loose coupling between the coils.

2.4.2 Leakage Inductance Compensation Method in Resonant IPT

To cancel out the reactance of inductors is crucial for high power delivery, and a viable solution is to insert capacitors in series or in parallel with the Tx and Rx coils, as shown in Figure 2.12.

The selection of capacitances is not straightforward and needs a lot of elaborated thought on its effects. Let us start with a simple idea that each leakage inductance of the primary and secondary sides is cancelled by each capacitance, as shown in Figure 2.13. The resonant conditions that make the impedances zero are as follows:

$$Z_1 = jX_{l1} - jX_{c1} = 0 \qquad (2.10a)$$
$$Z_2 = jX_{l2} - jX_{c2} = 0 \qquad (2.10b)$$

where

$$X_{l1} = \omega_s L_{l1}, \quad X_{c1} = \frac{1}{\omega_s C_1}, \quad X_{l2} = \omega_s L_{l2}, \quad X_{c2} = \frac{1}{\omega_s C_2}, \quad X_m = \omega_s L_m \qquad (2.11)$$

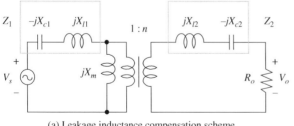

(a) Leakage inductance compensation scheme

Figure 2.13 Leakage inductance compensation scheme in the series–series resonant IPT.

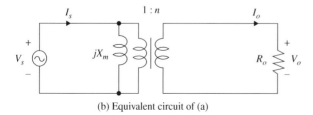

(b) Equivalent circuit of (a)

As identified from Figure 2.13, the output voltage and source current of the IPT become

$$V_o = nV_s \qquad (2.12a)$$

$$I_s = nI_o + \frac{V_s}{j\omega_s L_m} \qquad (2.12b)$$

The advantage of this compensation scheme is that the output voltage is an ideally voltage source determined solely by the turn ratio of the Tx and Rx coils. However, this scheme has a reactive current component for a resistive load, as identified from the right term of (2.12b). This reactive current may be very large for the weakly coupled inductor case, where L_m is very small. Therefore, this leakage compensation scheme is seldom used in practice.

2.4.3 Coil Compensation Method in Resonant IPT

Another method, which is widely used in practice, is to compensate the total inductance of each Tx and Rx coil by each capacitor, as shown in Figure 2.14. The resonance conditions that make each impedance zero become

$$Z_1 = jX_{l1} - jX_{c1} + jX_m = 0 \qquad (2.13a)$$

$$Z_2 = jX_m + (jX_{l2} - jX_{c2})/n^2 = 0 \qquad (2.13b)$$

First, let us analyze the circuit from the source side. As shown in Figure 2.14(b), the source side impedance of the right part of the circuit Z_3 becomes

$$
\begin{aligned}
Z_3 &= jX_m // (jX_{l2} - jX_{c2} + R_o)/n^2 \\
&= \frac{jX_m(jX_{l2} - jX_{c2} + R_o)/n^2}{jX_m + (jX_{l2} - jX_{c2} + R_o)/n^2} \\
&= \frac{jX_m(jX_{l2} - jX_{c2} + R_o)/n^2}{R_o/n^2}, \quad \because jX_m + (jX_{l2} - jX_{c2})/n^2 = 0 \\
&= \frac{jX_m(jX_{l2} - jX_{c2})}{R_o} + jX_m \\
&= \frac{jX_m(jX_{l2} - jX_{c2} + jX_m n^2 - jX_m n^2)}{R_o} + jX_m \\
&= \frac{jX_m(-jX_m n^2)}{R_o} + jX_m \\
&= \frac{X_m^2 n^2}{R_o} + jX_m \equiv R_{eq} + jX_m, \quad \because R_{eq} \equiv \frac{X_m^2 n^2}{R_o}
\end{aligned}
\qquad (2.14)
$$

As identified from (2.14), Z_3 becomes simply the sum of the reactance of the magnetizing inductance and an equivalent resistance that is inversely proportional to the output resistance. This concept that a short circuit in the output side looks like an open circuit from the input side and vice versa is difficult for beginners to understand.

The last step of the analysis from the source side is to find the equivalent circuit for Figure 2.14(c), which results in zero impedance due to the resonant condition of (2.13a). As

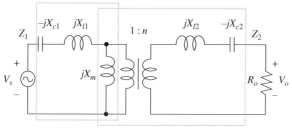

(a) Coil inductance compensation scheme

Figure 2.14 Coil compensation scheme in the series–series resonant IPT, viewed from the source side.

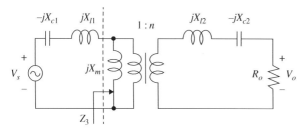

(b) Source side equivalent circuit of (a)

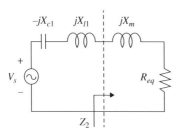

(c) Simplified source side equivalent circuit of (b)

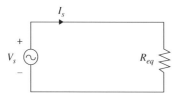

(d) Final simplified source side equivalent circuit of (c)

identified from Figure 2.14(d), the power factor of the source becomes unity for the coil compensation method when the load is resistive, and the source current is determined as follows:

$$I_s = \frac{V_s}{R_{eq}} = \frac{R_o V_s}{X_m^2 n^2} \tag{2.15}$$

Unfortunately, the output characteristics such as the output voltage, output-side impedance, and output power factor cannot be determined from the source-side equivalent circuit of Figure 2.14. This is why we need the load-side equivalent circuit, as shown in Figure 2.15. Because the Thevenin equivalent circuit of Figure 2.15(a) does not exist, the Norton equivalent circuit is obtained, as shown in Figure 2.15(b). (Try to obtain the Thevenin equivalent voltage and impedance by yourself!)

Figure 2.15 Coil compensation scheme in the series–series resonant IPT, viewed from the load side.

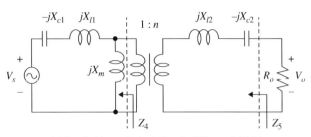

(a) Load-side equivalent circuit of Figure 2.14 (a)

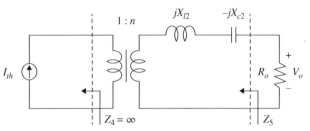

(b) Simplified load-side equivalent circuit of (a)

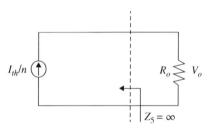

(c) Final simplified load-side equivalent circuit of (b)

The Norton current source and the impedance are determined as follows:

$$I_{th} = \frac{V_s}{jX_{l1} - jX_{c1}} = \frac{V_s}{jX_{l1} - jX_{c1} + jX_m - jX_m} = \frac{jV_s}{X_m} \tag{2.16a}$$

$$Z_4 = jX_m // (jX_{l1} - jX_{c1}) = \frac{jX_m(jX_{l1} - jX_{c1})}{jX_m + jX_{l1} - jX_{c1}}$$
$$= \frac{jX_m(jX_{l1} - jX_{c1} + jX_m - jX_m)}{0} = \frac{jX_m(-jX_m)}{0} = \frac{X_m^2}{0} = \infty \tag{2.16b}$$

In (2.16), the resonance condition of (2.13a) is used and the Norton equivalent circuit is found to be an ideal current source. Therefore, the impedance observed from load side Z_5 is infinite regardless of the secondary-side circuit impedance.

Finally, an ideal current source is observed from the load and the output voltage becomes

$$V_o = \frac{I_{th}}{n} R_o = \frac{jV_s}{nX_m} R_o \tag{2.17}$$

As identified from (2.17), the output voltage has a 90 degree phase difference from the source voltage.

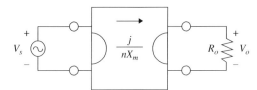

Figure 2.16 A gyrator as the equivalent circuit of the coil compensation scheme in the series–series resonant IPT.

Question 6 Why should we see a circuit from both the source side and the load side, as shown in Figures 2.14 and 2.15, instead of one side? Actually, you do not have to see a circuit from the source side if you are interested in the output voltage or current only.

As identified from Figures 2.14 and 2.15, it can be concluded that the coil compensated IPT is a sort of gyrator, as shown in Figure 2.16. From (2.16a), the gyrator gain is found to be an imaginary value as follows:

$$G \equiv \frac{I_o}{V_s} = \frac{I_{th}/n}{V_s} = \frac{j}{nX_m} = \frac{j}{n\omega_s L_m} \tag{2.18}$$

For example, the impedance viewed from the source side R_{eq} of (2.14) is also found from the gyrator characteristics as follows:

$$R_{eq} = \frac{1}{GG^*R_o} = \frac{X_m^2 n^2}{R_o} \tag{2.19}$$

A merit of the gyrator model is that you do not have to find equivalent circuits from the source side and load side, and now you can understand whole circuit characteristics from the model. A general and systematic gyrator model for IPT with a resonant circuit will be explained in Chapter 5.

It is remarkable to see the resemblance of gyrator equivalent circuits of Figures 2.16 and 2.6 even though there is no direct physical relationship between the two circuits. Do not forget that the gyrator equivalent circuits are only valid for static single frequency cases and are no longer valid for the dynamic transient state or multiple harmonic cases.

One of the drawbacks of the suggested coil-compensating IPT is that there are infinitely large source current flows for the no-load condition, which results in $R_{eq} = 0$. Therefore, the above-mentioned ideal resonant conditions of (2.13) are not used when the output side can be open. Furthermore, the coil parameters vary according to the distance and misalignment of Tx and Rx coils; hence, the resonant conditions cannot be maintained under these conditions.

2.4.4 Other Compensation Methods in Resonant IPT

Other resonant IPTs using two compensation capacitors are series–parallel (SP), parallel–series (PS), and parallel–parallel (PP) resonant circuits, as shown in Figure 2.17. It is also possible to use the current source for the four compensation circuits. Therefore, there are eight compensation circuits using two capacitors, which are analyzed in Chapters 5 and 14, so they are not further explained in detail in this chapter.

The two-capacitor scheme is the simplest configuration of IPT compensation because each Tx coil and Rx coil individually needs reactive compensation. Furthermore, no additional inductor is required for the two-capacitor scheme, which is quite beneficial to fabrication of a low-cost compact IPT system.

In cases where the number of compensation capacitors is not limited, there are a few resonant compensation circuits. One example is the three-capacitor compensation circuit, as shown in Figure 2.18. Like the two-capacitor compensation circuits, the characteristics of this

Figure 2.17 Other resonant IPTs with two capacitors, driven by the voltage source.

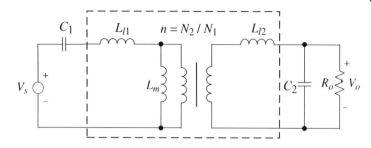

(a) Series–parallel (SP) resonant IPT

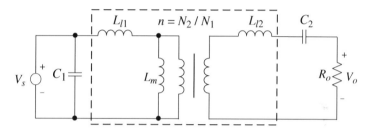

(b) Parallel–series (PS) resonant IPT

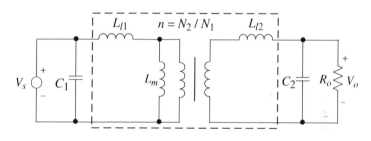

(c) Parallel–parallel (PP) resonant IPT

three-capacitor compensation circuit can be quite different from each other, depending on the resonant scheme, even though the circuit topology is the same. As shown in Figure 2.18(a), each inductance is compensated by a respective capacitor, and no reactive component remains after complete compensation, as shown in Figure 2.18(b).

Another category of compensation circuit is LCC, where an additional inductor is used together with two capacitors, as shown in Figure 2.19. You should be able to find other compensation circuits with more inductors and more capacitors; however, it is important to understand the basic two-capacitor compensation circuits and the LCC circuit because other compensation circuits quite often exploit the characteristics of them.

In order to understand this LCC resonant compensation, the LC resonant tank, as shown in Figure 2.20, is examined. As shown in Figure 2.20(b), the Norton equivalent circuit of Figure 2.20(a) is composed of a current source and impedance defined as follows:

$$I_{in} = \frac{V_s}{j\omega_s L_s} \tag{2.20a}$$

$$Z_{in} = j\omega_s L_s // \frac{1}{j\omega_s C_s} \tag{2.20b}$$

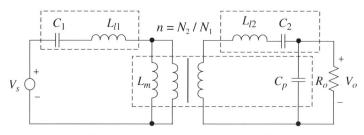

(a) Individual series and parallel resonant LC tanks

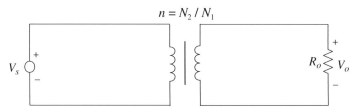

(b) An equivalent ideal transformer circuit without any reactive component when exactly tuned

Figure 2.18 Respective inductance compensation scheme in the three-capacitor resonant IPT.

In (2.20b), Z_{in} becomes infinite when the LC resonant tank is tuned to its resonant frequency, as shown in Figure 2.20(c).

Applying the Norton equivalent circuit of Figure 2.20 to the LCC circuit of Figure 2.19 results in Figure 2.21, where the equivalent current source is the same as (2.20a) and the equivalent voltage source is as follows:

$$V_{th} = j\omega_s L_m I_{in} = j\omega_s L_m \frac{V_s}{j\omega_s L_s} = \frac{L_m}{L_s} V_s = V_o \tag{2.21}$$

It is noted from (2.21) that the LCC circuit has a voltage source characteristic with in-phase source voltage when it is tuned to the secondary side resonant circuit, that is, $Z_{out} = 0$, as shown in Figure 2.21(c). A very useful feature of the LCC circuit, which is one of the rationales for popular use, is that this compensation circuit is robust to primary-side coil parameter change. As identified from Figure 2.21(a), the transfer function of the IPT system is insensitive to the changes of L_{l1}, L_m, and C_1 due to the current source behavior of the primary LC resonant tank. As long as the impedance of the secondary resonant tank Z_{out} is kept at zero, the IPT system is equivalent to an ideal voltage source. Of course, this wonderful characteristic is obtained by sacrificing compactness and cost of the additional primary-side LC tank. Furthermore, Z_{out} is not kept at zero and varies due to coil parameter change, which results in detuning of the resonant IPT.

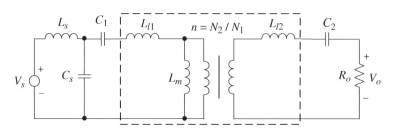

Figure 2.19 Primary LCC compensation circuit of resonant IPT.

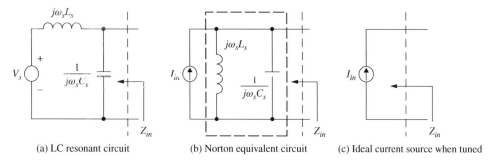

(a) LC resonant circuit   (b) Norton equivalent circuit   (c) Ideal current source when tuned

Figure 2.20 Norton equivalent circuit of the LC resonant circuit.

2.4.5 Discussion on Resonant Circuits

As discussed so far, the IPT system, adopting resonant compensation circuits, can deliver large power compared to the non-resonant IPT. By appropriate selection of the topology and resonant scheme of the compensation circuit, the IPT could be equivalent to either an ideal voltage

Figure 2.21 Circuit-based analysis of the primary LCC compensation circuit of Figure 2.19 in the static phasor domain.

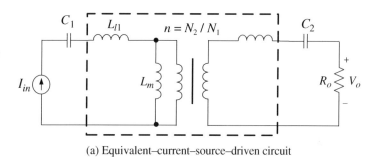

(a) Equivalent–current–source–driven circuit

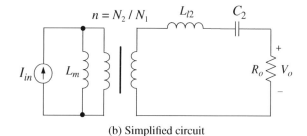

(b) Simplified circuit

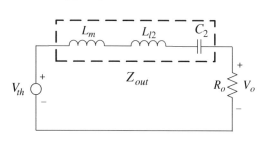

(c) Ideal voltage source when tuned ($Z_{out} = 0$)

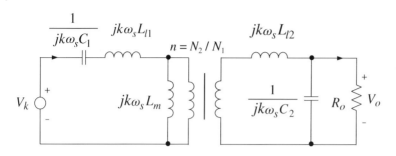

(a) Equivalent circuit for the kth harmonics from an inverter

Figure 2.22 Switching harmonic filtering characteristics of an example resonant compensation circuit of IPT.

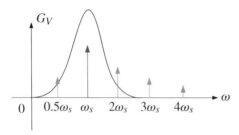

(b) Transfer function of the IPT system with the kth harmonics of an inverter

source (as shown in Figures 2.13, 2.14, 2.18, and 2.21) or an ideal current source (as shown in Figures 2.15 and 2.20).

It can be said that power flow in a WPT may be improved by nullifying the reactance of inductance or capacitance. However, the characteristics of compensation circuits vary due to distance and misalignment change of coils as well as load change.

Besides nullifying the reactance of Tx and Rx devices, resonant compensation circuits have numerous other characteristics:

1) Filtering out switching harmonics. First, the resonant compensation circuit is usually a good harmonics filter, which passes the fundamental component of a switching converter through LC resonance and inhibits higher or lower switching harmonics, as shown in Figure 2.22. The example series–parallel compensation circuit may effectively filter out harmonics, which are mainly odd harmonics. Usually, the currents of Tx and Rx are nearly sinusoidal due to good bandpass filtering of the resonant IPT.
2) Slowed dynamics and transient peaks. The penalties for using resonating compensation circuits are slowed system dynamics and transient peaks. Like other filters, the resonant compensation circuit has a finite dynamic response time. It is not easy, however, to calculate the response time and to analyze the system dynamics of an IPT system because the system order is high.

Furthermore, the system response for the sinusoidal voltage source requires special care. For example, an LC resonant tank, which is the simplest form of resonant compensation circuits, may be connected to a diode rectifier load, as shown in Figure 2.23(a). In case the source frequency is tuned to the resonant frequency, the envelope behavior of the resonant circuit can be analyzed by the phasor transformation [1,2]. As identified from Figure 2.23(b), the transient response of the LC resonant tank indicates that the peak current or voltage may be several times the steady-state value, which results in substantially increased current and voltage ratings of capacitors and coils of the IPT system.

Figure 2.23 An example of LC resonant compensation circuit connected to a rectifier load.

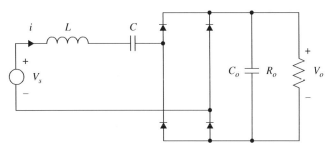

(a) An LC resonant circuit driven by a sinusoidal voltage source

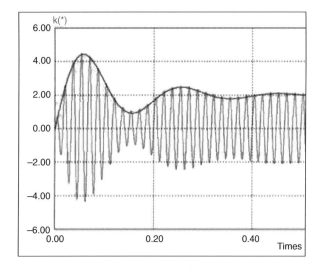

(b) Transient current waveform of the LC tank

2.5 Conclusion

Among various WPT, such as IPT, CPT, RF PT, and optical PT, to select the WPT method is often the first step we encounter. For long-distance applications, IPT and CPT may not be preferred but RF PT and optical PT may be considered. For submarine applications, galvanic isolation as well as penetration of water become important and only proximity power transfer is allowed for the four above-mentioned methods. For longer distance submarine applications, none of the methods will be appropriate but acoustic power transfer may be an alternative, although it is not introduced here.

As discussed so far, the Tx and Rx devices, compensators, and converters are key factors in the maximum power and high efficiency of a WPT system. These modules have specific system dynamics and often operate in nonlinear modes. The static behaviors of a WPT system in the steady state may not be the sufficient condition for a system design but may be the only necessary condition. The dynamic behaviors of the WPT system in the transient state always must be considered for a practical system design. Even though the dynamics of the WPT system is important, it has not been sufficiently dealt with in the introduction chapters because of the difficulties in understanding for beginners. In Chapters 13 and 14, the dynamic behaviors of WPT systems are introduced together with controller design issues.

References

1 Chun T. Rim, "Unified general phasor transformation for AC converters," *IEEE Trans. Power Electron.*, vol. 26, no. 9, pp. 2465–2475, September 2011.

2 S. Lee, B. Choi, and Chun T. Rim, "Dynamic characterization of the inductive power transfer system for online electric vehicles by Laplace phasor transform," *IEEE Trans. Power Electron.*, vol. 28, no. 12, pp. 5902–5909, December 2013.

3

Introduction to Electric Vehicles (EVs)

3.1 Overview of EVs

3.1.1 A History of EVs

Not surprisingly, the invention and commercialization of EVs was earlier than that of internal combustion engine (ICE) vehicles. The invention of ICE dates back to 1860, which is more than 150 years ago, and ICE vehicles became widely commercialized in 1900 [1, 2]. On the other hand, EV was first invented in 1827, more than 30 years earlier than that of ICE, and EV became commercialized before 1839 [3, 4]. At first, EV was preferred among steam, gasoline, and electric engines because of its comfort and ease of operation compared to gasoline cars of the time. To initiate a gasoline engine at that time, mechanical rotation torque had to be externally imposed by a hand crank, which is definitely difficult for a lady.

EVs were the fastest and longest vehicles in the early 1900s until the advent of powerful and light ICE cars. In 1900, about a third of the cars sold in the United States were known to be EVs. Before long, however, EVs were replaced by ICE cars due to a few historical events. In the early 1900s, road infrastructure was significantly improved, which made a vehicle possible for a longer distance drive than that offered by EVs of about 100 km at that time. Furthermore, large reserves of petroleum were discovered in Texas, California, and Oklahoma, which eventually made gasoline cars cheaper to operate than EVs. Cumbersome problems such as the hand crank for starting a gasoline engine and the noise generated by ICE cars became mitigated thanks to the inventions of the electric starter and muffler, respectively. At last, the mass production of gasoline cars by Henry Ford in 1913 made their cost cheaper than EVs, which became the ignitor for EVs to rapidly disappear from markets. Since then, ICE has been the dominant engine for automobiles for almost 100 years, but EVs did not disappear and still remained in a few areas such as trains and small vehicles.

Since 1990, General Motors started to develop modern EVs such as Impact and EV1. Then Chrysler, Ford, BMW, Nissan, Honda, and Toyota also produced limited numbers of EVs. Many of them, however, could not have succeeded in expanding their market share significantly due to the limited driving distance and relatively high price of EV, which have long been the same obstacles to commercialization of EVs. Recently, global warming and air pollution due to the petroleum-based transportations together with the peak oil price have drawn public attention again to EV. Furthermore, the price and energy density of the lithium battery for an EV have dramatically enhanced its desirability during the last two decades.

Though the global automobile market share of EV is still just a few percent in 2016, its growth rate is about 100% each year. In 2015, the world's best-selling pure-battery EV (BEV) was Nissan Leaf with global sales of 200 000 units and Tesla Model S with about 100 000 units [5]. Now the

Wireless Power Transfer for Electric Vehicles and Mobile Devices, First Edition. Chun T. Rim and Chris Mi.
© 2017 John Wiley & Sons Ltd. Published 2017 by John Wiley & Sons Ltd.

driving range of EV on a charge has extended to more than 300 km in 2016; however, the price of an EV without a tax incentive compared to an equivalent ICE car is still not very competitive. Including a plug-in hybrid EV (PHEV), EV has rapid growing market penetration, noticeably in Norway, China, the USA, and Japan.

3.1.2 Merits and Demerits of EVs

The EV has merits over fossil fuel-powered vehicles in terms of the electricity they consume, which can be generated from various energy sources such as fossil fuels (coals, petroleum, and natural gas), nuclear power, and renewable energy sources (wind power, solar power, hydraulic power, and tidal power).

The CO_2 and other emissions of an EV may be similar or less than ICE vehicles, depending on the fuel and technology used for electricity generation for the EV. Zero emission during drive is still a merit of an EV because it can significantly reduce air pollution in urban areas, which is a serious problem in modern large cities, especially in China.

Though the efficiency from battery to wheel of an EV is usually higher than 90%, power is lost during transmission, which varies from about 4% (South Korea) to about 30% (USA). This demerit of an EV is compensated by regenerative braking, which recovers vehicle momentum energy and stores it in the on-board battery.

Moreover, ICE vehicles also have similar problems concerning delivery of energy from the viewpoint of dwell-to-vehicle. Refueling or charging stations are an expensive social infrastructure, which should not be missed in cost comparisons. All these economic concerns are reflected in the price of refueling. It is hard to exactly compare the refueling price in general because it varies depending on countries, region, oil/gas prices, power generation and transmission costs. If I suggest a rough number on that, the refueling cost of an EV in South Korea is about 0.015 $/km, assuming 0.1 $/kWh and 0.15 kWh/km for a passenger car, whereas that of an ICE car in South Korea is about 0.060 $/km, assuming 1.2 $/liter and 20 km/liter. The calculated refueling cost of an ICE car is four times that of an EV in South Korea. If another assumption of 0.10 kWh/km for an EV and 15 km/liter for an ICE car is used, the relative refueling cost becomes as large as eight times. This relative cost of an ICE car could be less in other countries such as the USA, where the gas cost is cheaper and the electricity cost is more expensive than in South Korea.

Question 1 Calculate the refueling cost of EV and ICE buses in your country. For a fair comparison, assume that the number of passengers, air conditioning, and road conditions are the same in each case.

3.1.3 Configuration of EVs

Generally speaking, an EV comprises a conventional car body and EV specific components such as energy refuel equipment (typically, electric charger), on-board energy storage (typically, on-board battery), power converter (typically, inverter), and locomotive (typically, electric motor and transmission), as shown in Figure 3.1. It is instructive to find an analogy between EV and ICE cars: electric charger versus fueling station, on-board battery versus fuel tank, inverter versus fuel injector, and electric motor versus internal combustion engine.

Figure 3.1 A general configuration of an EV, charged by both wired and wireless chargers.

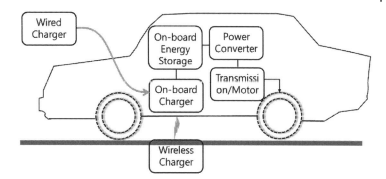

3.2 Classification of EVs

Depending on the energy source/refueling method, EV specific components mentioned above, and various applications, EVs can be classified into many categories, as will be explained in the following.

3.2.1 Classification by Energy Source or Refueling

One of the reference points of EV classification is the energy source or energy refueling method, such as external refueling and on-board generators.

External refueling methods can be further classified as follows:

- Wired-power EVs. Direct contact to long airway/ground power lines has been used for electric trains (subway, ground trams, high-speed trains) and trolley trucks/buses. Widely used EVs with a cable charger as well as the tethered aircraft can also be classified as wired-power EVs.
- Wireless-powered EVs (WEVs). Road-powered EVs (RPEVs) such as on-line electric vehicles (OLEVs) collect electric power contactless from buried high-frequency power cables under the road surface through magnetic induction. EVs with a stationary wireless charger fall into the category of WEVs. Small carrier vehicles with contactless power systems (CPSs) widely used in industry are also WEVs.
- Solar/RF-powered EVs. Other energy sources than electricity such as solar and RF energy can drive EVs, which could be ground vehicles, aerial vehicles, and water vehicles. Solar passenger cars, solar drones, solar ships, and RF stratosphere drones are good examples of potential future applications.
- Battery-swap EVs (BSEVs). This is a special case of an energy refueling method, where the discharged battery is replaced with a charged battery, usually by robots. The demerits of this BSEV are additional battery cost and expensive swap robot systems, whereas the merits are a quick recharging time and longer lifetime of batteries due to optimized slow charging.

On-board generators provide for EVs with electric power from many energy sources as follows:

- ICE-powered EVs. On-board ICEs such as gasoline, diesel, or liquid petroleum gas (LPG) engines can generate electricity or mechanical power for EVs, which are called hybrid EVs (HEVs). There are many possible combinations of ICE, battery, transmission, and motor/generator in HEVs.

- Fuel-cell-powered EVs. On-board fuel-cell EVs (FCEVs) utilize electricity directly generated from a fuel cell.
- Nuclear-powered EVs. On-board nuclear power plant or nuclear battery can generate long-endurance electricity without refueling for several months or several years, which is crucial for nuclear submarines and spacecraft.

Combinations of external refueling and on-board generator are also possible. Plug-in hybrid EV (PHEV) is an example, where an external rechargeable wire/wireless power equipment and an on-board ICE are installed together with a battery.

3.2.2 Classification by Components

EVs can be classified as major components such as on-board energy storage, power converter, and locomotive.

On-board energy storages used in EVs are as follows:

- Battery. Rechargeable chemical batteries such as an Li-ion battery are widely used in a battery EV (BEV), which is also referred to as a pure EV (PEV) or sometimes a full EV (FEV). Non-rechargeable batteries may be used in toy vehicles, which are not often referred to as EVs. The battery may be recharged while stationary or during moving.
- Supercap. Capacitors with a larger capacitance and higher energy density become available and used for quick charge and discharge EVs. Different from a BEV, a supercap EV (SEV) has an infinitely long charge/discharge cycle with a much higher (typically 20 times) instantaneous power capacity. Most advanced supercaps now being developed have nearly the same energy density (kWh/kg) as an Li-ion battery.
- Flywheel. For a short period of time, a flywheel can store kinetic energy efficiently and may be utilized to EVs. Of course, a flywheel alone cannot provide enough energy to drive for long distances.
- Hybrid energy storage. Because energy storages, mentioned above, have merits and demerits, hybrid energy storages are often used. A battery supercap is a good combination, where instantaneous/frequent high power and regeneration is provided by the supercap whereas long-endurance low power is provided by the battery. A converter usually connects the supercap and battery. In this way, battery life is extended and overall efficiency is improved.

Power converters and locomotives used in EVs are highly correlated with each other and can be classified as follows:

- Inverter–AC motor. Because DC power is available in most EVs, a DC–AC converter, called an inverter, is used in an EV, where AC motors such as three-phase induction motors (IM) and brushless DC motors (BLDC) are adopted. Speed and torque control is crucial for the inverter–motor system to respond to a driver's command. The inverter–motor together with a transmission module often becomes a combined system.
- Chopper–DC motor. Traditionally, DC motors have been used for some EVs where a large starting torque is required, that is, electric trains (often subway trains and trams) and forklift trucks. A DC–DC converter, called a chopper, is used to drive the DC motor.

3.2.3 Categories of EVs: Applications

EVs can be classified into applications such as ground, aerial, sea (ship/underwater), and space EVs.

Ground vehicles, as already discussed above, include BEVs, PEVs, PHEVs, FCEVs, and SEVs. They can be also classified as roadway EVs and railway EVs.

The roadway EVs can be further classified according to applications as electric passenger cars, electric buses, electric taxis, electric trolleys, electric trailers, electric trucks, electric motorcycles, electric bicycles, electric scooters, neighborhood electric cars, electric golf carts, electric milk floats, electric forklifts, and electric wheelchairs. A special application of ground vehicles would be space rovers, which include manned and unmanned space vehicles used for exploring the Moon and Mars with solar power.

Railway EVs can also be further classified according to applications such as light rail trains, subway trains, trams (street trains), and high-speed (rapid transit) trains. Usually they get power from the overhead wire through pantographs. Compared to conventional diesel locomotives, the EV trains have much better power-to-weight ratios and higher surge power for fast acceleration as well as regenerative brakes for higher energy efficiency. Another special railway EV will be the maglev train, where no mechanical rail exists but a magnetic levitation rail supports the floating vehicle, which is usually supplied by wireless power and driven by linear induction motors. However, the maglev train is very expensive and needs a lot of electric energy to sustain the floating vehicle until permanent magnets are adopted.

Airborne EVs once received great enthusiasm at the beginning era of aviation, which has now been revived. Recently, rechargeable battery drones are rapidly attracting people as a means of taking pictures, monitoring, and transporting goods. As mentioned above, stratosphere drones powered by RF from the ground have many opportunities to perform local satellite missions. Solar or fuel-cell-powered drones are also extensively being tested as potential future drones. Manned aircrafts can be flown by rechargeable battery, fuel cell, solar power, and RF power from the ground, which can be adequate for a short-distance flight of usually less than two hours or 1000 km.

Seaborne EVs include battery boats, solar boats, electric ferries, and submarines. Compared to airborne EVs, seaborne EVs have lot of space and weight available for energy storages but require a lot of energy due to low transportation efficiency. Therefore, the large charging time and power for battery boats could be one of the difficult problems for commercialization. For solar boats, a problem is that the required surface of a solar cell is typically about ten times that of a boat. Apart from this problem, electric boats are quiet and could have the infinite range of sail boats. Submarines often use batteries during many periods of their missions in order to keep quiet; these are sometimes charged by diesel or gasoline engines at the surface. Some of them are alternatively powered by nuclear power and fuel cells.

Spaceborne EVs have a long history of use. Batteries, solar cells, and sometimes a nuclear power system have long been used for the power sources of spacecraft. For propelling a spacecraft with electricity, the electrostatic ion thruster, the arc jet rocket, and the Hall effect thruster are under research.

Question 2 (1) Suggest a new combination of energy source, refueling method, energy storage, and application of EVs that has never been proposed. (2) Suggest any combination of energy source, refueling method, energy storage, and application of EVs that is not feasible or is inappropriate.

3.3 Technical and Other Issues on EVs

Though the number of EVs has been rapidly growing recently, there are still many technical problems as well as economic and environment issues to be resolved for widespread use.

If we restrict our discussions to typical ground EVs such as BEVs, PHEVs, and RPEVs, probably the biggest bottleneck is the battery related problem: the battery itself and battery charging problems. Battery size, cost, life, charging time, and freedom of recharging are expected to improve.

One of the important technical issues is electric safety, which includes electromagnetic field (EMF) emission, electromagnetic compatibility (EMC), and electric shock protection when the magnetic field is used for wireless power transfer and a high voltage of battery is adopted for higher power and better efficiency. Higher than 300 V DC is now widely used in EV passenger cars and buses.

Concerning energy efficiency, the "tank-to-wheel" efficiency of an EV is definitely a few times higher than an equivalent level ICE car but "well-to-wheel" efficiency of an EV is not always higher than the most efficient ICE cars. Well-to-wheel efficiency of an EV is related to not only the EV itself but also to the method of electricity production. For a fair comparison, the "well-to-wheel" efficiency of an ICE car should include energy spent on exploration, mining, refining, transporting, and fueling, as discussed earlier.

The recharging cost and charging infrastructure for EVs is an important economic issue, which needs to be also compared to ICE cars, where the refueling cost and charging stations are a few times more expensive than that of EVs.

Grid infrastructure is an important technical and economic issue concerning EVs. If not properly controlled, a few power plants and more grid infrastructure are required for massive commercialization of EVs. However, if EV charging is made mainly at night when there is a lot of unused electricity, this burden of the grid would be significantly mitigated. It could even help to stabilize the grid if vehicle-to-grid (V2G) connection is made when grid power is in deficiency. This V2G power may reduce the need for new power plants, but it reduces the battery life cycle and restricts the freedom of driving at any time. Therefore, V2G is not a viable solution until an innovative battery together with a high-efficiency converter become available.

A range of anxiety problems of EVs must also be resolved. An innovative battery with a few times higher energy density and a lower price with a few minutes of charging time for a thousand times of life cycles would be a solution, which still has the problems of explosion due to large energy storage and large capacity of recharging electric facilities. Another more viable solution is roadway wireless charging, which inherently requires no battery and utilizes any chances to get charged on the way.

Heating of EVs is a serious problem in cold climates because a substantial amount of energy is consumed to keep the interior of the vehicle warm and to defrost the windows, which is a simple task for conventional ICE cars, where heat already exists as waste. If an EV is connected to the grid, however, it can be preheated or precooled without using battery energy.

Question 3 (1) Evaluate the feasibility of an SEV in terms of solar energy available in a day and daily driving distance. (2) Design a deployable solar panel adequate for an EV to meet the requirement of (1).

One of the social effects of EVs lies in its high public transit efficiency, which is achieved by shifts from private to public transportation such as trains, trams, and buses in terms of individual km/kWh. Technically and economically, public transportation easily exploits EVs and is readily available in many countries, though private EVs are now struggling to compete with ICE cars.

Government incentives and promotion for EVs are globally widespread for reducing air pollution and fossil fuel consumption as well as encouraging technology innovation.

Question 4 Design a government incentive to promote widespread use of EVs. If possible, suggest an incentive requiring no government budget.

References

1 H.O. Hardenberg and O. Horst, *Samuel Morey and His Atmospheric Engine*, Warrendale, PA, 1992.

2 D. Clerk, *Gas and Oil Engines*, Longman Green & Co, 7th edn, 1897, pp. 3–5.

3 M. Guarnieri, "Looking back to electric cars," *Proc. HISTELCON 2012: The Origins of Electrotechnologies.* doi:10.1109/HISTELCON.2012.6487583.

4 A.P. Loeb, "Steam versus electric versus internal combustion: choosing the vehicle technology at the start of the automotive age," *Transportation Research Record, Journal of the Transportation Research Board of the National Academies*, No. 1885, at 1.

5 J. Cobb, *Plug-in Pioneers: Nissan Leaf and Chevy Volt Turn Five Years Old*, HybriCars.com, 2015. doi: http://www.hybridcars.com/plug-in-pioneers-nissan-leaf-and-chevy-volt-turn-five-years-old/.

Part II

Theories for Inductive Power Transfer (IPT)

Three fundamental theories useful for the analysis and design of inductive power transfer (IPT) are introduced.

The coupled inductor model is introduced as a basis theory of the IPT, which provides a generalized understanding of IPT principles, relating different models.

The gyrator circuit model has been recently introduced by the author as a circuit-based modeling technique for the IPT, which provides physical insights as a circuit friendly tool. It is expected that this model will be one of the best useful tools requisite for analyses and designs of IPT.

The magnetic mirror model has also been recently introduced by the author even though an ideal magnetic mirror model was available before. This enhanced magnetic mirror model is applicable to a core plate with finite length, which is quite useful in practice for calculation of the inductance and magnetic field.

The general unified dynamic phasor has also been recently introduced by the author. Static and dynamic performances of any linear AC circuits including a IPT system can be analyzed by the theory without much effort in manipulating equations. By inspecting an equivalent stationary circuit, readers can deal with AC converters and IPT systems like conventional DC circuits.

Wireless Power Transfer for Electric Vehicles and Mobile Devices, First Edition. Chun T. Rim and Chris Mi.
© 2017 John Wiley & Sons Ltd. Published 2017 by John Wiley & Sons Ltd.

4

Coupled Coil Model

4.1 Introduction to Coupled Coils

In order to model a generalized IPT, let us consider two coupled coils that are driven by two independent current sources, respectively, as shown in Figure 4.1. Note that the definition of currents and voltages at the primary and secondary coils is symmetrical in Figure 4.1(a); this is intentionally done to make the following derivations so general that the primary coil and secondary coil are not necessarily a transmitting coil and a receiving coil, respectively. The direction of wireless power flow can be arbitrary in coupled coils. It could be convenient and less confusing, however, if the direction of the secondary current source is reversely defined, as shown in Figure 4.1(b), where the definition of magnetic flux is unchanged. For this asymmetric definition of current sources, it appears that the power flows from the primary coil to the secondary coil; however, this is not true and there is no preference of power flow as discussed. The convention of current direction is just for convenience and does not change the physical performance at all.

Note the polarity of the two coupled coils, where the secondary side dot is shown at the bottom of Figure 4.1(a) in order to keep the notation such that all currents flowing into the dot generate the same direction of magnetic flux. The secondary side dot is at the upper side in Figure 4.1(b), where the secondary current direction is reversed but the direction of magnetic flux generated by the secondary current is unchanged. Keep in mind that the direction of interchanging magnetic fluxes ϕ_{12} and ϕ_{21} is reversely defined.

The currents and voltages of the coils are not necessarily sinusoidal but can be any waveform, unless otherwise specified in this chapter. Furthermore, the magnetic coupling between two coils is not necessarily high and can be arbitrarily small. However, it is assumed throughout this chapter that all coils are free from any parasitic capacitance and nonlinearity, and that the operating frequency of the IPT is low enough that we do not have to consider electromagnetic radiation and transmission delay. The coil may be coreless or with a core as long as it is not nonlinear.

According to Ampere's law, magnetic flux is generated from any conducting current at any place over a wire, as shown in Figure 4.1. Some of the generated magnetic flux of the primary coil intersects the secondary coil (ϕ_{11}) and the rest circulates (ϕ_{12}). This is true for the secondary coil, and the intersected magnetic flux is ϕ_{21} and the self-circulating magnetic flux is ϕ_{22}, as shown in Figure 4.1. Here, the parasitic inductance of wire of Figure 4.1 is not counted in the coupled coil model, but it can be separately considered as a lumped inductance. Therefore, let us focus on the magnetic flux generated from the coils only.

We will now focus on the voltage of primary and secondary sides. According to Faraday's law, the voltage of a coil can be determined by the time derivation (denoted by the upper dot)

Wireless Power Transfer for Electric Vehicles and Mobile Devices, First Edition. Chun T. Rim and Chris Mi.
© 2017 John Wiley & Sons Ltd. Published 2017 by John Wiley & Sons Ltd.

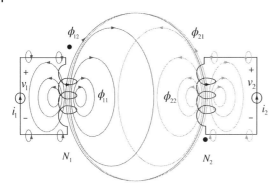

Figure 4.1 Two coupled coils driven by arbitrary independent current sources, respectively.

(a) Symmetric definition of current sources

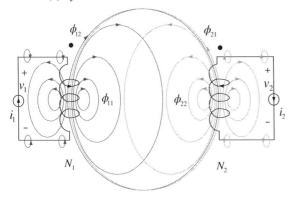

(b) Asymmetric definition of current sources

of magnetic flux intersecting the coil for Figure 4.1(a) as follows:

$$v_1 = N_1(\dot{\phi}_{11} + \dot{\phi}_{12} - \dot{\phi}_{21}) \tag{4.1a}$$

$$v_2 = N_2(\dot{\phi}_{22} + \dot{\phi}_{21} - \dot{\phi}_{12}) \tag{4.1b}$$

For the asymmetric case of Figure 4.1(b), the polarity of v_2 is reversed as follows:

$$v_1 = N_1(\dot{\phi}_{11} + \dot{\phi}_{12} - \dot{\phi}_{21}) \tag{4.1c}$$

$$v_2 = N_2(\dot{\phi}_{22} + \dot{\phi}_{21} - \dot{\phi}_{12}) \tag{4.1d}$$

The reason for polarity inversion for v_2 in (4.1d) is that Faraday's law for induced voltage is defined with respect to (w.r.t.) the current direction that generates magnetic flux. As the polarity of the secondary current source changes without changing the direction of the magnetic flux generated by the secondary current, the polarity of induced voltage v_2 should be changed to keep the same relationship of (4.1b).

Do not underestimate the difficulty of maintaining the correct sign in the definition of variables. It is very easy to get confused and readers must be cautious of the polarity of current, voltage, power, and field. I call this polarity problem the "-1x problem." Together with the "2x problem," which arises from the symmetric structure of phenomena in time, frequency, and space, these 2x problems are the biggest source of confusion and misunderstanding. Caution and patience are thus necessary when defining the polarity and symmetric configuration of phenomena.

In (4.1), it is assumed that all the magnetic fluxes intersect all windings N_1 and N_2 of each coil, and no magnetic flux intersects the fraction of the windings. Figure 4.1 intentionally shows the case when magnetic flux intersects only a part of the winding, which can be modeled by many coupled transformers. In order to avoid complicated discussions, only the simple case is considered, which would not deteriorate the generality of the proposed theory.

From Ampere's law, it is identified that each magnetic flux is proportional to its corresponding current for any waveform as follows:

$$\phi \ \ \alpha \ \ i, \quad \because \phi = \int_S \mathbf{B} \cdot d\mathbf{S}, \quad \mathbf{B} = \mu \mathbf{H}, \quad \oint_l \mathbf{H} \cdot d\mathbf{l} = Ni \tag{4.2}$$

The inductance is canonically defined by equating the induced voltage of an inductor as follows:

$$v = N\dot{\phi} = L\dot{i} \Rightarrow L \equiv \frac{N\phi}{i} \tag{4.3}$$

Applying (4.3) to (4.1) results in the following equations:

$$v_1 = N_1(\dot{\phi}_{11} + \dot{\phi}_{12} - \dot{\phi}_{21}) = L_{l1}\dot{i}_1 + (L_{12}\dot{i}_1 - L_{21}\dot{i}_2/n) \ \text{ for the symmetric case} \tag{4.4a}$$

$$v_2 = N_2(\dot{\phi}_{22} + \dot{\phi}_{21} - \dot{\phi}_{12}) = L_{l2}\dot{i}_2 + (L_{21}\dot{i}_2 - nL_{12}\dot{i}_1) \ \text{ for the symmetric case} \tag{4.4b}$$

$$v_1 = N_1(\dot{\phi}_{11} + \dot{\phi}_{12} - \dot{\phi}_{21}) = L_{l1}\dot{i}_1 + (L_{12}\dot{i}_1 - L_{21}\dot{i}_2/n) \ \text{ for the asymmetric case} \tag{4.4c}$$

$$v_2 = -N_2(\dot{\phi}_{22} + \dot{\phi}_{21} - \dot{\phi}_{12}) = -L_{l2}\dot{i}_2 - (L_{21}\dot{i}_2 - nL_{12}\dot{i}_1) \ \text{ for the asymmetric case} \tag{4.4d}$$

where the inductances and the turn ratio are defined as follows:

$$L_{l1} \equiv \frac{N_1\phi_{11}}{i_1}, \quad L_{l2} \equiv \frac{N_2\phi_{22}}{i_2}, \quad L_{12} \equiv \frac{N_1\phi_{12}}{i_1}, \quad L_{21} \equiv \frac{N_2\phi_{21}}{i_2} \tag{4.5a}$$

$$n = \frac{N_2}{N_1} \tag{4.5b}$$

From (4.4), a lumped-circuit-element model can be drawn, as shown in Figure 4.2. Note the polarity of the two coupled coils, where the secondary side dot is now shown at the bottom in order to keep the notation where all current flowing into the dot generates the same direction of magnetic flux, which is also applicable to Figure 4.1, although the dot is not shown.

In Figure 4.2, there is now no leak magnetic flux between the coupled coils, and the co-sharing magnetic flux becomes

$$\phi_m \equiv \phi_{12} - \phi_{21} \tag{4.6}$$

In (4.6), it is still not necessary for the two magnetic fluxes to be connected with each other even though we are now considering both of them together.

In the subsequent sections, we will find appropriate circuit models for the magnetic model of Figure 4.2, which is not analyzable by circuit theories due to the variables of magnetic flux.

4.2   Transformer Model

It is necessary to replace the magnetic coupling part of Figure 4.2 with an electric circuit to find an equivalent circuit of the IPT. Hence, the magnetic coupling part excluding the leakage inductance part of Figure 4.2 is redrawn to focus on, as shown in Figure 4.3, where the polarity

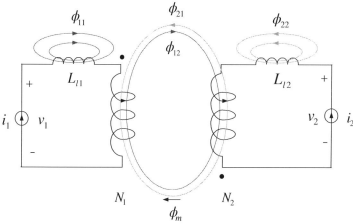

(a) Symmetric definition of current sources

Figure 4.2 A simplified lumped-circuit-element model for the two coupled coils.

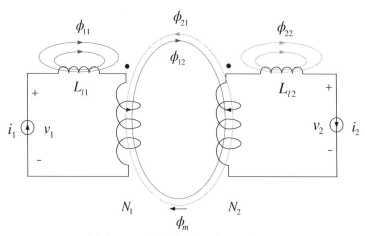

(b) Asymmetric definition of current sources

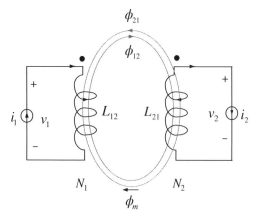

Figure 4.3 A simple magnetic coupling coil model without leakage inductances.

of the secondary current source is reversed but the direction of magnetic flux ϕ_{21} is unchanged because of the upper side dot, as identified from Figure 4.2(b).

Compared to Figure 4.2(a), the coupled coil, as shown in Figures 4.2(b) and 4.3, is often used because it matches well with the physical polarity of power transfer from the primary coil to the secondary coil. All the primary and secondary voltages and currents are in phase (positive) when it is powering from left to right. Note that the magnetic fluxes generated from the primary and secondary sides cancel each other when the primary and secondary currents are in phase, which results in a reduction of the magnitude of ϕ_m. The "physically right" model of Figure 4.3 may, however, mislead beginners to believe that the power flow should be from the primary coil to the secondary coil and that ϕ_{21} is always in the reverse direction of ϕ_{12}. As introduced in the previous section, there is no preference of polarity of power flow, and the phase and polarity of any voltage and current are arbitrary. Even the frequency and phase of the primary and secondary current sources could be different from each other.

Let us find the relationship between L_{12} and L_{21} of Figure 4.3, which are the inductances of primary and secondary sides, respectively, sharing magnetic flux ϕ_m. From (4.5a) and (4.6), L_{12} is found as follows:

$$L_{12} \equiv \frac{N_1\phi_{12}}{i_1} = \left.\frac{N_1\phi_m}{i_1}\right|_{i_2=0} = \frac{N_1}{i_1}\frac{N_1 i_1}{\mathscr{R}_m} = \frac{N_1^2}{\mathscr{R}_m} \tag{4.7}$$

where $\mathscr{R}_m$ is defined as the magnetic resistance of the magnetic coupling coil of Figure 4.3, which is the ratio of current and magnetic flux determined from (4.2) for a one-turn case. Similarly, L_{21} can be found from (4.5a) and (4.6) as follows:

$$L_{21} \equiv \frac{N_2\phi_{21}}{i_2} = \left.\frac{N_2\phi_m}{-i_2}\right|_{i_1=0} = \frac{N_2}{i_2}\frac{N_2 i_2}{\mathscr{R}_m} = \frac{N_2^2}{\mathscr{R}_m} \tag{4.8}$$

In (4.8), the polarity inversion of i_2 accounts for the same magnetic flux direction of ϕ_{21} as ϕ_m, as identified from (4.6).

Comparing (4.7) and (4.8), where the magnetic resistance $\mathscr{R}_m$ is common, the inductances have the following relationship:

$$L_{21} = n^2 L_{12} = n^2 L_m \quad \because L_{12} \equiv L_m \tag{4.9}$$

Applying (4.9) to (4.4c) and (4.4d) results in the following:

$$v_1 = L_{l1}\dot{i}_1 + (L_{12}\dot{i}_1 - L_{21}\dot{i}_2/n) = L_{l1}\dot{i}_1 + L_m(\dot{i}_1 - n\dot{i}_2) \equiv L_{l1}\dot{i}_1 + v_m \tag{4.10a}$$

$$v_2 = -L_{l2}\dot{i}_2 - (L_{21}\dot{i}_2 - nL_{12}\dot{i}_1) = (nL_m\dot{i}_1 - n^2 L_m\dot{i}_2)L_{l2}\dot{i}_2 = nv_m - L_{l2}\dot{i}_2 \tag{4.10b}$$

$$\because v_m = L_m(\dot{i}_1 - n\dot{i}_2) = L_m\dot{i}_m, \quad i_m \equiv i_1 - ni_2 \tag{4.10c}$$

The circuit reconstruction for (4.10) is shown in Figure 4.4, where the transformer is an ideal transformer with a turn ratio of n and L_m corresponds physically to the magnetizing inductance, which is defined as follows:

$$L_m \equiv \frac{N_1\phi_m}{i_m} = \frac{N_1^2}{\mathscr{R}_m} \tag{4.11}$$

For a core design, (4.11) is very useful to determine L_m and i_m for a given maximum magnetic flux density.

As identified from Figure 4.4, the equivalent circuit of IPT is indeed a conventional transformer. One of the major differences, however, is the values of leakage inductances L_{l1} and L_{l2}, which are relatively quite small compared to the magnetizing inductance L_m in the conventional

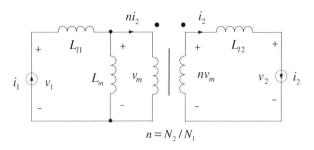

Figure 4.4 An explicit transformer model for the two coupled coils.

transformer (usually about 1%). The leakage inductances in the IPT are often larger than the magnetizing inductance (typically 100–10 000%). Furthermore, the voltages and currents of the primary and secondary coils are often not in phase although the frequency is the same relative to each other. Together with resonant circuits attached to coupled coils, the secondary current is in many cases quadratic against the primary current, which is completely different from a conventional transformer. The secondary voltage and current are not directly determined from the turn ratio n in IPT, and for this reason I have changed the polarity of the secondary current in order to differentiate the IPT from a canonical transformer.

Recall that the explicit transformer model, as shown in Figure 4.4, is generally applicable to an arbitrary waveform of source currents. In theory, it is possible to apply DC currents to the equivalent circuit of Figure 4.4.

The advantage of the explicit transformer model, also referred to as the "transformer model," is that the physical properties of the four circuit elements of two leakage inductances, magnetizing inductance, and ideal transformer are explicitly determined. Therefore, this model is convenient and useful for the design of coils for IPT.

4.3 M-Model

Another category of model that is widely used in IPT is a mutually coupled model, called the "M-model," as shown in Figure 4.5. Two coupled coils are described by three parameters instead of four parameters in the previous explicit transformer model as follows:

$$v_1 = L_1 \dot{i}_1 - M \dot{i}_2 \tag{4.12a}$$
$$v_2 = M \dot{i}_1 - L_2 \dot{i}_2 \tag{4.12b}$$

Note from (4.12) that there is neither a turn ratio nor a transformer, as shown in Figure 4.5(b). The polarity of the dependent voltage source at the primary side is reversed at the moment shown in Figure 4.5(b), but it can be inverted again if the polarity of i_2 is defined reversely, as shown in Figure 4.5(c), which is what the symmetrical configuration of Figures 4.1(a) and 4.2(a) is intended for. Actually, Figure 4.5(c) is more frequently used, but the polarity of i_2 should be carefully defined.

From (4.10) of the explicit transformer model, (4.12) can be deduced by rearranging terms as follows:

$$v_1 = L_{l1} \dot{i}_1 + L_m(\dot{i}_1 - n\dot{i}_2) = (L_{l1} + L_m)\dot{i}_1 - nL_m \dot{i}_2 \equiv L_1 \dot{i}_1 + M \dot{i}_2 \tag{4.13a}$$
$$v_2 = (nL_m \dot{i}_1 - n^2 L_m \dot{i}_2) - L_{l2} \dot{i}_2 = nL_m \dot{i}_1 - (n^2 L_m + L_{l2})\dot{i}_2 \equiv M \dot{i}_1 - L_2 \dot{i}_2 \tag{4.13b}$$
$$\because M \equiv nL_m, \quad L_1 \equiv L_{l1} + L_m, \quad L_2 \equiv L_{l2} + n^2 L_m \tag{4.13c}$$

Figure 4.5 M-model for two coupled coils.

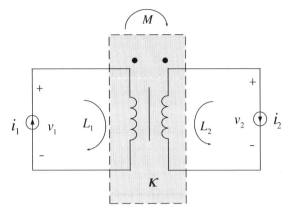

(a) Implicit transformer model

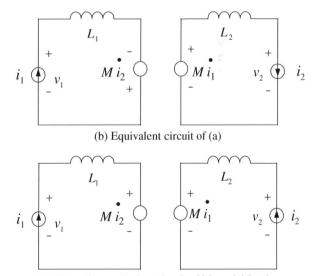

(b) Equivalent circuit of (a)

(c) Alternative equivalent circuit of M-model for the symmetric direction of current sources

As identified from (4.13), L_1 and L_2 are the inductances viewed from the primary and secondary sides, respectively, and M is the mutual inductance determined by the product of the turn ratio n and the magnetic inductance L_m. This is why the author would like to call the M-model an implicit transformer model, where some physical parameters do not explicitly appear. The value M can be generally calculated for a given geometry, as shown in the Appendix at the end of the chapter.

The M-model is exactly equivalent to the transformer model depicted in Figure 4.6.

Let us examine the power delivery through the dependent voltage sources in Figure 4.5(b). First, we will find the instantaneous input power of the primary side and the instantaneous output power of the secondary power, respectively, as follows:

$$p_1(t) = -i_1(M\dot{i}_2) \tag{4.14a}$$

$$p_2(t) = i_2(M\dot{i}_1) \tag{4.14b}$$

$$\Rightarrow p_1(t) = -Mi_1\dot{i}_2 \neq Mi_2\dot{i}_1 = p_2(t) \tag{4.14c}$$

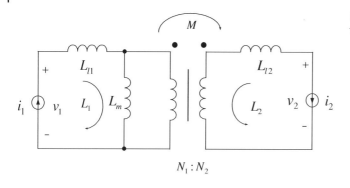

Figure 4.6 Equivalence of the transformer model and M-model.

For arbitrary currents of i_1 and i_2, the instantaneous input and output power is not preserved, as identified from (4.14). We may then consider the sinusoidal case of single frequency in the steady state, which is delineated as follows:

$$p_1(t) = -i_1(M\dot{i}_2) \Rightarrow P_1 = -\text{Re}\{I_1^*(Mj\omega_s I_2)\} \tag{4.15a}$$

$$p_2(t) = i_2(M\dot{i}_1) \Rightarrow P_2 = \text{Re}\{I_2^*(Mj\omega_s I_1)\} \tag{4.15b}$$

$$\because P_1 = -\text{Re}\{I_1^*(Mj\omega_s I_2)\} = -\text{Re}\{(I_1^*(Mj\omega_s I_2))^*\} = -\text{Re}\{I_1(-Mj\omega_s I_2^*)\}$$
$$= \text{Re}\{I_2^*(Mj\omega_s I_1)\} = P_2 \tag{4.15c}$$

It is identified from (4.15) that the power of the M-model is preserved in the static phasor domain. It can be said that wireless power is transferred through the dependent voltage sources of the M-model in the steady state of a sinusoidal source-driven case.

Question 1 Why is the instantaneous power not preserved but the static power is preserved in (4.14) and (4.15)?

An important parameter widely used in IPT as well as traditional transformers is the coupling coefficient or coupling factor κ, which is also represented as k. As identified from Figure 4.6, the primary and secondary open voltage gains of a transformer give us an idea of how much two coils are coupled to each other in the static phasor domain as follows:

$$\kappa^2 \equiv \left.\frac{V_2}{V_1}\right|_{I_2=0} \times \left.\frac{V_1}{V_2}\right|_{I_1=0} = \left(\frac{L_m}{L_{l1}+L_m}n\right) \cdot \left(\frac{n^2 L_m}{L_{l2}+n^2 L_m}\frac{1}{n}\right)$$
$$= \frac{nL_m}{L_{l1}+L_m} \cdot \frac{nL_m}{L_{l2}+n^2 L_m} = \frac{M}{L_1} \cdot \frac{M}{L_2} = \frac{M^2}{L_1 L_2} \tag{4.16a}$$

$$\Rightarrow \kappa = \frac{M}{\sqrt{L_1 L_2}} = \frac{nL_m}{\sqrt{(L_{l1}+L_m)(L_{l2}+n^2 L_m)}} \tag{4.16b}$$

For an ideal transformer, the coupling factor becomes unity as follows:

$$\kappa^2 = \left.\frac{V_2}{V_1}\right|_{I_2=0} \times \left.\frac{V_1}{V_2}\right|_{I_1=0} = \left(\frac{L_m}{0+L_m}n\right) \cdot \left(\frac{n^2 L_m}{0+n^2 L_m}\frac{1}{n}\right) = n \cdot \frac{1}{n} = 1 \tag{4.17a}$$

$$\Rightarrow \kappa = \frac{M}{\sqrt{L_1 L_2}} = 1 \quad \text{or} \quad M = \sqrt{L_1 L_2} \tag{4.17b}$$

Figure 4.7 T-model for two coupled coils.

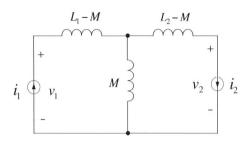

For IPT where the coupling factor is very small, however, the mutual inductance is relatively very small as follows:

$$M = \kappa\sqrt{L_1 L_2} \ll \sqrt{L_1 L_2} \quad \text{for} \quad \kappa \ll 1 \tag{4.18}$$

The advantage of the M-model is its simplicity, but circuit-based analyses are not permitted due to dependent voltage sources.

4.4 T-Model

Another useful model for the coupled coils is the T-model, which comprises three inductors, resembling the letter "T," as shown in Figure 4.7.

This T-model can be derived from the M-model of (4.13) by rearranging terms as follows:

$$v_1 = L_1\dot{i}_1 - M\dot{i}_2 = (L_1 - M)\dot{i}_1 + M(\dot{i}_1 - \dot{i}_2) \tag{4.19a}$$
$$v_2 = M\dot{i}_1 - L_2\dot{i}_2 = -(L_2 - M)\dot{i}_2 + M(\dot{i}_1 - \dot{i}_2) \tag{4.19b}$$

The circuit reconstruction of (4.19) is Figure 4.7; however, Figure 4.7 is not an exact equivalent circuit of the coupled coils because there is no galvanic isolation. Except for this isolation issue, Figure 4.7 is an electrically exact equivalent circuit for the coupled coils.

It is remarkable that the inductance values of the T-model except for M are not necessarily positive. Actually, they become negative for the following conditions:

$$L_1 - M < 0 \Rightarrow L_1 < M = \kappa\sqrt{L_1 L_2} \Rightarrow L_1^2 < \kappa^2 L_1 L_2 \Rightarrow L_1/L_2 < \kappa^2 \tag{4.20a}$$
$$L_2 - M < 0 \Rightarrow L_2 < M = \kappa\sqrt{L_1 L_2} \Rightarrow L_2^2 < \kappa^2 L_1 L_2 \Rightarrow L_2/L_1 < \kappa^2 \tag{4.20b}$$

In the case of a symmetric configuration, $L_1 = L_2$ and both conditions of (4.20) do not hold; this means the two inductances of the T-model are both positive. However, in an asymmetric case, for example, $L_1 < L_2$, (4.20a) may be met, which results in a negative inductance value.

The advantage of the T-model is its simplicity and versatility for circuit-based analyses. The disadvantage of the T-model is that circuit parameters such as the turn ratio and coupling factor do not appear when their effects are sought.

Question 2 (1) In case the inductance of (4.20) becomes negative, what happen to its physical energy and system stability? (2) What are the similarity and difference of the negative inductor from a capacitor with positive capacitance? (3) What are the similarity and difference of the negative inductor from a resistor with negative resistance?

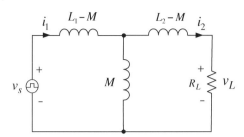

Figure 4.8 An example of an arbitrary source and load, where the T-model is valid.

4.5 Further Discussions and Conclusion

It has been proven that the transformer model, M-model, and T-model are equivalent to each other. Their strengths and drawbacks are also addressed; therefore, the preference of models may be arbitrary. Thus any model that provides a solution can be chosen.

For circuit-based analyses, the transformer model and T-model are preferred, while the M-model is preferred for equation-based analyses. When the M-model is used, only the inductances of L_1 and L_2 are seen and the dependent voltage sources look like ideal voltage sources with zero impedances; however, the dependent voltage sources may be equivalent to any reactance through primary and secondary circuit interactions.

Recall that the above-mentioned three models are not only valid for static sinusoidal current sources but also for arbitrary current sources in the transient state. Furthermore, the models are valid for voltage sources and are compatible with any circuits, as shown in the example in Figure 4.8.

Though not included in the derivation of the models, resistances accounting for conduction loss and core losses (hysteresis loss and eddy current loss) as well as parasitic capacitances can be included in the detailed transformer model, as shown in Figure 4.9. Of course, this transformer model is compatible with arbitrary sources and external circuits, like a conventional transformer.

For the designs of IPT, the effects of the air gap and lateral displacements on the leakage inductances, magnetizing inductance, and the number of turns of coupled coils have been extensively explored. Because the transformer model gives us detailed information on all the circuit parameters, the author preferably uses that model.

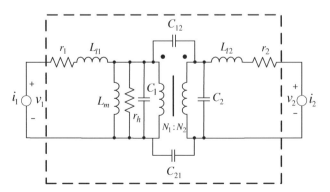

Figure 4.9 Detailed transformer model for coupled coils including parasitics.

Appendix

The mutual inductance between two loops with a single turn, as shown in Figure 4.10, can generally be obtained. The shape and orientation of the loops may be arbitrary and the current and magnetic flux can be either DC or AC (though demonstrated for the DC case here), which does not affect the mutual inductance. To show clearly that the following derivation is valid for real-time variables as well as phasor variables, arrow-type vector notation is adopted here.

Using the double integral Neumann formula [1, 2], the mutual inductance of Figure 4.10 can be calculated as follows:

$$M_{ij} \equiv \frac{\phi_i}{I_j} = \frac{\mu_0}{4\pi} \oint_{C_i} \oint_{C_j} \frac{\vec{dl_i} \cdot \vec{dl_j}}{|\vec{R_{ij}}|} \tag{4.21}$$

where the magnetic flux is calculated from Stokes' theorem and the concept of the vector potential is used:

$$\phi_i = \int_{S_i} \vec{B} \cdot \vec{ds_i} = \int_{S_i} (\nabla \times \vec{A}) \cdot \vec{ds_i} = \int_{C_i} \vec{A} \cdot \vec{dl_i} = \int_{C_i} \left(\frac{\mu_0 I_j}{4\pi} \int_{C_j} \frac{\vec{dl_j}}{|\vec{R_{ij}}|} \right) \cdot \vec{dl_i} \tag{4.22}$$

The parameters are explained in the following list:

M_{ij} is the mutual inductance between the ith loop coil and the jth loop coil with a single-turn.
ϕ_i is the magnetic flux through the ith loop coil surface S_i.
I_j is the current of the jth loop coil contour C_j.
μ_0 is the permeability of free space ($=4\pi \times 10^{-7}$).
$\vec{B}$ is the magnetic flux density, generated by the current I_j.
$\vec{A}$ is the vector potential, generated by the current I_j.
$\vec{dl_i}$ is the infinitesimal length vector on the contour C_i.
$\vec{dl_j}$ is the infinitesimal length vector on the contour C_j.
$\vec{R_{ij}}$ is the distance vector between two points at $\vec{dl_i}$ and $\vec{dl_j}$.
$\vec{ds_i}$ is the infinitesimal surface vector on the contour C_i.

Needless to say, the mutual inductance for multiple turns of N_i and N_j becomes

$$M_{ij,t} = N_i N_j M_{ij,t} = N_i N_j \frac{\mu_0}{4\pi} \oint_{C_i} \oint_{C_j} \frac{\vec{dl_i} \cdot \vec{dl_j}}{|\vec{R_{ij}}|} \tag{4.23}$$

Note from (4.21) that the total magnetic flux is proportional to both N_i and N_j.

Figure 4.10 Mutual inductance between two arbitrary loops.

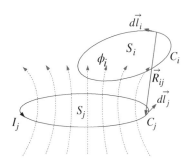

Problems

4.1 It is often misunderstood about the nature of a transformer that AC power can be transferred through the transformer but DC power cannot. Let us examine this myth using
the following explicit transformer model of Figure 4.11(a), having no internal resistances
and parasitic capacitances but having a finite magnetizing current. Therefore, the transformer loses its magnetic induction capability when its magnetizing current i_m reaches the
saturation current I_m. Assume that $V_s = 100$ V, $L_{l1} = 10$ µH, $L_m = 90$ µH, $L_{l2} = 80$ µH,
$I_m = 3$ A, $n = 4$. Also assume that $R_L = \infty$.

   (a) The output voltage v_2 can be described, as shown in Figure 4.11(b). Explain why the
voltage waveform is kept constant for a while and then eventually drops sharply to
zero.

   (b) Determine the output voltage level V_o and the non-saturation time T_1 of Figure 4.11(b).

   (c) Draw the output voltage v_2 for the given AC input voltage of Figure 4.11(c), assuming
two cycles with $T_2 = 20$ ms. Find the maximum input voltage V_m that eventually makes
the transformer become saturated.

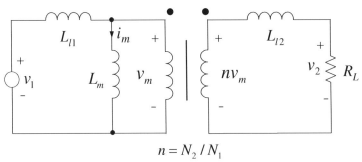

$$n = N_2 / N_1$$

(a) An explicit transformer model without any losses

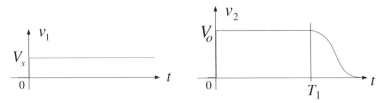

(b) Step input and its output response in time

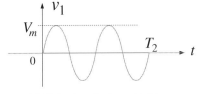

(c) Finite length sinusoidal input

Figure 4.11 Transient response of a transformer having a finite saturation current of magnetizing inductance.

(**d**) Suggest the method to mitigate the saturation of the transformer. (*Hint.* Change the starting phase of the AC voltage.) How much can we increase V_m in this way?

4.2 Examine the effect of load current on the transformer of Figure 4.11(a), assuming that the circuit parameters are the same as 4.1 except for $R_L = 10\ \Omega$.

(**a**) Draw the output voltage v_2 for the step input voltage of Figure 4.11(b). Explain the voltage waveform, especially the difference from the zero load current case of 4.1.

(**b**) Determine the output voltage level V_o and the non-saturation time T_1 of Figure 4.11(b).

(**c**) Draw the output voltage v_2 for the given AC input voltage of Figure 4.11(c), assuming two cycles with $T_2 = 20$ ms. Find the maximum input voltage V_m that eventually makes the transformer become saturated.

(**d**) Compare the above results of (a) to (c) with the zero load current case of 4.1. Does the load current deteriorate the saturation characteristics of the transformer?

4.3 Calculate the mutual inductance between two circular loop coils that are on the same plane. Assume that the inner and outer coil diameters are 1 m and 2 m, respectively, and the number of turns of the two coils are both 4.

References

1 J.D. Jackson, *Classical Electrodynamics*, John Wiley & Sons, 1975, pp. 176 and 263.
2 C.R. Paul, *Inductance Loop and Partial*, John Wiley and Sons, Inc., Hoboken, NJ, USA, 2010.

5

Gyrator Circuit Model

5.1 Introduction

Inductive power transfer systems (IPTSs) are becoming widely used in electric vehicles, consumer electronics, medical devices, lighting, factory automation, defense systems, and remote sensors [1–14]. Compensation circuits are essential in most IPTSs to increase load power by canceling out reactive powers generated from coupled coils. There have been various compensation circuits reported in the literature such as voltage-source-type primary series–secondary series (V-SS), voltage-source-type primary series–secondary-parallel (V-SP), and inductor–capacitor–inductor (LCL) [15–42]. Each compensation circuit has its own electrical characteristics at a given resonance frequency such as source-to-load voltage/current gain, load R_L-independent output characteristics, power factor at the source, sign of the source phase angle, and allowances for open/short loads [15–35, 43]. Hence, it is crucial to select the most appropriate compensation circuit for given applications at the beginning of the IPTS design.

A typical configuration for compensation circuits is shown in Figure 5.1. It basically includes coupled inductors, which transfer the wireless power to the load, and two compensation capacitors. More capacitors or inductors can be used [36–42, 44, 59–62]. Analyses of these compensation circuits have been mostly performed by the equational approaches, which usually include matrix manipulation. By solving the equations derived from the circuits, the characteristics of the circuits can be determined; hence, the equational approaches are extremely general, unified, and definite. However, the manipulation of circuit equations can be tedious and time-consuming for compensation circuits involving a larger number of reactive components than four. As an alternative to the problems, a graphical approach for modeling IPTSs is explained in this chapter. Practical merits of the graphical approaches over the equational approaches are widely known [46, 63–73]. Circuit averaging techniques, which have been widely used for the modeling of switching converters, allow manipulations on the circuit diagram [63–71]. By virtue of the graphic-based approach, high-order systems can be easily manipulated by hand, giving fruitful physical insights, and unified general equivalent circuits for switching converters can be found as electronic transformers [46].

In this chapter, the fact that all IPTSs inherently have the nature of a gyrator is explained and verified through the graphical approach. A gyrator is one of the basic circuit elements, which was introduced by Tellegen in 1948 based on the idea that there should be a complement to the ideal transformer [45]. In the field of power electronics, a gyrator was first utilized by Dr Rim for the analysis of alternating current (AC) converters by representing D-Q transformed AC subcircuits with gyrators [46]. Magnetic components are modeled in the electrical domain by modeling the windings as gyrators that convert electrical current to a magnetomotive force, and vice versa [47]. Power-processing properties of fundamental switching converters are modeled

Wireless Power Transfer for Electric Vehicles and Mobile Devices, First Edition. Chun T. Rim and Chris Mi.
© 2017 John Wiley & Sons Ltd. Published 2017 by John Wiley & Sons Ltd.

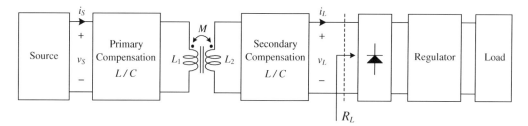

Figure 5.1 Typical configuration of compensation circuits in an IPTS.

using gyrators [48, 49]. By utilizing the proposed approach, analyses of compensation circuits can be greatly simplified. The feature of the proposed approach is that it is suitable for practical use due to its simple and systematic aspects for the following reasons.

1. It is simple because the proposed analysis is carried out with graphical steps and does not involve complicated equations.
2. It is systematic because the proposed analysis is applied to different compensation schemes in a consistent manner. First, the compensation circuit is equivalently transformed to a gyrator-based circuit while the source and the load R_L remain unchanged. Second, source-to-load and load-to-source electrical characteristics are analyzed using a gyrator's circuit properties.

The above aspects are shown with some compensation circuits that are widely used in IPTSs.

The proposed approach is so general that it is applicable to not only a resonant frequency but also any frequencies. The effects of the equivalent series resistances (ESRs) can also be evaluated although it is not covered in detail due to the page limit. In particular, the proposed analysis is compatible with a compensation circuit whose tuning frequency is magnetic coupling independent. This is due to the characteristic of the gyrator model of coupled inductors whereby all the magnetic coupling-dependent behaviors are included in the gyrator's gain, and tuning is then achieved by canceling out the remaining reactive components whose values are magnetic coupling independent. This implies that the proposed analysis is suitable for low-cost and highly reliable IPTSs that do not require complicated control algorithms that track the desired tuning frequency altered by airgap changes [15, 16, 22, 30, 31].

Source-to-load voltage/current gain, load R_L-independent output characteristics, power factor at the source, sign of the source phase angle, and allowances for open/short R_L, which are important characteristics of a compensation circuit between its source and load, are evaluated using the proposed analysis. Widely used compensation circuits, V-SS, I-SP, and V-LCL-P in their magnetic coupling-independent tuning conditions are analyzed as examples, and the results all agree with previous studies in the literature. In addition, effects of mistuning on these characteristics are also evaluated, while maintaining the simple and systematic aspects of the proposed analysis [74].

5.2 Representation of Compensation Circuits with Gyrators

In this section, gyrators in resonant circuits and coupled inductors are derived and important properties of the gyrator are given. Throughout this chapter, it is assumed that the IPTS is in the steady state and that high-frequency harmonics involved in v_S or i_S in Figure 5.1, which are generated by the source, have negligible impacts on the fundamental frequency component

Figure 5.2 Electric circuit model of a gyrator.

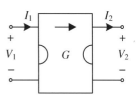

due to the bandpass filter-like frequency response of the compensation circuits. The rectifier of Figure 5.1 is modeled as R_L considering only the fundamental frequency components of v_L and i_L under the continuous conduction mode operation in the steady state [15–44, 58].

As shown in Figure 5.2, a gyrator is a linear, time-invariant, and lossless two-port network that converts the voltage of one port to the current of another port [45]. It is a bilateral network, whose signal propagates back from the secondary port to the primary port, as well as forward from the primary port to the secondary port, and its voltage–current relationships are given as follows:

$$I_2 = G \times V_1 \tag{5.1a}$$
$$I_1 = G^* \times V_2, \tag{5.1b}$$

where I_1, V_1, I_2, and V_2 are the phasors of the primary-port current and voltage and the secondary-port current and voltage, respectively. Their sign conventions are given in Figure 5.2. G and G^* are defined as the gyrator forward gain and reverse gain, respectively, and are complex numbers. Note that the concept of complex gyrator is first introduced here [74] and that the meaning of "complex" needs to be carefully examined.

Note from (5.1) that the secondary-port current I_2 is uniquely determined by the primary-port voltage V_1, and vice versa. In addition, because the reverse gain G^* is complex conjugate of the forward gain G, the real powers on the primary side P_1 and secondary side P_2 are the same and the reactive powers of them are in opposite polarity, as follows:

$$P_1 \equiv \mathrm{Re}(V_1 I_1^*) = \mathrm{Re}(GV_1 V_2^*) = \mathrm{Re}\{(G^* V_1^* V_2)^*\} = \mathrm{Re}(G^* V_1^* V_2) = \mathrm{Re}(V_2 I_2^*) \equiv P_2 \tag{5.2a}$$

$$Q_1 \equiv \mathrm{Im}(V_1 I_1^*) = \mathrm{Im}(GV_1 V_2^*) = \mathrm{Im}\{(G^* V_1^* V_2)^*\} = -\mathrm{Im}(G^* V_1^* V_2) = -\mathrm{Im}(V_2 I_2^*) \equiv -Q_2 \tag{5.2b}$$

Note that the power relation of (5.2) is true only when the complex gyrator is so defined as in (5.1).

Question 1 (1) Define the complex gyrator as $I_2 = G \times V_1$ and $I_1 = G \times V_2$, and apply this definition to (5.2) to see the physical meaning of this definition. Find out the condition on G that the real power is preserved. (2) Find another definition of the complex gyrator $I_2 = G^* \times V_1$ and $I_1 = G \times V_2$. Does this definition fulfill (5.2)?

5.2.1 Realization of a Gyrator with Passive Components

It should be emphasized that a gyrator can be realized using only passive components. Figure 5.3 shows examples of a gyrator that consists of impedances Z and $-Z$, which are complex values in the steady state. Figure 3(a) and (b) are equivalent and can be derived from each other through a T–Π transformation.

Voltage and current relationships of the circuits in Figure 5.3(a) and (b) can be derived using simple calculations. Applying Kirchhoff's voltage law on V_1 and V_2 in Figure 5.3(a) gives the

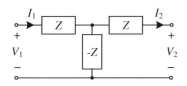

Figure 5.3 Examples of gyrators that consist of purely imaginary impedances Z and $-Z$.

(a) Example of a gyrator in T-network

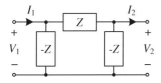

(b) Example of a gyrator in Π-network

following equations:

$$V_2 = (-Z) \times (I_1 - I_2) + Z \times (-I_2) = -Z \times I_1 \tag{5.3a}$$

$$V_1 = (-Z) \times (I_1 - I_2) + Z \times I_1 = Z \times I_2 \tag{5.3b}$$

Similarly, applying Kirchhoff's current law on I_1 and I_2 in Figure 5.3(b) gives the following equations:

$$I_2 = \left(-\frac{1}{Z}\right) \times (-V_2) + \frac{1}{Z} \times (V_1 - V_2) = \frac{1}{Z} \times V_1 \tag{5.4a}$$

$$I_1 = \left(-\frac{1}{Z}\right) \times V_1 + \frac{1}{Z} \times (V_1 - V_2) = -\frac{1}{Z} \times V_2 \tag{5.4b}$$

Note that (5.3) and (5.4) do not require any specific assumptions and apply to all cases. Rearranging (5.3) in terms of I_1 and I_2, then comparing with (5.4) reveals that Figure 5.3(a) and (b) have identical voltage–current relationships (V_1, V_2, I_1, I_2), which are as follows:

$$I_2 = \frac{1}{Z} \times V_1 \tag{5.5a}$$

$$I_1 = -\frac{1}{Z} \times V_2 \tag{5.5b}$$

Comparing (5.5) with (5.1) shows that the necessary and sufficient condition for the circuits in Figure 5.3(a) and (b) to be a gyrator in Figure 5.2 is that impedance Z is purely imaginary, as follows:

$$-\frac{1}{Z} = \left(\frac{1}{Z}\right)^* \quad \Leftrightarrow \quad \text{Re}(Z) = 0 \tag{5.6}$$

Question 2 Prove (5.6) by yourself, assuming $Z = R + jX$. Note that X could be either positive or negative.

Under this condition, Figure 5.3(a) and (b) can be equivalently represented as a gyrator in Figure 5.4, whose forward gain G is the admittance $1/Z$ is as follows:

$$G = \frac{1}{Z} = \frac{1}{jX} \tag{5.7}$$

Figure 5.4 Equivalent gyrator model of Figure 5.3(a) and (b) if and only if Z is purely imaginary.

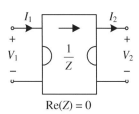

When Z in Figure 5.3 is not purely imaginary, representation of a T- or Π-network with a gyrator requires an extended representation from the lossless gyrator described in (5.1) to one with a power loss:gyrator's reverse gain of $-1/Z$ is not complex conjugate of the forward gain $1/Z$; thus, (5.2) does not hold. Note that such a gyrator is still linear and time-invariant, and that I_2 and I_1 are uniquely determined by V_1 and V_2, respectively. This concept is required when ESRs are taken into account. General modeling of resonant circuits that involve ESRs are briefly discussed in Section 5.2.4.

Question 3 (1) In case the gyrator is defined as (5.5) with a non-zero real part of $Z = R + jX$, what about the input and output real/reactive powers of (5.2) for Figure 5.3(a) and (b) if the AC input voltage V_s and resistive load R_L are applied to the circuits? (2) What will be the power efficiency? Use the gyrator relationship of (5.5) when you calculate the power and efficiency, which enables easier calculation.

Note from (5.6) and (5.7) that the gyrator in Figure 5.4 is symmetric due to its purely imaginary forward gain G; when the primary port and the secondary port are reversed, the forward gain remains unchanged. Hence, the voltage–current relationships between the two ports are not changed, regardless of the direction of the power flow. For the rest of this chapter, forward gain G is referred to as gyrator gain.

Purely imaginary impedances Z and $-Z$ can be implemented with reactive components inductor L and capacitor C that resonate at the angular frequency of interest, ω_0, as follows:

$$\omega_0 = \frac{1}{\sqrt{LC}} \tag{5.8}$$

This is illustrated in Figure 5.5. Note that the admittance of L, $1/(jX_0)$, becomes the gyrator gain where X_0 is the characteristic impedance of the L-C resonant circuit, as follows:

$$X_0 \equiv \sqrt{\frac{L}{C}} = \omega_0 L = \frac{1}{\omega_0 C} \tag{5.9}$$

In Figure 5.5, L and C can be interchanged without loss of generality.

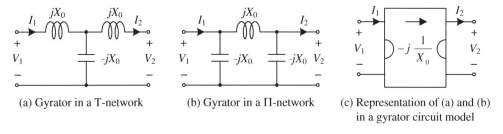

(a) Gyrator in a T-network   (b) Gyrator in a Π-network   (c) Representation of (a) and (b) in a gyrator circuit model

Figure 5.5 Implementation of a gyrator with reactive components.

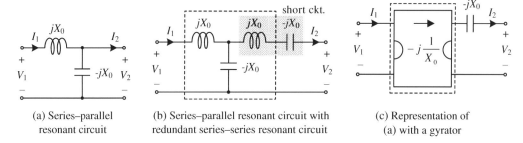

(a) Series–parallel
resonant circuit

(b) Series–parallel resonant circuit with
redundant series–series resonant circuit

(c) Representation of
(a) with a gyrator

Figure 5.6 Representation of a series–parallel resonant circuit with a gyrator.

Based on Figure 5.5, resonant circuits and coupled inductors that are mainly used as the components in compensation circuits can be represented by gyrators whose gain G is purely imaginary.

It is an amazing fact, as first discovered by the authors of [74], that the symmetric T- and Π-networks in resonance become an equivalent gyrator with imaginary gain.

5.2.2 Gyrators in an IPTS: Resonant Circuits

Figure 5.6(a) shows a resonant circuit in series–parallel connection where inductance L and capacitance C resonate at the angular frequency of interest, ω_0. This resonant circuit is commonly found in many compensation circuits; for example, L and C are the secondary-side self-inductance L_2 and compensation capacitance C_2 of the secondary–parallel compensation circuit [15–35]. As shown in Figure 5.6(b), the series–series resonant circuit L-C, which is effectively a short circuit (ckt.), can be added in series to the circuit in Figure 5.6(a) without affecting its external port voltage–current relationships (V_1, V_2, I_1, I_2). Then the gyrator T-network in Figure 6.5(a) appears; thus, the circuit in Figure 5.6(a) can be equivalently transformed using the gyrator, as shown in Figure 5.6(c). Note that parallel-connected C in Figure 5.6(a) becomes series-connected, and its admittance becomes the gyrator gain in Figure 5.6(c).

The same analogy is applied to the parallel–series resonant circuit in Figure 5.7(a), which is also often found in compensation circuits; for instance, capacitance C and inductance L are the primary-side compensation capacitance C_1 and self-inductance L_1 of the primary parallel compensation circuit [15–35]. Again, this assumes that L-C resonates at the angular frequency of interest, ω_0. Note from Figure 5.7(b) that the redundant parallel–parallel resonant circuit L-C, which is effectively an open circuit, is added in parallel, and the gyrator Π-network in Figure 5.5(b) appears. Thus, the original circuit in Figure 5.7(a) becomes the one in Figure 5.7(c),

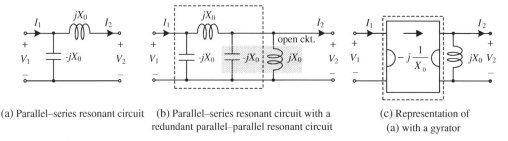

(a) Parallel–series resonant circuit

(b) Parallel–series resonant circuit with a
redundant parallel–parallel resonant circuit

(c) Representation of
(a) with a gyrator

Figure 5.7 Representation of a parallel–series resonant circuit with a gyrator.

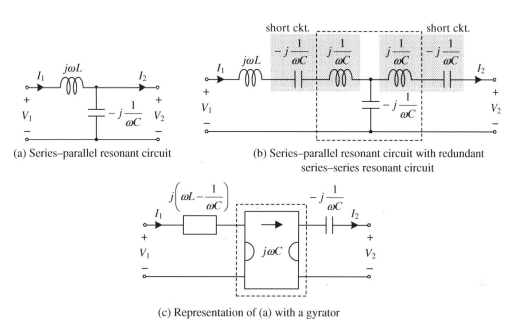

(a) Series–parallel resonant circuit

(b) Series–parallel resonant circuit with redundant
series–series resonant circuit

(c) Representation of (a) with a gyrator

Figure 5.8 Representation of a series–parallel resonant circuit with a gyrator.

where series-connected L in Figure 5.7(a) becomes parallel-connected in Figure 5.7(c), and the admittance of C becomes the gyrator gain.

When resonant circuits are not fully resonated, unlike those of Figures 5.6(a) and 5.7(a), they can also be represented with a gyrator. In this case, addition of redundant series-series or parallel-parallel resonant circuits, as illustrated in Figures 5.6(b) and 5.7(b), is also utilized to make explicit T- or Π-networks. Detailed explanations are omitted due to similar analogy to those in Figures 5.6 and 5.7, which are described in Figures 5.8 and 5.9. Note that this gyrator

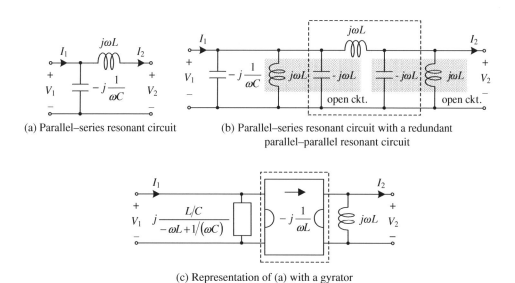

(a) Parallel–series resonant circuit

(b) Parallel–series resonant circuit with a redundant
parallel–parallel resonant circuit

(c) Representation of (a) with a gyrator

Figure 5.9 Representation of a parallel–series resonant circuit with a gyrator.

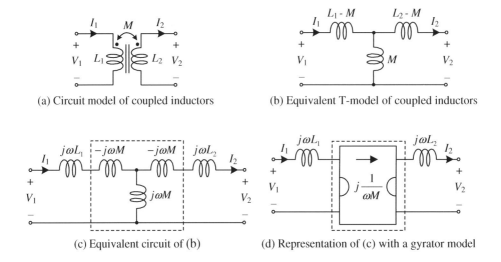

(a) Circuit model of coupled inductors

(b) Equivalent T-model of coupled inductors

(c) Equivalent circuit of (b)

(d) Representation of (c) with a gyrator model

Figure 5.10 Representation of coupled inductors with a gyrator.

modeling enables the representation of a compensation circuit in the general angular frequency domain.

5.2.3 Gyrators in IPTS: Coupled Inductors

Figure 5.10 represents coupled inductors using a gyrator. Figure 5.10(a) shows coupled inductors L_1 and L_2 with a mutual inductance of M, and Figure 5.10(b) illustrates its well-known T-model, which provides equivalent voltage and current relationships (V_1, V_2, I_1, I_2) although the galvanic isolation is not considered [58]. It can be found from Figure 5.10(b) that coupled inductors implicitly contain a gyrator T-network as described in Figure 5.5(a), which is illustrated in Figure 5.10(c). Here, $-M$ can be regarded as a capacitor due to its negative inductance, where effective capacitance C_{eff} is given as follows [50]:

$$\frac{1}{j\omega C_{eff}} = -j\omega M \Rightarrow C_{eff} = \frac{1}{\omega^2 M} \tag{5.10}$$

Hence, the equivalent circuit becomes one as seen in Figure 5.10(d) using a gyrator. It should be noted that the equivalent circuit in Figure 5.10(d) is valid at all angular frequencies.

5.2.4 Gyrators in IPTS: Including ESRs

Resonant circuits that include ESRs can also be represented by a gyrator model. This requires extension of the lossless gyrator described in (5.1) to one that is not lossless. Figure 5.11(a) shows series–parallel resonant circuits that involve ESRs. For generality, it is assumed that two ESRs r_L and r_C have different values and L and C are not fully resonated. Similar to that described in Figure 5.8(b), a redundant series–series resonant circuit can be added with $-r_C$, as illustrated in Figure 5.11(b). Considering (5.5) and the T-network of Figure 5.3(a), it can be noticed that the T-network in the dotted box of Figure 5.11(b) can be replaced with a gyrator having the forward gain of $\{r_C - j/(\omega C)\}^{-1}$, as illustrated in Figure 5.11(c). Note that the forward gain of the gyrator in Figure 5.11(c) is not purely imaginary, unlike that of Figure 5.8(c). Hence, it is not

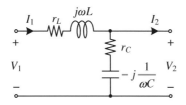

(a) Series–parallel resonant circuit including ESRs

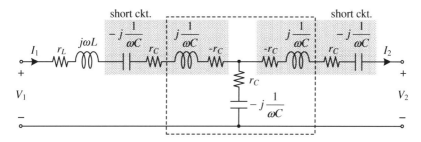

(b) Series–parallel resonant circuit with a redundant series–series resonant circuit including ESRs

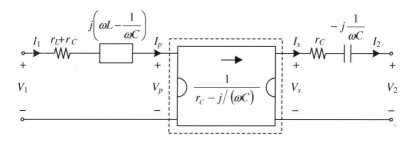

(c) Representation of (a) with a gyrator that has complex forward gain and power loss

Figure 5.11 Representation of a series–parallel resonant circuit with a gyrator.

lossless while its voltage–current relationships (V_p, V_s, I_p, I_s) still hold in the same manner as described in (5.5) with $Z = -r_C + j/(\omega C)$ as follows:

$$I_2 = \frac{1}{-r_C + j/(\omega C)} \times V_1 \tag{5.11a}$$

$$I_1 = -\frac{1}{-r_C + j/(\omega C)} \times V_2 \tag{5.11b}$$

The same procedure addressed so far can be applied to a parallel–series resonance circuit, though it is not explained here.

Question 4 Note that the imaginary gain gyrator model can be used in the special case of $r_C = 0$ with non-zero r_L in Figure 5.11, which is a highly reasonable assumption in many practical applications. Is this also valid for the parallel–series resonance circuit case?

From the discussions in this section, it can be seen that a gyrator with a purely imaginary gain G can be implemented with lossless reactive components in resonance; such gyrators are

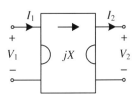

Figure 5.12 Gyrator with a purely imaginary gain *jX*.

found in resonance circuits and coupled inductors. This implies that compensation circuits in an IPTS inherently have the nature of a gyrator and hence can be regarded as a combination of gyrators.

5.3 Circuit Characteristics of the Proposed Purely Imaginary Gyrator

In this section, circuit characteristics of a gyrator that are mainly used for the analysis of compensation circuits are given. As discussed in the previous section, it is assumed that the gyrator gain G is purely imaginary; $G = jX$, as shown in Figure 5.12. Hence, circuit characteristics of a gyrator in this section are valid, regardless of whether the port of interest is primary or secondary. The external-port voltage–current relationship of the gyrator in Figure 5.12 is as follows:

$$I_2 = jX \times V_1 \tag{5.12a}$$
$$I_1 = -jX \times V_2 \tag{5.12b}$$

The voltage–current relationship in (5.12) is fundamental for deriving the circuit characteristics of a gyrator given in this section.

5.3.1 Source-Type Conversion Rule: Voltage-to-Current and Current-to-Voltage

A gyrator converts an ideal voltage source to an ideal current source, and vice versa. This is illustrated in Figure 5.13, where the equivalent circuit is modeled from right to left at the secondary port of the gyrator. This model can be directly derived from (5.12).

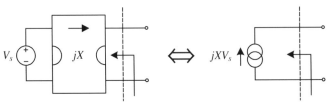

(a) Voltage source to current source conversion

Figure 5.13 Source-type conversion rule of a gyrator.

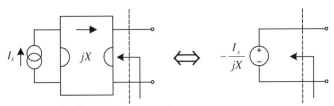

(b) Current source to voltage source conversion

Figure 5.14 Terminated impedance inversion rule of a gyrator: terminated secondary port by Z_2 (left) and its inverse impedance at the primary port (right).

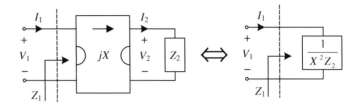

5.3.2 Terminated Impedance Inversion Rule

When the secondary port of the gyrator is terminated with $Z_2 = R_2 + jX_2$, as shown in the left circuit of Figure 5.14, the impedance $Z_1 = R_1 + jX_1$, that is, the inverted Z_2 at the primary port, can be derived from (5.12) as follows:

$$Z_1 = \frac{V_1}{I_1} = \frac{I_2/(jX)}{-jXV_2} = \frac{I_2}{X^2 V_2} = \frac{1}{X^2 Z_2} \tag{5.13}$$

From (5.13), four important features can be found when Z_2 is inverted from the secondary port to the primary port, which are as follows:

1. A short circuit becomes an open circuit, and vice versa:

$$|Z_1|\big|_{|Z_2|=0} = \infty \tag{5.14a}$$

$$|Z_1|\big|_{|Z_2|=\infty} = 0 \tag{5.14b}$$

2. The magnitude of impedance $|Z|$ is divided by $X^2|Z|^2$ and is described as follows:

$$|Z_1| = |Z_2| \times \frac{1}{X^2|Z_2|^2} \tag{5.15}$$

3. The power factor on impedance Z_1 is the same as that on impedance Z_2 and is described as follows:

$$\cos(\angle Z_1) = \frac{R_1}{\sqrt{R_1^2 + X_1^2}} = \frac{\dfrac{R_2}{X^2(R_2^2 + X_2^2)}}{\sqrt{\left\{\dfrac{R_2}{X^2(R_2^2 + X_2^2)}\right\}^2 + \left\{\dfrac{X_2}{X^2\left(R_2^2 + X_2^2\right)}\right\}^2}} \tag{5.16}$$

$$= \frac{R_2}{\sqrt{R_2^2 + X_2^2}} = \cos(\angle Z_2)$$

4. Inductive Z_2 becomes capacitive Z_1, and vice versa, and is described as follows:

$$\text{Im}(Z_2) > 0 \Rightarrow \text{Im}(Z_1) < 0 \tag{5.17a}$$

$$\text{Im}(Z_2) < 0 \Rightarrow \text{Im}\left(Z_1\right) > 0 \tag{5.17b}$$

5.3.3 Unterminated Impedance Inversion Rule

It was found in the previous subsection that terminated impedance Z at the secondary port of a gyrator can be inverted to the primary port, and vice versa. This is extended to unterminated impedance Z in this subsection. It is shown that unterminated impedance Z at the secondary port can be moved to the primary port while not affecting the external port voltage–current relationships (V_1, V_2, I_1, I_2).

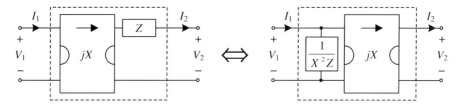

(a) Series-connected unterminated impedance Z at the secondary port (left) and its equivalent circuit (right)

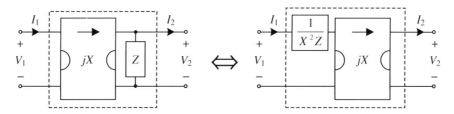

(b) Parallel-connected unterminated impedance Z at the secondary port (left) and its equivalent circuit (right)

Figure 5.15 Unterminated impedance inversion rule of a gyrator.

When Z is connected in series to the secondary port of the gyrator, as shown in the left circuit of Figure 5.15(a), its external port voltage–current relationships can be derived as follows:

$$V_2 = \left(-\frac{1}{jX}\right) \times I_1 + Z \times (-I_2) \tag{5.18a}$$

$$V_1 = \frac{1}{jX} \times I_2 \tag{5.18b}$$

Rearranging (5.18) in terms of I_1 and I_2 gives the following equations:

$$I_2 = jXV_1 \tag{5.19a}$$
$$I_1 = X^2ZV_1 - jXV_2 \tag{5.19b}$$

Identical voltage–current relationships can be derived from the right circuit in Figure 5.15(a), demonstrating equivalence between the two circuits in Figure 5.15(a). The same analogy applies when Z is connected in parallel to the secondary port, as shown in Figure 5.15(b); thus, detailed explanations are omitted.

From the equivalences illustrated in Figure 5.15, it is found that the series-connected impedance Z at the secondary port can be equivalently inverted to the primary port as parallel-connected impedance $1/(X^2Z)$, and vice versa. Note that the terminated impedance inversion rule described in Section 5.3.1 is a special case with $V_2 = 0$ in Figure 5.15(a) (or $I_2 = 0$ in Figure 5.15(b)).

With this property, several intermediate impedances Z between two gyrators can be moved to the external port. This eases the analysis by enabling the use of the gyrators' merge rule described in Section 5.3.4. Figure 5.16 shows an example where impedances Z_1 and Z_2 are inserted in parallel–series, respectively, between two gyrators with gains of jX_1 and jX_2. By continuously inverting Z_1 and Z_2, as shown in ckt. 1 and ckt. 2 of Figure 5.16, the equivalent circuit of Figure 5.16(c) is achieved that has the same external port voltage–current relationships (V_1, V_2, I_1, I_2) as that of Figure 5.16(a).

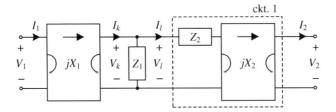

(a) Impedances Z_1 and Z_2 between two gyrators

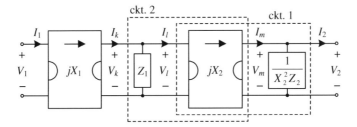

(b) Equivalent circuit of (a), where the circuits in the dashed boxes are equivalently replaced

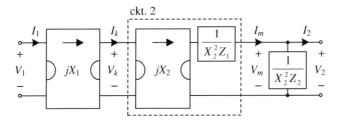

(c) Equivalent circuit of (b), where the circuits in the dashed boxes are equivalently replaced

Figure 5.16 Utilization of the unterminated impedance inversion rule of a gyrator.

5.3.4 Gyrators Merge Rule: Series-Connected Multiple Gyrators

As shown in Figure 5.17, series connections of multiple gyrators can be merged into a single ideal transformer or a single gyrator depending on whether the number of series-connected gyrators is even or odd. This stems from the gyrator's intrinsic property that it converts the voltage of one port to the current of another port, and vice versa.

When n gyrators with the gain of jX_1, jX_2, ..., jX_n are connected in series and n is even-numbered, the conversion between voltage and current caused by each gyrator occurs even-numbered times. Then the n gyrators can be merged into a single ideal transformer whose effective turn ratio N_{eff} is given as follows:

$$N_{eff} = \prod_{i=1}^{n/2} \left(-\frac{X_{2i-1}}{X_{2i}} \right) \tag{5.20}$$

which is real-numbered. At the same time this implies that an ideal transformer is a special kind of gyrator; however, the same is not true in reverse. Similarly, when n is odd-numbered,

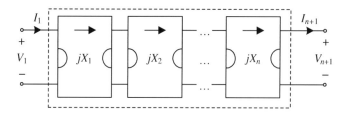

(a) Original n series-connected gyrators

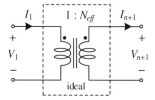

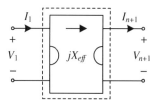

(b) Equivalent ideal transformer circuit when n is even-numbered

(c) Equivalent gyrator circuit when n is odd-numbered

Figure 5.17 Gyrator merge rule: n series-connected gyrators and the equivalent circuits.

the conversion between voltage and current occurs odd-numbered times. Hence, n series-connected gyrators can be merged into a single gyrator with the forward gain of jX_{eff} as follows:

$$jX_{eff} = jX_n \prod_{i=1}^{(n-1)/2} \left(-\frac{X_{2i-1}}{X_{2i}} \right) \tag{5.21}$$

which is purely imaginary; hence, it still has reverse gain, which is the same as the forward gain.

5.4 Analyses of Perfectly Tuned Compensation Circuits with the Proposed Method

Based on the discussions in the previous sections, compensation circuits are analyzed and important electrical characteristics between the source and load are derived: source-to-load gain, load R_L-independent output characteristics, power factor at the source, sign of the source phase angle, and allowance of open/short R_L. To show the validity of the proposed gyrator-based analysis, three of the most widely used compensation circuits are discussed as examples [15–43]: voltage-source-type primary-series–secondary-series (V-SS), current-source-type primary-series–secondary-parallel (I-SP), and voltage-source-type LCL secondary-parallel (V-LCL-P). The analyses of different compensation circuits have two common steps:

1. A compensation circuit is equivalently transformed to a gyrator-based circuit while the source and the load remain unchanged (Section 5.2).
2. Then the source-to-load and load-to-source electrical characteristics are analyzed using a gyrator's circuit properties (Section 5.3).

Throughout this section, it is shown that the proposed gyrator-based analysis is applied to different compensation schemes in a consistent manner and provides a simple and systematic way to analyze compensation circuits in an IPTS.

Figure 5.18 V-SS and the equivalent gyrator-based models.

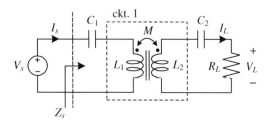

(a) V-SS circuit model

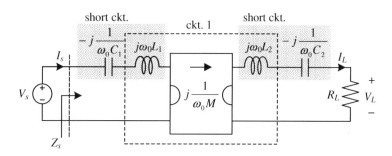

(b) Equivalent V-SS circuit of (a) with a gyrator

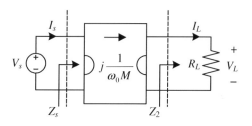

(c) Equivalent V-SS circuit of (b) with a gyrator

5.4.1 V-SS

V-SS is composed of a voltage source, two compensation capacitors, and coupled inductors, as illustrated in Figure 5.18(a). Its own characteristics are well known from the literature, such as load R_L-independent voltage-to-current gain, unity power factor at the source, and unsafe operation of the source when R_L is open-circuited [15–21,[26–31, 43]. These characteristics require that the source angular frequency ω_S of the V-SS be compensated as follows:

$$\omega_S = \frac{1}{\sqrt{L_1 C_1}} = \frac{1}{\sqrt{L_2 C_2}} = \omega_0 \tag{5.22}$$

Note that the condition in (5.22) is magnetic coupling coefficient k-independent.

As described in Section 5.2.3, V-SS in Figure 5.18(a) can be equivalently transformed to the circuit in Figure 5.18(b) by replacing the coupled inductors with the gyrator in Figure 5.10. Series–series resonant circuits in Figure 5.18(b),

L_1-C_1 and L_2-C_2, then become short-circuits providing the simplified equivalent circuit in Figure 5.18(c).

From the gyrator model in Figure 5.18(c) and the properties of the gyrator in the previous section, source-to-load gain and power factor at the source are calculated in a simple manner directly from the circuit diagram. Source-to-load gain A_G is derived by utilizing the properties of the gyrator outlined in Section 5.3.2, where the combination of a voltage source and gyrator is equivalent to a current source. Thus, the load current I_L is determined solely by source voltage V_s and gyrator gain $-1/(j\omega_0 M)$, which implies at the same time that I_L is independent of R_L as follows:

$$\frac{\partial I_L}{\partial R_L} = 0 \tag{5.23}$$

A_G is directly the gyrator gain in Figure 5.18(c), as follows:

$$A_G \equiv \frac{I_L}{V_s} = -\frac{1}{j\omega_0 M} \tag{5.24}$$

The power factor at the source is obtained using the properties of the gyrator in the previous Section 5.3.1. Because the power factor is maintained when the impedance on the secondary side of the gyrator is inverted to the primary side, the power factor at the source is equivalent to that of the secondary side, as follows:

$$PF|_{Z_s} = PF|_{Z_2} = \cos 0 = 1 \tag{5.25}$$

where Z_s and Z_2 are defined as the impedance seen at the source and at the secondary side, respectively, as expressed in Figure 5.18(c). Thus, the source has no reactive power rating.

Polarity of the source phase angle, $\angle Z_s$, can also be found by applying the properties of the gyrator in the previous Section 5.3.2. As shown in Figure 5.18(c), Z_2 is purely resistive; thus, Z_s is also purely resistive, as follows:

$$\text{Im}(Z_s) = 0 \tag{5.26}$$

Thus, when the source voltage, V_s, is implemented with a bridge-type converter such as the full-bridge and half-bridge converters commonly used to drive the compensation circuit source, current I_s does not lag V_s and active switches that constitute the bridge-type converter may not achieve zero voltage switching (ZVS) turn-on.

Safety of the source when R_L is an open circuit or a short circuit is also easily identified. As the gyrator converts an open circuit on the secondary side to a short circuit on the primary side, and vice versa, the source current I_s becomes zero when $R_L = 0$ and infinity when $R_L = \infty$, indicating unsafe operation of V-SS on open R_L as follows:

$$\lim_{R_L \to \infty} |I_s| = \infty \tag{5.27}$$

Characteristics of V-SS derived from the proposed gyrator-based analysis in this subsection all agree with studies in the literature, showing the validity of the gyrator-based analysis [15–21, 26–31, 43].

5.4.2 I-SP

I-SP, shown in Figure 5.19(a), consists of a current source, two compensation capacitors, and coupled inductors. The current source could be implemented by regulating source current I_s with feedback control of a voltage-type inverter [51–56]. To achieve its desirable characteristics, such as load R_L-independent current-to-current gain, its source angular frequency ω_S should

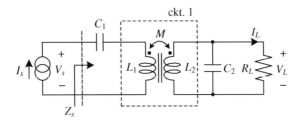

(a) I-SP circuit model

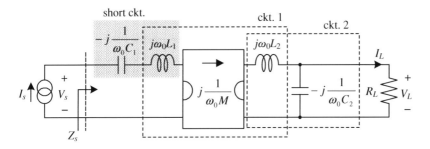

(b) Equivalent I-SP circuit of (a) with a gyrator

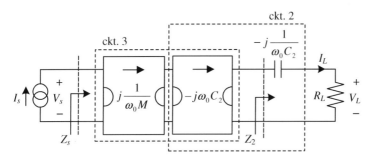

(c) Equivalent I-SP circuit of (b) with a gyrator

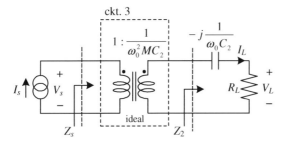

(d) Equivalent I-SP circuit of (c) with a gyrator

Figure 5.19 I-SP and the equivalent gyrator-based models.

be tuned at the secondary-resonance angular frequency, which is magnetic coupling coefficient k-independent as follows [43]:

$$\omega_S = \frac{1}{\sqrt{L_2 C_2}} = \omega_0 \tag{5.28}$$

In Figure 5.19(a), it is noticed that the coupled inductors can be replaced by the gyrator model in Figure 5.10, and the equivalent circuit becomes as seen in Figure 5.19(b). Then the series–series resonant circuit L_1-C_1 becomes a short circuit and the series–parallel resonant circuit L_2-C_2 is replaced by the gyrator in Figure 5.6, providing the equivalent circuit of Figure 5.19(c). Since two-series–connected gyrators are effectively a single transformer, as described in Figure 5.17(b), the circuit in Figure 5.19(c) can be further simplified to the circuit in Figure 5.19(d).

As analyzed in the previous subsection, the source-to-load gain A_G and power factor at the source can easily be found from the equivalent circuit in Figure 5.19(d), referring to the properties of a gyrator. Because the current source I_s is connected to the primary side of the transformer, the load current I_L is solely determined by I_s and the turn ratio. This shows the R_L-independent I_L characteristic of an I-SP as follows:

$$\frac{\partial I_L}{\partial R_L} = 0 \tag{5.29}$$

The source-to-load gain A_I is then given from the effective turn ratio of the transformer in Figure 5.19(d) as follows:

$$A_I \equiv \frac{I_L}{I_s} = \omega_0^2 M C_2 \tag{5.30}$$

The power factor at the source is equivalent to that of Z_2 in Figure 5.19(d), as discussed in Section 5.2.1 and as follows:

$$PF|_{Z_S} = PF|_{Z_2} = \frac{R_L}{\sqrt{R_L^2 + \left(\frac{1}{\omega_0 C_2}\right)^2}} \tag{5.31}$$

It should be noted that (5.31) provides two parameters, R_L and $1/(\omega_0 C_2)$, that fully determine the reactive power rating at the source, and this is derived in a simple manner.

Polarity of the source phase angle, $\angle Z_s$, is found by applying the properties of the gyrator described in the previous Sections 5.3.2 and 5.3.3. As the transformer in Figure 519(d) is a series connection of an even number of gyrators, the capacitive Z_2 is inverted to the capacitive Z_s as follows:

$$\text{Im}(Z_s) < 0 \tag{5.32}$$

Thus, the source voltage V_s lags I_s in phase.

Safety of the source when R_L is open or short is also identified. As the gyrator converts an open circuit on the secondary side to a short circuit on the primary side and the transformer in Figure 5.19(d) is a series connection of an even number of gyrators, the source voltage V_s becomes zero when $R_L = 0$. V_s becomes infinity when $R_L = \infty$, indicating unsafe operation of I-SP with an open load as follows:

$$\lim_{R_L \to \infty} |V_s| = \infty \tag{5.33}$$

Again, the characteristics of the I-SP analyzed by the gyrator-based analysis in this subsection all agree with studies in the literature [43].

5.4.3 V-LCL-P

V-LCL-P is composed of two inductors, L_0 and L_1, and capacitor C_1 on the primary side of the coupled inductors, and the secondary side is parallel-compensated, as shown in Figure 5.20(a).

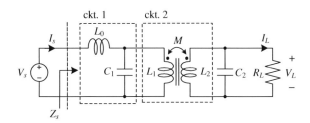

(a) V-LCL-P circuit model

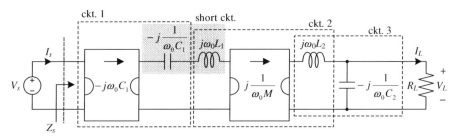

(b) Equivalent V-LCL-P circuit of (a) with a gyrator

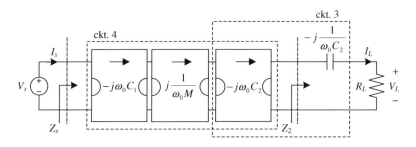

(c) Equivalent V-LCL-P circuit of (b) with a gyrator

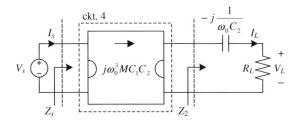

(d) Equivalent V-LCL-P circuit of (c) with a gyrator

Figure 5.20 V-LCL-P and the equivalent gyrator-based models.

The magnetic coupling coefficient k-independent tuning condition for V-LCL-P is as follows [36–42]:

$$\omega_S = \frac{1}{\sqrt{L_0 C_1}} = \frac{1}{\sqrt{L_1 C_1}} = \frac{1}{\sqrt{L_2 C_2}} = \omega_0 \tag{5.34}$$

It can be seen from Figure 5.20(a) that the series–parallel resonant circuit L_0-C_1 and the coupled inductors can be equivalently replaced by the gyrators in Figures 5.6 and 5.10, respectively, and the equivalent circuit becomes as seen in Figure 5.20(b). Then the series–series resonant circuit C_1-L_1 becomes a short circuit and the series–parallel resonant circuit L_2-C_2 can be replaced by the gyrator model in Figure 5.6. Thus the circuit in Figure 5.20(b) is further simplified as seen in Figure 5.20(c), providing three-series–connected gyrators. With the property of the gyrator described in Figure 5.17(b), three gyrators can be replaced by a single gyrator, as in Figure 5.20(d).

From the gyrator circuit model in Figure 5.20(d) and the property of the gyrator in Section 5.2.2, the source-to-load gain and power factor at the source are easily found. Because the voltage source at the primary port of the gyrator is equivalent to a current source on the secondary side, it is directly found from Figure 5.20(d) that I_L is independent of R_L as follows:

$$\frac{\partial I_L}{\partial R_L} = 0 \tag{5.35}$$

The source-to-load gain A_G is directly the gyrator gain in Figure 5.20(d), as follows:

$$A_G \equiv \frac{I_L}{V_s} = j\omega_0^3 M C_1 C_2 \tag{5.36}$$

The power factor at the source is also easily obtained from the circuit model in Figure 5.20(d) and the property discussed in Section 5.3.1. It is equivalent to the power factor on the secondary side of the gyrator in Figure 5.20(d) and is given as follows:

$$PF|_{V_s} = PF|_{Z_2} = \frac{R_L}{\sqrt{R_L^2 + \left(\frac{1}{\omega_0 C_2}\right)^2}} \tag{5.37}$$

In addition, the sign of the source phase angle, $\angle Z_s$, can be found by noting that the capacitive Z_2 implies inductive Z_s as described in (5.17). Thus, the source can be operated under ZVS turn-on operation when it is configured as a bridge-type converter:

$$\text{Im}(Z_s) > 0 \tag{5.38}$$

Safety of the source when R_L is an open circuit or a short circuit is also identified. As the gyrator converts an open circuit on the secondary side to a short circuit on the primary side, the source current I_s becomes infinity when $R_L = \infty$, indicating unsafe operation of V-LCL-P with the open R_L as follows:

$$\lim_{R_L \to \infty} |I_s| = \infty \tag{5.39}$$

The characteristics of V-LCL-P analyzed by the gyrator-based analysis in this subsection all agree with studies in the literature [36–42].

5.4.4 Discussion

Although only three compensation circuits, V-SS, I-SP, and V-LCL-P, are selected as examples among all compensation circuits, they have demonstrated that the transformation of any compensation circuit to a gyrator-based circuit is possible. This can be justified because a compensation circuit with any number of reactive components is decomposed to series or parallel redundant LC resonant circuits, as shown in Figures 5.8 and 5.9.

The application procedure of the proposed gyrator-based graphical approach to any IPTSs can be summarized as follows:

1. Replace the coupled inductors of an IPTS with a gyrator, as shown in Figure 5.10.
2. Add redundant *LC* resonant tanks if needed so that an appropriate T- or Π-network can be constituted, which results in an equivalent gyrator, as shown in Figures 5.8 and 5.9.
3. Move intermediate impedances between gyrators to the end port, as shown in Figure 5.16, and merge gyrators into a single gyrator or transformer, as in Figure 5.17.

In this way, the most simplified equivalent circuit composed of a gyrator or transformer with passive circuit components is obtained, which can be further analyzed for deriving useful equations and determining characteristics of the IPTSs. When this criteria is taken together with the gyrator's circuit properties described in Section 5.3, topological design as well as the analysis of the compensation circuit is possible: one can determine which kind of compensation is required to achieve certain characteristics.

5.5 Analyses of Mistuned Compensation Circuits with the Proposed Method

Although the compensation circuit is perfectly tuned under the nominal condition, it is easily mistuned by disturbances like misalignment of coupled coils and the appearance of foreign objects between the coupled coils [57]. To achieve a high reliability in IPTSs, evaluating the effects of mistuning on electrical characteristics of the compensation circuit is essential at the design step. This is especially true when the compensation circuit has a high load quality factor Q_L, because it makes the IPTS very sensitive to parameter changes [43]. However, quantifying these effects commonly involves a series of complicated equations and this makes identifying significant parameters difficult. In this section, it is shown that the proposed gyrator-based analysis can also be applied to evaluations of mistuning effects without any approximations, while maintaining its simple and systematic features. This reveals the practical aspects of the proposed gyrator-based analysis in that it does not necessarily require perfectly tuned reactive components for its application.

Most considerably mistuned parameters from a perfectly tuned condition may be the coupled inductor parameters: M, L_1, and L_2. These parameters are altered by misalignment of coupled coils or the appearance of foreign objects near the coupled coils that changes the magnetic reluctance around the coupled coils. For example, L_1 and L_2 as well as M increase from their nominal values when the distance between the coupled coils is reduced and both coils involve the magnetic core. L_1, L_2, and M decrease from their nominal values when conductive foreign objects appear near the coupled coils. Other parameters, compensation capacitances and inductances, are relatively insensitive to these disturbances. Hence, in this chapter, coupled inductor parameters M, L_1, and L_2 are considered to be mistuned parameters.

Mistuned coupled inductor parameters M, L_1, and L_2 can be represented with deviations from their nominal values M_0, L_{10}, and L_{20} as ΔM, ΔL_1, and ΔL_2, respectively, as follows:

$$M = M_0 + \Delta M \tag{5.40a}$$

$$L_1 = L_{10} + \Delta L_1 \tag{5.40b}$$

$$L_2 = L_{20} + \Delta L_2 \tag{5.40c}$$

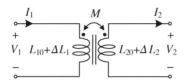

(a) Circuit model of mistuned coupled inductors

Figure 5.21 Representation of mistuned coupled inductors with a gyrator.

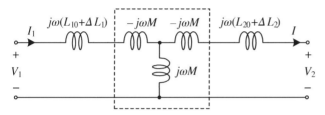

(b) Equivalent T-model of (a)

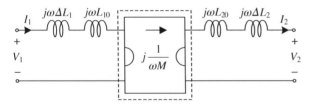

(c) Equivalent gyrator model of (b)

Based on (5.40), the coupled inductors and their equivalent T-models can be represented as seen in Figure 21(a) and 5.21(b). Then the equivalent gyrator model can be derived like Figure 5.10 and is illustrated in Figure 5.21(c). Comparing Figure 5.10(d) and Figure 5.21(c) shows that the deviation of M is all included in the deviation of the gyrator gain $-1/\{j\omega(M_0+\Delta M)\} = -1/(j\omega M)$, and deviations of L_1 and L_2 are all expressed as deviations of inductances $L_{10} + \Delta L_1$ and $L_{20} + \Delta L_2$ that are in series with the gyrator.

In this section, mistuned V-LCL-P, as shown in Figure 5.22(a), is analyzed as an example. It is assumed that V-LCL-P is perfectly tuned with nominal L_{10} and L_{20} as follows:

$$\omega_S = \frac{1}{\sqrt{L_0 C_1}} = \frac{1}{\sqrt{L_{10} C_1}} = \frac{1}{\sqrt{L_{20} C_2}} = \omega_0 \tag{5.41}$$

The mistuned coupled inductors are replaced with the gyrator model in Figure 5.21(c) and the equivalent circuit becomes as seen in Figure 5.22(b). Because all the deviations in the self-inductances of the coupled inductors are modeled as ΔL_1 and ΔL_2, L_{10} and L_{20} in Figure 5.22(b) still represent perfect tuned inductances. Thus, resonant circuits L_{20}-C_2 and L_{10}-C_1 can be replaced with the gyrator circuit in Figure 5.6 and a short circuit, respectively, and the equivalent circuit becomes as seen in Figure 5.22(c). By applying the gyrator's property described in Section 5.3.3, the mistuning disturbance ΔL_2 in Figure 5.22(c) that is in series with the primary port of the right-most gyrator can be equivalently moved to the secondary port like the parallel-connected capacitance $C_2 \Delta L_2/L_{20}$ in Figure 5.22(d). Mistuned disturbance ΔL_1 can also be moved to the secondary side of the right-most gyrator in the same manner, and the

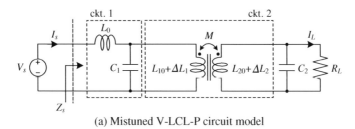

(a) Mistuned V-LCL-P circuit model

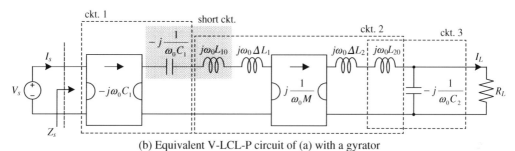

(b) Equivalent V-LCL-P circuit of (a) with a gyrator

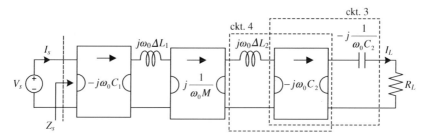

(c) Equivalent V-LCL-P circuit of (b) with a gyrator

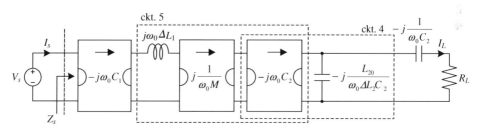

(d) Equivalent V-LCL-P circuit of (c) with a gyrator

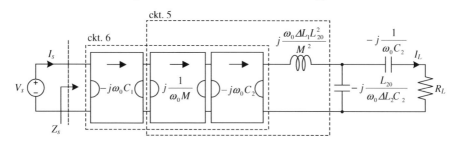

(e) Equivalent V-LCL-P circuit of (d) with a gyrator

Figure 5.22 Mistuned LCL-P and the equivalent circuits with gyrators.

Figure 5.22 (*Continued*)

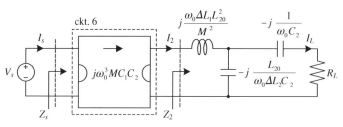

(f) Equivalent V-LCL-P circuit of (e) with a gyrator

equivalent circuit becomes as seen in Figure 5.22(e). Then three-series-connected gyrators are found, whichy can be replaced with a single gyrator according to Section 5.3.4, giving the further simplified equivalent circuit in Figure 5.22(f). Note that Figure 5.22(f) is equivalent to Figure 5.20(d) when there are no mistuned disturbances, $\Delta M = \Delta L_1 = \Delta L_2 = 0$.

From the simplified gyrator model in Figure 5.22(f), the effects of mistuning can be evaluated in the same manner as that of the perfectly tuned condition described in Section 5.4. According to the property of a gyrator described in Section 5.3.2, the voltage source at the primary port of the gyrator becomes a current source at the secondary port. Thus, current I_2 in Figure 5.22(f) of the secondary port of the gyrator is fully determined by source voltage V_s and the gyrator gain. Then load current I_L can be calculated by considering the current dividing ratio I_L/I_2 between $C_2\Delta L_2/L_{20}$ and $C_2 \cdot R_L$, and the source-to-load gain A_G is given as follows:

$$
\begin{aligned}
A_G &\equiv \frac{I_2}{V_s} \times \frac{I_L}{I_2} \\
&= j\omega_0^3 \left(M_0 + \Delta M\right) C_1 C_2 \times \frac{L_{20}}{\left(L_{20} + \Delta L_2\right) + j\omega_0 R_L \Delta L_2 C_2}
\end{aligned}
\tag{5.42}
$$

Note from (5.42) that mistuning ΔL_1 does not affect the source-to-load gain. This is because the inductance $\Delta L_1 L_{20}^2/(M_0 + \Delta M)^2$ is redundant when connected in series with current source I_2.

The power factor at the source V_s is derived by utilizing the property of a gyrator described in Section 5.3.1, where the power factor is maintained when the secondary-port impedance is inverted to the primary port of a gyrator. Thus, the effect of mistuning on the power factor at the source can be explicitly considered with that of Z_2 in Figure 5.22(f). Moreover, a condition where the source loses its ZVS operation is obtained by identifying whether Z_2 is inductive or capacitive. If Z_2 is inductive, Z_s becomes capacitive and the source voltage V_s does not lead the source current I_s, losing the ZVS turn-on operation in the source. Thus, it is qualitatively found that $\Delta L_1 \gg 0$ or $\Delta M \ll 0$ can lead to a high switching loss in the source.

5.6 Example Design and Experimental Verifications

The proposed gyrator-based analysis method was verified with the V-LCL-P compensation circuit for not only the perfectly tuned case but also for the mistuned case. Figures 5.23 and 5.24 show the experimental kit and its circuit schematic, respectively. The resonance frequency f_0 in (5.41) was set at $f_0 = 100$ kHz by carefully selecting the capacitance and inductance of the

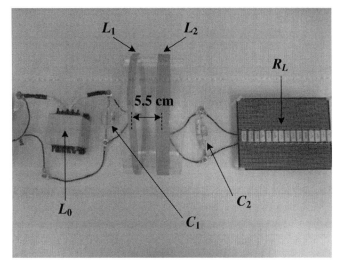

(a) Prototype fabrication for the perfectly tuned case

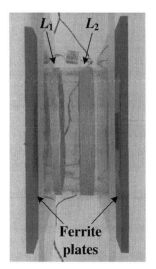

(b) Mistuned coupled coils

Figure 5.23 Prototype fabrication for experiments.

compensation components C_1, C_2, and L_0. The targeted load power was 85 W and the distance between the coupled coils L_1 and L_2 was 5.5 cm. To implement the mistuned coupled coils, ferrite core plates were deliberately placed near the coupled coils, as shown in Figure 5.23(b). All the circuit parameters are tabulated in Table 5.1.

The source-to-load gain A_G and the source phase angle $\angle Z_s$ were measured with respect to frequencies from 80 kHz to 120 kHz and compared with the calculation results for both perfectly tuned and mistuned conditions in Figure 5.25, where the solid lines represent the calculation results that assume that reactive components L_0, L_1, L_2, C_1, and C_2 had negligible parasitic resistance. To demonstrate that the discrepancy between the experimental results and lossless calculation results are mainly due to the ESRs, the calculation results including the measured ESRs are also represented in Figure 5.25, plotted with computer-aided simulations. The calculation results match well with the experimental results at the resonance frequency $f_0 = 100$ kHz demonstrating the proposed analysis method. The discrepancy between the solid lines and the dotted lines represents the effects of the power losses in the reactive components on the accuracy of the proposed analysis. The accuracy depends on the power efficiency from V_s to the load R_L, and the measured power efficiencies at f_0 were 91% and 81% for the perfectly

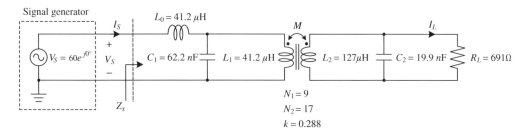

Figure 5.24 V-LCL-P circuit schematic for the experiments: perfectly tuned condition.

Table 5.1 Experimental circuit parameters at the source frequency of 100 kHz

Circuit components	Value
Linear amplifier A	MP108FD (200 V_{max}, 100 W_{max})
Inductively coupled coil set: perfectly tuned ($k = 0.288$)	$N_1 = 9$ turns $L_1 = 41.2$ μH $R_1 = 256$ mΩ
	$N_2 = 17$ turns $L_2 = 127$ μH $R_2 = 483$ mΩ
Inductively coupled coil set: mistuned ($k = 0.340$)	$N_1 = 9$ turns $L_1 = 48.3$ μH $R_1 = 263$ mΩ
	$N_2 = 17$ turns $L_2 = 142$ μH $R_2 = 510$ mΩ
Primary compensation inductor L_0	$L_0 = 41.2$ μH $R_{L0} = 43.2$ mΩ
Primary compensation capacitor C_1	$C_1 = 62.2$ nF $R_{C1} = 33.5$ mΩ
Secondary compensation capacitor C_2	$C_2 = 19.9$ nF $R_{C2} = 21.2$ mΩ

tuned and mistuned conditions, respectively. Including these effects, however, makes the analysis quite complicated and formidable to handle in practice. The experimental results match well with the dotted lines and the small discrepancy is mainly due to the modeling errors caused by measurement errors of the parasitic resistance and the nonlinear aspect of the core losses in L_0.

5.7 Conclusion

It was discussed that not only the magnetically coupled inductors but also all IPTSs inherently have the nature of a gyrator. A graphical approach that utilizes the gyrator's nature was proposed for modeling IPTSs. The proposed method utilizes that nature and provides a simple and systematic way to analyze important electrical characteristics of compensation circuits. Through analyses of widely used compensation circuits, V-SS, V-SP, and V-LCL-P, it was shown that the proposed method does not involve complicated high-order equations and applies to different compensation circuits in the same manner. In addition, it was also shown that the effects of mistuning can be evaluated with the proposed method while maintaining its simple and systematic aspects.

Question 5 (1) One of the limitations of the gyrator model is that it is applicable to only static circuit analysis. Can you suggest any idea to extend the gyrator model applicable to dynamic circuit analysis? (2) What is the relationship between the gyrator model and the phasor transformation (or dynamic phasor)? You may get some idea from this comparison and you can be a pioneer on this new modeling of the so-called "dynamic gyrator model," which has not yet

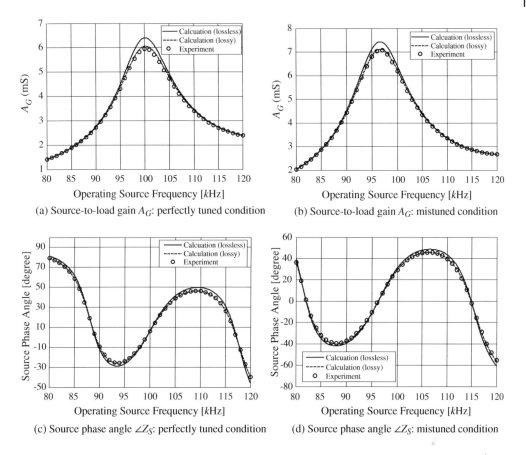

(a) Source-to-load gain A_G: perfectly tuned condition

(b) Source-to-load gain A_G: mistuned condition

(c) Source phase angle $\angle Z_S$: perfectly tuned condition

(d) Source phase angle $\angle Z_S$: mistuned condition

Figure 5.25 Experimental results for the prototype.

appeared. (3) What is the relationship between the gyrator model and the circuit DQ transformation [72, 73]?

Problems

5.1 Determine the efficiency of the experiment circuit under perfectly tuned conditions, as shown in Figure 5.24, with the parameters in Table 5.1. You need to analyze the gyrator circuit with internal resistances of coils, which is similar to Figure 5.20.

5.2 Determine the efficiency of the experiment circuit under the mistuned condition, as shown in Figure 4.24, with the parameters in Table 5.1. You need to analyze the gyrator circuit with internal resistances of coils, which is similar to Figure 5.22.

References

1 G.A. Covic and J.T. Boys, "Inductive power transfer," *Proceedings of the IEEE*, vol. 101, no. 6, pp. 1276–1289, June 2013.

2 E. Abel and S. Third, "Contactless power transfer-An exercise in topology," *IEEE Trans. Magn.*, vol. MAG-20, no. 5, pp. 1813–1815, September–November 1984.

3 A.W. Green and J.T. Boys, "10 kHz inductively coupled power transfer-concept and control," in *Proc. 5th Int. Conf. on Power Electron. Variable-Speed Drives*, October 1994, pp. 694–699.

4 Y. Hiraga, J. Hirai, A. Kawamura, I. Ishoka, Y. Kaku, and Y. Nitta, "Decentralised control of machines with the use of inductive transmission of power and signal," in *Proc. IEEE Ind. Appl. Soc. Annual Meeting*, 1994, vol. 29, pp. 875–881.

5 A. Kawamura, K. Ishioka, and J. Hirai, "Wireless transmission of power and information through one high-frequency resonant AC link inverter for robot manipulator applications," *IEEE Trans. Ind. Applic.*, vol. 32, no. 3, pp. 503–508, May/June 1996.

6 J.M. Barnard, J.A. Ferreira, and J.D. van Wyk, "Sliding transformers for linear contactless power delivery," *IEEE Trans. Ind. Electron.*, vol. 44, no. 6, pp. 774–779, December 1997.

7 D.A.G. Pedder, A.D. Brown, and J.A. Skinner, "A contactless electrical energy transmission system," *IEEE Trans. Ind. Electron.*, vol. 46, no. 1, pp. 23–30, February 1999.

8 J.T. Boys, G.A. Covic, and A.W. Green, "Stability and control of inductively coupled power transfer systems," *IEE Proceedings – Electric Power Applications*, vol. 147, no. 1, pp. 37–43, 2000.

9 K.I. Woo, H.S. Park, Y.H. Choo, and K.H. Kim, "Contactless energy transmission system for linear servo motor," *IEEE Trans. Magn.*, vol. 41, no. 5, pp. 1596–1599, May 2005.

10 G.A.J. Elliott, G.A. Covic, D. Kacprzak, and J.T. Boys, "A new concept: asymmetrical pick-ups for inductively coupled power transfer monorail systems," *IEEE Trans. Magn.*, vol. 42, no. 10, pp. 3389–3391, October 2006.

11 P. Sergeant and A. Van den Bossche, "Inductive coupler for contactless power transmission," *IET Electr. Power Applic.*, vol. 2, pp. 1–7, 2008.

12 J.T. Boys and A.W. Green, "Intelligent road studs – lighting the paths of the future," *IPENZ Trans.*, vol. 24, no. 1, pp. 33–40, 1997.

13 H.H. Wu, G.A. Covic, and J.T. Boys, "An AC processing pickup for IPT systems," *IEEE Trans. Power Electron. Soc.*, vol. 25, no. 5, pp. 1275–1284, May 2010.

14 H.H. Wu, G.A. Covic, J.T. Boys, and D. Robertson, "A series tuned AC processing pickup," *IEEE Trans. Power Electron. Soc.*, vol. 26, no. 1, pp. 98–109, January 2011.

15 W. Chwei-Sen, G.A. Covic, and O.H. Stielau, "Power transfer capability and bifurcation phenomena of loosely coupled inductive power transfer systems," *IEEE Trans. Ind. Electron.*, vol. 51, no. 1, pp. 148–157, February 2004.

16 W. Chwei-Sen, O.H. Stielau, and G.A. Covic, "Design considerations for a contactless electric vehicle battery charger," *IEEE Trans. Ind. Electron.*, vol. 52, no. 5, pp. 1308–1314, October 2005.

17 W. Chwei-Sen, G.A. Covic, and O.H. Stielau, "General stability criterions for zero phase angle controlled loosely coupled inductive power transfer systems," in *2001 IECON Conference*, pp. 1049–1054.

18 K. Aditya and S.S. Williamson, "Comparative study of series–series and series–parallel compensation topologies for electric vehicle charging," in *2014 ISIE Conference*, pp. 426–430.

19 K. Aditya and S. S. Williamson, "Comparative study of series–series and series-parallel compensation topology for long track EV charging application," in *2014 ITEC Conference*, pp. 1–5.

20 C. Yuan-Hsin, S. Jenn-Jong, P. Ching-Tsai, and S. Wei-Chih, "A closed-form oriented compensator analysis for series–parallel loosely coupled inductive power transfer systems," in *2007 PESC Conference*, pp. 1215–1220.

21 S. Chopra and P. Bauer, "Analysis and design considerations for a contactless power transfer system," in *2011 INTELEC Conference*, pp. 1–6.

22 G.B. Joung and B.H. Cho, "An energy transmission system for an artificial heart using leakage inductance compensation of transcutaneous transformer," *IEEE Trans. Power Electron.*, vol. 13, no. 6, pp. 1013–1022, November 1998.

23 A.J. Moradewicz and M.P. Kazmierkowski, "Contactless energy transfer system with FPGA-controlled resonant converter," *IEEE Trans. Ind. Electron.*, vol. 57, no. 9, pp. 3181–3190, September 2010.

24 J. How, Q. Chen, Siu-Chung Wong, C.K. Tse, and X. Ruan, "Analysis and control of series/series–parallel compensated resonant converters for contactless power transfer," *IEEE Journal of Emerging and Selected Topics in Power Electronics*, vol. PP, no. 99, p. 1, 2014.

25 S.-Y. Cho, I.-O. Lee, S. Moon, G.-W. Moon, B.-C. Kim, and K.Y. Kim, "Series–series compensated wireless power transfer at two different resonant frequencies," in *2013 ECCE Conference*, pp. 1052–1058.

26 W. Zhang, S.-C. Wong, C.K. Tse, and Q. Chen, "Analysis and comparison of secondary series- and parallel-compensated inductive power transfer systems operating for optimal efficiency and load-independent voltage-transfer ratio," *IEEE Trans. Power Electron.*, vol. 29, no. 6, pp. 2979–2990, June 2014.

27 W. Zhang, S.-C. Wong, and C.K. Tse, "Compensation technique for optimized efficiency and voltage controllability of IPT systems," in *2012 ISCAS Conference*, pp. 225–228.

28 W. Zhang, S.-C. Wong, C.K. Tse, and Q. Chen, "Load-independent current output of inductive power transfer converters with optimized efficiency," in *2014 IPEC Conference*, pp. 1425–1429.

29 W. Zhang, S.-C. Wong, C.K. Tse, and Q. Chen, "Analysis and comparison of secondary series- and parallel- compensated IPT systems," in *2013 ECCE Conference*, pp. 2898–2903.

30 Z. Wei, S.-C. Wong, C.K. Tse, and C. Qianhong, "Design for efficiency optimization and voltage controllability of series–series compensated inductive power transfer systems," *IEEE Trans. Power Electron.*, vol. 29, no. 1, pp. 191–200, January 2014.

31 X. Ren, Q. Chen, L. Cao, X. Ruan, S.-C. Wong, and C.K. Tse, "Characterization and control of self-oscillating contactless resonant converter with fixed voltage gain," in *2012 IPEMC Conference*, pp. 1822–1827.

32 O.H. Stielau and G.A. Covic, Design of loosely coupled inductive power transfer systems," *International Conference on Power System Technology*, 2000, pp. 85–90.

33 H. Abe, H. Sakamoto, and K. Harada, "A noncontact charger using a resonant converter with parallel capacitor of the secondary coil," *IEEE Trans. Ind. Applic.*, vol. 36, no. 2, pp. 444–451, March–April 2000.

34 J.L. Villa, J. Sallan, J.F. Sanz Osorio, and A. Llombart, "High-misalignment tolerant compensation topology for ICPT systems," *IEEE Trans. Ind. Electron.*, vol. 59, no. 2, pp. 945–951, February 2012.

35 J. Sallan, J.L. Villa, A. Llombart, and J.F. Sanz, "Optimal design of ICPT systems applied to electric vehicle battery charge," *IEEE Trans. Ind. Electron.*, vol. 56, no. 6, pp. 2140–2149, June 2009.

36 M.L.G. Kissin, H. Chang-Yu, G.A. Covic, and J.T. Boys, "Detection of the tuned point of a fixed-frequency LCL resonant power supply," *IEEE Trans. Power Electron.*, vol. 24, no. 41, pp. 1140–1143, April 2009.

37 H. Hao, G.A. Covic, and J.T. Boys, "An approximate dynamic model of LCL-T-based inductive power transfer power supplies," *IEEE Trans. Power Electron.*, vol. 29, no.10, pp. 5554–5567, October 2014.

38 C.Y. Huang, J.T. Boys, and G.A. Covic, "LCL pickup circulating current controller for inductive power transfer systems," *IEEE Trans. Power Electron.*, vol. 28, no. 4, pp. 2081–2093, April 2013.

39 C.Y. Huang, J.E. James, and G.A. Covic, "Design considerations for variable coupling lumped coil systems," *IEEE Trans. Power Electron.*, vol. 30, no. 2, pp. 680–689, February 2015.

40 N.A. Keeling, G.A. Covic, and J.T. Boys, "A unity-power-factor IPT pickup for high-power applications," *IEEE Trans. Ind. Electron.*, vol. 57, no. 2, pp. 744–751, February 2010.

41 C.W. Wang, G.A. Covic, and O.H. Stielau, "Investigating an LCL load resonant inverter for inductive power transfer applications," *IEEE Trans. Power Electron.*, vol. 19, no. 4, pp. 995–1002, July 2004.

42 H. Hao, G.A. Covic, and J.T. Boys, "A parallel topology for inductive power transfer power supplies," *IEEE Trans. Power Electron.*, vol. 29, no. 3, pp. 1140–1151, March 2014.

43 Y.H. Sohn, B.H. Choi, E.S. Lee, G.C. Lim, G.H. Cho, and Chun T. Rim, "General unified analyses of two-capacitor inductive power transfer systems," *IEEE Trans. Power Electron.*, vol. 30, no. 1, pp. 6030–6045, November 2015.

44 Z. Pantic, B. Sanzhong, and S. Lukic, "ZCS LCC-compensated resonant inverter for inductive-power-transfer application," *IEEE Trans. Ind. Electron.*, vol. 58, no. 9, pp. 3500–3510, August 2011.

45 B.D.H. Tellegen, "The gyrator, a new electric network element," *Philips Res. Rep.*, vol. 3, pp. 81–101, April 1948.

46 Chun T. Rim, Dong Y. Hu, and G.-H. Cho, "Transformers as equivalent circuits for switches: general proofs and D-Q transformation-based analyses," *IEEE Trans. Ind. Applic.*, vol. 26, no. 4, pp. 777–785, July/August 1990.

47 D.C. Hamill, "Lumped equivalent circuits of magnetic components: the gyrator-capacitor approach," *IEEE Trans. Power Electron.*, vol. 8, no. 2, pp. 97–103, April 1993.

48 S. Singer and R.W. Erickson, "Canonical modeling of power processing circuits based on the POPI concept," *IEEE Trans. Power Electron.*, vol. 7, no. 1, pp. 37–43, January 1992.

49 A. Cid-Pastor, L. Martinez-Salamero, C. Alonso, R. Leyva, and S. Singer, "Paralleling DC–DC switching converters by means of power gyrators," *IEEE Trans. Power Electron.*, vol. 22, no. 6, pp. 2444–2453, November 2007.

50 K. Woronowicz, A. Safaee, T. Dickson, M. Youssef, and S. Williamson, "Boucherot bridge based zero reactive power inductive power transfer topologies with a single phase transformer," in *2014 IEVC Conference*, pp. 1–6.

51 J. Huh, S.W. Lee, W.Y. Lee, G.H. Cho, and Chun T. Rim, "Narrow-width inductive power transfer system for online electrical vehicles," *IEEE Trans. Power Electron.*, vol. 26, no. 12, pp. 3666–3679, December 2011.

52 S.Y. Choi, B.W. Gu, J. Huh, W.Y. Lee, J.G. Cho, and Chun T. Rim, "Asymmetric coil sets for wireless stationary EV chargers with large lateral tolerance by dominant field analysis," *IEEE Trans. Power Electron.*, vol. 29, no. 12, pp. 6406–6420, December 2014.

53 C.B. Park, S.W. Lee, and Chun T. Rim, "Innovative 5 m-off-distance inductive power transfer systems with optimally shaped dipole coils," *IEEE Trans. Power Electron.*, vol. 30, no. 2, pp. 817–827, February 2015.

54 C.B. Park, S.W. Lee, G.-H. Cho, S.-Y. Choi, and Chun T. Rim, "Two-dimensional inductive power transfer system for mobile robots using evenly displaced multiple pick-ups," *IEEE Trans. Ind. Applic.*, vol. 50, no. 1, pp. 558–565, January–February 2014.

55 S.W. Lee, B. Choi, and Chun T. Rim, "Dynamics characterization of the inductive power transfer system for on-line electric vehicles by Laplace phasor transform," *IEEE Trans. Power Electron.*, vol. 28, no. 12, pp. 5902–5909, December 2013.

56 B.H. Choi, J.P. Cheon, J.H. Kim, and Chun T. Rim, "7 m-off-long-distance extremely loosely coupled inductive power transfer systems using dipole coils," in *2014 ECCE Conference*, pp. 858–863.

57 S.Y. Choi, B.W. Gu, S.Y. Jeong, and Chun T. Rim, "Advances in wireless power transfer systems for road powered electric vehicles," *IEEE Journal of Emerging and Selected Topics in Power Electronics*, vol. 3, no. 1, pp. 18–36, March 2015.

58 R.L. Steigerwald, "A comparison of half-bridge resonant converter topologies," *IEEE Trans. Power Electron.*, vol. 3, no. 2, pp. 174–182, April 1988.

59 S. Li, W. Li, J. Deng, and C. Mi, "A Double-sided LCC compensation network and its tuning method for wireless power transfer," *IEEE Transactions on Vehicle Technology*, issue 99, pp. 1–12, 2014.

60 F. Lu, H. Hofmann, J. Deng, and C. Mi, "Output power and efficiency sensitivity to circuit parameter variations in double-sided LCC-compensated wireless power transfer system," in *2015 APEC Conference*, pp. 597–601.

61 J. Deng, F. Lu, W. Li, R. Ma, and C. Mi, "ZVS double-side LCC compensated resonant inverter with magnetic integration for electric vehicle wireless charger," in *2015 APEC Conference*, pp. 1131–1136.

62 J. Deng, F. Lu, S. Li, T. Nguyen, and C. Mi, "Development of a high efficiency primary side controlled 7 kW wireless power charger," *IEEE International Electric Vehicle Conference*, 2014, pp. 1–6.

63 G.W. Wester and R.D. Middlebrook, "Low-frequency characterization of switched DC–DC converters," *IEEE Transactions an Aerospace and Electronic Systems*, vol. AES-9, pp. 376–385, May 1973.

64 R. Tymerski and V. Vorperian, "Generation, classification and analysis of switched-mode DC-to-DC converters by the use of converter cells," in *1986 INTELEC Conference*, pp. 181–195.

65 V. Vorperian, R. Tymerski, and F.C. Lee, "Equivalent circuit models for resonant and PWM switches," *IEEE Trans. Power Electron.*, vol. 4, no. 2, pp. 205–214, April 1989.

66 V. Vorperian, "Simplified analysis of PWM converters using the model of the PWM switch: Parts I and II," *IEEE Transactions on Aerospace and Electronic Systems*, vol. AES-26, pp. 490–505, May 1990.

67 S. Freeland and R.D. Middlebrook, "A unified analysis of converters with resonant switches," *in 1987 PESC Conference*, pp. 20–30.

68 A. Witulski and R.W. Erickson, "Extension of state-space averaging to resonant switches and beyond," *IEEE Trans. Power Electron.*, vol. 5, no, 1, pp. 98–109, January 1990.

69 D. Maksimovic and S. Cuk, "A unified analysis of PWM converters in discontinuous modes," *IEEE Trans. on Power Electron.*, vol. 6, no. 3, pp. 476–490, July 1991.

70 D.J. Shortt and F.C. Lee, "Extensions of the discrete-average models for converter power stages," in *1983 PESC Conference*, pp. 23–37.

71 O. AL-Naseem and R.W. Erickson, "Prediction of switching loss variations by averaged switch modeling," in *2000 APEC Conference*, pp. 242–248.

72 Chun T. Rim, D.Y. Hu, and G.-H. Cho, "The graphical D-Q transformation of general power switching converters," in *1998 Industry Applications Society Annual Meeting*, pp. 940–945.

73 Chun T. Rim, N.S. Choi, G.C. Cho, and G.-H. Cho, "A complete DC and AC analysis of three-phase controlled-current PWM rectifier using circuit D-Q transformation," *IEEE Trans. Power Electron.*, vol. 9, no. 4, pp. 390–396, July 1994.

74 Y.H. Son, B.H. Choi, G.-H. Cho, and Chun T. Rim, "Gyrator-based analysis of resonant circuits in inductive power transfer systems," *IEEE Trans. on Power Electronics*, vol. 31, no. 10, pp. 6824–6843, October 2016.

6

Magnetic Mirror Model

6.1 Introduction

Wireless power transfer technology has strongly attracted the attention of the public due to its convenience, safety, and reasonably high power efficiency. Wireless battery chargers for mobile electronic devices and robots [1–9], wireless electric vehicles (EVs) such as the on-line electric vehicle (OLEV) [10–27], and biomedical implant devices such as pacemakers, magnetic nerve stimulators, and blood pressure sensors [28–35] that require no surgical operation are all examples of the application of this new technology. In a wireless power transfer system using inductive coupling, the primary and pick-up coils include cores of high permeability to enhance the transferred power and system efficiency. As an example, ferromagnetic core plates can be used to generate strong magnetic flux in magnetic resonance imaging (MRI) and magnetic flux pumps [36–39]. In wireless battery chargers for mobile electronic devices and robots [1–9], very light core plates are used to transfer uniform power regardless of the device position. Ferrite cores are also used in wireless battery chargers for EVs [40–45] and in road-powered EVs to transfer high power of more than several tens of kW [10–23].

To improve the performance of the coils, the magnetic resistance between two coils, the hysteresis loss of the core, and the maximum allowable magnetic flux in the core should be clearly identified. Therefore, the magnetic field distribution of the coils is crucial for an efficient coil design. Besides the electric field calculation, however, it is very difficult to calculate the magnetic field where the core is involved. The core distorts the magnetic flux path, meaning that there is no simple means of finding the magnetic resistance for an arbitrary core shape. A simple analysis method is crucial for magnetic field calculations such as those pertaining to the electric field.

An ideal magnetic mirror model for a plane core with infinite permeability and an infinite width is commonly used to calculate the magnetic field distribution, where a mirror current of the same magnitude as the source current at the opposite point from the core surface replaces the complicated core effect on the magnetic field [39]. The model is similar to the image charge method of calculating the electric field for a perfect conductor. It is very convenient and useful to use this model when a current source is positioned on an ideal core plate, where the magnetic flux density on the core surface is exactly doubled by the perfect core as the magnetic resistance in the free space is halved. For a loop coil with a cylindrical core having finite permeability and a finite coil size, it has been identified by finite element method (FEM) simulations that the mirror current value is reduced [39], but this has been calculated for only several specific cases. For practical applications, more generalized magnetic mirror models applicable to the plane core case with a finite width are required. Most papers dealing with inductive power transfer systems (IPTSs) [1–23, 36–45], however, rely only on simulations and experiments without

Wireless Power Transfer for Electric Vehicles and Mobile Devices, First Edition. Chun T. Rim and Chris Mi.
© 2017 John Wiley & Sons Ltd. Published 2017 by John Wiley & Sons Ltd.

formulating proper equational models to determine the magnetic flux density for a given coil structure.

In this chapter, improved magnetic mirror models (IM^3) of a closed equational form for a core plate with a finite width are explained. The permeability of the core is, however, assumed to be infinite in this case due to the difficulties and complexities in modeling. By introducing an appropriate mirror current whose magnitude is smaller than the source current, the magnetic flux density over a core plate can be determined. The ratio of the mirror current and the source current, which is a function of the width of the core plate and the distance between the source current and core plate, is rigorously verified by FEM simulations. By applying the proposed IM^3 to the mono and dual coils used for wireless electric vehicles, the magnetic flux density over an open core plate is analyzed and its maximum points on or in the plates, which is crucial in the design of the coils to avoid local magnetic saturation, can also be found. Furthermore, the magnetic flux density when a pick-up core plate is positioned over a primary core plate is analyzed by introducing successive mirror currents. The proposed IM^3 was extensively verified by experiments, including site tests, to demonstrate the practical usefulness of these models.

Nomenclature

I_s Magnitude of the source current (A)
I_m Magnitude of the mirror current (A)
B_1 Magnetic flux density for a mono coil with an open core plate (T)
B_2 Magnetic flux density for a dual coil with an open core plate (T)
B_3 Magnetic flux density for a mono coil with parallel core plates (T)
B_4 Magnetic flux density for a dual coil with parallel core plates (T)
d_s Vertical distance between the surface of a core plate and the center of the source current (m)
d_m Vertical distance between the surface of a core plate and the center of the mirror current (m)
d Vertical distance between the surface of a core plate and the center of either the source current or mirror current (m)
d_c Thickness of a core plate (m)
w_s Horizontal displacement of the center of the source current from the center of the surface of a core plate (m)
w_o Width of a core plate (m)
l_o Length of a core plate (m)
r_s Radius from the center of the surface of a core plate to the center of the source current (m)
r_m Radius from the center of the surface of a core plate to the center of the mirror current (m)
x Horizontal value of the measurement point from the center of the surface of a core plate (m)
y Vertical value of the measurement point from the center of the surface of a core plate (m)
h Air gap between the facing surfaces of the primary and pick-up core plates (m)
μ_o Permeability in free space, that is, $4\pi \times 10^{-7}$ (H/m)
μ_r Relative permeability of a core plate
a_x Unit vector in the horizontal direction
a_y Unit vector in the vertical direction
γ_1 Ratio of the mirror current and the source current for a mono coil
γ_2 Ratio of the mirror current and the source current for a dual coil
α_1 Curve-fitting coefficient of the mirror current model γ_1 for a mono coil
α_2 Curve-fitting coefficient of the mirror current model γ_2 for a dual coil

6.2   Improved Magnetic Mirror Models for Coils with Open Core Plates

The magnetic flux density for coils having a core plate with a finite width, an infinite length, and infinite permeability is calculated and thus the magnitude of the mirror current for IM^3 can be determined as having an equational form. It is assumed that the core has no loss and that the source current can be arbitrary, that is, DC or AC. Both mono and dual coils are considered here because they are being implemented in OLEVs [17–19] requiring explicit design models. There is no pick-up coil over the primary coils; therefore, half of the free space of the core plate is open. FEM simulations are performed by Ansoft Maxwell (Ver. 12) for a two-dimensional case to verify the proposed models.

6.2.1   Mono Coil with a Finite Width

The mono coil with a pick-up coil used in OLEVs is shown in Figure 6.1, where the length of the mono coil l_o is infinitely long but its width w_o is typically less than 1 m [17]. The current flowing on the core plate is divided into two and returned out of the core plate, and the width of the pick-up coil is slightly larger than that of the mono coil in order to provide lateral tolerance. To deal with this practical problem, we simplify the problem and solve it incrementally. In this section, the case of a mono coil with an open core plate is considered first, as shown in Figure 6.2. The magnetic flux is severely distorted by the core plate; hence, it seems analytically difficult to determine it. The magnetic field on the surface of the core is, however, close

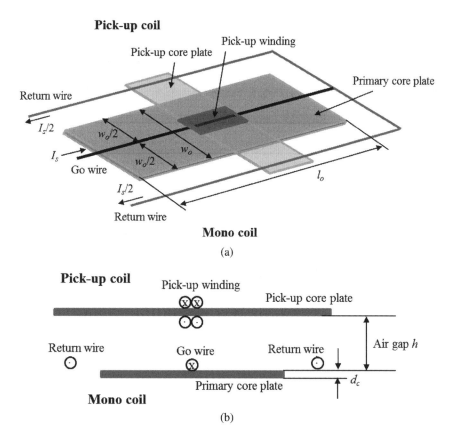

Figure 6.1 Mono coil with a pick-up coil for OLEV: (a) bird's view, (b) cross-sectional view.

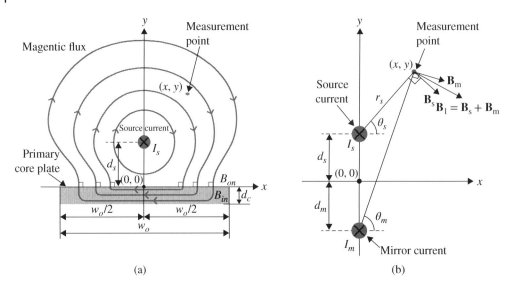

Figure 6.2 Proposed IM³ of a mono coil with an open core plate. (a) Mono coil consists of a source current and a core plate with finite width and infinite permeability. (b) Proposed improved magnetic mirror model (IM³) of the mono coil.

to perpendicular when the permeability is infinitely large, as shown is Figure 6.3. Therefore, it can be solved by applying a modified magnetic mirror model in which the mirror current generates a canceling magnetic field against the source current for the horizontal direction x on the surface of the core plate. Figure 6.3 shows the FEM simulation results for the angle of the magnetic flux density on the surface of the core plate as a function of the relative permeability. It was found that the magnetic flux is close to perpendicular when the relative permeability exceeds approximately 1000. Therefore, it is applicable for most practical core materials where this value typically ranges from 2000 to 5000.

IM³ for the mono coil is, however, not readily available because the distance from the surface of the core d_m as well as the magnitude of the mirror current I_m have yet to be determined. For the case of a mono coil with a finite width, there is no rationale for the distance d_m to be equal to the distance from the source current to the surface of the core d_s, as shown in Figure 6.2. Furthermore, the mirror current I_m should be found from the closed-form equation rather than specific numbers, as in [39]. They are verified by magnetic field simulations under various conditions in this chapter because no theoretical analysis is available at present.

From Ampere's law, the following equation holds in general:

$$\oint \mathbf{H} \cdot d\mathbf{l} = I \quad \text{and} \quad \oint \mathbf{B} \cdot d\mathbf{l} = \mu_o I \tag{6.1}$$

The magnetic flux density of the open mono coil B_1 for a mirror current I_m and a source current I_s at an arbitrary measurement point (x, y) is obtained as follows:

$$B_1 \equiv |\mathbf{B}_1| = |\mu_o \mathbf{H}_1| = \left| \frac{\mu_o I_s}{2\pi} \frac{\alpha_x (y - d_s) - \alpha_y x}{r_s^2} + \frac{\mu_o I_m}{2\pi} \frac{\alpha_x (y + d_m) - \alpha_y x}{r_m^2} \right|$$

$$= \frac{\mu_o I_s}{2\pi} \sqrt{\left\{ \frac{y - d_s}{r_s^2} + \gamma_1 \frac{y + d_m}{r_m^2} \right\}^2 + \left\{ \frac{x}{r_s^2} + \gamma_1 \frac{x}{r_m^2} \right\}^2} \quad \text{for} \quad |x| < \frac{w_o}{2}, \ 0 \le y \tag{6.2a}$$

$$r_s = \sqrt{x^2 + (y - d_s)^2}, \ r_m = \sqrt{x^2 + (y + d_m)^2} \tag{6.2b}$$

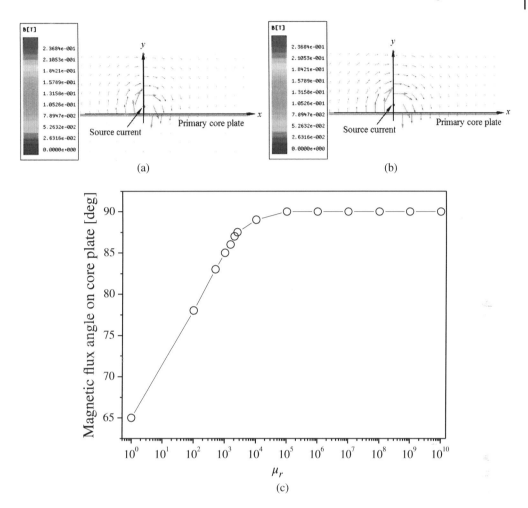

Figure 6.3 FEM simulation results of the magnetic flux density for the mono coil with an open core plate along the relative permeability of the core plate at $x = 1$ m under the conditions $I_s = 100$ A, $d = 0.01$ m, and $w_o = 5$ m. (a) The magnetic flux density vector for $\mu_r = 100$, (b) the magnetic flux density vector for $\mu_r = 10^{10}$, and (c) the magnetic flux density vector angle at $x = 1\,m$ on the core plate.

Through magnetic field FEM simulations, as shown in Figures 6.4 and 6.5, it was found that the simulation results for the magnetic flux density B_1 are the best fit for the IM^3 of (6.2) under the following conditions:

$$d_m = d_s = d, \quad \gamma_1 \equiv \frac{I_m}{I_s} \cong \left(1 - e^{-w_o/(\alpha_1 y)}\right) \quad \because \alpha_1 \cong 1.7 \tag{6.3}$$

Fortunately, the distances of the mirror current and the source current from the core surface were found to be identical regardless of any of the other parameters. This can be explained via the boundary condition on the surface of the core plate, that is, $y = 0$, where the magnetic flux is perpendicular to the core surface for an infinitely large permeability. The boundary condition is met for any measurement point $(x, 0)$ only when $d_m = d_s$ and $I_m = I_s$, which is fulfilled by (6.3) because $\gamma_1 = 1$ when $y = 0$. However, γ_1 is a function of the core width w_o, as shown in Figure 6.4, where the magnitude of the mirror current becomes zero for the case without a core plate and approaches the ideal case, that is, $\gamma_1 = 1$, exponentially as the core plate width w_o increases.

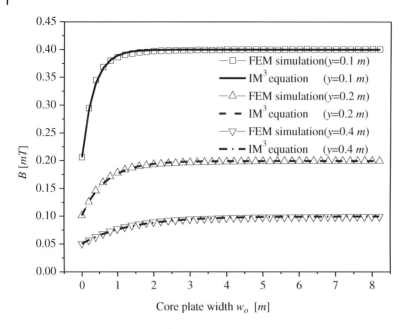

Figure 6.4 Comparison of the IM[3] with FEM simulation results for the magnetic flux density of the mono coil with an open core plate for different y along w_o at $(0, y)$ under the conditions $\mu_r = 10^7$, $I_s = 100$ A, and $d = 0.01$ m.

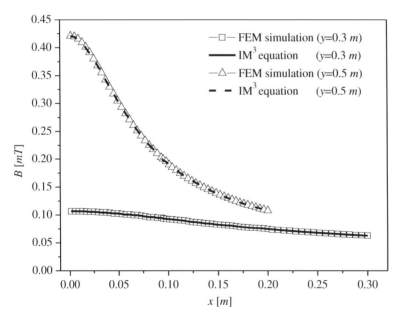

Figure 6.5 Comparison of IM[3] with FEM simulation results for the magnetic flux density of the mono coil with an open core plate for different y along x at (x, y) under the conditions $\mu_r = 10^7$, $I_s = 100$ A, $d = 0.45$ m, and $w_o = 3$ m.

Therefore, the magnetic flux density is eventually doubled when $w_o = \infty$, as expected. Oddly, γ_1 is not a function of d but a function of the measurement point y, implying that the mirror current is little changed as the distance d varies but decreases when measured far from the core plate, as identified by (6.3). In other words, even for a fixed configuration of a coil, the mirror current should be changed accordingly for different measurement points. This can be explained physically because the contribution of the magnetic flux induced by the finite width core to the observer at the measurement point increases as a function of the viewing factor w_o/y, as shown in (6.3). The proposed model is quite different from the ideal mirror model where the mirror current is unchanged by the observer. In the proposed IM^3, however, the mirror current of each observer can be different because the viewing factors are different from each other.

The IM^3 of (6.2) and (6.3) is also well fitted to the simulation results in the x direction for the different measuring heights y, as shown in Figure 6.5. In addition to the comparison results shown in Figures 6.4 and 6.5, extensive simulations were performed to confirm the curve fitting of (6.3). Caution is required regarding the use of (6.2) and (6.3) because the proposed model is valid for the region inside the core plate, that is, $|x| < w_o/2$. It was found that the model is quite accurate for $|x| < w_o/4$ within 5% error. For the edges of a core plate, the fringe effect of the magnetic field dominates and the proposed magnetic mirror model IM^3 deviates from the simplified ideal condition, which results in roughly 20% error. This is not a practical problem for a coil design in which the wire is on the core plate, that is, where the region of interest is bounded to the inside of a core plate.

Question 1 (1) What is the rationale for (6.3)? In other words, is there any possible explanation or physical meaning of the exponential function? (2) Why $\alpha_1 \cong 1.7$? Is it actually an approximated number of a specific number such as $\sqrt{3}$?

6.2.2 Dual Coil with a Finite Width

A dual coil [17], where the length of the core plate l_o is infinitely long but the width w_o is finite, is modeled here. The dual coil consists of a core plate with a go wire and a return wire, as shown in Figure 6.6. The proposed IM^3 for the mono coil cannot be used for the dual coil because the geometric properties and configurations are quite different from each other. Therefore, similar works of the mono coil should be done for this case.

Figure 6.6 Dual coil with finite width and infinite permeability.

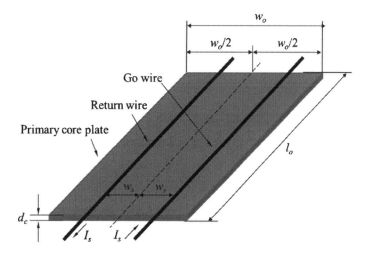

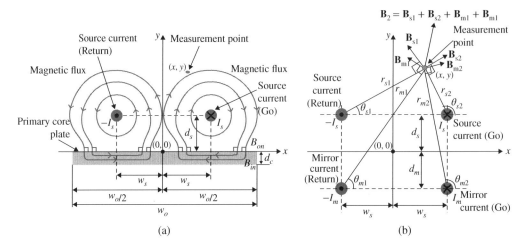

Figure 6.7 Proposed IM³ of a dual coil with an open core plate. (a) Dual coil consists of a two-source current and a core plate with finite width and infinite permeability. (b) Proposed improved magnetic mirror model (IM³) of the dual coil.

Because the dual coil has two source currents that flow in opposite directions, it is postulated that the core plate effect on the magnetic flux can be modeled as two mirror currents, as shown in Figure 6.7. As explained in the previous section, the distances between the mirror current and the source current from the core surface are identical regardless of any of the other parameters. The magnetic flux density at any point (x, y) for the source currents and the mirror currents is determined by applying Ampere's law, as shown in (6.4a) and (6.4b):

$$B_2 \equiv |\mathbf{B}_2| = \left| \frac{\mu_o I_s}{2\pi} \left\{ \left(-\frac{y-d}{r_{s1}^2} + \frac{y-d}{r_{s2}^2} \right) \alpha_x + \left(\frac{x+w_s}{r_{s1}^2} - \frac{x-w_s}{r_{s2}^2} \right) \alpha_y \right\} \right.$$
$$\left. + \frac{\mu_o I_m}{2\pi} \left\{ \left(-\frac{y+d}{r_{m1}^2} + \frac{y+d}{r_{m2}^2} \right) \alpha_x + \left(\frac{x+w_s}{r_{m1}^2} - \frac{x-w_s}{r_{m2}^2} \right) \alpha_y \right\} \right|$$

$$(6.4a)$$

$$= \frac{\mu_o I_s}{2\pi} \sqrt{ \begin{array}{l} \left(-\dfrac{y-d}{r_{s1}^2} + \dfrac{y-d}{r_{s2}^2} - \gamma_2 \dfrac{y+d}{r_{m1}^2} + \gamma_2 \dfrac{y+d}{r_{m2}^2} \right)^2 \\ + \left(\dfrac{x+w_s}{r_{s1}^2} - \dfrac{x-w_s}{r_{s2}^2} + \gamma_2 \dfrac{x+w_s}{r_{m1}^2} - \gamma_2 \dfrac{x-w_s}{r_{m2}^2} \right)^2 \end{array} } \quad for \quad |x| < \frac{w_o}{2}, \ 0 \leq y$$

$$r_{s1} = \sqrt{(x+w_s)^2 + (y-d)^2}, \ r_{s2} = \sqrt{(x-w_s)^2 + (y-d)^2}$$
$$r_{m1} = \sqrt{(x+w_s)^2 + (y+d)^2}, \ r_{m2} = \sqrt{(x-w_s)^2 + (y+d)^2}$$

$$(6.4b)$$

Here, r_{s1}, r_{s2}, r_{m1}, and r_{m2} are the distances between the measurement point and the center of each current source.

By extensive magnetic field FEM simulations, including the results shown in Figures 6.8 to 6.10, it was found that the simulation results for the magnetic flux density B_2 are the best fit for the IM³ of (6.4) under the following conditions:

$$\gamma_2 \equiv \frac{I_m}{I_s} \cong \left(1 - e^{-(w_o+2w_s)/(\alpha_2 y)} \right) \quad \because \alpha_2 \cong 2.4$$

$$(6.5)$$

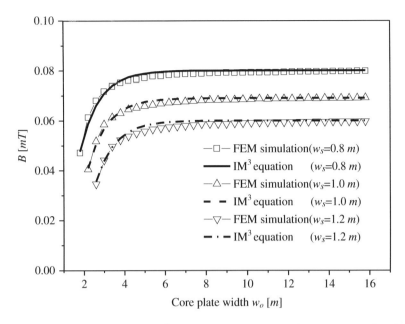

Figure 6.8 Comparison of the IM³ with FEM simulation results for the magnetic flux density of the dual coil with an open core plate for different w_s along w_o at $(0, y)$ under the conditions $\mu_r = 10^7$, $I_s = 100$ A, $d = 0.01$ m, and $y = 0.4$ m.

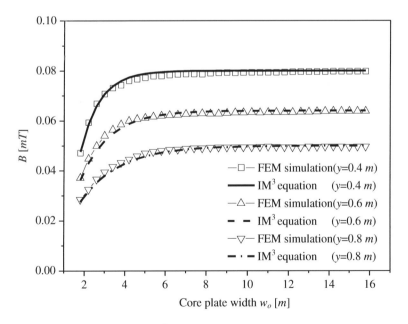

Figure 6.9 Comparison of the IM³ with FEM simulation results for the magnetic flux density of the dual coil with an open core plate for different y along w_o at $(0, y)$ under the conditions $\mu_r = 10^7$, $I_s = 100$ A, $d = 0.01$ m, and $w_s = 0.8$ m.

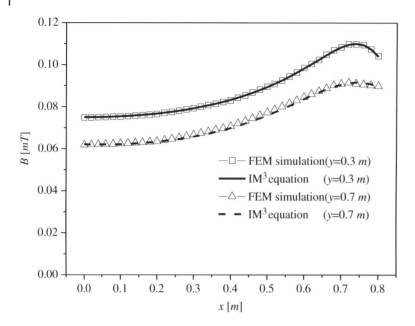

Figure 6.10 Comparison of the IM³ with FEM simulation results for the magnetic flux density of the dual coil with an open core plate for different y along x under the conditions $\mu_r = 10^7$, $I_s = 100$ A, $d = 0.45$ m, $w_s = 0.8$ m, and $w_o = 10$ m.

In (6.5), γ_2 are functions of the core width w_o, the wire displacement w_s, and the height of the measurement point y only. As shown in Figures 6.8 and 6.9, the magnitude of the mirror current approaches the ideal case, that is, $\gamma_2 = 1$, exponentially as the core plate width w_o increases. It is noteworthy that the curve-fitting parameter α_2 of (6.5) is different from that of (6.3), which confirms that the dual coil cannot be modeled as a combination of two mono coils. The IM³ of (6.4) and (6.5) is well fitted to the simulation results for the region inside the core plate, that is, $|x| < w_o/2$.

Question 2 What is the relationship between α_1 and α_2? There should be good rationales that (6.3) and (6.5) are of the same exponential form and $(\alpha_2 - 1) \cong 2(\alpha_1 - 1)$. If you can answer Questions 1 and 2 correctly, in the author's opinion you probably deserve to write good papers on that to be published in qualified journals.

6.2.3 Mono-Coil Local Saturations

In many IPTSs using ferrite cores, local saturation of the core plates deteriorates the power transfer because the resonant frequency of the IPTS changes due to the resultant magnetic resistance change [23]. Therefore, it is crucial in the design of an IPTS coil to keep the peak magnetic flux density of the cores below an allowable limit.

By virtue of the analytical forms of the proposed IM³, it is possible to find the maximum magnetic flux density of a core. There are two possible local saturation points; one is on the core plate and the other is in the core plate.

For the case of the mono coil with an open core plate, it is found from (6.3) that $\gamma_1 = 1$ when $y = 0$; therefore, a very simple form of the magnetic flux density equation on the core plate

$B_{1,on}$ is obtained, as follows:

$$B_{1,on} = B_1|_{y=0} = \frac{\mu_o I_s}{\pi} \frac{x}{d^2 + x^2} \quad for \quad |x| < \frac{w_o}{2} \tag{6.6}$$

The point of maximum magnetic flux density is then found as follows:

$$\left.\frac{\partial B_{1,on}}{\partial x}\right|_{x=x_{1m}} = \left.\frac{\mu_o I_s}{\pi} \frac{d^2 - x^2}{(d^2 + x^2)^2}\right|_{x=x_{1m}} = 0 \quad \Rightarrow x_{1m} = \pm d$$

$$\therefore B_{1m,on} \equiv \left.\frac{\mu_o I_s}{\pi} \frac{x}{d^2 + x^2}\right|_{x=x_{1m}} = \frac{\mu_o I_s}{2\pi d} \tag{6.7}$$

It is also found that local saturation on the core plate occurs at the $(d, 0)$ and $(-d, 0)$ points and that the saturation becomes severe as the distance d decreases.

On the other hand, the magnetic flux density equation in the core plate $B_{1,in}$ is obtained as follows:

$$B_{1,in} \equiv B_1|_{-d_c < y < 0} = \frac{1}{l_o d_c} \int_x^\infty l_o B_{1,on} dx \cong \frac{1}{d_c} \int_x^{w_o/2} \frac{\mu_o I_s}{\pi} \frac{x}{d^2 + x^2} dx$$

$$= \frac{\mu_o I_s}{2\pi d_c} \ln \frac{d^2 + w_o^2/4}{d^2 + x^2} \quad for \quad |x| < \frac{w_o}{2} \tag{6.8}$$

In (6.8), it is assumed that all of the magnetic flux on the core plate is absorbed into the core plate and that the internal magnetic flux density is even for a given value of x, as shown in Figure 6.11. It is found from (6.8) that local saturation in the core occurs just below the source current, as follows:

$$B_{1m,in} \equiv B_{1,in}|_{x=0} = \frac{\mu_o I_s}{2\pi d_c} \ln\left(1 + \frac{w_o^2}{4d^2}\right) \tag{6.9}$$

Figure 6.11 Comparison of the IM³ with FEM simulation results for the magnetic flux density in the core plate of the mono coil with an open core plate for different d along x under the conditions $\mu_r = 10^7$, $I_s = 100$ A, $d_c = 0.02$ m, and $w_o = 3$ m.

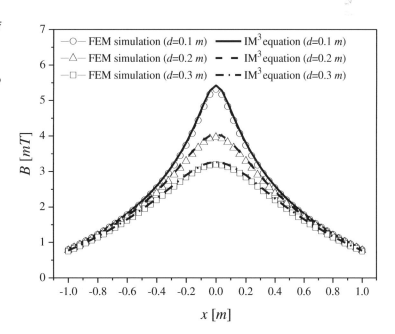

Local saturation can occur either on or in the core plate; hence, the magnetic flux densities determined by (6.7) and (6.9) should be simultaneously lower than the allowable limit. In an OLEV application as an example [23], where $I_s = 200$ A, $B_{1m} = 0.1$ T, and $w_o = 0.8$ m, it can be determined that $d = 0.40$ mm from (6.7), which is much smaller than a conventional litz wire 20 mm in diameter. In other words, local saturation on the core plate is negligible for a conventional wire as long as the current is less than 5000 A. If a thinner litz wire is used in this example, that is, $d = 10$ mm, it is calculated from (6.9) that $d_c = 2.95$ mm. This means that the core is not locally saturated if its thickness is larger than 3 mm, which is the case for OLEVs [23]. It should be noted that the thickness of the core plate is the most prime parameter to be manipulated so as to avoid local saturation.

6.2.4 Dual-Coil Local Saturations

For the case of the dual coil with an open core plate, it was found from (6.5) that $\gamma_2 = 1$ when $y = 0$; therefore, a very simple form of the magnetic flux density equation on the core plate $B_{2,on}$ is obtained, as shown below:

$$B_{2,on} \equiv B_2|_{y=0} = \frac{\mu_o I_s}{\pi} \left\{ \frac{w_s + x}{d^2 + (w_s + x)^2} + \frac{w_s - x}{d^2 + (w_s - x)^2} \right\} \quad for \quad |x| < \frac{w_o}{2} \tag{6.10}$$

The point of maximum magnetic flux density is then found as shown below:

$$\frac{\partial B_{2,on}}{\partial x}\bigg|_{x=x_{2m}} = \frac{\mu_o I_s}{\pi} \left\{ \frac{d^2 - (w_s + x)^2}{\{d^2 + (w_s + x)^2\}^2} - \frac{d^2 - (w_s - x)^2}{\{d^2 + (w_s - x)^2\}^2} \right\}\bigg|_{x=x_{2m}} = 0$$

$$\Rightarrow x_{2m} = \pm\sqrt{w_s^2 + d^2 - 2d\sqrt{w_s^2 + d^2}} \cong \pm(w_s - d) \quad for \quad d \ll w_s \tag{6.11}$$

Equation (6.11) shows that the maximum point on the core surface is close to the mono-coil case of (6.7) if d is much less than w_s. This represents the practical case [23].

The magnetic flux density equation in the core plate $B_{2,in}$ is obtained by considering symmetry at $x = 0$, as shown below:

$$B_{2,in} \equiv B_2|_{-d_c < y < 0} = \frac{1}{l_o d_c} \int_x^\infty l_o B_{2,on} dx$$

$$= \frac{1}{d_c} \int_0^x \frac{\mu_o I_s}{\pi} \left\{ \frac{w_s + x}{d^2 + (w_s + x)^2} + \frac{w_s - x}{d^2 + (w_s - x)^2} \right\} dx$$

$$= \frac{\mu_o I_s}{2\pi d_c} \left\{ \ln \frac{d^2 + (w_s + x)^2}{d^2 + w_s^2} - \ln \frac{d^2 + (w_s - x)^2}{d^2 + w_s^2} \right\} \tag{6.12}$$

$$= \frac{\mu_o I_s}{2\pi d_c} \ln \left\{ \frac{d^2 + (w_s + x)^2}{d^2 + (w_s - x)^2} \right\} \quad \because 0 \le x < w_o/2$$

Figure 6.12 Comparison of the IM³ with FEM simulation results for the magnetic flux density in the core plate of the dual coil with an open core plate for different d along x under the conditions $\mu_r = 10^7$, $I_s = 100$ A, $d_c = 0.02$ m, $w_s = 0.5$ m, and $w_o = 3$ m.

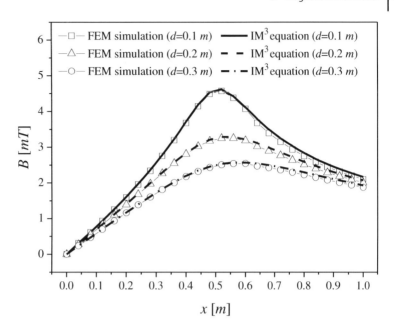

To determine the local saturation point from (6.12), its derivative is explored. The maximum value is found by

$$
\left. \frac{\partial B_{2,in}}{\partial x} \right|_{x=x_{3m}} = 0 \quad \Rightarrow x_{3m} = \sqrt{w_s^2 + d^2} \cong w_s \quad for \quad d \ll w_s
$$

$$
\therefore B_{2m,in} \equiv \left. |B_{2,in}| \right|_{x=x_{3m}} \cong \left. |B_{2,in}| \right|_{x=w_s} = \frac{\mu_o I_s}{2\pi d_c} \ln\left(1 + \frac{4w_s^2}{d^2} \right)
$$

(6.13)

As with the mono-coil case, local saturation is severe in the core plate just below the source current, as identified by (6.13). The local saturation results are verified by simulations, as shown in Figure 6.12, showing good agreement with the theoretical results.

6.3 Improved Magnetic Mirror Models for a Coil with Parallel Core Plates

6.3.1 Mono Coil and Pick-up Coil with a Finite Width

The proposed IM³ is extended to the analysis of an IPTS in which a pick-up coil is placed on a primary mono coil and the coils are magnetically coupled to each other. This problem can be dealt with in a manner similar to that used with the line charge located between two parallel conducting planes, as shown in Figure 6.13. In the image charge model, the line charge placed between two conducting planes is infinitely reflected by each side of the conducting planes with alternating polarity, as shown in Figure 6.13(a). It is postulated in this chapter that the same infinite reflection may occur for two parallel core plates with the same polarity of currents at $\pm d$, $\pm(2h \pm d)$, $\pm(4h \pm d)$, …, as shown in Figure 6.13(b), where the IM³ for a mono coil with a finite width and a pick-up core plate is modeled. This can be justified in that all of the reflected mirror currents are required for each core plate to ensure that the horizontal magnetic field

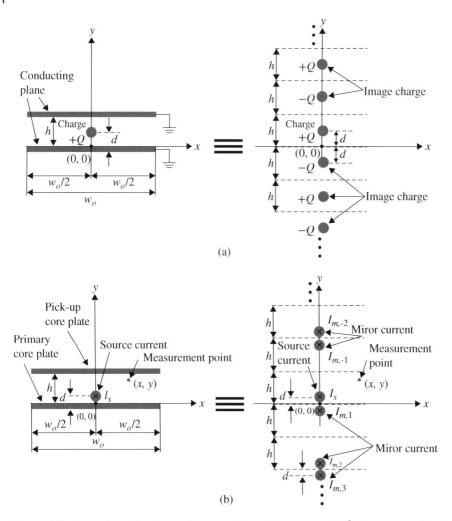

Figure 6.13 The analogy of an image charge model and the proposed IM[3] for a mono coil and a pick-up core plate with finite width. (a) A line charge between two parallel conduction planes and its image charge model and (b) a source current between two parallel core plates with infinite permeability and its mirror currents model.

on its core plate becomes zero, as discussed in Section 6.2.1 above. All of the upside mirror currents and downside mirror currents have values identical to each other. This comes from the fact that the magnitude of the mirror currents is independent of the distance from the core surface and is solely determined by an observation point, as identified by (6.3). The magnetic flux density at any point (x, y) between the parallel core plates of the gap h considering $2n + 1$ reflections is determined from (6.2), as follows:

$$B_3 \equiv |\mathbf{B}_3| = \left| \frac{\mu_o I_s}{2\pi} \sum_{k=-n}^{+n} \left\{ \gamma_{1,2k} \frac{\alpha_x(y - d + 2hk) - \alpha_y x}{x^2 + (y - d + 2hk)^2} + \gamma_{1,2k+1} \frac{\alpha_x(y + d + 2hk) - \alpha_y x}{x^2 + (y + d + 2hk)^2} \right\} \right|$$

$$for \ |x| < \frac{w_o}{2}, \ 0 \leq y \leq h \tag{6.14}$$

In (6.14), $\gamma_{1,k}$ is determined by the successive application of IM^3 of the mono coil with an open core plate to this two-plate case as expressed, as shown below:

$$
\begin{aligned}
&\gamma_{1,0} = 1, \\
&\gamma_{1,1} - \gamma_1(y) - \left(1 - e^{-w_o/(\alpha_1 y)}\right), \quad \gamma_{1,-1} - \gamma_1(h - y), \\
&\gamma_{1,2} = \gamma_{1,-1}\big|_{y=0}\gamma_1(y) = \gamma_1(h)\gamma_1(y), \quad \gamma_{1,-2} = \gamma_{1,1}\big|_{y=h}\gamma_1(h - y) = \gamma_1(h)\gamma_1(h - y), \\
&\gamma_{1,3} = \gamma_{1,-2}\big|_{y=0}\gamma_1(y) = \gamma_1^2(h)\gamma_1(y), \quad \gamma_{1,-3} = \gamma_{1,2}\big|_{y=h}\gamma_1(h - y) = \gamma_1^2(h)\gamma_1(h - y), \\
&\cdots \\
&\gamma_{1,k} = \gamma_1^{k-1}(h)\gamma_1(y), \quad \gamma_{1,-k} = \gamma_1^{k-1}(h)\gamma_1(h - y), \quad k = 1, 2, 3, \dots, 2n + 1
\end{aligned}
\tag{6.15}
$$

In (6.15), $\gamma_{1,0}$ represents the source current and $\gamma_{1,1}$ represents the first downward mirror current, as shown in Figure 6.13. In addition, $\gamma_{1,-1}$ represents the first upward mirror current, whose magnitude is reduced as the measurement point moves away from the pick-up plate. $\gamma_{1,2}$ represents the second downward mirror current, which is the reflection of the first upward mirror current; it becomes equal to $\gamma_{1,-1}$ when $y = 0$ to meet the boundary condition and decreases as y increases. Similarly, this iterative process can be continued forever. Applying (6.15) to (6.14) results in the following complete solution, representative of a vector form:

$$
\begin{aligned}
\mathbf{B}_3 =\ &\frac{\mu_o I_s}{2\pi}\left\{ \frac{\alpha_x(y - d) - \alpha_y x}{x^2 + (y - d)^2} + \gamma_1(y)\frac{\alpha_x(y + d) - \alpha_y x}{x^2 + (y + d)^2} \right\} \\
&+ \frac{\mu_o I_s}{2\pi}\sum_{k=1}^{+n}\gamma_1(y)\left\{ \gamma_1^{2k-1}(h)\frac{\alpha_x(y - d + 2hk) - \alpha_y x}{x^2 + (y - d + 2hk)^2} + \gamma_1^{2k}(h)\frac{\alpha_x(y + d + 2hk) - \alpha_y x}{x^2 + (y + d + 2hk)^2} \right\} \\
&+ \frac{\mu_o I_s}{2\pi}\sum_{k=-1}^{-n}\gamma_1(h - y)\left\{ \gamma_1^{-2k-1}(h)\frac{\alpha_x(y - d + 2hk) - \alpha_y x}{x^2 + (y - d + 2hk)^2} + \gamma_1^{-2k-2}(h)\frac{\alpha_x(y + d + 2hk) - \alpha_y x}{x^2 + (y + d + 2hk)^2} \right\} \\
&\text{for } |x| < \frac{w_o}{2},\ 0 \le y \le h
\end{aligned}
\tag{6.16}
$$

Equation 6.15 shows that the mirror currents of multiple reflections diminish exponentially for larger values of k. The number of reflections required to diminish $\gamma_{1,k}$ to e^{-1} is then determined as follows:

$$
\gamma_{1,2n+1}\big|_{y=0} = \gamma_1^{2n}(h)\gamma_1(0) = \left(1 - e^{-w_o/(\alpha_1 h)}\right)^{2n} = e^{-1} \Rightarrow n = \frac{1}{-2\ln\left(1 - e^{-w_o/(\alpha_1 h)}\right)}
\tag{6.17}
$$

For example, the number of reflections n predicted by (6.17) is as high as 8.3×10^9 when $w_o/h = 40$ in Figure 6.14(a); it becomes 5 when $w_o/h = 4$ in Figure 6.15. This means that the reflection ratio becomes nearly unity for the case of a small air gap and the ideal mirror model can be used instead of the proposed model, whereas the proposed model is adequate for a large air gap when n is small.

As shown in Figures 6.14 and 6.15, the magnetic flux densities of (6.16) and (6.18), in the next subsection, are verified by FEM simulations of various conditions, respectively. They become zero near wires because the reflected currents measured at $y = h$ are identical to the counterpart currents in terms of strength, as identified by (6.15) and (6.20), in the next subsection, and thus cancel each other out. It was found that the proposed IM^3 results are similar to the FEM

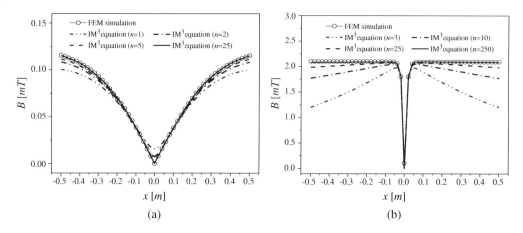

Figure 6.14 Comparison of the IM³ with FEM simulation results for the magnetic flux density of the mono coil with a pick-up core plate for different n along x at (x, h) under the conditions $\mu_r = 10^7$, $I_s = 100$ A, $d = 0.01$ m, and $w_o = 20$ m for (a) $h = 0.5$ m and (b) $h = 0.03$ m.

simulations for a larger value of n, as determined either by (6.17) or w_o/h; the latter corresponds to the mirror current at the height of w_o, where its contribution becomes small.

6.3.2 Dual Coil and Pick-up Coil with Finite Width

As in the previous mono-coil case, a dual coil and a pick-up coil with a finite width can be modeled by the proposed IM³. The magnetic flux density at any point (x, y) between the parallel

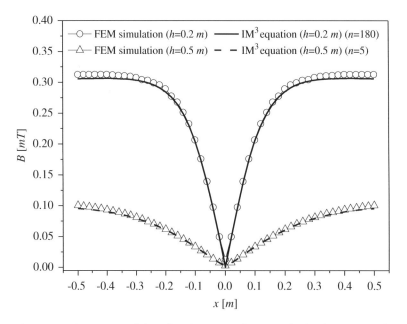

Figure 6.15 Comparison of the IM³ with FEM simulation results for the magnetic flux density of the mono coil with a pick-up core plate for different h along x at (x, h) under the conditions $\mu_r = 10^7$, $I_s = 100$ A, $d = 0.01$ m, and $w_o = 3$ m.

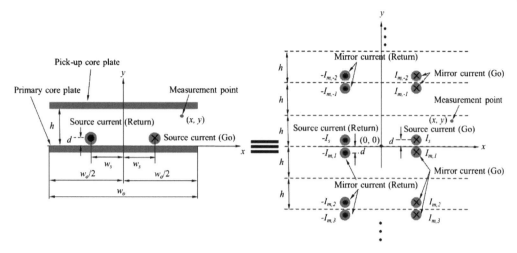

Figure 6.16 The proposed IM^3 for a dual coil and a pick-up plate with finite width.

core plates of the gap h considering $2n+1$ reflections, as shown in Figure 6.16, is determined from (6.4), as shown below:

$$\mathbf{B}_4 = \frac{\mu_0 I_s}{2\pi} \sum_{k=-n}^{+n} \gamma_{2,2k} \left\{ \left(-\frac{1}{\gamma_{s1,k}^2} + \frac{1}{\gamma_{s2,k}^2} \right) (y - d + 2hk)\alpha_x + \left(\frac{x + w_s}{\gamma_{s1,k}^2} - \frac{x - w_s}{\gamma_{s2,k}^2} \right) \alpha_y \right\}$$
$$+ \frac{\mu_0 I_s}{2\pi} \sum_{k=-n}^{+n} \gamma_{2,2k+1} \left\{ \left(-\frac{1}{\gamma_{m1,k}^2} + \frac{1}{\gamma_{m2,k}^2} \right) (y + d + 2hk)\alpha_x + \left(\frac{x + w_s}{\gamma_{m1,k}^2} - \frac{x - w_s}{\gamma_{m2,k}^2} \right) \alpha_y \right\}$$
$$for \quad |x| < \frac{w_o}{2}, \; 0 \leq y \leq h, \tag{6.18}$$

Here (6.4) and (6.5) are appropriately adjusted to consider the multiple reflections, as shown below:

$$\gamma_{s1,k} = \sqrt{(x + w_s)^2 + (y - d + 2hk)^2}, \; \gamma_{s2,k} = \sqrt{(x - w_s)^2 + (y - d + 2hk)^2}$$
$$\gamma_{m1,k} = \sqrt{(x + w_s)^2 + (y + d + 2hk)^2}, \; \gamma_{m2,k} = \sqrt{(x - w_s)^2 + (y + d + 2hk)^2} \tag{6.19}$$

$$\gamma_{2,0} = 1, \quad \gamma_2(y) = \left(1 - e^{-(w_o + 2w_s)/(\alpha_2 y)} \right),$$
$$\gamma_{2,k} = \gamma_2^{k-1}(h)\gamma_2(y), \quad \gamma_{2,-k} = \gamma_2^{k-1}(h)\gamma_2(h - y), \quad k = 1, 2, 3, \ldots, 2n + 1 \tag{6.20}$$

The magnetic flux density of (6.18) is verified by FEM simulations with good agreement for various conditions, as shown in Figure 6.17.

6.4 Example Design and Experimental Verifications

6.4.1 Magnetic Flux Density of the Mono and Dual Coils with an Open Core Plate

The experimental setup involving the mono coil with an open core plate consists of a litz wire and a ferrite core of 2 cm in thickness, as shown in Figure 6.18. The frequency and magnitude of the source current were selected as 20 kHz and 10 A, respectively, and the permeability of the core was about 2000. The magnetic flux density of the mono coil was measured in two

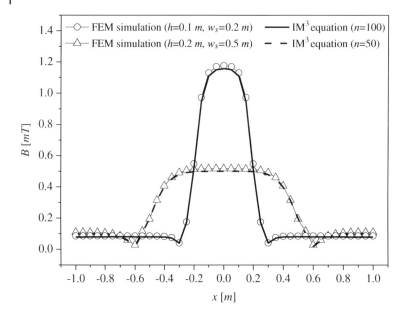

Figure 6.17 Comparison of the IM[3] with FEM simulation results for the magnetic flux density of the dual coil with a pick-up core plate for different h and w_s along x at $(0, h)$ under the conditions $\mu_r = 10^7$, $I_s = 100$ A, $d = 0.01$ m, and $w_o = 5$ m.

cases, when $y = 0.25$ m and $y = 0.45$ m, and compared to the results of the proposed model of (6.2) and to the FEM simulation, as shown in Figure 6.19. Using the same open core plate, a dual coil was set and the magnetic flux density for different values of w_s was also measured and then compared to the results of the proposed model of (6.4) and the FEM simulation, as shown in Figure 6.20. The small errors between the IM[3] and the FEM simulation in Figures 6.19 and 6.20 stem from the difference in the core permeability, as the proposed models of (6.2) and (6.4) assumed infinite permeability, whereas this is set to 2000 in the FEM simulation. The experimental results deviate from the IM[3] and the FEM simulation for a large value of x due to

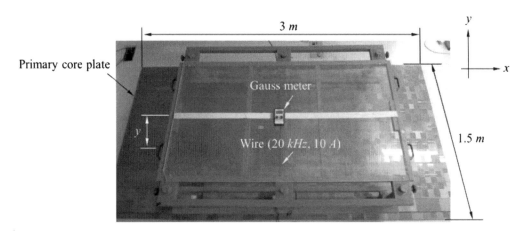

Figure 6.18 An experiment set for measuring the magnetic flux density of the mono coil with an open core plate.

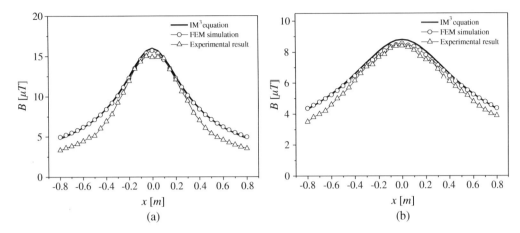

Figure 6.19 Comparison of the IM^3, FEM simulation, and experiment results for the magnetic flux density of the mono coil with an open core plate for different y along x under the conditions $\mu_r = 2000$, $I_s = 10$ A, $d = 0.01$ m, and $w_o = 3$ m for (a) $y = 0.25$ m and (b) $y = 0.45$ m.

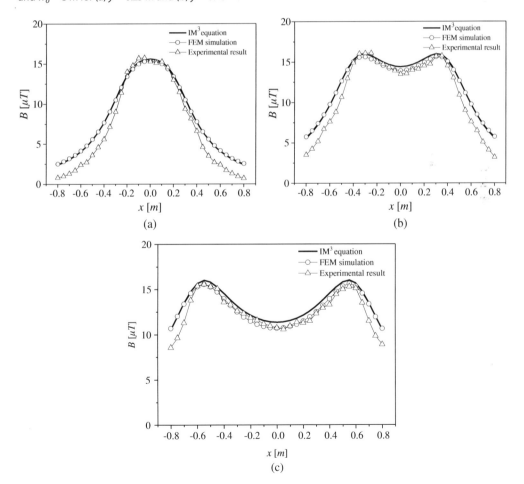

Figure 6.20 Comparison of the IM^3, FEM simulation, and experiment results for the magnetic flux density of the dual coil with an open core plate for different w_s along x at $y = 0.25$ m under the conditions $\mu_r = 2000$, $I_s = 10$ A, $d = 0.01$ m, and $w_o = 3$ m for (a) $w_s = 0.2$ m, (b) $w_s = 0.4$ m, and (c) $w_s = 0.6$ m.

the finite length of the wire, $l_o = 1.5$ m. Given a wire with a finite length, in general, the magnetic flux density is degenerated by the factor of $\sin \phi$, as follows [46]:

$$B = \frac{\mu_o I_s}{2\pi x} \sin \phi, \quad \because \sin \phi \equiv \frac{l_o/2}{\sqrt{x^2 + (l_o/2)^2}} \qquad (6.21)$$

For example, the degeneration factor is $1/\sqrt{2}$ when $x = l_o/2 = 0.75$ m, and it was found that (6.21) explains the deviations in Figures 6.19 and 6.20 well. Permitting small errors, the proposed IM^3 can be useful in the practical design of these types of coils.

To verify the maximum magnetic flux density points on the core plate, as predicted by (6.7) and (6.11), experimental measurements were performed for the test set shown in Figure 6.18.

6.4.2 Maximum Magnetic Flux Density on the Open Core Plate of the Mono and Dual Coils

As shown in Figures 6.21 and 6.22, the experimental results are reasonably well fitted to the FEM simulations for different values of d and also match the proposed IM^3 model of (6.7) and (6.11). The slight mismatch can also be explained by the finite permeability and finite length of the core plate; the proposed IM^3 assumes infinite permeability and an infinite length and the FEM simulation was done while also assuming finite permeability and an infinite length. A systematic measurement error was also involved due to the measurement gap of the Gauss meter of about 1 cm, as shown in Figure 6.23(b); therefore, the measured magnetic flux density does not go to zero near the wires, in contrast to the discussion in Section 6.3.1. Regardless of the small errors, the proposed model can be used for predictions of the local saturation points.

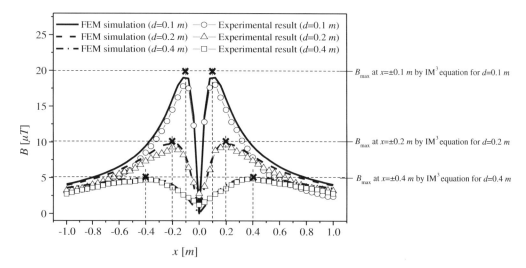

Figure 6.21 Comparison of the IM^3, FEM simulation, and experiment results for the maximum magnetic flux density and its point on the core plate of the mono coil with an open core plate, that is, $y = 0$ for different d along x under conditions $\mu_r = 2000$, $I_s = 10$ A, and $w_o = 3$ m.

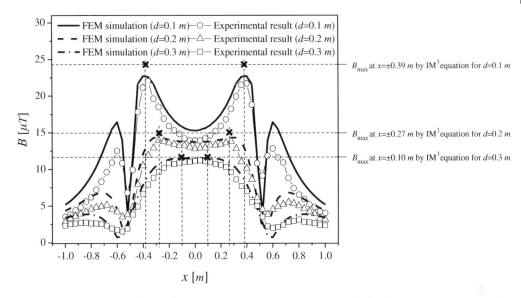

Figure 6.22 Comparison of the IM^3, FEM simulation, and experiment results for the maximum magnetic flux density and its point on the core plate of the dual coil with an open core plate, that is $y = 0$ for different d along x under the conditions $\mu_r = 2000$, $I_s = 10$ A, $w_s = 0.5$ m, and $w_o = 3$ m.

6.4.3 Magnetic Flux Density of the Mono and Dual Coils with Parallel Core Plates

The experimental setup in this case was implemented by placing a pick-up core plate on the primary core plate, as shown in Figure 6.23. The frequency and magnitude of the source current were set to 20 kHz and 10 A, respectively. The magnetic flux densities of the mono coil and dual coils with a pick-up coil were measured and compared to the proposed models of (6.16) and (6.18) and the FEM simulations, showing reasonably good agreement, as shown in Figures 6.24 and 6.25, respectively. The mismatches can be explained by the finite permeability and the finite length of the core plate and by the measuring gap of the Gauss meter. Hence, the proposed

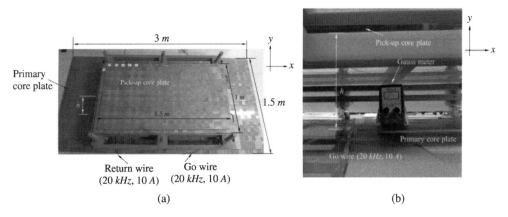

Figure 6.23 An experiment set for measuring the magnetic flux density of the dual coil with a pick-up core plate: (a) the dimensions of an experiment set and (b) front view.

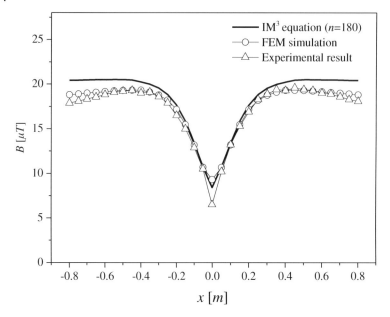

Figure 6.24 Comparison of the IM^3, FEM simulation, and experiment results for the magnetic flux density of the mono coil with a pick-up core plate along x at $y = 0.2$ m under the conditions $\mu_r = 2000$, $I_s = 10$ A, $d = 0.01$ m, $h = 0.3$ m, and $w_o = 3$ m.

model can be used for wireless power transfer coil designs, bearing in mind the small errors near the edges.

6.4.4 Site Test for the On-Line Electric Vehicle

A site test involving an OLEV was performed using the mono coil with an open core plate. This test used a litz wire and a core plate with a width of 0.8 m, thickness of 0.01 m, and a length of 5 m, as shown in Figure 6.26. The core plate has ferrite cores with permeability of about 2000. The frequency and magnitude of the source current for the OLEV are 20 kHz and 200 A,

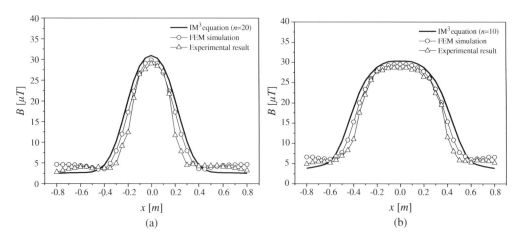

Figure 6.25 Comparison of the IM^3, FEM simulation, and experiment results for the magnetic flux density of the dual coil with a pick-up core plate for different w_s along x at $y = 0.2$ m under the conditions $\mu_r = 2000$, $I_s = 10$ A, $d = 0.01$ m, $h = 0.3$ m, and $w_o = 3$ m for (a) $w_s = 0.2$ m and (b) $w_s = 0.4$ m.

Figure 6.26 An experimental mono coil for the OLEV test.

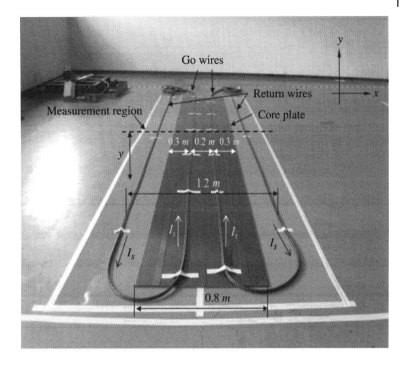

respectively. Two pairs of wires are placed on the core plate, where the space between the wires is 0.2 m, as shown in Figure 6.26. Each wire carries 100 A of current in the same direction in order to generate a uniform magnetic flux density near the center of the core plate. The magnetic flux density of the mono coil for the OLEV was measured along the measurement line shown in Figure 6.26 and compared with the results of (6.2), assuming a single current source of 200 A at the center, as shown in Figure 6.27. The experimental results show a slightly

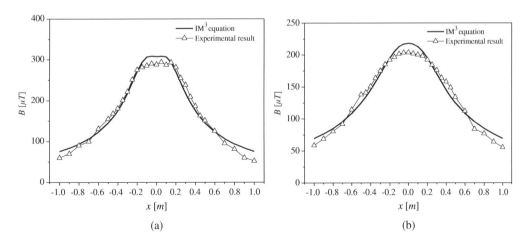

Figure 6.27 Comparison of IM³ with experiment results for the magnetic flux density of the mono coil for OLEV along x for different y under the conditions $\mu_r = 2000$, $I_s = 100$ A, $d = 0.03$ m, and $w_o = 0.8$ m for (a) $y = 0.2$ m and (b) $y = 0.3$ m.

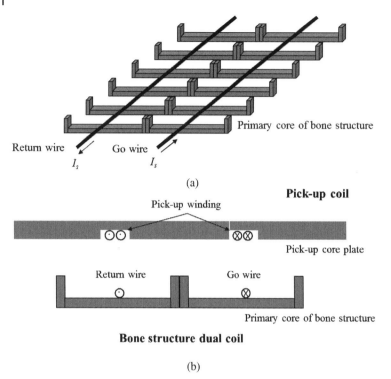

Figure 6.28 A bone structure dual coil with a pick-up coil for OLEV: (a) bird's view and (b) cross-section view.

broadened field distribution compared to the proposed IM^3 due to the divided current sources with a space of 0.2 m.

A bone structure dual coil using sparse cores was introduced for an OLEV to improve the durability of the coil by concrete pouring and to reduce the construction cost by reducing the amount of cores, as shown in Figure 6.28. However, even with the sparse structure, the magnetic flux density of the bone structure dual coil is reduced by only 8% as compared to the case of the ideal core plate [17]. To obtain a strong magnetic flux density at the center and both edges, three vertical poles were set up, as shown in Figures 6.28(b) and 6.29(a). Due to the thickness of the fiber-reinforced plastic (FRP) pipes used to protect the wires, as shown in Figure 6.29(a), the distance between the wires and the core d increased slightly, thereby preventing local saturation of the core plate, as denoted in (6.7), (6.9), (6.11), and (6.13). The magnetic flux density of the bone-structure dual coil with an open core plate for an OLEV was measured along the measurement line over the test road, as shown in Figure 6.29(b). As shown in Figure 6.30, the experimental results are still in good agreement with the proposed IM^3 in terms of the averaged magnetic flux density distribution. The deviations are mainly due to the three vertical poles where the magnetic flux density is somewhat higher than the results of (6.4).

Thus, it was experimentally verified that the proposed IM^3 is applicable to the coil design of an OLEV if small errors are permitted.

6.5 Conclusions

The improved magnetic mirror models (IM^3) of the closed equational form for the core plate with finite width and infinite permeability have been explained in this chapter. The proposed models have been rigorously verified by the FEM simulations and extensive experiments. It

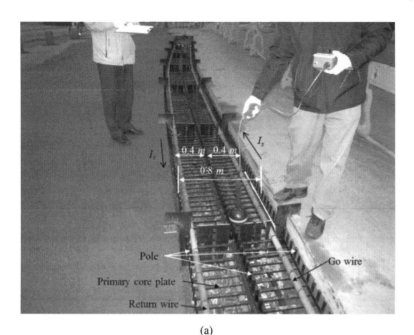

(a)

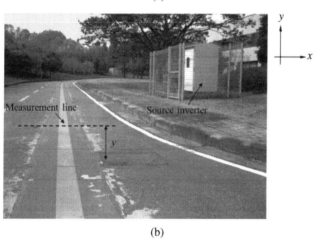

(b)

Figure 6.29 The dual coil and test road for OLEV: (a) installation of the dual coil in the test road and (b) apparent view of the test road.

was shown that the models can be used for the prediction of the magnetic flux density of the mono and dual coils with an open plate or with parallel plates if permitting a reasonably small error, which is less than 5% for the permeability of 2000. Magnetic local saturation points and their magnitudes can also be calculated, which are crucial for the coil design of wireless power transfer such as OLEV. More generalized magnetic mirror models that can be applicable even to the low permeability is left for further work.

Question 3 As a theory in engineering, IM^3 looks imperfect in terms of accuracy and physical rationales. Discuss the validity of the proposed model as an engineering theory with such imperfection. Note that no theory is perfectly proven by all means and can be true for all time and applicable areas.

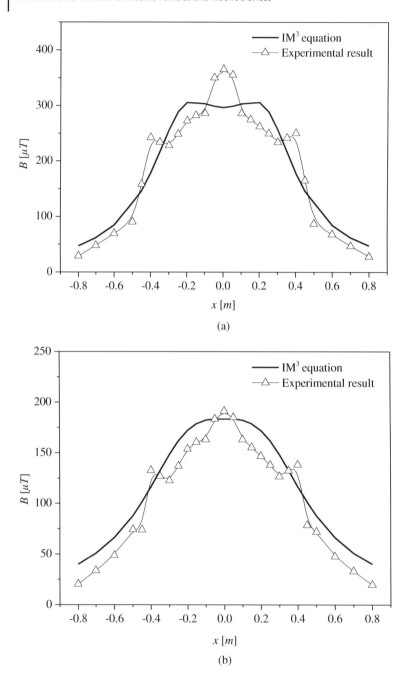

Figure 6.30 Comparison of the IM³ with experiment results for the magnetic flux density of the dual coil for OLEV along *x* for different *y* under the conditions $\mu_r = 2000$, $I_s = 200$ A, $d = 0.05$ m, $w_s = 0.3$ m, and $w_o = 0.8$ m for (a) *y* = 0.2 m and (b) *y* = 0.3 m.

Problems

6.1 **(a)** Draw the graph of (6.6) for an infinitely long round cable with a given current.

(b) Repeat (a) for two identical infinitely long round cables with a distance of $d/2$. Note that two graphs for each cable can be simply summed due to the superposition theorem, which can be applicable to these magnetic mirror currents. Find the maximum magnetic field value.

(c) Discuss the local magnetic saturation of a single cable and that of two cables if the total ampere-turn is the same for each. Discuss the effect of the shape of a cable (round versus rectangular) and the winding method of a cable (single versus bundle) on local magnetic saturation.

6.2 As shown in Figure 6.30, IM^3 can be approximately identified as being applicable to the core of the fish-borne structure of Figure 6.28. It is, however, uncertain whether the mirror current of the sparse structure behaves in a similar way to the canonical core of the plane structure. Identify how Figure 6.30 is obtained through FEM simulation of Figure 6.28.

References

1 B.H. Lee, H.J. Kim, S.W. Lee, C.B. Park, and C.T. Rim, "Resonant power shoes for humanoid robots," in *IEEE Energy Conversion Congress and Exposition (ECCE)*, 2011, pp. 1791–1794.

2 W.X. Zhong, X. Liu, and S.Y. Hui, "A novel single-layer winding array and receiver coil structure for contactless battery charging systems with free-positioning and localized charging features," *IEEE Trans. on Ind. Electron.*, vol. 58, no. 9, pp. 4136–4144, September 2011.

3 L. Jingkun, X. Guizhi, H. Minchai, and L. Lijuan, "Application of wireless energy transfer based on BCI to the power system of intelligent wheelchair," in *IEEE International Conference on Networking and Digital Society (ICNDS)*, 2010, pp. 35–38.

4 W.P. Choi, W.C. Ho, X. Liu, and S.Y.R. Hui, "Bidirectional communication technique for wireless battery charging systems and portable consumer electronics," in *IEEE Applied Power Electronics Conference and Exposition (APEC)*, 2010, pp. 2251–2259.

5 P. Arunkumar, S. Nandhakumar, and A. Pandian, "Experimental investigation on mobile robot drive system through resonant induction technique," in *IEEE International Conference on Computer and Communication Technology (ICCCT)*, 2010, pp. 699–705.

6 T. Imura and Y. Hori, "Maximizing air gap and efficiency of magnetic resonant coupling for wireless power transfer using equivalent circuit and Neumann formula," *IEEE Trans. on Ind. Electron.*, vol. 58, no. 10, pp. 4746–4752, October 2011.

7 S. Rajagopal and F. Khan, "Multiple receiver support for magnetic resonance based wireless charging," in *IEEE International Conference on Communications (ICC)*, 2011, pp. 1–5.

8 H. Jabbar, Y.S. Song, and T.T. Jeong, "RF energy harvesting system and circuits for charging of mobile devices," *IEEE Trans. on Consumer Electron.*, vol. 56, no. 1, pp. 247–253, February 2010.

9 Y. Yao, H. Zhang, and Z. Geng, "Wireless charger prototype based on strong coupled magnetic resonance," in *IEEE Electronic and Mechanical Engineering and Information Technology (EMEIT)*, 2011, pp. 2252–2254.

10 G.A.J. Elliott, S. Raabe, G.A. Covic, and J.T. Boys, "Multiphase pickups for large lateral tolerance contactless power-transfer systems," *IEEE Trans. on Ind. Electron.*, vol. 57, no. 5, pp. 1590–1598, May 2010.

11 C.Y. Huang, G.A. Covic, J.T. Boys, and S. Ren, "LCL pick-up circulating current controller for inductive power transfer systems," in *IEEE Energy Conversion Congress and Exposition (ECCE)*, 2010, pp. 640–646.

12 H.L. Li, A.P. Hu, and G.A. Covic, "Current fluctuation analysis of a quantum AC–AC resonant converter for contactless power transfer," in *IEEE Energy Conversion Congress and Exposition (ECCE)*, 2010, pp. 1838-1843.

13 M.L.G. Kissin, G.A. Covic, and J.T. Boys, "Steady-state flat-pickup loading effects in polyphase inductive power transfer systems," *IEEE Trans. on Ind. Electron.*, vol. 58, no. 6, pp. 2274–2282, June. 2011.

14 N.A. Keeling, G.A. Covic, and J.T. Boys, "A unity-power-factor IPT pickup for high-power applications," *IEEE Trans. on Ind. Electron.*, vol. 57, no. 2, pp. 744–751, February 2010.

15 H.H. Wu, J.T. Boys, and G.A. Covic, "An AC processing pickup for IPT systems," *IEEE Trans. on Power Electron.*, vol. 25, no. 5, pp. 1275–1284, May 2010.

16 H.H. Wu, G.A. Covic, J.T. Boys, and D.J. Robertson, "A series-tuned inductive-power-transfer pickup with a controllable AC-voltage output," *IEEE Trans. on Power Electron.*, vol. 26, no. 1, pp. 98–109, January 2011.

17 S.W. Lee, J. Huh, C.B. Park, N.S. Choi, G.H. Cho, and C.T. Rim, "On-line electric vehicle using inductive power transfer system," in *IEEE Energy Conversion Congress and Exposition (ECCE)*, 2010, pp. 1598–1601.

18 J. Huh and C.T. Rim, "KAIST wireless electric vehicles – OLEV," in *JSAE Annual Congress*, 2011, invited paper.

19 N.P. Suh, D.H. Cho, and C.T. Rim, "Design of on-line electric vehicle (OLEV)," in *Plenary Lecture at the 2010 CIRP Design Conference*, 2010.

20 J. Huh, S.W. Lee, C.B. Park, G.H. Cho, and C.T. Rim, "High performance inductive power transfer system with narrow rail width for on-line electric vehicles," in *IEEE Energy Conversion Congress and Exposition (ECCE)*, 2010, pp. 647–651.

21 S.W. Lee, W.Y. Lee, J. Huh, H.J. Kim, C.B. Park, G.H. Cho, and C.T. Rim, "Active EMF cancellation method for I-type pick-up of on-line electric vehicles," in *IEEE Applied Power Electronics Conference and Exposition (APEC)*, 2011, pp. 1980–1983.

22 J. Huh, W.Y. Lee, G.H. Cho, B H. Lee, and C.T. Rim, "Characterization of novel inductive power transfer systems for on-line electric vehicles," in *IEEE Applied Power Electronics Conference and Exposition (APEC)*, 2011, pp. 1975–1979.

23 J. Huh, S.W. Lee, W.Y. Lee, G.H. Cho, and C.T. Rim, "Narrow-width inductive power transfer system for on-line electric vehicles," *IEEE Trans. on Power Electron.*, vol. 26, no. 12, pp. 3666–3679, December 2011.

24 M. Pahlevaninezhad, P. Das, J. Drobnik, P.K. Jain, and A. Bakhshai, "A new control approach based on the differential flatness theory for an AC/DC converter used in electric vehicles," *IEEE Trans. on Power Electron.*, vol. 27, no. 4, pp. 2085–2103, April 2012.

25 M. Budhia, G.A. Covic, and J.T. Boys, "Design and optimization of circular magnetic structures for lumped inductive power transfer systems," *IEEE Trans. on Power Electron.*, vol. 26, no. 11, pp. 3096–3108, Nov. 2011.

26 H.L. Li, A.P. Hu, and G.A. Covic, "A direct AC–AC converter for inductive power transfer systems," *IEEE Trans. on Power Electron.*, vol. 27, no. 2, pp. 661–668, February 2012.

27 H. Matsumoto, Y. Neba, K. Ishizaka, and R. Itoh, "Model for a three-phase contactless power transfer system," *IEEE Trans. on Power Electron.*, vol. 26, no. 9, pp. 2676–2678, September 2011.

28 C. Zheng and D. Ma, "Design of monolithic CMOS LDO regulator with D^2 coupling and adaptive transmission control for adaptive wireless powered bio-implants," *IEEE Trans. on Circuit and Systems*, vol. 58, no. 10, pp. 1–11, October 2011.

29 C. Zheng and D. Ma, "Design of monolithic low dropout regulator for wireless powered brain cortical implants using a line ripple rejection technique," *IEEE Trans. on Circuit and Systems*, vol. 57, no. 9, pp. 686–690, September 2010.

30 S.J.A. Majerus, P C. Fletter, M.S. Damaser, and S.L. Garverick, "Low-power wireless micro nanometer system for acute and chronic bladder-pressure monitoring," *IEEE Trans. on Biomedical Engineering*, vol. 58, no. 3, pp. 763–767, March 2011.

31 A.K. Ramrakhyani, S. Mirabbasi, and M. Chiao, "Design and optimization of resonance-based efficient wireless power delivery systems for biomedical implants," *IEEE Trans. on Biomedical Engineering*, vol. 5, no. 1, pp. 48–63, February 2011.

32 M. Kianiand and M. Ghovanloo, "An RFID-based closed-loop wireless power transmission system for biomedical applications," *IEEE Trans. on Circuit and Systems*, vol. 57, no. 4, pp. 260–264, April 2010.

33 P. Cong, W.H. Ko, and D.J. Young, "Wireless batteryless implantable blood pressure monitoring microsystem for small laboratory animals," *Sensors and Materials J.*, vol. 10, no. 2, pp. 327–340, February 2010.

34 X. Luo, S. Niu, S.L. Ho, and W.N. Fu, "A design method of magnetically resonating wireless power delivery systems for bio-implantable devices," *IEEE Trans. on Magnetics*, vol. 47, no. 10, pp. 3833–3836, October 2011.

35 P. Cong, N. Chaimanonart, W.H. Ko, and D.J. Young, "A wireless and batteryless 10-bit implantable blood pressure sensing microsystem with adaptive RF powering for real-time laboratory mice monitoring," *IEEE J. on Solid-State Circuits*, vol. 44, no. 12, pp. 3631–3644, December 2009.

36 Z. Bai, S. Ding, C. Li, C. Li, and G. Yan, "A newly developed pulse-type microampere magnetic flux pump," *IEEE Trans. on Applied Superconductivity*, vol. 20, no. 3, pp. 1667–1670, June 2010.

37 D. Zhang, *et al.*, "Research on M_gB_2 superconducting magnet with iron core for MRI," *IEEE Trans. on Applied Superconductivity*, vol. 20, no. 3, pp. 764–768, June 2010.

38 D. Zhang, *et al.*, "Research on stability of M_gB_2 superconducting magnet for MRI," *IEEE Trans. on Applied Superconductivity*, vol. 21, no. 3, pp. 2100–2103, June 2011.

39 C.H. Moon, H.W. Park, and S.Y. Lee, "A design method for minimum-inductance planar magnetic-resonance-imaging gradient coils considering the pole-piece effect," *Journal of Measurement Science and Technology*, vol. 10, pp. 136–141, August 1999.

40 U.K. Madawala and D.J. Thrimawithana, "Current sourced bi-directional inductive power transfer system," *IET Trans. on Power Electron.*, vol. 4, no. 4, pp. 471–480, September 2010.

41 H.H. Wu, A. Gilchrist, K. Sealy, P. Israelsen, and J. Muhs, "A review on inductive charging for electric vehicles," in *IEEE International Electric Machines and Drives Conference (IEMDC)*, 2011, pp. 143–147.

42 M. Budhia, G.A. Covic, and J.T. Boys "A new IPT magnetic coupler for electric vehicle charging systems," in *36th Annual Conference on IEEE Industrial Electronics Society (IECON)*, 2010, pp. 2487–2492.

43 U.K. Madawala and D.J. Thrimawithana, "A bidirectional inductive power interface for electric vehicles in V2G systems," *IEEE Trans. on Indust. Electron.*, vol. 58, no. 10, pp. 4789–4796, October 2011.

44 Y. Hori, "Future vehicle society based on electric motor, capacitor and wireless power supply," in *IEEE International Power Electronics Conference (IPEC)*, 2010, pp. 2930–2934.

45 D.J. Thrimawithana, U.K. Madawala, and Y. Shi, "Design of a bi-directional inverter for a wireless V2G system," in *IEEE International Conference on Sustainable Energy Technologies (ICSET)*, 2010, pp. 1–5.

46 D.K. Cheng, *Field and Wave Electromagnetics*, Pearson Education, New Jersey, 1994.

7

General Unified Dynamic Phasor

7.1 Introduction

Let me explain first why we need a general unified linear time-invariant model for any AC converter in general.

Power switches in a converter change the circuit configuration over time; hence, all switching converters are inherently time-varying, and switching harmonics are inevitably generated from the switches. In the history of power electronics, how to deal properly with the time-varying nature of switching converters has been, perhaps, the most important issue of numerous models and analyses [1–23]. Among them, the switching function-based Fourier analysis techniques [1,3,4] for fundamental and harmonics analyses in the steady state, the averaging techniques [2, 10] for static and dynamic analyses of DC converters, the D–Q transformations [5,7,11,13,14] for three-phase AC converter analyses, and the circuit transformations [7,8,11,13] for static and dynamic analyses of converters have drawn great attention from power electronics specialists.

The phasor transformation, first introduced in [8] in 1990 for single-phase AC converter analyses, has evolved to various areas [24–30]. The dynamic phasor [28–30], applied even to harmonic analysis, is an integral form of the phasor transformation. The conventional phasor concept that the magnitude and phase of a sinusoid are constant [31] was substituted with the generalized time-varying phasor by the phasor transformation. It is applicable to not only single-phase DC–AC, AC–DC, and AC–AC converter analyses but also to any resonant converter analyses. Recently, a unified general circuit-oriented phasor transformation encompassing polyphase AC converters, which was previously covered by the circuit D–Q transformation [7, 11, 13] in a complicated manner, was proposed [32]. This new phasor transformation drastically simplifies the AC converter analysis so that any balanced polyphase AC converters can be degenerated into single-phase converters and multiple switches in AC converters can be replaced with an equivalent transformer having a complex turn ratio regardless of the number of switches.

Now a phasor transformation oriented analysis procedure can be established in general, as shown in Figure 7.1. Any switching circuit of a converter for analysis can be substituted with corresponding electronic transformer(s) according to the principle that "a switch set is exactly equivalent to a transformer having the time-varying turn-ratio determined by the switching function of the switch set" [7]. If the switching function is affected by the currents or the voltages of the converter, the system is in the DCM (discontinuous conduction mode) and becomes a nonlinear system. If the switching function is solely determined by external commands, that is, turn-on/off duty cycle, the system is in the CCM (continuous conduction mode) and becomes a linear system so far as linear circuit elements are used. The switched transformer in a linear switching system can be averaged for the fundamental component analysis to determine the

Wireless Power Transfer for Electric Vehicles and Mobile Devices, First Edition. Chun T. Rim and Chris Mi.
© 2017 John Wiley & Sons Ltd. Published 2017 by John Wiley & Sons Ltd.

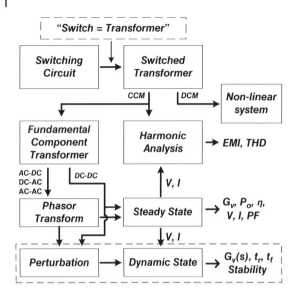

Figure 7.1 Phasor transformation-based analysis flow of switching converters.

steady-state characteristics, that is voltage gain G_V, output power P_o, efficiency η, DC operating points, power factor (PF), and dynamic state characteristics, that is transfer function $G_V(s)$, rising/falling times, and system stability. Harmonic analysis for EMI (electromagnetic interference) and THD (total harmonic distortion) design can be performed by the harmonic voltage/current sources derived from the switched transformer model.

The dynamic state analyses for phasor transformed AC circuits have, however, not yet been possible in general. By the conventional phasor transformation, an AC converter is transformed into a circuit that contains imaginary resistors and electronic transformers with complex turn ratios [8, 32]. Conventional linear transformations such as Laplace transformations, Fourier transformations, and z-transformations, which deal with only real variables, cannot be directly applied to complex circuits whose voltages and currents are complex variables.

In this chapter, the complex Laplace transformation was introduced first for the dynamic analyses of phasor transformed complex circuits and was verified as a quite useful mathematical tool for AC converter manipulations; hence, a complete analysis flow for any switching converters, including the IPT system, has been established. Note that the unified general dynamic phasor is applicable not only to the IPT system but also to any linear AC circuits. This very general technique is introduced in this book because I feel that there are quite a few examples in the literature that deal with the dynamics of IPT systems. Without the help of this very simple and powerful theory, it is formidably difficult to find a dynamic response of an IPT system of high order.

7.2 Complex Laplace Transformation for AC circuits

7.2.1 Theory of Complex Laplace Transformation

The Laplace transformation in the complex form was first proposed as a generalized form of the previous Laplace transformation by Poincare [33, 34] as follows:

$$\boldsymbol{F}(s) = \int_0^\infty \boldsymbol{f}(t)e^{-st}\mathrm{d}t, \ \boldsymbol{f}(t)\text{: complex variable} \tag{7.1}$$

The time-domain variable of (7.1) is no longer a real number but a complex number. In recent years, however, this concept has been gradually used in engineering areas such as digital signal processing [35] and time-varying control [36–38]. Now this theory is to be applied to power electronics in which complex circuits are involved. It is not straightforward to apply (7.1) to complex circuits in a way that can give us meaningful physical meaning, but a method to do just that will be made clear in this chapter.

Question 1 Compared to the conventional real variable, what happens to the Laplacian s of (7.1), which is believed to be complex number? In other words, the complex variable together with complex s may result in another complex number.

7.2.2 Unified General Phasor Transformation

A three-phase rectifier is chosen to show the procedure of a complex Laplace transformation, as shown in Figure 7.2(a). It is assumed that the three-phase rectifier is well balanced and the LC input filters and the output filter can sufficiently diminish switching harmonics so as not to meaningfully affect the fundamental voltage or current components. The converter in the rotary time frame can be transformed into a circuit in the stationary time frame, as shown in Figure 7.2(b), by the recently proposed power-invariant phasor transformation [32]. The system order is degenerated from 7 to 3, and imaginary resistors appear though the circuit is still in the

Figure 7.2 Example of phasor transformation for a three-phase rectifier.

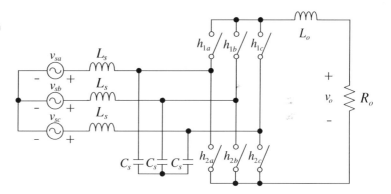

(a) Original power circuit in the real-time domain

(b) Phasor transformed circuit in the time domain

time domain, which should be differentiated from the conventional reactance of AC inductors and capacitors.

7.2.3 Application of Complex Laplace Transformation to Complex Circuit Elements

Phasor transformed AC circuits, also called complex circuits, can be decomposed into nine circuit elements in the phasor domain, as shown in Figure 7.3. They are linear time-invariant circuit elements, that is, phasor voltage sources, phasor current sources, phasored inductors, phasored capacitors, phasored real resistors, phasored imaginary resistors, complex matrix transformers, complex VSI (voltage source inverter) transformers, and complex CSI (current source inverter) transformers. Therefore, if the complex Laplace transformation can be applicable to each complex circuit element, then any complex circuit composed of the complex circuit elements can be analyzed, in general, by the complex Laplace transformation.

7.2.3.1 Phasored inductor
First of all, the application of the complex Laplace transformation to a phasor transformed inductor, called a phasored inductor in this chapter, is performed by using the conventional Laplace transformation of real form.

The governing time-domain equation for a phasored inductor, as shown in Figure 7.3(a) ①, is as follows:

$$v_L = L\frac{d\,i_L}{dt} \tag{7.2}$$

where v_L and i_L are the complex voltage and current of a phasored inductor in the time domain, respectively, and L is the inductance of the real value. Since a complex variable can be decomposed into a real part and an imaginary part, (7.2) can be rewritten as follows:

$$v_{Lr} + jv_{Li} = L\left(\frac{di_{Lr}}{dt} + j\frac{di_{Li}}{dt}\right), \quad \because j \equiv \sqrt{-1} \tag{7.3}$$

where v_L and i_L are defined using the real-time variables, respectively, as

$$v_L \equiv v_{Lr} + jv_{Li}, \quad i_L \equiv i_{Lr} + ji_{Li} \tag{7.4}$$

Decomposing (7.3) into a real part equation and an imaginary part equation, two sets of independent equations in the real-time domain are obtained, as follows:

$$\begin{aligned} v_{Lr} &= L\frac{di_{Lr}}{dt} \\ v_{Li} &= L\frac{di_{Li}}{dt} \end{aligned} \tag{7.5}$$

A conventional Laplace transformation of the form (7.7) can be applied to (7.5) without any mathematical problems as follows:

$$\begin{aligned} V_{Lr}(s) &= L\{sI_{Lr}(s) - i_{Lr}(0)\} \\ V_{Li}(s) &= L\{sI_{Li}(s) - i_{Li}(0)\} \end{aligned} \tag{7.6}$$

$$F(s) \equiv \int_0^\infty f(t)e^{-st}dt, \quad f(t): \text{ real variable} \tag{7.7}$$

Figure 7.3 Complex Laplace transformation for basic circuit elements of complex circuits.

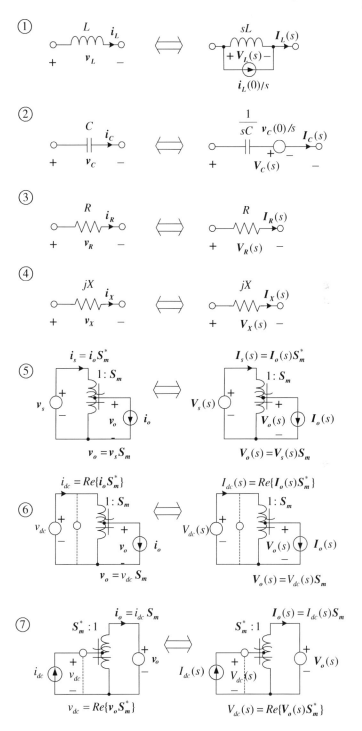

(a) Complex time domain (left) (b) Complex Laplace transform domain (right)

It is found that applying the complex Laplace transformation (7.1) to (7.4) and using (7.6) results in the following relationship:

$$V_L(s) \equiv \int_0^\infty v_L e^{-st} dt = \int_0^\infty (v_{Lr} + jv_{Li}) e^{-st} dt = \int_0^\infty v_{Lr} e^{-st} dt$$

$$+ j \int_0^\infty v_{Li} e^{-st} dt = V_{Lr}(s) + jV_{Li}(s)$$

$$= L\{sI_{Lr}(s) - i_{Lr}(0)\} + jL\{sI_{Li}(s) - i_{Li}(0)\} = sL\{I_{Lr}(s) + jI_{Li}(s)\} - L\{i_{Lr}(0) + ji_{Li}(0)\} \quad (7.8)$$

$$= sL\,I_L(s) - L\,i_L(0)$$

$$= \int_0^\infty L \frac{d\,i_L}{dt} e^{-st} dt$$

According to (7.8), it can be concluded that direct application of the complex Laplace transformation (7.1) and (7.2) gives us the same results as the conventional Laplace transformation-based approach of (7.3) to (7.7). This means that the complex Laplace transformation is valid for the phasored inductor; hence, its equivalent circuit in the complex Laplace transformed domain can be obtained, as shown in Figure 7.3(b) ①, based on the following equation derived from (7.8):

$$I_L(s) = \frac{V_L(s)}{sL} + \frac{i_L(0)}{s} \tag{7.9}$$

It is important to note that the complex Laplace transformed voltage and current of (7.8) and (7.9) can be decomposed into real parts and imaginary parts regarding the Laplacian s as a real value, which was a complex value in the conventional Laplace transformation.

Question 2 Why should we regard the Laplacian s as a real value when we derive (7.8) from (7.2)? What happens to (7.9) if we regard s as a complex number? Note that the real and imaginary parts of a complex variable can be independently Laplace-transformed only when regarding s as indifferent to the Laplace transformation.

7.2.3.2 Phasored Capacitor

Application of the complex Laplace transformation to a phasored capacitor can also be performed in a fashion similar to that of the phasored inductor case above. The governing time-domain equation for a phasored capacitor, as shown in Figure 7.3(a) ②, is as follows:

$$i_C = C \frac{d\,v_C}{dt} \tag{7.10}$$

Performing the complex Laplace transformation to (7.10) results in the following Laplace equation.

$$I_C(s) = sCV_C(s) - Cv_C(0) \quad \text{or} \quad V_C(s) = \frac{I_C(s)}{sC} + \frac{v_C(0)}{s} \tag{7.11}$$

The equivalent circuit of (7.11) in the complex Laplace transformed domain can be obtained, as shown in Figure 7.3(b) ②, where the voltage source term represents the initial capacitor voltage and can be removed if it is zero.

7.2.3.3 Phasored Real Resistor

The application of the complex Laplace transformation to a phasored real resistor, as shown in Figure 7.3(a) ③, is straightforward since its governing time-domain equation is as follows:

$$v_R = R\,i_R \tag{7.12}$$

Performing the complex Laplace transformation to (7.12) results in the following Laplace equation:

$$V_R(s) = R\,I_R(s) \tag{7.13}$$

The equivalent circuit of (7.13) in the complex Laplace transformed domain is shown in Figure 7.3(b) ③.

7.2.3.4 Phasored Imaginary Resistor

The application of the complex Laplace transformation to a phasored imaginary resistor, as shown in Figure 7.3(a) ④, is performed for the governing time domain equation, as follows:

$$v_X = jX\,i_X \tag{7.14}$$

Performing the complex Laplace transformation to (7.14) results in the following Laplace equation:

$$V_X(s) = jX\,I_X(s) \tag{7.15}$$

The equivalent circuit of (7.15) in the complex Laplace transformed domain is shown in Figure 7.3(b) ④.

7.2.3.5 Complex Matrix Transformer

A complex matrix transformer, as shown in Figure 7.3(a) ⑤, is a phasor transformed equivalent circuit element of an AC–AC converter, that is, a matrix converter, whose governing time-domain equation is as follows [32]:

$$v_o = v_s S_m, \quad i_s = i_o S_m^* \tag{7.16}$$

where S_m, that is, $S_m e_S^{j\varphi}$ last s is subscript, is a complex turn ratio representing the voltage conversion ratio S_m and the phase difference ϕ_S between the source and output voltages of the fundamental components. $S_m^*(= S_m e^{-j\phi_S})$ is a complex conjugate of the complex turn ratio and represents the relationship between the source and the output currents. It should be remarked that the complex turn ratio includes neither switching harmonics nor time-varying components; hence, it is just a time-invariant complex number.

The complex turn ratios of the complex transformer models in Figure 7.3 are assumed to have already been perturbed [8]; hence, the transformer models have fixed complex turn ratios; perturbed voltage sources and current sources are excluded from the models. Under this condition, the application of the complex Laplace transformation to (7.16) results in the following Laplace equation:

$$V_o(s) = V_s(s)S_m, \quad I_s(s) = I_o(s)S_m^* \tag{7.17}$$

The equivalent circuit of (7.17) in the complex Laplace transformed domain is shown in Figure 7.3(b) ⑤.

7.2.3.6 Complex VSI Transformer

A complex VSI transformer, as shown in Figure 7.3(a) ⑥, is a phasor transformed equivalent circuit element of a DC–AC voltage source converter, that is, a voltage source inverter or a current source rectifier, whose governing time-domain equation is as follows [32]:

$$\boldsymbol{v_o} = v_{dc}\boldsymbol{S_m}, \quad i_{dc} = Re\left\{i_o\boldsymbol{S_m^*}\right\} \tag{7.18}$$

where the cumbersome real part operation is used to describe the fact that there is no complex voltage and current in the real time-domain circuits, that is, the DC side circuits. The dotted line, together with the small circle in Figure 7.3(a) ⑥, represents a border line that divides the real domain (left side) and the complex domain (right side), where the small circle denotes the dummy current source [32] to nullify the imaginary current of the complex transformer flowing into the DC side. Except for this real part operator, which is composed of the dotted line and the small circle, the complex transformers in Figure 7.3 ⑥ and ⑦ are exactly the same as that in Figure 7.3 ⑤.

The $\boldsymbol{S_m}$ of (7.18) is a constant complex turn ratio, as discussed in the above complex matrix transformer case; hence, the application of the complex Laplace transformation to (7.18) results in the following Laplace equation:

$$\boldsymbol{V_o}(s) = V_{dc}(s)\boldsymbol{S_m},$$
$$I_{dc}(s) = \int_0^\infty Re\left\{i_o\boldsymbol{S_m^*}\right\}e^{-st}\mathrm{d}t = Re\left\{\int_0^\infty i_o e^{-st}\mathrm{d}t\boldsymbol{S_m^*}\right\} = Re\left\{\boldsymbol{I_o}(s)\boldsymbol{S_m^*}\right\} \tag{7.19}$$

From (7.19), it should be remarked that the real part operator $Re\{\}$ is valid for the complex Laplace transformation regarding the Laplacian s as a real number. The equivalent circuit of (7.19) in the complex Laplace transformed domain is shown in Figure 7.3(b) ⑥.

Question 3 Is the real part operator linear? Prove it by yourself, showing that $Re\{a\boldsymbol{x_1} + b\boldsymbol{x_2}\} = aRe\{\boldsymbol{x_1}\} + bRe\{\boldsymbol{x_2}\}$.

7.2.3.7 Complex CSI Transformer

A complex CSI transformer, as shown in Figure 7.3(a) ⑦, is a phasor transformed equivalent circuit element of a current source DC–AC converter, that is a current source inverter or a voltage source rectifier, whose governing time-domain equation is as follows [32]:

$$\boldsymbol{i_o} = i_{dc}\boldsymbol{S_m}, \quad v_{dc} = Re\left\{\boldsymbol{v_o}\boldsymbol{S_m^*}\right\} \tag{7.20}$$

which is analogous to (7.18); the real part operation was used for the same purpose as the complex VSI transformer case. The dotted line, together with the small circle in Figure 7.3 ⑦, also divides the real domain (left side) and the complex domain (right side), where the small circle denotes the dummy voltage source [32], which is used to nullify the imaginary voltage of the complex transformer on the DC side.

The $\boldsymbol{S_m}$ of (7.20) is also a constant complex turn ratio; hence, the application of the complex Laplace transformation to (7.20) results in the following Laplace equation:

$$\boldsymbol{I_o}(s) = I_{dc}(s)\boldsymbol{S_m},$$
$$V_{dc}(s) = \int_0^\infty Re\left\{\boldsymbol{v_o}\boldsymbol{S_m^*}\right\}e^{-st}\mathrm{d}t = Re\left\{\int_0^\infty \boldsymbol{v_o}e^{-st}\mathrm{d}t\boldsymbol{S_m^*}\right\} = Re\left\{\boldsymbol{V_o}(s)\boldsymbol{S_m^*}\right\} \tag{7.21}$$

From (7.21), the real part operator $Re\{\}$ is also valid for the complex Laplace transformation regarding the Laplacian s as a real number. The equivalent circuit of (7.21) in the complex Laplace transformed domain is shown in Figure 7.3(b) ⑦.

The applications of the complex Laplace transformation to the phasor voltage source and the phasor current source are not shown separately here; however, they are inherently included in the above discussions, for example, in (7.16) and (7.17).

Figure 7.4 Complex Laplace transformed three-phase rectifier.

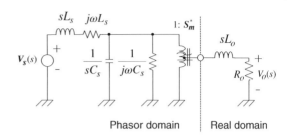

Phasor domain | Real domain

7.2.4 Application of Complex Laplace Transformation to Complex Circuits

The application of a complex Laplace transformation to the individual complex circuit elements shown in Figure 7.2(b) assuming zero initial conditions and the reconstruction of them result in a circuit in the frequency domain, as shown in Figure 7.4. Apparently, the complex Laplace transformation is very similar to a conventional Laplace transformation, as can be seen in Figure 7.4, in which inductors and a capacitor have impedances represented in Laplacian terms. It should be noted, however, that the well-known and very convenient Laplace transformation can fortunately be used for the complex domain circuit, which would not be possible but for the complex Laplace transformation in this chapter. Furthermore, it is not easy to deal with the real part operator, which was dealt with for the steady state [32] and which will be clarified in the next section.

7.3 Analyses of Complex Laplace Transformed Circuits

Dynamic responses, as well as static responses, can be analyzed by the proposed complex Laplace transform. The example three-phase rectifier, shown in Figure 7.4, is used in this chapter to demonstrate the analysis process in detail.

7.3.1 Static Analysis of Complex Laplace Transformed Circuit

In the steady state, inductors are shorted and capacitors are opened for a complex Laplace transformed circuit, as shown in Figure 7.5(a). This is the same as the conventional Laplace transformation case, where $s \to 0$ in the steady state.

The source voltage V_s is a complex value representing the magnitude and phase of the source voltages, as shown in Figure 7.2(a), and the angular frequency ω is the same as that of the source voltages. A Thevenin equivalent circuit for the left part of the complex transformer, as shown in Figure 7.5(b), can drastically simplify the analysis, where the Thevenin equivalent voltage and impedance are as follows:

$$V_{t1} = \frac{\dfrac{1}{j\omega C_s}}{j\omega L_s + \dfrac{1}{j\omega C_s}} V_s = \frac{1}{1 - \omega^2 L_s C_s} V_s \tag{7.22}$$

$$Z_{t1} = \frac{j\omega L_s \dfrac{1}{j\omega C_s}}{j\omega L_s + \dfrac{1}{j\omega C_s}} = \frac{j\omega L_s}{1 - \omega^2 L_s C_s} \tag{7.23}$$

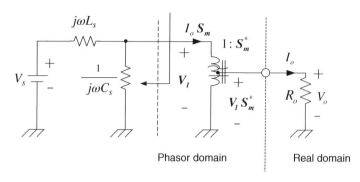

Figure 7.5 Static equivalent circuits of the three-phase rectifier.

(a) Original static circuit.

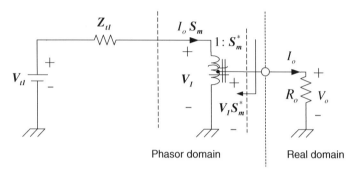

(b) The first Thevenin equivalent circuit for the source side.

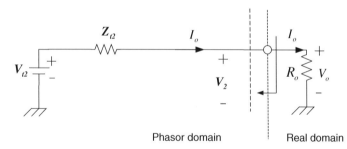

(c) The second Thevenin equivalent circuit, removing the complex transformer.

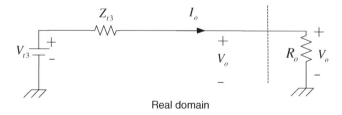

(d) The third Thevenin equivalent circuit, removing the real part operator.

The complex transformer in Figure 7.5(b) can be eliminated when another Thevenin equivalent circuit for the right part is sought, and considering that the complex transformer is linear, as shown in Figure 7.5(c). The Thevenin equivalent voltage is the open-circuit voltage shown in Figure 7.5(b), in which no current flows, as follows:

$$V_{t2} = V_2\big|_{I_o=0} = V_1 S_m^* \big|_{I_o=0} = V_{t1} S_m^* \tag{7.24}$$

The Thevenin equivalent resistance can be calculated from Figure 7.5(b), as follows:

$$Z_{t2} \equiv \frac{V_1 S_m^*}{-I_o}\bigg|_{V_{t1}=0} = \frac{(-I_o S_m Z_{t1}) S_m^*}{-I_o} = Z_{t1} S_m S_m^* = Z_{t1}|S_m|^2 \equiv Z_{t1} S_m^2 \tag{7.25}$$

Finally, Figure 7.5(c) can be further simplified to remove the cumbersome real part operator with a little caution on the phasor domain and real domain as follows:

$$V_o = Re\{V_{t2} - Z_{t2}I_o\} = Re\{V_{t2}\} - Re\{Z_{t2}\}I_o \equiv V_{t3} - Z_{t3}I_o \tag{7.26}$$

where

$$V_{t3} \equiv Re\{V_{t2}\}, \quad Z_{t3} \equiv Re\{Z_{t2}\} \tag{7.27}$$

From (7.26) and (7.27), it can be seen that the final circuit of Figure 7.5(d) includes only the real part variables and that a conventional circuit analysis is now possible; hence, the DC output voltage can be calculated from (7.22) to (7.27), as follows:

$$V_o = \frac{R_o}{Z_{t3} + R_o} V_{t3} = \frac{R_o}{Re\{Z_{t2}\} + R_o} Re\{V_{t2}\} = \frac{R_o}{Re\{Z_{t1}S_m^2\} + R_o} Re\{V_{t1}S_m^*\}$$

$$= \frac{R_o}{Re\left\{\dfrac{j\omega L_s}{1 - \omega^2 L_s C_s}\right\} S_m^2 + R_o} Re\left\{\frac{V_s S_m^*}{1 - \omega^2 L_s C_s}\right\} = \frac{R_o}{0 + R_o} \frac{Re\{V_s S_m^*\}}{1 - \omega^2 L_s C_s} = \frac{Re\{V_s S_m^*\}}{1 - \omega^2 L_s C_s}$$

$$\tag{7.28}$$

Note that Z_{t3} becomes zero because the real part of (7.23) or (7.25) is zero; hence, the analytical result of (7.28) is very simple. The procedure for a DC phasor analysis seems somewhat complicated for this example, which is intentionally illustrated in detail; however, the circuit-oriented phasor analysis requires very few equations and can even be drastically simple, as shown in [32].

7.3.2 Dynamic Analysis of a Complex Laplace Transformed Circuit

The complex Laplace transformed three-phase rectifier shown in Figure 7.4 can be analyzed in a fashion similar to that of the above static analysis case for the dynamics characterization, as shown in Figure 7.6. Because the complex Laplace transformed circuit shown in Figure 7.6(a) is linear, a Thevenin equivalent circuit for the left part of the complex transformer is obtained, as shown in Figure 7.6(b), in which the Thevenin equivalent voltage and impedance are as follows:

$$V_{t1}(s) = \frac{1/(sC_s + j\omega C_s)}{sL_s + j\omega L_s + 1/(sC_s + j\omega C_s)} \cdot V_s(s) = \frac{1}{1 + (sL_s + j\omega L_s)(sC_s + j\omega C_s)} \cdot V_s(s) \tag{7.29}$$

$$Z_{t1}(s) = \frac{(sL_s + j\omega L_s)\dfrac{1}{sC_s + j\omega C_s}}{(sL_s + j\omega L_s) + \dfrac{1}{sC_s + j\omega C_s}} = \frac{sL_s + j\omega L_s}{1 + (sL_s + j\omega L_s)(sC_s + j\omega C_s)} \tag{7.30}$$

Figure 7.6 Dynamic equivalent circuits of the three-phase rectifier in the complex Laplace domain.

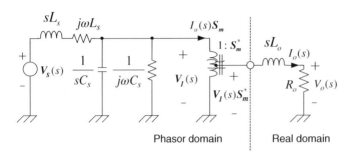

Phasor domain | Real domain

(a) Original dynamic circuit.

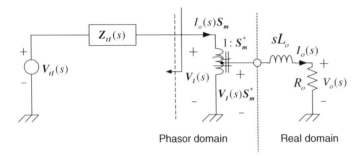

Phasor domain | Real domain

(b) The first Thevenin equivalent circuit for the source side.

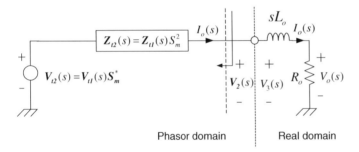

Phasor domain | Real domain

(c) The second Thevenin equivalent circuit, removing the complex transformer.

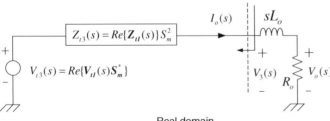

Real domain

(d) The third Thevenin equivalent circuit, removing the real part operator.

The complex transformer in Figure 7.6(b) can be eliminated by finding a Thevenin equivalent circuit for the left side of the real part operator, as shown in Figure 7.6(c). The Thevenin equivalent voltage is the open-circuit voltage shown in Figure 7.6(b), at which no current flows, as follows:

$$V_{t2}(s) = V_2(s)\big|_{I_o(s)=0} = V_1(s)S_m^*\big|_{I_o(s)=0} = V_{t1}(s)S_m^* \tag{7.31}$$

The Thevenin equivalent resistance is calculated from Figure 7.6(b) as follows:

$$Z_{t2}(s) \equiv \frac{V_1(s)S_m^*}{-I_o(s)}\bigg|_{V_{t1}(s)=0} = \frac{-I_o(s)S_m Z_{t1}(s)S_m^*}{-I_o(s)} = Z_{t1}(s)S_m S_m^* = Z_{t1}(s)S_m^2 \tag{7.32}$$

Figure 7.6(c) is further simplified by removing the real part operator as follows:

$$V_3(s) = Re\{V_2(s)\} = Re\{V_{t2}(s) - Z_{t2}(s)I_o(s)\} = Re\{V_{t2}(s)\} - Re\{Z_{t2}(s)\}I_o(s) \equiv V_{t3}(s) - Z_{t3}(s)I_o(s) \tag{7.33}$$

where

$$V_{t3}(s) \equiv Re\{V_{t2}(s)\} = Re\{V_{t1}(s)S_m^*\}, \quad Z_{t3}(s) \equiv Re\{Z_{t2}(s)\} = Re\{Z_{t1}(s)\}S_m^2 \tag{7.34}$$

From (7.33) and (7.34), the final circuit of Figure 7.6(d) can be drawn; this circuit is in the conventional Laplace transformation domain and no complex variables exist. Thus, the output voltage transfer function can be calculated from (7.29) to (7.34) as follows:

$$V_o(s) = \frac{R_o}{Z_{t3}(s) + sL_o + R_o}V_{t3}(s) = \frac{R_o}{Re\{Z_{t1}(s)\}S_m^2 + sL_o + R_o}Re\{V_{t1}(s)S_m^*\} \tag{7.35}$$

where the real part operation of the dynamic case is quite complicated in comparison with the static case of (7.28), as follows:

$$
\begin{aligned}
Re\{V_{t1}(s)S_m^*\} &= Re\left\{\frac{V_s(s)S_m^*}{1+(sL_s+j\omega L_s)(sC_s+j\omega C_s)}\right\} = Re\left\{\frac{V_s(s)S_m^*}{1+(s^2-\omega^2)L_sC_s+j2\omega sL_sC_s}\right\} \\
&= Re\left\{\frac{V_s(s)S_m^*\left\{1+(s^2-\omega^2)L_sC_s-j2\omega sL_sC_s\right\}}{\left\{1+(s^2-\omega^2)L_sC_s+j2\omega sL_sC_s\right\}\left\{1+(s^2-\omega^2)L_sC_s-j2\omega sL_sC_s\right\}}\right\} \\
&= \frac{Re\left\{V_s(s)S_m^*\left\{1+(s^2-\omega^2)L_sC_s-j2\omega sL_sC_s\right\}\right\}}{\left\{1+(s^2-\omega^2)L_sC_s\right\}^2+4\omega^2s^2L_s^2C_s^2} \\
&= \frac{Re\left\{V_s(s)S_m^*\right\}\left\{1+(s^2-\omega^2)L_sC_s\right\}+Im\left\{V_s(s)S_m^*\right\}2\omega sL_sC_s}{\left\{1+(s^2-\omega^2)L_sC_s\right\}^2+4\omega^2s^2L_s^2C_s^2}
\end{aligned} \tag{7.36}
$$

$$
\begin{aligned}
Re\{Z_{t1}(s)\} &= Re\left\{\frac{sL_s+j\omega L_s}{1+(sL_s+j\omega L_s)(sC_s+j\omega C_s)}\right\} \\
&= \frac{Re\left\{(sL_s+j\omega L_s)\left\{1+(s^2-\omega^2)L_sC_s-j2\omega sL_sC_s\right\}\right\}}{\left\{1+(s^2-\omega^2)L_sC_s\right\}^2+4\omega^2s^2L_s^2C_s^2} \\
&= \frac{sL_s\left\{1+(s^2-\omega^2)L_sC_s\right\}+(\omega L_s)2\omega sL_sC_s}{\left\{1+(s^2-\omega^2)L_sC_s\right\}^2+4\omega^2s^2L_s^2C_s^2} \\
&= \frac{sL_s\left\{1+(s^2+\omega^2)L_sC_s\right\}}{\left\{1+(s^2-\omega^2)L_sC_s\right\}^2+4\omega^2s^2L_s^2C_s^2}
\end{aligned} \tag{7.37}
$$

It should be noted that the imaginary part operation $Im\{\}$ is used in (7.36) and that the real part operation is not directly applicable to the denominator. Equations (7.36) and (7.37) constitute the key parts of this chapter, because how to apply the real part operation for the complex Laplace transformed circuit is illustrated as an example.

As emphasized in the previous section, the Laplacian s is regarded as a non-complex number in the real part operations of (7.36) and (7.37). This may be very strange for readers who are familiar with the idea that the Laplacian s should in general be a complex number. Furthermore, the proposed complex Laplace transformation of (7.1) is applied to the complex variables, as shown in Figure 7.3. It is worthwhile, at the moment, to remark on the validity of the proposed real part operation in the complex domain related with the complex Laplace transformation. We now begin with a complex Laplace transformed function $F(s)$, assuming that it can be decomposed into two conventional real Laplace transformed functions, $F_r(s)$ and $F_i(s)$, as follows:

$$\boldsymbol{F}(s) = F_r(s) + jF_i(s) \tag{7.38}$$

where

$$\boldsymbol{F}(s) \equiv \int_0^\infty \boldsymbol{f}(t)\mathrm{e}^{-st}\mathrm{d}t, \quad F_r(s) \equiv \int_0^\infty f_r(t)\mathrm{e}^{-st}\mathrm{d}t, \quad F_i(s) \equiv \int_0^\infty f_i(t)\mathrm{e}^{-st}\mathrm{d}t \tag{7.39}$$

and

$$\boldsymbol{f}(t) = f_r(t) + jf_i(t), \quad \because f_r(t), f_i(t) \in \mathrm{R}^1: \text{real} \tag{7.40}$$

Note that $F_r(s)$ and $F_i(s)$ of (7.39) are of real value if the Laplacian s is a real number because their corresponding time-domain functions $f_r(t)$ and $f_i(t)$ are real. Therefore, $F_r(s)$ and $F_i(s)$ become complex values only if the Laplacian s is a complex number. In other words, the nature of complex variables in the conventional Laplace transformed function stems not from the time-domain function but from postulation of the complex Laplacian s. It can be seen from (7.38) and (7.40) that the real part operation in the complex Laplace transformation is valid for the real Laplacian s, as follows:

$$Re\{\boldsymbol{F}(s)\} = Re\left\{\int_0^\infty \boldsymbol{f}(t)\mathrm{e}^{-st}\mathrm{d}t\right\} = \int_0^\infty Re\{\boldsymbol{f}(t)\}\mathrm{e}^{-st}\mathrm{d}t \quad \text{if } s \in \mathrm{R}^1$$
$$= \int_0^\infty f_r(t)\mathrm{e}^{-st}\mathrm{d}t = F_r(s) = Re\{F_r(s) + jF_i(s)\} \quad \text{if } s \in \mathrm{R}^1 \tag{7.41}$$

It should not be misunderstood, however, from (7.41) that the complex Laplace transformation explained in this chapter is valid for a real Laplacian s only. What (7.41) shows is just a way of finding the real part of a complex Laplace function, that is, it does not necessarily impose the Laplacian s on a real number in practical applications. To further clarify this arguable statement, a conventional real Laplace transformed function $F(s)$ of (7.7), which represents one of $F_r(s)$ and $F_i(s)$, is introduced, as follows:

$$F(s) \equiv \frac{b_0 + b_1 s^1 + b_2 s^2 + \cdots + b_m s^m}{a_0 + a_1 s^1 + a_2 s^2 + \cdots + a_n s^n}, \quad \because \{a_k\} \in \mathrm{R}^1, 1 \le k \le n,$$
$$\{b_k\} \in \mathrm{R}^1, 1 \le k \le m, a_n \ne 0, b_m \ne 0$$
$$= \frac{G_z(s)}{G_p(s)}, \quad G_p(s) = a_0 + a_1 s^1 + a_2 s^2 + \cdots + a_n s^n,$$
$$G_z(s) = b_0 + b_1 s^1 + b_2 s^2 + \cdots + b_m s^m \tag{7.42}$$

It is assumed that $F(s)$ is composed of polynomial functions $G_p(s)$ and $G_z(s)$ with real coefficients, which is the case in a linear system with real variables such as an ordinary electrical circuit. Permitting complex poles p_k and zeros z_k, $F(s)$ of (7.42) can be rewritten in the following form:

$$F(s) = \frac{b_m(s-z_0)(s-z_1)\cdots(s-z_m)}{a_n(s-p_0)(s-p_1)\cdots(s-p_n)}, \quad G_p(s) = a_n(s-p_0)(s-p_1)\cdots(s-p_n),$$

$$G_z(s) = b_m(s-z_0)(s-z_1)\cdots(s-z_m) \tag{7.43}$$

Note that the complex poles in $G_p(s)$ or the complex zeros in $G_z(s)$ always have their corresponding complex conjugate pairs so that the coefficients can be real, as shown for the following complex pole pair case:

$$(s-p_k)\left(s-p_k^*\right) = s^2 - s\left(p_k + p_k^*\right) + p_k p_k^* = s^2 - s \cdot 2Re\{p_k\} + |p_k|^2, \quad 1 \le k \le n \tag{7.44}$$

The time-domain function $f(t)$, which is the inverse Laplace transformation of (7.43), includes no complex value, as defined in (7.7); hence, the complex pole pairs or zero pairs do not generate any complex time-domain value. In this way, the real Laplace transformed functions $F_r(s)$ and $F_i(s)$, though having complex poles or zeros, are inversely Laplace transformed to the time-domain functions of real value only. In a word, what matters is not the nature of Laplacian s but the real coefficients in $F(s)$; hence, the decomposition of a complex Laplace transformed function $\boldsymbol{F}(s)$ into $F_r(s)$ and $F_i(s)$ can be conveniently performed by finding the real and imaginary parts of $\boldsymbol{F}(s)$ regarding s as real, as shown in (7.36) and (7.37). From this viewpoint, s can be called a pseudo real variable for the real or imaginary part operations in the proposed complex Laplace transformation.

7.3.3 Perturbation Analysis of a Complex Laplace Transformed Circuit

In the previous section, the source voltage $V_s(s)$ was considered only as an input for the purpose of large signal dynamic analysis; however, the complex transformer, as shown in Figure 7.2(b) or Figure 7.6(a), is the control driver in most applications. The complex transformer is no longer linear for a time-varying complex turn ratio s_m, though it is linear for a constant S_m, as has been discussed so far.

Therefore, a perturbed complex transformer model in the complex Laplace domain neglecting the product of two perturbed variables is as follows:

$$v_o \equiv V_o + \hat{v}_o = v_s s_m \equiv (V_s + \hat{v}_s)(S_m + \hat{s}_m) = V_s S_m + V_s \hat{s}_m + S_m \hat{v}_s + \hat{v}_s \hat{s}_m$$
$$\cong V_s S_m + V_s \hat{s}_m + S_m \hat{v}_s$$
$$i_s \equiv I_s + \hat{i}_s = i_o s_m^* \equiv (I_o + \hat{i}_o)\left(S_m^* + \hat{s}_m^*\right) = I_o S_m^* + I_o \hat{s}_m^* + S_m^* \hat{i}_o + \hat{i}_o \hat{s}_m^*$$
$$\cong I_o S_m^* + I_o \hat{s}_m^* + S_m^* \hat{i}_o \tag{7.45}$$

The small signal perturbed variables, excluding the large signal constant variables from (7.45), as shown in Figure 7.7(a), are as follows:

$$\hat{v}_o \cong V_s \hat{s}_m + S_m \hat{v}_s$$
$$\hat{i}_s \cong I_o \hat{s}_m^* + S_m^* \hat{i}_o \tag{7.46}$$

The complex Laplace transformation applied to (7.46), as shown in Figure 7.7(b), results in the following:

$$\hat{V}_o(s) \cong V_s \hat{S}_m(s) + S_m \hat{V}_s(s)$$
$$\hat{I}_s(s) \cong I_o \hat{S}_m^*(s) + S_m^* \hat{I}_o(s) \tag{7.47}$$

Figure 7.7 Perturbed complex transformer model.

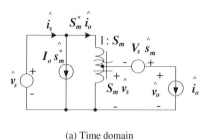

(a) Time domain

(b) Complex Laplace domain

Considering $s_m = s_m e^{j\phi_s}$ and $s_m^* = s_m e^{-j\phi_s}$, the perturbed complex turn ratios in (7.46) can be further resolved as follows:

$$\hat{s}_m = \hat{s}_m e^{j\phi_s} + j\hat{\phi}_s S_m e^{j\phi_s} = (\hat{s}_m/S_m + j\hat{\phi}_s)S_m$$
$$\hat{s}_m^* = \hat{s}_m e^{-j\phi_s} - j\hat{\phi}_s S_m e^{-j\phi_s} = (\hat{s}_m/S_m - j\hat{\phi}_s)S_m^* \tag{7.48}$$

The complex Laplace transformation for (7.48) results in the following:

$$\hat{S}_m(s) = \{\hat{S}_m(s)/S_m + j\hat{\phi}_s(s)\}S_m$$
$$\hat{S}_m^*(s) = \{\hat{S}_m(s)/S_m - j\hat{\phi}_s(s)\}S_m^* \tag{7.49}$$

The perturbed complex Laplace transformed transformer model of Figure 7.7(b) can be applied to Figure 7.6(a) for the perturbation analysis, as shown in Figure 7.8(a). The source voltage is also perturbed; hence, there are three independent voltage and current sources in Figure 7.8(a), in which the source side LC filter is substituted with the Thevenin resistance $Z_{t1}(s)$, which is the same as (7.30), and the perturbed Thevenin voltage $\hat{V}_{t1}(s)$ is as follows:

$$\hat{V}_{t1}(s) = \frac{1}{1 + (sL_s + j\omega L_s)(sC_s + j\omega C_s)} \cdot \hat{V}_s(s) \tag{7.50}$$

Removing the complex transformer and obtaining the Thevenin open voltage for the three independent sources, a more simplified equivalent circuit is found in Figure 8(c), in which the Thevenin resistance $Z_{t2}(s)$ is the same as (7.32) and the perturbed Thevenin voltage $\hat{V}_{t2}(s)$ is as follows:

$$\hat{V}_{t2}(s) = S_m^* \hat{V}_{t1}(s) - S_m^* I_o Z_{t1}(s)\hat{S}_m(s) + V_1 \hat{S}_m^*(s) \tag{7.51}$$

Finally, removing the real part operator, the Thevenin equivalent circuit is found in Figure 7.8(d), where the Thevenin resistance $Z_{t3}(s)$ is the same as (7.34) and the perturbed Thevenin voltage $\hat{V}_{t3}(s)$ is as follows:

$$\hat{V}_{t3}(s) = Re\{\hat{V}_{t2}(s)\} \tag{7.52}$$

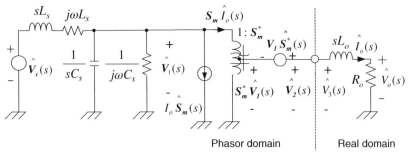

(a) Original perturbed circuit.

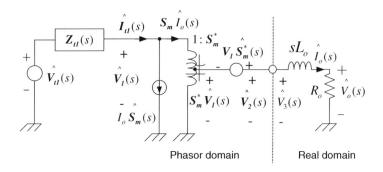

(b) The first Thevenin equivalent circuit for the source side.

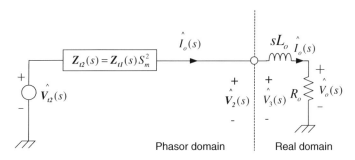

(c) The second Thevenin equivalent circuit, removing the complex transformer.

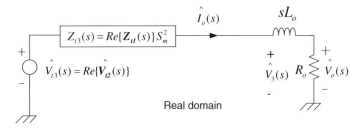

(d) The third Thevenin equivalent circuit, removing the real part operator.

Figure 7.8 Perturbed circuits of the 3-phase rectifier in the complex Laplace domain.

Now the perturbed output voltage can be obtained by applying conventional real domain circuit analysis to Figure 7.8(d), which is very similar to (7.35), as follows:

$$\hat{V}_o(s) = \frac{R_o}{Z_{t3}(s) + sL_o + R_o} \hat{V}_{t3}(s) = \frac{R_o}{Re\{Z_{t1}(s)\}S_m^2 + sL_o + R_o} Re\{\hat{V}_{t2}(s)\} \tag{7.53}$$

where $Re\{Z_{t1}(s)\}$ is the same as (7.37) and $Re\{\hat{V}_{t2}(s)\}$ can be obtained using (7.49) to (7.51):

$$\begin{aligned}
Re\{\hat{V}_{t2}(s)\} &= Re\left\{S_m^* \hat{V}_{t1}(s) - S_m^* I_o Z_{t1}(s)\hat{S}_m(s) + V_1 \hat{S}_m^*(s)\right\} \\
&= Re\left\{S_m^* \hat{V}_{t1}(s)\right\} - I_o Re\left\{S_m^* Z_{t1}(s)\{\hat{S}_m(s)/S_m + j\hat{\phi}_S(s)\}S_m\right\} \\
&\quad + Re\left\{V_1\{\hat{S}_m(s)/S_m - j\hat{\phi}_S(s)\}S_m^*\right\} \\
&= Re\left\{S_m^* \hat{V}_{t1}(s)\right\} - I_o S_m^2 \left[Re\{Z_{t1}(s)\}\hat{S}_m(s)/S_m - Im\{Z_{t1}(s)\}\hat{\phi}_S(s)\right] \\
&\quad + \left[Re\left\{V_1 S_m^*\right\}\hat{S}_m(s)/S_m + Im\left\{V_1 S_m^*\right\}\hat{\phi}_S(s)\right]
\end{aligned} \tag{7.54}$$

In similar ways to those used for obtaining (7.36) and (7.37), $Re\{S_m^* \hat{V}_{t1}(s)\}$ and $Im\{Z_{t1}(s)\}\hat{\phi}_S(s)$ can be found as follows:

$$Re\left\{S_m^* \hat{V}_{t1}(s)\right\} = \frac{Re\left\{S_m^* \hat{V}_s(s)\right\}\{1 + (s^2 - \omega^2)L_s C_s\} + 2Im\left\{S_m^* \hat{V}_s(s)\right\}\omega s L_s C_s}{\{1 + (s^2 - \omega^2)L_s C_s\}^2 + 4\omega^2 s^2 L_s^2 C_s^2} \tag{7.55}$$

$$\begin{aligned}
Im\{Z_{t1}(s)\} &= \frac{Im\{(sL_s + j\omega L_s)\{1 + (s^2 - \omega^2)L_s C_s - j2\omega s L_s C_s\}\}}{\{1 + (s^2 - \omega^2)L_s C_s\}^2 + 4\omega^2 s^2 L_s^2 C_s^2} \\
&= \frac{\omega L_s\{1 - (s^2 + \omega^2)L_s C_s\}}{\{1 + (s^2 - \omega^2)L_s C_s\}^2 + 4\omega^2 s^2 L_s^2 C_s^2}
\end{aligned} \tag{7.56}$$

7.4 Verifications of Complex Laplace Transformed Circuits by Simulation

Several controversial concepts and elements of mathematics were newly introduced in the proposed complex Laplace transformation discussed so far; hence, it is crucial to verify the theory rigorously. Experimental verification, though it gives us practical insight in most cases, is not preferred at this time since it is not enough to be convincing, even if the experimental results are found to coincide well with the theoretical estimations. Instead, real time-domain numerical simulations using the state equation for Figure 7.2(a), neglecting switching harmonics, were performed for the following circuit parameters, unless otherwise specified:

$$\begin{aligned}
&V_s = 440\angle\frac{\pi}{3} \quad S_m = 0.90\angle\frac{\pi}{4} \quad f_s = 60\,\text{Hz} \\
&L_s = 5\,\text{mH} \quad C_s = 300\,\mu\text{F} \quad L_o = 3\,\text{mH} \quad R_L = 10\,\Omega
\end{aligned} \tag{7.57}$$

A time domain transient response for all zero initial conditions and the circuit parameters of (7.57) is shown in Figure 7.9, in which the converter is stabilized within 2–3 cycles but undergoes large transient peak voltage stress due to an abrupt step source voltage.

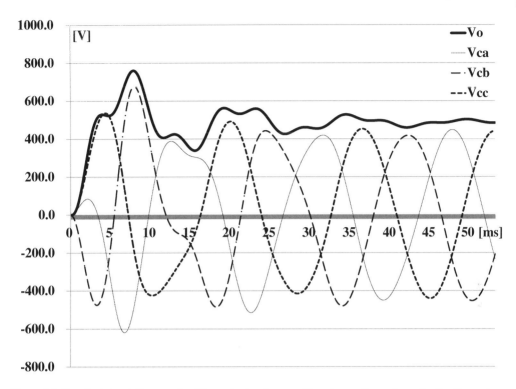

Figure 7.9 Transient simulation results of the output voltage and input capacitor voltages for the three-phase rectifier examples of Figure 7.2(a) with the parameters of (7.57).

7.4.1 Verification for the Static Analysis

The validity of the DC output voltage analysis result of (7.28) was verified by the time-domain simulations for different switching functions, that is, S_m and ϕ_S. The DC voltage gain should be as follows:

$$G_V \equiv \frac{V_o}{V_s} = \frac{1}{V_s} \frac{Re\left\{V_s S_m^*\right\}}{1 - \omega^2 L_s C_s} = \frac{1}{V_s} \frac{Re\{V_s e^{j\phi_V} S_m e^{-j\phi_S}\}}{1 - \omega^2 L_s C_s} = \frac{S_m \cos(\phi_V - \phi_S)}{1 - \omega^2 L_s C_s} \equiv \frac{S_m \cos \phi_{VS}}{1 - \omega^2 L_s C_s}$$

$$(7.58)$$

It is verified by the time-domain simulation, as shown in Figure 7.10, that the DC voltage gain of (7.58), as a function of ϕ_{VS}, is well within the 0.1% error, which is regarded as a simulation error. The DC gain G_V could be negative if the four quadrant AC switches are used in the example circuit shown in Figure 7.2(a).

7.4.2 Verification for the Dynamic Perturbation Analysis

The small signal transfer functions can be found from (7.53) to (7.56), (7.37), and (7.49). For simplicity, it is assumed that the source voltage phasor has a zero phase angle, by which the analysis results do not lose generality because the responses will be relative with respect to the

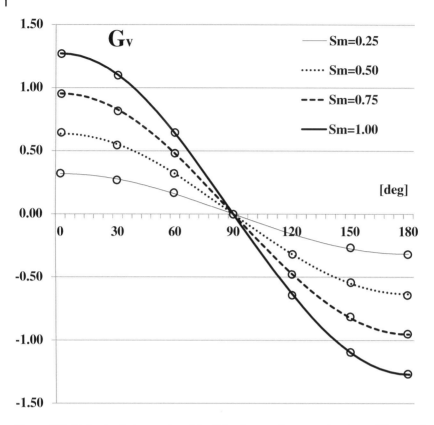

Figure 7.10 Static simulation results of the DC voltage gain versus the phase difference between the source voltage and the switching function for different turn ratios shown in Figure 7.2(a).

$$\left(V_s = 440\angle\frac{\pi}{3}, \quad S_m: \text{variable}, \quad f_s = 60\text{ Hz}, \quad L_s = 5\text{ mH}, \quad C_s = 300\text{ μF}, \quad L_o = 3\text{ mH}, \quad R_L = 10\,\Omega\right)$$

source voltage. In this way, the small signal source voltage transfer function can be found as follows:

$$
G_v(s) \equiv \frac{\hat{V}_o(s)}{\hat{V}_s(s)} = \frac{1}{\hat{V}_s(s)} \frac{R_o}{Re\{Z_{t1}(s)\}S_m^2 + sL_o + R_o} Re\{\hat{V}_{t2}(s)\}\Big|_{\hat{S}_m(s)=\hat{\phi}_S(s)=0}
$$

$$
= \frac{R_o}{\hat{V}_s(s)} \frac{Re\{S_m^* \hat{V}_{t1}(s)\}}{Re\{Z_{t1}(s)\}S_m^2 + sL_o + R_o}
$$

$$
= \frac{R_o}{\hat{V}_s(s)} \frac{Re\{S_m^* \hat{V}_s(s)\}\{1 + (s^2 - \omega^2)L_sC_s\} + 2Im\{S_m^* \hat{V}_s(s)\}\omega sL_sC_s}{sL_s\{1 + (s^2 + \omega^2)L_sC_s\}S_m^2 + (sL_o + R_o)\left[\{1 + (s^2 - \omega^2)L_sC_s\}^2 + 4\omega^2 s^2 L_s^2 C_s^2\right]}
$$

$$
= S_m R_o \frac{\cos\phi_S\{1 + (s^2 - \omega^2)L_sC_s\} - 2\sin\phi_S\omega sL_sC_s}{sL_s\{1 + (s^2 + \omega^2)L_sC_s\}S_m^2 + (sL_o + R_o)\left[\{1 + (s^2 - \omega^2)L_sC_s\}^2 + 4\omega^2 s^2 L_s^2 C_s^2\right]}
$$

$$(7.59)$$

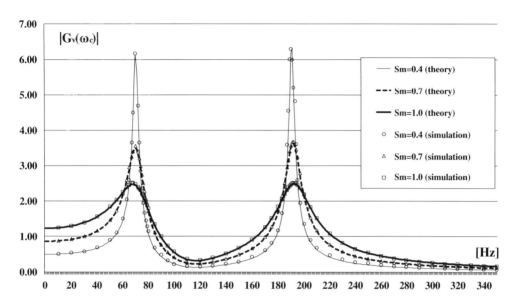

Figure 7.11 Simulation results for the small signal frequency response of the amplitude gain for the example converter of Figure 7.2(a).

$$\left(V_s = 440\angle 0, \quad \phi_{VS} = \frac{\pi}{12}, \quad f_s = 60\,\text{Hz}, \quad L_s = 5\,\text{mH}, \quad C_s = 300\,\mu\text{F}, \quad L_o = 3\,\text{mH}, \quad R_L = 10\,\Omega \right)$$

The amplitude frequency response of the example converter can be obtained from (7.59), as follows:

$$|G_v(\omega_c)| \equiv |G_v(s)|_{s=j\omega_c, \omega=\omega_s}$$

$$= S_m R_o \frac{\left| \cos\phi_S \left\{ 1 - (\omega_c^2 + \omega_s^2) L_s C_s \right\} - j2\sin\phi_S \omega_s \omega_c L_s C_s \right|}{\left| j\omega_c L_s \left\{ 1 + (\omega_s^2 - \omega_c^2) L_s C_s \right\} S_m^2 + (j\omega_c L_o + R_o) \left[\left\{ 1 - (\omega_c^2 + \omega_s^2) L_s C_s \right\}^2 - 4\omega_s^2 \omega_c^2 L_s^2 C_s^2 \right] \right|}$$

$$(7.60)$$

Note that ω_c is a variable control angular frequency, whereas ω_s is the fixed source voltage angular frequency, and that the frequency response of a complex Laplace transfer function can be calculated by letting $s = j\omega_c$, just as in the conventional Laplace transformation case. Other small signal transfer functions, though not shown here, can also be obtained in ways similar to those used in (7.59) and (7.60).

As shown in Figure 7.11, this theoretical calculation of (7.60) is well within ±0.1 from the time-domain simulation result, in which a small perturbation signal was inserted into the source voltage using the circuit parameters of (7.57) except for the turn ratio. The errors stem from the inaccurate measurements of the simulated output voltages, in which some harmonic ripples remained. It is found that there are two poles at 68–70 Hz and 191–193 Hz, which roughly correspond to the difference between and the addition of the input filter resonance frequency 130 Hz and the source frequency 60 Hz. This can be explained by noting that the modulating control signal over the source voltages generates frequencies of the difference and the addition of each frequency, and these frequencies resonate with the input filter.

7.5 Conclusion

The application of a complex Laplace transformation to the dynamic characterization of a phasor transformed circuit including complex variables was successfully done and rigorously verified by simulations. Together with the static analysis theory for AC converters [32], the general unified phasor transformation theory is now completely established by the proposed complex Laplace transformation in this chapter. Any linear time-varying AC converters with multiple phases and switches can be substituted with a complex circuit including a complex transformer and can be analyzed by the general unified phasor transformation theory.

Question 4 The key concept of the proposed complex Laplace transformation to AC converters can be summarized as a conventional Laplace transformation technique that can be applied to complex AC circuits regarding *s* as a real number. Can you tell the difference between the proposed complex Laplace transformation and the conventional dynamic phasor?

Problems

7.1   Determine the dynamics of a current source inverter with a capacitor–resistor load.

7.2   Determine the dynamics of a series–series compensated IPT circuit with an output full-bridge diode rectifier and a capacitor–resistor load. For simplicity, assume that the primary and secondary sides are fully resonated, independently.

7.3   Repeat 7.2 for an output full-bridge active rectifier with phase angle control.

References

 1 D.W. Novotny, "Switching function representation of polyphase inverters," in *IEEE Ind. Applic. Society Conf. Rec.* 1975, pp. 823–831.
 2 R. Middlebrook and S. Cuk, "A general unified approach to modeling switching power converter stages," in *IEEE PESC* 1976, pp. 18–34.
 3 P. Wood, "General theory of switching power converters," in *IEEE PESC* 1979, pp. 3–10.
 4 A. Alesina and M.G.B. Venturini, "Solid-state power conversion: a Fourier analysis approach to generalized transformer synthesis," *IEEE Trans. Circuits Syst.*, vol. CAS-28, no. 4, pp. 319–330, April 1981.
 5 K.D.T. Ngo, "Low frequency characterization of PWM converter," *IEEE Trans. Power Electron.*, vol. PE-1, pp. 223–230, October 1986.
 6 V. Vorperian, R. Tymersky, and F.C. Lee, "Equivalent circuit models for resonant and PWM switches," *IEEE Trans. Power Electron.*, vol. PE-4, no. 2, pp. 205–214, April 1989.
 7 C.T. Rim, D.Y. Hu, and G.H. Cho, "Transformers as equivalent circuits for switches: general proofs and D–Q transformation-based analyses," *IEEE Trans. Ind. Applic.*, vol. 26, no. 4, pp. 777–785, July/August 1990.
 8 C.T. Rim and G.H. Cho, "Phasor transformation and its application to the DC/AC analyses of frequency phase-controlled series resonant converters (SRC)," *IEEE Trans. Power Electron.*, vol. 5, pp. 201–211, April 1990.
 9 V. Vorperian, "Simplified analysis of PWM converters using the model of the PWM switch, part I and part II," *IEEE Trans. Aerospace Electron. Syst.*, vol. 26, no. 3, pp. 490–505, 1990.
10 S.R. Sanders, J.M. Noworolski, X.Z. Liu, and G.C. Verghese, "Generalized averaging method for power conversion circuits," *IEEE Trans. Power Electron.*, vol. 6, pp. 251–259, April 1991.

11 C.T. Rim, N.S. Choi, G.C. Cho, and G.H. Cho, "A complete DC and AC analysis of three-phase controlled-current rectifier using circuit D–Q transformation," *IEEE Trans. Power Electron.*, vol. 9, no. 4, pp. 390–396, July 1994.

12 H.C. Mao, D. Boroyevich, and C.Y. Lee, "Novel reduced-order small signal model of a three-phase rectifier and its application in control design and system analysis," *IEEE Trans. Power Electron.*, vol. 13, no. 3, pp. 511–521, May 1998.

13 J. Chen and K.D.T. Ngo, "Graphical phasor analysis of three-phase PWM converters," *IEEE Trans. Power Electron.*, vol. 16, no. 5, pp. 659–666, September 2001.

14 P. Szczesniak, Z. Fedyczak, and M. Klytta, "Modeling and analysis of a matrix-reactance frequency converter based on buck-boost topology by DQ0 transformation," in *13th International Power Electronics and Motion Control Conference (EPE-PEMC 2008)*, pp. 165–172.

15 S. Kwak, and T. Kim, "An integrated current source inverter with reactive and harmonic power compensators," *IEEE Trans. Power Electron.*, vol. 24, no. 2, pp. 348–357, February 2009.

16 B. Yin, R. Oruganti, S.K. Panda, and A.K.S. Bhat, "A simple single-input–single-output (SISO) model for a three-phase rectifier," *IEEE Trans. Power Electron.*, vol. 24, no. 3, pp. 620–631, March 2009.

17 J. Sun, "Small-signal methods for AC distributed power systems – a review," *IEEE Trans. Power Electron.*, vol. 24, no. 11, pp. 2545–2554, November 2009.

18 V. Valdivia, A. Barrado, A. Laazaro, P. Zumel, C. Raga, and C. Fernandez, "Simple modeling and identification procedures for 'black-box' behavioral modeling of power converters based on transient response analysis," *IEEE Trans. Power Electron.*, vol. 24, no. 12, pp. 2776–2790, December 2009.

19 J. Sun, Z. Bing, and K.J. Karimi, "Input impedance modeling of multi pulse rectifiers by harmonic linearization," *IEEE Trans. Power Electron.*, vol. 24, no. 12, pp. 2812–2820, December 2009.

20 J. Dannehl, F. Fuchs, and P. Thøgersen, "PI state space current control of grid-connected PWM converters with LCL filters," *IEEE Trans. Power Electron.*, vol. 25, no. 9, pp. 2320–2330, September 2010.

21 R. Bucknall, and K. Ciaramella, "On the conceptual design and performance of a matrix converter for marine electric propulsion," *IEEE Trans. Power Electron.*, vol. 25, no. 6, pp. 1497–1508, June 2010.

22 S. Kim, Y. Yoon, and S. Sul, "Pulse width modulation method of matrix converter for reducing output current ripple," *IEEE Trans. Power Electron.*, vol. 25, no. 10, pp. 2620–2629, October 2010.

23 R. Barazarte, G. González, and M. Ehsani, "Generalized gyrator theory," *IEEE Trans. Power Electron.*, vol. 25, no. 7, pp. 1832–1837, July 2010.

24 C.T. Rim, *"A complement of imperfect phasor transformation,"* in Korea Power Electronics Conference, Seoul, 1999, pp. 159–163.

25 S. Ben-Yaakov, S. Glozman, and R. Rabinovici, "Envelope simulation by spice-compatible models of linear electric circuits driven by modulated signals," *IEEE Trans. Ind. Applic.*, vol. 37, no. 2, pp. 527–533, March/April 2001.

26 Y. Yin, R. Zane, J. Glaser, and R.W. Erickson, "Small-signal analysis of frequency-controlled electronic ballasts," *IEEE Trans. Circuits Syst. – I: Fundamental Theory and Applications*, vol. 5, no. 8, August 2003.

27 Z. Ye, P.K. Jain, and P.C. Sen, "Phasor-domain modeling of resonant inverters for high-frequency AC power distribution systems, " *IEEE Trans. Power Electron.*, vol. 24, no. 4, pp. 911–924, April 2009.

28 P. Mattavelli, A.M. Stanković, and G.C. Verghese, "SSR analysis with dynamic phasor model of thyristor-controlled series capacitor," *IEEE Trans. Power Syst.*, vol. 14, pp. 200–208, February 1999.

29 A.M. Stanković, S.R. Sanders, and T. Aydin, "Dynamic phasors in analysis of unbalanced polyphase ac machines," *IEEE Trans. Energy Conversion*, vol. 17, no. 1, pp. 107–113, March 2002.

30 J.A. de la O Serna, "Dynamic phasor estimates for power system oscillations," *IEEE Trans. Instrum. Meas.*, vol. 56, no. 5, pp. 1648–1657, October 2007.

31 C.P. Steinmetz, *"Complex quantities and their use in electrical engineering,"* in Proc. AIEE Int. Elect. Congress, Chicago, IL, 1894, pp. 33–74.

32 C.T. Rim, "Unified general phasor transformation for AC converters," *IEEE Trans. Power Electron.*, accepted for publication.

33 R.C. Paley and N. Wiener, *Fourier Transforms in the Complex Domain*, American Mathematical Society Colloquium Publications, vol. 19, 1934.

34 M.A.B. Deakin, "The ascendancy of the Laplace transform and how it came about", *Archive for History of Exact Sciences*, vol. 44, no. 3, pp. 265–286, 1992.

35 W.S. Steven, *The Scientists and Engineers Guide to Digital Signal Processing*, California Technical Publishing, San Diego, CA, 1997, pp. 567–604.

36 N. Bayan and S. Erfani, *"Frequency analysis of linear time-varying systems: a new perspective,"* in Proc. of IEEE Midwest Symposium on Circuits and Systems, 2005, pp. 1494–1497.

37 N. Bayan and S. Erfani, *"Laplace transform approach to analysis and synthesis of Bessel type linear time-varying systems,"* in Proc. of IEEE Midwest Symposium on Circuits and Systems, 2007, pp. 919–923.

38 S. Erfani, *"Extending Laplace and Fourier transforms and the case of variable systems: a personal perspective,"* in Proc. of IEEE Signal Processing, Long Island Section, 2007. Available at: http://www.ieee.li/pdf/viewgraphs/laplace.pdf.

Part III

Dynamic Charging for Road-Powered Electric Vehicles (RPEVs)

In this part of the book, dynamic charging issues are dealt with, which would be the most integral part of the book.

Starting from the introduction of dynamic charging, the history of RPEVs is summarized. Narrow-width power rails of I-type and ultra-slim power rails of S-type are explained, where the width of the power rail is important in road construction of the infrastructure for RPEVs.

Controllers and compensation circuits for IPT are also explained. Electromagnetic field (EMF) cancel methods including active cancel are explained. Large tolerance design is provided using double dual coils and power rail segmentation and deployment issues are addressed.

Wireless Power Transfer for Electric Vehicles and Mobile Devices, First Edition. Chun T. Rim and Chris Mi.
© 2017 John Wiley & Sons Ltd. Published 2017 by John Wiley & Sons Ltd.

8

Introduction to Dynamic Charging

8.1 Introduction to RPEV

According to the Fourth Assessment Report of the Intergovernmental Panel on Climate Change (IPCC), the ambient temperature of the Earth may rise by more than 2 °C relative to the preindustrial level unless the average CO_2 concentration of the Earth's atmosphere is reduced by 50% and that of the industrialized nations by close to 100% [1]. If the temperature rise is unchecked, we may invite many adverse ecological consequences such as heat waves, drought, tropical cyclones, and extreme tides. To prevent such ecological calamity, many nations are now imposing limits on greenhouse gas emission.

Historically major new technological advances have become the engine for economic growth. With economic growth, the use of energy has also increased. Since the primary source of energy has been fossil fuels, the concentration of greenhouse gases in the atmosphere, especially CO_2, has increased. Today, the United States and China are two of the major emitters of CO_2 on a per capita basis, while on a GDP basis, Russia and China are the leading CO_2 generators. The International Energy Agency [2] clearly states that the current energy trend is not sustainable environmentally, economically, and socially. Therefore, we must devise solutions to achieve the future economic growth without adverse environmental effects.

There are some 800 million automobiles with internal combustion engines (ICEs) in use today worldwide. These automobiles are a major source of greenhouse gases, especially CO_2. Thus, an effective way of dealing with the global warming problem is to replace ICE-powered automobiles with all electric vehicles (EVs). The use of electric cars will also improve the quality of air around major cities.

To replace ICEs, many automobile companies are developing "plug-in" electric cars, which use lithium ion (or polymer) batteries that can be recharged at home or at charging stations. However, the basic premise for plug-in electric cars raises many questions. First, the cost of lithium batteries is high. Second, the batteries are heavy. Third, the charging time for the battery is so long that it requires an expensive infrastructure for charging stations. Finally, perhaps the most important of all, is the finite supply of lithium on Earth. Earth has only about ten million tons of lithium that can be mined economically, which is enough for about 800 million cars, almost the same as the number of cars in use today.

Battery replacement for EVs can drastically reduce the time to switch over the battery but costs a lot due to a backup battery and expensive robot replacement facilities. Spontaneous charging at each stop of EVs may relieve the battery capacity and reduces additional charging time but a high-power charging station is crucial, which is a substantial burden to the grid.

Wireless Power Transfer for Electric Vehicles and Mobile Devices, First Edition. Chun T. Rim and Chris Mi.
© 2017 John Wiley & Sons Ltd. Published 2017 by John Wiley & Sons Ltd.

An RPEV can be a good solution for this battery and charging problem, if only road infrastructure cost can be mitigated. An OLEV has drastically simplified the primary coil buried in the road, so that the road infrastructure cost has been drastically reduced [3–13].

8.2 Functional Requirements (FRs) and Design Parameters (DPs) of OLEV

A high-level design of OLEV is provided here as a brief of [3] to [5]. You will be able to see how the concept of OLEV has been conceived.

8.2.1 FRs, DPs, and Constraints of OLEV

The performance of OLEV is expected to be approximately the same as vehicles with ICEs. The highest-level functional requirements (FRs) of OLEV are as follows [3–5]:

FR1 = propel the vehicle with electric power
FR2 = transfer electricity from underground electric cable to the vehicle
FR3 = steer the vehicle
FR4 = brake the vehicle
FR5 = reverse the direction of motion
FR6 = change the vehicle speed
FR7 = provide the electric power when there is no external electric power supply
FR8 = supply electric power to the underground cable

Constraints are as follows:

C1 = safety regulations governing electric systems
C2 = price of OLEV (should be competitive with cars with IC engines)
C3 = no emission of greenhouse gases
C4 = long-term durability and reliability of the system
C5 = vehicle regulations for space clearance between the road and the bottom of the vehicle

The design parameters (DPs) of OLEV may be chosen as follows:

DP1 = electric motor
DP2 = underground coil
DP3 = conventional steering system
DP4 = conventional braking system
DP5 = electric polarity
DP6 = motor drive
DP7 = rechargeable battery
DP8 = electric power supply system

Question 1 The selection of FRs, Cs, and DPs above is not straightforward and can be arbitrary. Can you add two more FRs, Cs, and DPs according to your own engineering judgment?

8.2.2 The Second-Level FRs, and DPs

The first-level FRs and DPs given in the preceding section must be decomposed until the design is completed with all the details required for full implementation.

The second-level FRs are the FRs for the highest–level DPs and, at the same time, the children FRs of the first-level FRs [3–5]. For example, FR1 can be decomposed to lower-level FRs, for example, FR11, FR12, etc., which are FRs for DP1. Then DP11 can be selected to satisfy FR11, etc. These lower-level FRs and DPs provide further details of the design. There are many patents [8–13] that describe the details of the OLEV system [6,7], including the lower-level FRs and DPs [5].

8.2.3   Design Matrix (DM)

The design matrix (DM) relates the FR vector, {FRs}, to the DP vector, {DPs}, which can be formulated after DPs are selected to satisfy the FRs. DM is used to check if there is any coupling of FRs caused by the specific DPs selected for the design. According to the independence axiom, FRs must be independent of each other.

An integration team of the OLEV project constructed the DM for the OLEV system to identify and eliminate coupling between the FRs at several levels [6, 7]. The final design was either an uncoupled or a decoupled design. When there was coupling, its effect was minimized by making the magnitude of the element of the design matrix that caused coupling much smaller than other elements through design changes.

8.2.4   Modeling of the FRs and DPs

A given FR may have several different DPs. In this case, the final DPs were selected through modeling and simulation of the design using different DPs. The final values of DPs were also determined through modeling and simulations before the hardware was actually built.

8.3   Discussion: Future Prospect of RPEV

A high-level prospect on future vehicles and their impact on energy, environment, economy, etc., is provided as a brief of [3] to [5].

8.3.1   Energy and Environment

Two basic reasons for developing OLEV buses and cars are for better air quality in large cities and the reduction of CO_2 in the Earth's atmosphere to slow global warming. If we remove cars with ICEs from the streets of major cities, the quality of air will improve. However, the total reduction of greenhouse gases depends on the specific means of electricity generation, which may change more gradually. The use of OLEV may not affect the world's primary energy demand in the short term.

According to IEA [2], the world energy use will grow at the rate of 1.6% per year on average in 2006–2030 from 11 730 million tons of oil equivalent (MTOE) to 17 000 MTOE, an increase of 45%, non-OECD countries accounting for 87% of the increase. About a half of the overall increase will be because of the economic growth of China and India. The energy consumption by non-OECD countries exceeded that of the OECD in 2005. Global demand for natural gas grows more quickly, by 1.8% per year, its share in total energy demand rising to 22%. World demand for coal increases by 2% a year on average, its share in global energy demand climbing from 26% in 2006 to 29% in 2030, which is a major generator of CO_2. The use of nuclear power will decrease from 6% to 5% relative to the increased use of energy, although the number of nuclear power plants will increase in all regions except some European OECD countries. Modern renewable technologies are growing rapidly, overtaking gas to become the second-largest source of electricity, behind coal, within a decade.

Question 2 (1) Select a country other than South Korea and calculate the number of power plants of 1 GW to replace all ground vehicles, currently operated. (2) Compare total CO_2 emission of the country before and after the replacement. You may introduce appropriate assumptions if required.

8.3.2   RPEV and Supply of Electricity

To replace all ICE cars being used in South Korea in 2009 with OLEV cars, for example, South Korea needs to dedicate two nuclear power plants for electricity generation. At this time, South Korea produces about 40% of its electricity using nuclear power plants. The cost of electricity is only 22.7% of fossil fuel in South Korea.

Many countries in the world do not have any oil. These countries will have to rely on nuclear power if they are to replace all ICE-driven cars with RPEV cars and buses, without causing global warming.

To reduce the emission of CO_2, we must use more nuclear power plants and renewable energies to generate electricity. In Denmark, windmills produce about 20% of the electricity used in the country. Until we develop other green technologies for generating electric power, many countries will have to rely on nuclear power during the next 50 years. Countries like South Korea are not best situated to make use of renewable energy sources. According to the International Energy Commission, the nations around the world need to build 1750 new nuclear power plants (or equivalent other power plants) until 2050 to meet the energy needs of the world, about 35 new plants a year.

8.3.3   RPEV versus Plug-In All-Battery Cars

The developers of plug-in all-battery cars are banking on low-cost light-weight batteries. However, there is a fundamental limit to the reduction of size and weight of any battery, because batteries need physical structures and space that do not contribute to electric power generation. Furthermore, the total known deposit of lithium is limited. Although there is lithium in seawater, the cost of removing it will be prohibitive unless a new low-cost technology is developed.

Although these all-battery cars have advantages over RPEV in the regions where the population density is so sparse that the cost of laying underground coils for RPEV cannot be justified, they require many charging stations, which may add significant cost. In these regions, cars with ICE may be the best alternative.

There are many significant problems associated with implementing the all-battery plug-in car system, which include the long charging time, the high power capacity needed at charging stations, and the reduced efficiency with the increase in the charging rate.

8.3.4   Electromagnetic Safety

To be sure that there is no question at all about the perceived safety of RPEV, the system must be so designed that people may be minimally exposed to the electromagnetic field (EMF) within the allowable limits. Where the exposure to the electric field is unavoidable, the magnitude of EMF is controlled to be well below the allowable level. A segmented power supply system for OLEV and specially designed coils will further reduce the EMF level to enhance safety.

8.4 Concluding Remarks: The Need for Dynamic Charging

As clearly identified from previous sections, cost-effective solutions of dynamic charging are crucial for the success of RPEVs, where OLEVs show the potential of RPEVs as the future transportation among many candidates. Though RPEVs may not replace most vehicles now, a drastic increase of worldwide market share will be the sign of success.

One of fundamental difficulties of dynamic charging is that the inductive power transfer (IPT) system should not only be robust to electrical stress but also strong against mechanical shocks simultaneously, which is a quite new challenge in road engineering society. Furthermore, it should be cheap and simple enough for deployment over a wide area. Moreover, EVs are ever moving and changing sensitive parameters such as lateral displacement and airgap, which is much more difficult than a static charger whose parameters are fixed once parked.

It looks to be formidable work at a glance to develop such an RPEV system, meeting all the tough requirements at the same time. However, it has been verified by the KAIST team that RPEV, that is, OLEV, is sufficiently possible to develop and could be a promising candidate for future transportations.

References

1 Intergovernmental Panel on Climate Change (IPCC), Fourth Assessment Report (AR4): Climate Change, 2007.
2 International Energy Agency, OECD, *World Energy Outlook*, 2008.
3 N.P. Suh, *The Principles of Design*, Oxford University Press, New York, 1990.
4 N.P. Suh, *Axiomatic Design: Advances and Applications*, Oxford University Press, New York, 2001.
5 N.P. Suh, D.H. Cho, and Chun T. Rim, "Design of on-line electric vehicle (OLEV)," *Plenary Lecture at the 2010 CIRP Design Conference in Nantes*, France, April 19–21, 2010, pp. 3–8.
6 Chun T. Rim, "Wireless charging research activities around the world: KAIST Tesla Lab (Part)," *IEEE Power Electronics Magazine*, vol. 1, no. 2, pp. 32–37, June 2014.
7 Chun T. Rim, "A review of recent developments of wireless power transfer systems for road powered electric vehicles," *IEEE Transportation Electrification Newsletter*, September/October 2014.
8 N.P. Suh, D.H. Cho, Chun T. Rim, S.J. Jeon, J.H. Kim, and S. Ahn, "Method and device for designing a current supply and collection device for a transportation system using an electric vehicle," US Patent Application 13/810,066, 2011, USA.
9 S.H. Chang, J.G. Cho, G.H. Cho, D.H. Cho, Chun T. Rim, and N.P. Suh, "Ultra slim power supply device and power acquisition device for electric vehicle," US Patent Application 13/262,879, 2010, USA.
10 N.P. Suh, D.H. Cho, Chun T. Rim, S.J. Jeon, J.H. Kim, and S. Ahn, "Power supply device, power acquisition device and safety system for electromagnetic induction-powered electric vehicle," US Patent Application 13/202,753, 2010, USA.

11 N.P. Suh, S.H. Chang, D.H. Cho, J.G. Cho, and Chun T. Rim, "Power supply apparatus for on-line electric vehicle, method for forming same and magnetic field cancelation apparatus," US Patent Application 13/501,691, 2010, USA.

12 N.P. Suh, D.H. Cho, Chun T. Rim, J.W. Kim, K.M. Park, and B.Y. Song, "Modular electric-vehicle electricity supply device and electrical wire arrangement method," US Patent Application 13/510,218, 2010, USA.

13 S.J. Jeon, D H. Cho, Chun T. Rim, and G.H. Jeong, "Load-segmentation-based full-bridge inverter and method for controlling same," US Patent Application 13/518,213, 2009, USA.

9

History of RPEVs

9.1 Introduction

As conventional transportations that heavily rely on internal combustion engines face world-wide growing pressure to reduce emissions of greenhouse gases such as CO_2 and to mitigate air pollution in urban areas, electric vehicles (EVs) are becoming more attractive than ever, though the "petroleum era" has been extended due to the economic exploitation of shale gas. Therefore, automobile manufacturers have been developing various EVs such as pure-battery EV (BEV) [1–4], hybrid EV (HEV) [5–8], plug-in hybrid EV (PHEV) [9–11], battery-replace EV (BREV) [12–15], and road-powered EV (RPEV) [16–99]. Probably the biggest challenge to the commercialization of the EV is the battery, which is still heavy, bulky, and expensive even though it was commercialized for EVs more than 130 years ago [100]. Moreover, it is made of scarce materials, such as lithium, buried in only a few countries and may be explosive in car accidents. The charging of the battery is another obstacle in the commercialization of EVs because they should be frequently recharged after operating for a short range due to the low energy density of batteries. The currently available quick charging time of 20 minutes [101–103] is still too long for drivers accustomed to rapid fueling and deteriorates battery lifetime severely, requiring quite expensive and large-size charging facilities. Promising battery technologies of extremely quick charging, that is, less than 5 minutes, have been reported [104], but they make the quick charging problems worse; nevertheless, they will become technically stabilized and economically available. Other challenges in developing EV components, such as light and robust motors, efficient and compact inverters, and miscellaneous powertrain units, are relatively trivial and are no longer technical problems but economic ones.

Unfortunately, EVs, such as PEVs and BREVs, heavily rely on a large battery; therefore, innovations in the battery are crucial for the commercialization of these vehicles. Relatively, HEVs, PHEVs, and RPEVs do not require the battery innovations for commercialization; in other words, they can be readily available in markets using currently affordable EV batteries. Thus, HEVs are becoming more popular in worldwide markets among all EVs, even though the role of the battery is quite limited to short-term energy recovery. When the power supply rails for transmitting power to RPEVs are fully deployed under the road, RPEVs will not require battery energy storage for their traction because they directly get required power from a road while they are moving on it. Hence, RPEVs are mostly free from the battery-related problems among EVs and quite promising candidates for future transportation of small cars, passenger cars, taxies, buses, trams, trucks, trailers, and trains, even in competition with internal combustion engines. Despite the fact that RPEVs are free from battery problems, RPEVs have not been widely used so far. The biggest challenge of the RPEV in commercialization is to transfer high power from the road in an efficient, economic, and safe way. The power transfer can either be wired or

Wireless Power Transfer for Electric Vehicles and Mobile Devices, First Edition. Chun T. Rim and Chris Mi.
© 2017 John Wiley & Sons Ltd. Published 2017 by John Wiley & Sons Ltd.

wireless. Traditionally, the former was preferred because there were no appropriate means for wireless power transfer. Even though wired electric buses [16–18] are no longer widely used in urban areas, it is amazing that still the highest speed train is powered through pantographs [19,20], which are a kind of wired power transfer device. Because of the wearing of pantographs and maintenance problems, wired power transfer is gradually being replaced by wireless one as hundreds of kW of power become available. Thus, various wireless power transfer systems (WPTSs) [21–99] have been widely developed for RPEVs. Therefore, it is worthwhile to focus on the wireless RPEVs and exclude the wired ones from further discussions.

In this chapter, a full history of the WPTSs for RPEVs is described from its advent developments in the 1890s to cutting-edge technologies now. Important technical issues in the development of inductive power transfer systems (IPTSs), the majority of WPTSs, are addressed, and major milestones of the developments of RPEVs are summarized, focusing on the development of on-line electric vehicles (OLEVs) that have been recently commercialized.

9.2 Fundamentals of Wireless Power Transfer Systems for RPEV

9.2.1 Overall Configuration of the WPTS

The WPTS for RPEV should be capable of delivering high power efficiently through a moderate airgap for avoiding collisions between the RPEVs and the road. The WPTS are composed of two subsystems: one is the roadway subsystem for providing power, which includes a rectifier, a high-frequency inverter, a primary capacitor bank, and a power supply rail, and another is the on-board subsystem for receiving power, which includes a pick-up coil, a secondary capacitor bank, a rectifier, and a regulator for the battery, as shown in Figure 9.1(a). The roadway subsystem should be so robust and cheap that it may withstand severe road environments for a long time and should be economic to install over a long distance, whereas the on-board subsystem should be compact in size and light in weight so that it may be adopted into the RPEV.

WPTSs, in general, can be classified into inductive power transfer systems (IPTSs) [21–99], coupled magnetic resonance systems (CMRSs) [105–107], and capacitive power transfer systems (CPTSs) [110]. These three systems were regarded as quite different from one another; however, CMRS are found to be just a special form of IPTS whose quality factor Q is extremely high, and resonating repeaters extend power-delivering distance [107]. Moreover, it is no longer true that CMRS are appropriate for a long-distance power delivery whereas IPTS are adequate for a short-distance high-power delivery because, recently, a new world record of 5-m-off long-distance power delivery has been demonstrated by using IPTS without a very high Q factor [108,109]. Considering the difficulties in maintaining resonance for multiple resonant repeaters with a very high Q factor and the bulky configuration of pick-ups, conventional CMRSs are not quite appropriate candidates for RPEV, in general. In addition, CPTS is not appropriate for RPEVs because it needs a huge area of conductor to transfer several kW of power with an airgap of 20 cm, which may be bigger than the bottom space of a vehicle. Therefore, IPTSs will be dealt with in detail in subsequent sections, though CMRSs and CPTSs are not completely excluded from review of papers.

The RPEV mission system, as shown in Figure 9.1(b), includes not only the WPTS system but also the control system and EV system. The control system is unique and crucial for RPEVs because it senses and identifies the EV and then appropriately turns the inverters on and off. Moreover, it monitors the health of the IPTSs and RPEVs and provides an accounting service and communication links.

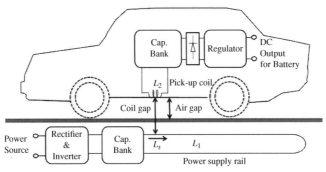

(a) The configuration of a wireless power transfer system for RPEV [86]

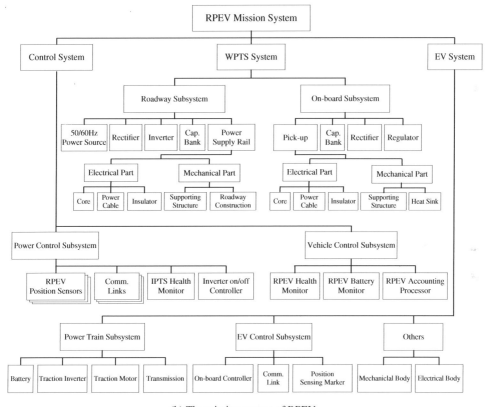

(b) The mission system of RPEV

Figure 9.1 Overall configurations of RPEV and IPTS.

9.2.2 Fundamental Principles of the IPTS

Now the fundamental principles of IPTSs, instead of all WPTSs, for RPEVs will be briefly explained. The IPTSs are governed by Ampere's law and Faraday's law among four Maxwell equations, as shown in Figure 9.2(a). It can be briefly explained as follows:

1. Time-varying magnetic flux is generated from the AC current of a power supply rail in accordance with Ampere's law.

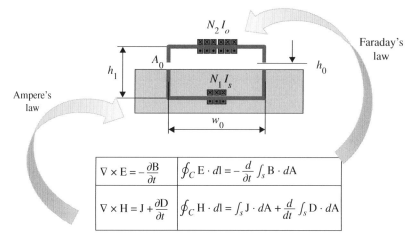

(a) The governing equations of the IPTS [73]

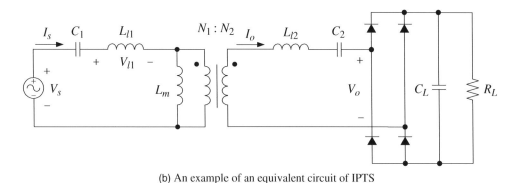

(b) An example of an equivalent circuit of IPTS

Figure 9.2 The fundamental principle and an equivalent circuit of IPTS for RPEV.

2. Voltage is induced from the pick-up coil, coupled with the power supply rail, in accordance with Faraday's law.
3. Power is wirelessly delivered through magnetic coupling, where capacitor banks are used to nullify inductive reactance.

The governing equations of IPTS for sinusoidal magnetic field, voltage, and current in the steady state are as follows:

$$\nabla \times \mathbf{H} = \mathbf{J} \quad \text{(Ampere's law)} \tag{9.1a}$$
$$\nabla \times \mathbf{E} = -j\omega\mathbf{B} \quad \text{(Faraday's law)} \tag{9.1b}$$

In order to provide moderately high-frequency AC current to the power supply rail, a high-frequency switching inverter is introduced, and a rectifier is attached to the pick-up coil side to obtain a DC voltage for providing power to an on-board battery of RPEV, as shown in Figures 9.1(a) and 9.2(b). Meanwhile, the electromagnetic force (EMF) for pedestrians should be under the constraints such as ICNIRP guidelines [111,112]. This is why passive and active EMF cancelation techniques have been so developed [113–125].

It is quite often misunderstood that the purpose of resonating LC circuits in the IPTS, as shown in Figure 9.2(b), is to amplify power by the Q factor. Actually, resonance is not mandatory

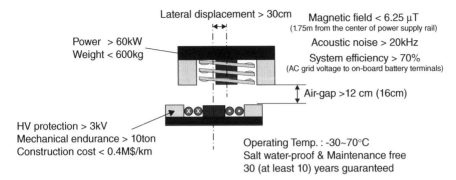

Lateral displacement > 30cm

Magnetic field < 6.25 µT
(1.75m from the center of power supply rail)

Power > 60kW
Weight < 600kg

Acoustic noise > 20kHz

System efficiency > 70%
(AC grid voltage to on-board battery terminals)

Air-gap >12 cm (16cm)

HV protection > 3kV
Mechanical endurance > 10ton
Construction cost < 0.4M$/km

Operating Temp. : -30~70°C
Salt water-proof & Maintenance free
30 (at least 10) years guaranteed

Figure 9.3 An example of desired requirements of the IPTSs for RPEVs [73].

if the required power delivery is small so that the voltage drop of leakage inductances of coils may not be large due to the small current [86–88]. Moreover, the switching frequency of the inverter f_s for IPTS is not necessarily exactly tuned to the resonant frequency of LC circuits f_r. Very often, f_s is intentionally increased a little bit higher than f_r to guarantee the zero voltage switching (ZVS) of inverters [60,61]. Depending on the source and load types, the compensating capacitors can be connected in series or parallel with the source or loads. Figure 9.2(b) shows a series–series compensation example, which is adequate for high-power applications of IPTS whose output characteristic is an ideal voltage source [60,61].

Question 1 Besides the two equations, what other Maxwell equations can you suggest that are related or useful for IPT?

9.2.3 Basic Requirements of the IPTS

As shown in Figure 9.3, the IPTSs for RPEVs are differentiated from conventional IPTSs for stationary charging because additional requirements such as larger lateral tolerance, higher airgap, and lower construction cost should be considered for dynamic charging. Moreover, the IPTSs for RPEVs should survive harsh road conditions such as extremely high and low temperatures, high humidity, and repetitive mechanical shocks for at least 10 years. Basically, the high voltage power rail under a road should be electrically and mechanically well protected, but electricity under the conditions of a wet and fragile structure is not compatible with a road, in general. Therefore, it is quite challenging to build the roadway subsystem robustly. On the other hand, the on-board subsystem should also survive the harsh road conditions as well as vibrating RPEV operating conditions, where strict technical and legislative regulations for vehicles should be met.

9.2.4 Design Issues of the IPTS

The design goals of the IPTS are summarized as follows:

1. To increase magnetic coupling as much as possible so that a higher induced voltage can be obtained.
2. To increase power efficiency for a given power capacity, device ratings, and cost.
3. To make modules as compact as possible to accommodate a given space and weight.
4. Not to increase or cancel out EMF.

5. To manage resonance frequency variation and coupling factor change due to misalignment of pick-up position, airgap change, and even temperature change.

In order to meet the design goals, several important design issues of the IPTSs, which are unique for RPEVs, should be resolved as follows:

1. *Switching frequency.* The switching frequency of the inverter and rectifier of an RPEV governs overall performances. As the switching frequency f_s increases, the coil and capacitor size decreases for the same required power, but the switching loss and core loss as well as the conduction loss of wires due to the skin effect increase. In the case of a switching frequency lower than 20 kHz, acoustic noise may be a problem [61]. As shown in Figure 9.2(b), another unique feature of RPEVs concerning frequency is the increased voltage stress in a distributed power supply rail V_{l1} for a higher switching frequency, which is proportional to the frequency and line current I_s as follows:

$$V_{l1} = j\omega_s L_{l1} I_s \tag{9.2a}$$

$$\frac{\partial V_{l1}}{\partial x} = j\omega_s I_s \frac{\partial L_{l1}}{\partial x} \tag{9.2b}$$

Because of these restrictions, the switching frequency tends to be no less than 20 kHz but not far beyond 50 kHz. For the on-line electric vehicles (OLEVs), the frequency was finally selected as 20 kHz after examining for 20 kHz, 25 kHz, and 30 kHz.

The interoperability of an IPTS between RPEVs and stationary wireless EVs is an important issue that needs further study because the design of IPTSs for RPEVs is not a direct extension of that of stationary chargers. The design goals of the IPTSs of RPEVs include low infra cost, high power, continuity of power delivery, and low fluctuation during in-motion, which are not so important for stationary chargers.

2. *High power and large current.* In addition to the high operating frequency of RPEVs, high power of hundreds of kW with a large current of hundreds of amperes [60, 61] makes it difficult to handle cables, converters, and devices. For example, a litz wire of 300 A rating with 20 kV insulation capability is not commercially mass produced. Capacitors and IGBTs, which have high voltage and large current ratings with an operating frequency higher than 20 kHz, become scarce. Furthermore, all the inverters, rectifiers, coils, capacitors, and cables used in the IPTS of RPEV are operated outdoors, where debris and salty and wet material may damage them.

3. *Power efficiency.* In order to make RPEVs competitive against internal combustion engine vehicles (ICEVs), the power efficiency or energy efficiency should not be too low. The bottom line of the overall power efficiency, which is defined from the AC power source to the DC output power for the battery, would be about 50% [61], considering grid loss and fuel cost. Fortunately, modern IPTSs have fairly good power efficiency larger than 80% [60, 61]. More importantly, power loss rather than power efficiency is a serious problem in designing the inverter and pick-up due to over temperature caused by excessive heat generation.

4. *Coil design.* In order to focus magnetic field from the power supply rail to the pick-up and mitigate leakage flux so that a large induced voltage as well as a low EMF level can be achieved, a novel coil design is crucial for the IPTS. This design is unique for the roadway subsystem in that the cost per km is critical; therefore, the roadway power supply rail should be low enough not to deteriorate the economic feasibility of the overall RPEV solution, whereas the pick-up should be neither heavy nor thick so that it can be successfully implemented on the bottom of the RPEV.

5. *Insulation.* In order to guarantee stable operation of IPTS, a few kV level of insulation of the roadway subsystem and on-board subsystem should be provided. The insulation of a roadway power supply is of great importance because high voltage is induced along through

the distributed power rail over a road rather than over a point and is increased by the back-emf of the pick-up coil abruptly displaced on the power supply rail. The insulation of the pick-up coil is also an important issue when a high power output is sought because the large output current of the pick-up coil induces a high voltage, which could be as much as 10 kV if not appropriately mitigated.

6. *Segmentation of the power supply rail.* RPEVs typically require many IPTS along a road because a power supply rail cannot be infinitely deployed. It should be segmented so that each segment can be independently turned on and off. The length of a power supply rail is an important design issue because it would be too expensive if the length is very short due to an increased number of inverters and switch boxes, whereas the power loss would be too large if the length is very long due to increased resistance. Besides these issues, the amount of cable use, EMF level, and car length should be considered when determining the segmentation scheme.

7. *Roadway construction.* In order to minimize traffic obstruction, the roadway construction time of the IPTS should be as short as possible. Moreover, there are numerous difficulties in roadway work because of debris and dirt. Keeping all of the electric components of the IPTS clean during roadway construction for a few days or weeks is a serious problem in practice. There should be a smart way to overcome these problems.

8. *Resonant frequency variation.* The resonant frequencies of the roadway subsystem and on-board subsystem are significantly varied as the magnetic coupling between a power supply rail and a pick-up changes due to airgap change, lateral misalignment, and longitudinal movement. Moreover, the frequency changes are different from time to time for different cars and different numbers of cars on a power supply rail. There should be either the smart coil design and inverter design to cope with these frequency changes or in situ frequency tuning capability.

9. *Control of dynamic resonating circuits.* The resonant circuits, as shown in Figure 9.2(b), have finite dynamic response times and may give rise to high voltage and current peaks that can destroy electric devices. They should be moderately controlled so that the voltage and current level can be mitigated, where the frequency response of AC circuits can be managed with the recently developed Laplace phasor transformation [126]. A resonating LC circuit has a first-order response in the dynamic phasor domain [127]; hence, a high-order IPTS with more than two resonating circuits can be managed with the general phasor transforms [52, 56, 60, 126, 127].

9.3 Early History of RPEV

9.3.1 The Origin of the RPEV: The Concept of a "Transformer System for Electric Railways"

The concept of the RPEV stems from the first patent, "Transformer system for electric railways" by M. Hustin and M. Leblanc in France in 1894 [21], where a large airgap transformer was displaced under a train for electric power transfer, as shown in Figure 9.4.

In the patent, they claimed several design issues of IPTSs on the deployment of power supply rails, high-power transfer to pick-up coils, and reduction of conduction and eddy current losses, which are still important design issues.

9.3.2 The First Development of RPEVs

During the oil crisis of the 1970s, interest in RPEVs increased in the USA, where several research teams started to investigate RPEVs for reducing the use of petroleum in highway vehicles

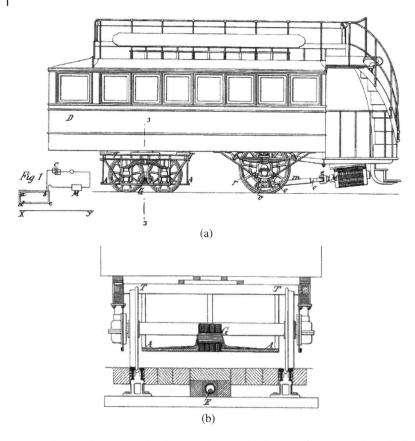

(a)

(b)

Figure 9.4 The first patent of an RPEV in 1894 [21]. (a) Side view of the electric train. (b) Cross-section view cut on the line 3-3 of (a).

[21–44]. The first development of RPEVs was begun in 1976 by the Lawrence Berkeley National Laboratory in order to confirm the technical feasibility of RPEVs [24, 25, 27]. A prototype IPTS was built and tested for 8 kW wireless power transfer [25]; however, it was not a fully operational system [37]. In 1979, the Santa Barbara Electric Bus Project was started and another prototype RPEV was also developed [30–35].

After the two frontier projects of RPEV, the Partners for Advanced Transit and Highways (PATH) program began in 1992 to determine the technical viability of RPEV in the University of California, Berkeley [42, 43]. Throughout the PATH program, broad research and field tests on RPEV [43], including designs of IPTS, installations of an IPTS to a bus, road constructions of power supply rails, and environmental impact studies, were performed. The PATH team achieved an efficiency of 60% at an output power of 60 kW with a 7.6 cm airgap. The PATH program, however, had not been successfully commercialized due to high-power rail construction costs of around 1 million $/km, heavy coils, and acoustic noises, as well as relatively low power efficiency and a large primary current of thousands of amperes owing to a low operating frequency of 400 Hz. Furthermore, the small airgap does not meet road safety regulations and lateral displacement of less than 10 cm is not acceptable for practical use [43]. Despite the limitations for practical applications, the PATH team's work was well documented [42, 43] and stimulated subsequent research and development on modern RPEVs.

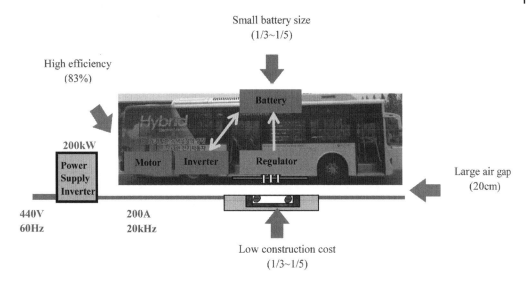

Figure 9.5 The development concepts of the OLEV bus developed by KAIST [73]. Reproduced with permission of IEEE.

9.4   Developments of On-Line Electric Vehicles

A stream of modern RPEVs is the OLEV, which has solved most of the remaining problems of the PATH team's work, as shown in Figures 9.5 to 9.7. The OLEV project was started in 2009 by a research team led by KAIST (including the author), South Korea [45–78]. Innovative coil designs and roadway construction techniques as well as all of the systems operating at a reasonably high frequency of 20 kHz made it possible to achieve the highest power efficiency of 83% at an output power of 60 kW with a large airgap of 20 cm and a fairly good lateral tolerance of 24 cm [61]. Moreover, the power rail construction cost of the OLEV, which is responsible for more than 80% of the total deployment cost for RPEVs [45], has been dramatically reduced to at least a third of that of the PATH project. The primary current has also been reasonably

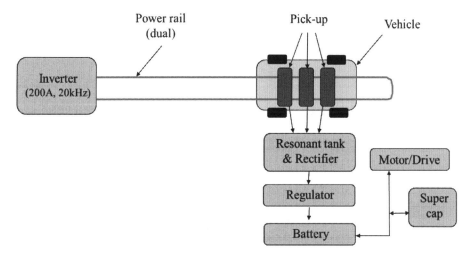

Figure 9.6 The overall configuration of IPTS of the OLEV (plane view).

Figure 9.7 Roadway construction of IPTS of the OLEV.

mitigated to as low as 200 A, and the battery size has been significantly reduced to 20 kW h, which can be further reduced by increasing the length of the power supply rails.

As shown in Figures 9.8 and 9.9, the first-generation (1G) concept demonstration car of OLEV, the second-generation (2G) OLEV buses, and the third-generation (3G) OLEV passenger car were developed and extensively tested at the test sites of KAIST in 2009, and three OLEV trains (3^+G) were successfully deployed at the Seoul Grand Park, South Korea, in 2010. Two upgraded OLEV buses (3^+G) were deployed at the 2012 Yeosu EXPO, South Korea, and another two OLEV buses (3^+G) have been in full operation at the main campus of KAIST since

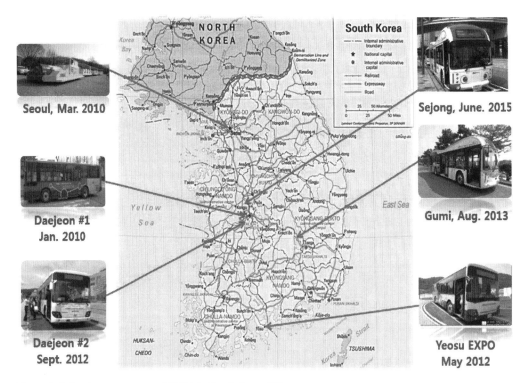

Figure 9.8 Deployment status of OLEVs in South Korea [73].

	1 G (Car)	2 G (Bus)	3 G (SUV)	3⁺ G (Bus)	3ᵗG (Train)	4 G (Bus)
Date	Feb. 27, 2009	July 14, 2009	Aug. 14, 2009	Jan. 31, 2010	Mar. 9, 2010	2010~ (under development)
Vehicle						
System Spec.	air-gap= 1cm efficiency= 80%	air-gap= 17cm efficiency= 72%	air-gap= 17cm efficiency= 71%	air-gap= 20cm Efficiency= 83%	air-gap= 12cm efficiency= 74%	air-gap= 20cm, efficiency= 80%
	All the efficiencies are measured by AC grid voltage to on-board battery terminals					
EMF	10mG	51mG	50mG	50mG	50mG	<10mG
Power Rail (width)						
	20cm	140cm	80cm	80cm	80cm	10cm
Pick-up						
Power	3kW / pick-up	6kW / pick-up	15kW / pick-up	15kW / pick-up	15kW / pick-up	25kW / pick-up
Weight (Pick-up)	20kg	80kg	110kg	110kg	110kg	80kg
Size	55x18x4cm³	160x60x11cm³	170x80x8cm³	170x80x8cm³	170x80x8cm³	80x100x8cm³

Figure 9.9 A summary on the developments of OLEV including their IPTS [73].

2012 [73]. Recently, two OLEV buses (3⁺G) were firstly commercialized at a 48 km route in Gumi, South Korea.

The fourth-generation (4G) OLEV, showing more practical performances, such as a very narrow rail width of 10 cm, a larger lateral displacement of 40 cm, a lower EMF level, and lower power rail construction costs compared to the previous generation OLEV, has also been developed. Now the development of the fifth-generation (5G) OLEV is in progress, where an ultraslim S-type power supply coil of 4 cm rail width is proposed for power rail construction with much less cost and time [76].]

Question 2 Discuss the heatsink of the power supply rail of each generation. In a hot summer, the temperature increase due to the continuous and large current of the power supply rail may be a disaster.

Figure 9.10 The 1G golf cart platform of OLEV [73].

9.4.1 The First-Generation (1G) OLEV

The 1G OLEV, which was announced on February 27, 2009, is a golf cart equipped with a mechanically controlled pick-up to automatically align to the power supply rail of 45 m within a 3 mm lateral displacement, as shown in Figure 9.10. The 1G OLEV adopted E-type cores for both the power supply rail and pick-up coil, as shown in Figure 9.9, and achieved an overall system efficiency of 80% at 20 kHz switching frequency with an output power of 3 kW at a 1 cm airgap [73]. It successfully demonstrated the wireless power delivery to an EV and became the basis of confidence in succeeding developments.

9.4.2 The Second-Generation (2G) OLEV

The 2G OLEV, announced on July 14, 2009, focused on drastically improving the airgap of the 1G OLEV without mechanical moving parts, and finally achieved a 17 cm airgap, which meets road regulations, 12 cm in South Korea and 16 cm in Japan. At the same time, it achieved a maximum efficiency of 72% and a maximum output power of 60 kW using 10 pick-ups. The power supply rail, where the width is 1.4 m and the total length is 240 m, is paved with asphalt to provide the same friction force as normal roads, as shown in Figure 9.11 [50]. In order to realize the 17 cm airgap, the U-type power supply rail and the flat pick-up coil of IPTS have

Figure 9.11 The 2G OLEV bus at the KAIST Munji Campus with four test tracks of power supply rails (60 m each) [73].

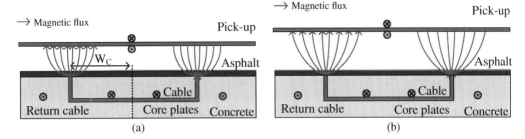

Figure 9.12 Cross-section of the U-type power supply rail and I-type pick-up coil [73] for (a) a small airgap and (b) a large airgap.

been developed, where the name "U-type" stems from the cross-section of the power supply rail, as shown in Figure 9.12. A pair of return power cables is used to mitigate the EMF from the power supply rail for the ICNIRP guidelines [111, 112]. With the U-type power supply rail and I-type pick-up coil, it is found that an effective area of the magnetic path between the power supply rail and the pick-up coil increases as the airgap increases due to the fringe effect, as shown in Figures 9.12 and 9.13.

The lateral displacement, at which the output power drops to half of the maximum output power and the induced voltage or magnetic flux density becomes 70.7% of their maximum, was achieved to be 23 cm, as shown in Figure 9.13(b). The cross-section of the core plates of the power rail, as shown in Figure 9.12, is significantly reduced by using the high operating frequency of 20 kHz, which is 50 times that of the PATH team [43]. The reason for selecting 20 kHz is that it is the lowest inaudible frequency, which mitigates the line voltage stress, as given in (9.2). The upward magnetic flux generated from the pick-up coil was appropriately shielded by an aluminum plate with a 5 cm distance.

9.4.3 The Third-Generation (3G) OLEV

The 3G OLEV SUV (sports utility vehicle), announced on August 14, 2009, as shown in Figure 9.9, adopted the W-type power supply rail and the flat pick-up with overlapped double coils for higher power and increased lateral tolerance. Thus, the upward magnetic leakage flux from the pick-up of 2G OLEV had been drastically mitigated by using the flat pick-up core, which prohibits the magnetic flux between the power supply rail and pick-up from leaking. A magnetic field shield, such as an aluminum plate, or additional space are no longer required. The overall efficiency and airgap of the 3G OLEV were 71% and 17 cm, respectively, which were respectable but slightly disappointing numbers; therefore, overall systems including a roadway rectifier and inverter, power supply rail, pick-up, and on-board regulator were redesigned to achieve a maximum efficiency of 83% with a 20 cm airgap, and this new design was called the 3G⁺ OLEV. Four 3G⁺ OLEV buses, as shown in Figure 9.9, were built for test purposes, and six more 3G⁺ OLEV buses were built for full operation purposes, whereas three 3G⁺ OLEV trains were deployed at the Seoul Grand Park, as shown in Figure 9.14.

The W-type power supply rail, as shown in Figure 9.15, has many cores of "W" shape, where the total magnetic resistance becomes three-quarters smaller than that of the 2G OLEV for the same airgap, which eventually leads to the increase of the output power from 6 kW to 15 kW for each pick-up. Moreover, the width of the power supply rail is decreased to 70 cm, which is just half that of the 2G OLEV, and can reduce the deployment cost of OLEV for commercialization.

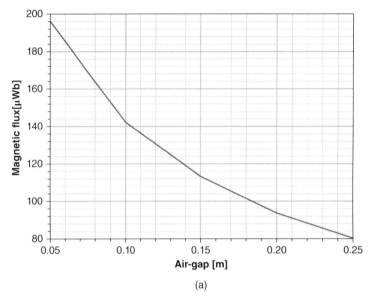

(a)

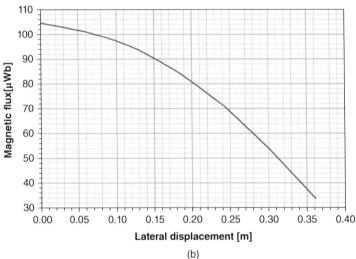

(b)

Figure 9.13 Magnetic flux characteristics of the 2G IPTS [73]: (a) along the airgap and (b) along the lateral displacement.

The EMF at 1.75 m distance meets the ICNIRP guideline because the two power cables of opposite polarity of magnetic flux cancel each other out [73].

One of the problems found during the development of the 2G OLEV was the inherently weak structure of the power supply rail, where the core separates the concrete into two and undergoes severe mechanical stress from heavy vehicles. As a remedy for this mechanical weakness, bone structure cores of the power supply rail, as shown in Figure 9.15(b), have been proposed and registered as a patent [63], where the bone structure cores are installed with the bone coregap X_D. Despite the large gap between cores, the magnetic flux does not decrease considerably, as shown in Figure 9.16.

In addition, during power rail road construction, concrete can percolate down through the bone structure cores. Therefore, the power supply rail, which is reinforced with two iron bars and in which power cables are protected by FRP (fiber reinforced plastic) pipes, has almost

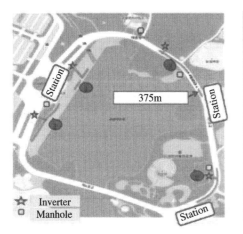

Figure 9.14 The 3G$^+$ OLEV train on the 2.2 km road at the Seoul Grand Park, where 375 m was paved with 24 m power supply rails [73].

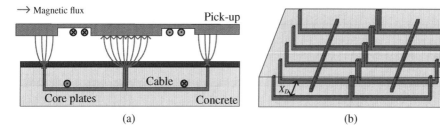

Figure 9.15 Views of the W-type power supply rail and flat pick-up coil [73]. (a) cross-section view and (b) bird's eye view of a bone structure.

Figure 9.16 Simulated magnetic flux characteristics for different bone coregaps X_D [73].

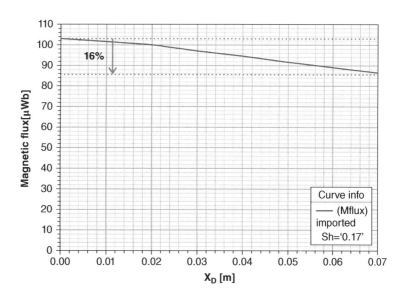

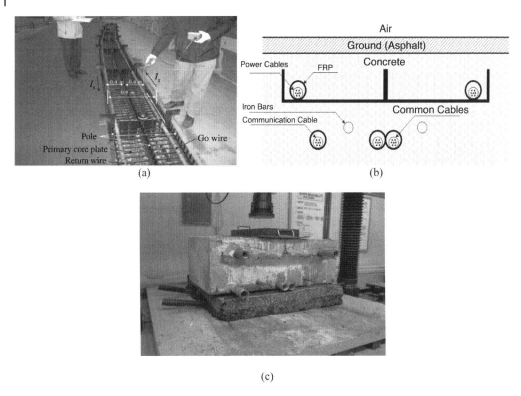

Figure 9.17 The W-type power supply rail of the 3G OLEV [73]. (a) under road construction, (b) cross-section view [74], and (c) being tested. Reproduced with permission of IEEE.

the same endurance of concrete, as shown in Figure 9.17. In practice, roadway construction of the power supply rail takes a few weeks, being affected by weather and debris, and this delay obstructs the commercialization of OLEV.

9.4.4   The Fourth-Generation (4G) OLEV

As shown in Figure 9.18, the 4G OLEV bus, announced in 2010, has an innovative I-type structure of a power supply rail with a maximum output power of 27 kW for a double pick-up coil having a 20 cm airgap and 24 cm lateral displacement [60, 61]. As shown in Figure 9.19(a), the

Figure 9.18 The 4G OLEV bus at KAIST Munji Campus [73].

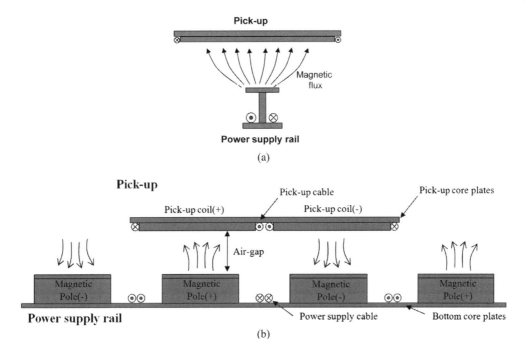

Figure 9.19 The I-type power supply rail and double flat pick-up coil [61]. (a) front view and (b) side view. Reproduced with permission of IEEE.

I-type power supply rail, where the name "I-type" stems from its front shape, has only a 10 cm width, which leads to a deployment cost reduction of 20% compared to the 3G OLEV. Unlike previous generations of OLEV, a very low EMF for pedestrians, as low as 1.5 μT at a distance of 1 m from the center of the power supply rail, can be obtained [61] because the power supply rail has alternating magnetic poles along the road, as shown in Figure 9.19(b). As shown in Figure 9.20, the 4G OLEV adopted a module concept for the I-type power supply rail to reduce the deployment time within a few hours, which had been one of the drawbacks of the 3G OLEV. The I-type power supply module, which includes the power supply rail and capacitor banks inside, as shown in Figure 9.20(a), should be robust to high humidity and external mechanical impacts at least for 10 years in accordance with the system requirements of Figure 9.3.

Figure 9.20 The I-type power supply module of the 4G OLEV [73] (a) a prototype module and (b) deployed modules at the 24 m test site.

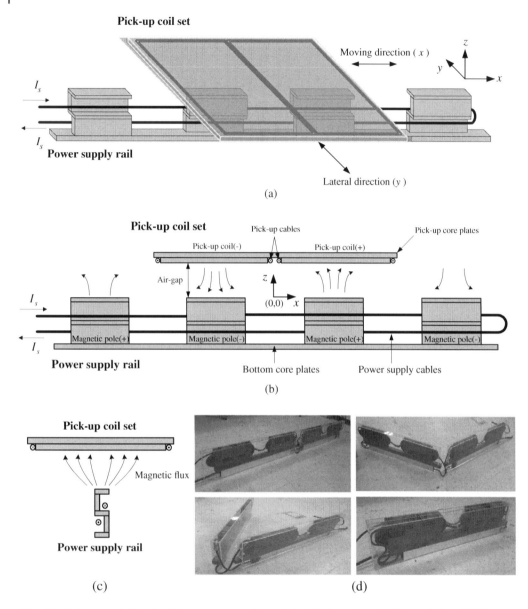

Figure 9.21 Fabricated ultra-slim S-type power supply modules [71]. Reproduced with permission of IEEE.

9.4.5 The Fifth-Generation (5G) OLEV

To further reduce the construction cost and increase the robustness of power supply rails, the 5G OLEV with an ultra-slim S-type core was recently proposed, which has an S-shape when viewed from the front, as shown in Figure 9.21(c) [76]. The S-type power supply module has a width of only 4 cm, which has been decreased from the I-type width of 10 cm; hence, the S-type model leads to less construction cost and deployment time. Moreover, the S-type model makes it easier to fold itself, which means connecting power cables is no longer necessary after being deployed. The impact of deploying the S-type power supply modules on the surface of roads is now minimal, which fortunately does not change existing road operation conditions.

9.5 A Few Technical and Economic Issues of OLEV

A few practical technical issues, such as EMF cancelations and segmentations as well as some economic issues such as component costs and regional economic analyses, are addressed here. Fundamental operating principles on OLEV, including steady-state operations [47, 48, 60, 61] and dynamic characteristics [56, 62, 75, 126, 127] of the IPTS, are not dealt with here due to the page limit.

9.5.1 Generalized Active EMF Cancelation Methods

In the IPTS of OLEV, the total EMF, which is the summation of EMF generated by the power supply rail and the pick-up coil, should be lower than the ICNIRP guideline for the safety of pedestrians. Among EMF cancelation methods, passive methods rely on the use of ferromagnetic materials, conductive materials, and selective surfaces for protecting against radio frequency interference [113–125], which is found to be adequate for OLEV. An active EMF cancelation design was applied to the RPEV of the PATH team [43], where a canceling coil mitigates the EMF without RPEV. Another active EMF cancelation design was made for the pick-up coil of the OLEV [125], where counter current is appropriately controlled. However, the EMF is generated from both the power supply rail and pick-up coil of the IPTS, as shown in Figure 9.22; therefore, they must be canceled together without accurate sensing of EMF and complicated control circuits, which should be avoided for practical use.

A generalized active EMF design principle, which can be extended to any IPTS, has been developed [78], where there are three design methods: independent self-EMF cancelation (ISEC), 3 dB dominant EMF cancelation (3DEC), and leakage-free EMF cancelation (LFEC). By adding an active EMF cancelation coil to each power supply rail and pick-up coil without EMF sensing and control circuits, the EMF generated from each main coil can be independently canceled by their corresponding cancelation coils, as shown in Figure 9.22.

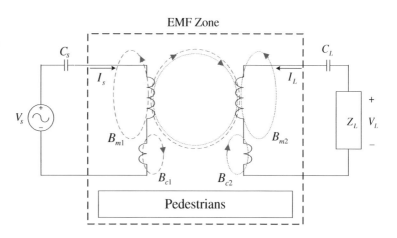

Figure 9.22 Independent self-EMF cancelation (ISEC) method for primary and secondary sides, fetching its cancelation current from each main coil [78]. Reproduced with permission of IEEE.

Because the current of each cancelation coil is fetched from its corresponding main coil, the magnetic flux of a cancelation coil is in the phase of that of a main coil; hence, the EMF $\boldsymbol{B}_t$ becomes independent of both phase and load conditions as follows:

$$\boldsymbol{B}_t = (\boldsymbol{B}_{m1} + \boldsymbol{B}_{c1}) + (\boldsymbol{B}_{m2} + \boldsymbol{B}_{c2}) \cong 0 \quad \because \quad \boldsymbol{B}_{c1} \cong -\boldsymbol{B}_{m1}, \quad \boldsymbol{B}_{c2} \cong -\boldsymbol{B}_{m2} \text{ and}$$
$$\boldsymbol{B}_k \equiv \boldsymbol{B}_{kx}\boldsymbol{x}_0 + \boldsymbol{B}_{ky}\boldsymbol{y}_0 + \boldsymbol{B}_{kz}\boldsymbol{z}_0, k \text{ is } t, m1, m2, c1, c2 \tag{9.3}$$

Together with ISEC, the 3DEC method is quite useful in practice for designing the IPTS because it makes it possible to completely isolate the primary and secondary cancelation designs from each other. For the full resonant IPTS that is very common in practice [60,61], the primary and secondary currents are in quadrature, that is, $B_{1x} \perp B_{2x}, \quad B_{1y} \perp B_{2y}, \quad B_{1z} \perp B_{2z}$. Then the total magnetic field B_t can be constrained to the ICNIRP guidelines in so far as a dominating magnetic field B_1 is less than 3 dB below the guideline as follows [78]:

$$B_t \equiv |\boldsymbol{B}_t| = \sqrt{|\boldsymbol{B}_1|^2 + |\boldsymbol{B}_2|^2} = \sqrt{B_1^2 + B_2^2} \leq \sqrt{B_1^2 + B_1^2} = \sqrt{2}B_1 \leq B_{ref}$$
$$\because B_1 \equiv |\boldsymbol{B}_1|, \ B_2 \equiv |\boldsymbol{B}_2|, \quad B_2 \leq B_1 \tag{9.4}$$

A design example for the 4G I-type IPTS based on the ISEC and 3DEC is shown in Figure 9.23 , where no cancelation coil was used for the power supply rail due to the low EMF generated from itself [60,61].

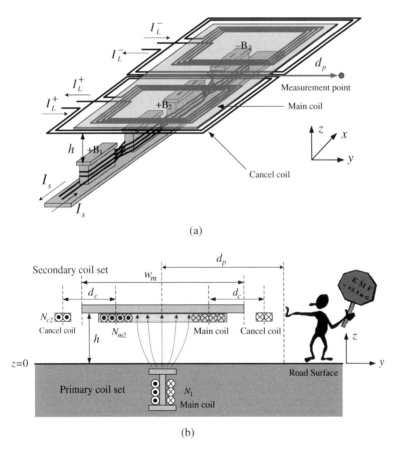

(a)

(b)

Figure 9.23 An active EMF cancelation design example for the I-type IPTS [78]: (a) a bird's eye view and (b) front view. Reproduced with permission of IEEE.

Figure 9.24 A bad design case for cancelation coils, which are strongly coupled with the main magnetic linkage [78].

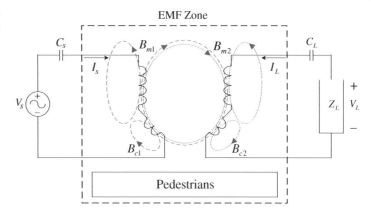

In case the EMF cancelation coils are involved in an unwanted magnetic linkage, as shown in Figure 9.24, the induced load voltage drops; therefore, it is highly recommended to design the cancelation coils based on the LFEC method so that the magnetic linkage does not intersect them. It was verified that the LFEC method, shown in Figure 9.25, can improve the load voltage by as much as 21% [78].

9.5.2 The Cross-Segmented Power Rails (X-Rail)

Unlike conventional EV, the RPEV need their own roadways, where power supply rails should be activated when RPEVs are on the roadway, but they should be deactivated not to waste electric power and generate unwanted EMF for pedestrians; this is why the power supply rail is sometimes segmented into a few subrails. Each subrail can be activated by providing it with high-frequency current through a switch box from an inverter, as shown in Figure 9.26. The first segmented power supply rail developed for OLEV, as shown in Figure 9.26(a), was the centralized switching type composed of a few subrails, a bundle of supply cables, and a centralized switch box, where an inverter is connected to one of several pairs of supply cables through the switch box one at a time. One of the demerits of the centralized one is that too many bundles of power cables are needed. This is why the distributed switching power rail, as shown in Figure 9.26(b), was proposed, which is composed of a few subrails, a pair of common power supply cables, and multiple switch boxes located between two subrails. Thus, the total length of the

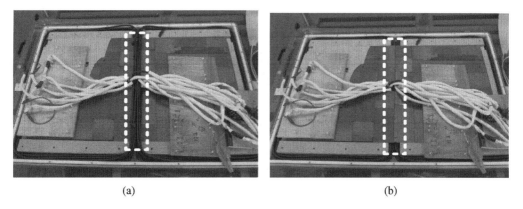

(a)   (b)

Figure 9.25 The application of the LFEC design to a pick-up set [78]: (a) applied and (b) not applied.

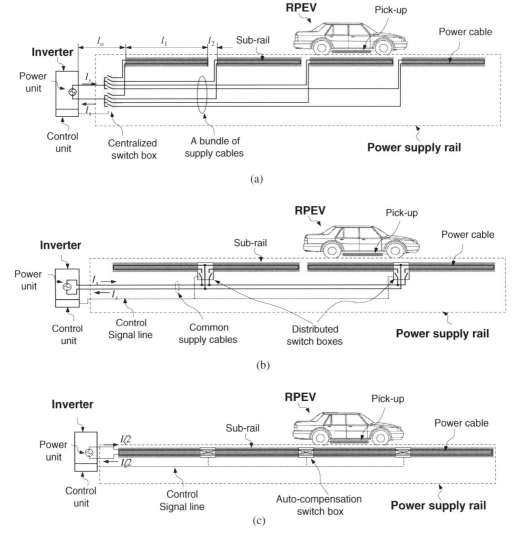

Figure 9.26 Segmentation of power supply rails for RPEV [74]: (a) centralized switching, (b) distributed switching, and (c) X-segmented switching.

cable could be reduced compared to the centralized one, especially for a larger number of cables. The distributed one, however, requires the common power supply cables, which increases the power rail construction cost and conductive power loss, as shown in Figure 9.17(b). A recently proposed cross-segmented power supply rail (X-rail) [74], consisting of segmented subrails, autocompensation switch boxes, control signal lines, and roadway harnesses, as shown in Figure 9.26(c), has the shortest length of power cable among the segmentations developed so far, and the inverter can drive multiple RPEVs, unlike the previous segmentations that activate only one subrail at a time.

As shown in Figure 9.27, the cable cost of the X-rail is nearly halved compared to the distributed switching, and the cable length reduction effect becomes dominant as the number of segmented subrails increases.

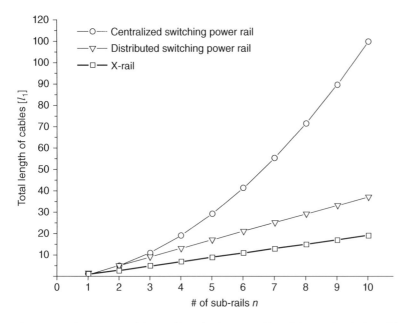

Figure 9.27 Comparison of the total length of power cables of the X-rail with that of the conventional switching rails [74].

A prototype X-rail demonstrated for the I-type subrails that are 2 m long and its magnetic flux density without pick-ups are shown in Figure 9.28. The activated subrail generates enough magnetic flux whereas the EMF is mitigated by the X-rail, which is measured at the center of the power supply rail [74].

9.5.3 Brief Economic Analyses of the OLEV

The component costs of the prototype W-type and I-type roadway subsystems, directly obtained from the real developments [45], are compared with each other, as shown in Table 9.1.

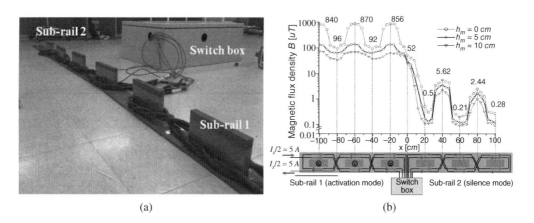

(a)   (b)

Figure 9.28 The experimental set of the X-rail [74]: (a) overall experimental setup and (b) magnetic flux measurement on the rail.

Table 9.1 Component cost comparison of the W-type and I-type roadway subsystems [45]

Module	Components	Unit Price Cost ($/km)	W-type (3G OLEV) Number/ km	W-type (3G OLEV) Sum ($/km)	I-type (4G OLEV) Number/ km	I-type (4G OLEV) Sum ($/km)	Remarks
Inverter (100 kW)	–	55 000	4	220 000	4	220 000	
Cable	Supply cable	54 000	2	108 000	4.4	237 600	250 A
	Common cable	54 000	2	108 000	2	108 000	250 A
Cable protector	FRP pipe	15 000	4	60 000	0	0	
	FRP package	26 000	0	0	1	25 920	
Core	Rib type	250 000	1	250 000	0	0	20 EA
	Square type	96 000	0	0	1	96 000	
Capacitor	Cap. box	5 000	40	200 000	0	0	
	Module type	300	0	0	500	150 000	
Road construction	Concrete	60 000	1	60 000	0	0	
	Asphalt uncover	2 000	1	2 000	0.5	1 000	
	Asphalt cover	11 000	1	11 000	0	0	3.5 m width, 5 cm thickness
Miscellaneous	–	50 000	1	50 000	0.2	10 000	
Total				**1 069 000**		**848 520**	

The total component cost of the developed I-type one is found to be 0.85 million $/km, which is 79% of that of the W-type one. The component costs in mass production, which are of great concern in practical applications, would be 2–5 times less than these component costs at the R&D stage when considering the cost reduction rate in general.

The total cost analysis for the regional area of Seoul, the capital of South Korea, is performed to compare the economic feasibility of the OLEV with other existing vehicles, as shown in Figure 9.29(a). It is assumed that each vehicle runs 20 000 km each year and that the major roads in Seoul, which are about 600 km, are paved with the OLEV-IPTS, which costs about 0.8 million $/km for two-way roads in mass production. The necessary output power of the OLEV is determined as 100 kW, considering not only weight and speed in downtown areas but also air resistance and air conditioner consumption of vehicles. Furthermore, the capacity of the battery of the OLEV and the average energy consumption per distance downtown are assumed to be 20 kWh and 1 kWh/km, respectively. The life and price of the OLEV battery are estimated as 10 years and $440/kWh, respectively. Parameters of other vehicles such as PHEV, BEV, and ICE are conservatively assumed. As identified from Figure 9.29(a), the unit vehicle price and total infracost of the OLEV are estimated to be much lower than any other candidate. The overall costs for deploying the vehicles including infrastructure and operation for 10 years are also analyzed, as shown in Figure 9.29(b). The PEV is the most expensive solution regardless of the number of vehicles, whereas the OLEV is the cheapest solution for any vehicle number. The OLEV is as much as 2–4 times cheaper than any other vehicle, which is because of the relatively low vehicle price and operation cost as well as the economic OLEV-IPTS.

Figure 9.29 Cost analysis for the deployment of vehicles in Seoul [45]: (a) investment cost comparison and (b) total cost versus number of cars.

Costs	BEV	PHEV	ICE	OLEV
One vehicle	50 k$ (45–60 k$)	35 k$ (22–55 k$)	20 k$	20–25 k$
Energy (for 10 years)	3.6 k$	7.8 k$	20 k$	4 k$
Home charger	5 k$	5 k$	0	0–5 k$
Unit vehicle price	**58.6 k$**	**47.8 k$**	**40.0 k$**	**24–34 k$**
Infra (charger)	3500 EA	700 EA	700 EA	600 km
Price/infra	5 M$	5 M$	5 M$	0.8 M$/km
Total infra cost	**17.5 B$**	**3.5 B$**	**3.5 B$**	**0.48 B$**

(a)

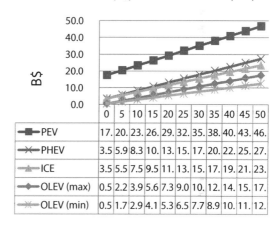

(b)

The cost and benefit of the commercialization of the OLEV system throughout South Korea is summarized [45], as shown in Figure 9.30. The cost includes the costs of power rail construction, R&D investment, maintenance of power rails, inverters, and infrastructures as well as the road usage fees and the cost of emergency charging stations. It assumes that about 30% of the total roads in South Korea are paved with the OLEV-IPTS. Even though the investment cost for the OLEV is huge, the benefit for 30 years is more than 13-fold.

As the benefit–cost (B/C) ratio and the net present value (NPV) are much higher than 1 and 0, respectively, as shown in Figure 9.31, the investment cost for the commercialization of OLEV can be compensated in 2024. As a result, the commercialization of OLEV in South Korea has high economic feasibility because the B/C ratio is 5.8 and the NPV is $79.4 billion USD in 2038 [45].

Question 3 Select another city in the world and calculate the economy of EVs.

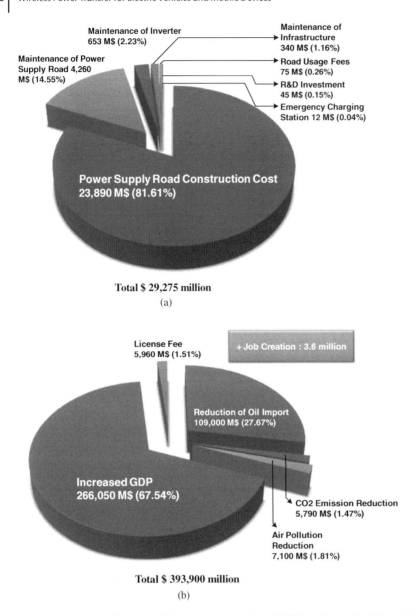

Figure 9.30 Cost and benefit of the commercialization of OLEV for the whole of South Korea for 30 years [45]: (a) investment cost and (b) overall benefit.

9.6 Research Trends of Road powered Electric Vehicles by Other Research Teams

9.6.1 The Auckland University Research Team

Since the 1990s, a research team in Auckland University, New Zealand, known as the Auckland team, has been proposing various IPTSs for wireless charging [79–89]. Among them, the circular coils that adopted ferrite bars instead of ferrite plates need to be addressed because

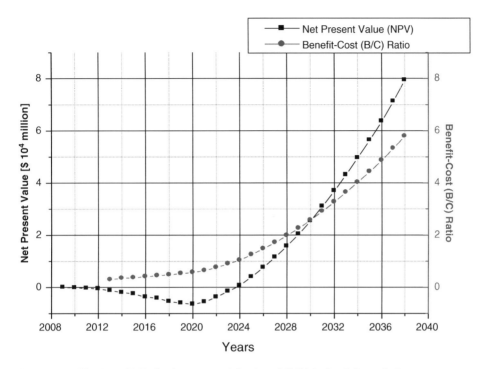

Figure 9.31 B/C ratio and NPV for the commercialization of OLEVs in South Korea [45].

of their compact structure, low weight, and low EMF [79], as shown in Figure 9.32. The circular type coils, however, are inadequate for high-power RPEVs because of their low-power transfer capability and small lateral tolerance. In principle, the lateral tolerance can be increased with a larger diameter of the circular coils, but it is not applicable in practice due to the limited space at the bottom of a car and increased EMF for pedestrians.

In 2010, the double-sided coils that consist of a rectangular core plate and a vertically wounded cable were proposed, as shown in Figure 9.33, which have high lateral tolerance and a high coupling factor [80]. The magnetic resistance between the primary and secondary coils could be greatly reduced by the proposed core structure. As mentioned in the previous section of the 2G OLEV, the double-sided coil has similar problems, such as upward magnetic flux leakage, because its structure is similar to the I-type pick-up, except for the ends.

Figure 9.32 The expanded view of a circular-type coil using ferrite bars by the Auckland team [79].

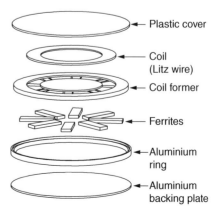

Figure 9.33 The double-sided coils used by the Auckland team [80].

After that, single-sided polarized coils comprising horizontally wound cables on core plates, as shown in Figure 9.34, were introduced [81–84]. A larger coupling coefficient as well as better lateral and longitudinal tolerances compared to the circular coil were obtained. In addition, the modified single-sided polarized coils, where another coil is added between the single-sided coils, as shown in Figure 9.34(c), or coils overlapping each other, as shown in Figure 9.34(d), had improved performances, as shown in Figure 9.35.

In detail, Figure 9.35 shows the uncompensated power transfer when a circular-type coil is used as the primary coil with the various secondary coils, such as a circular-type coil, double-sided coil, and single-sided polarized coil. As shown in Figure 9.35(a), a circular-type coil is used as the secondary side and the maximum power transfer is obtained when they are centered but the power transfer band is very narrow to any misalignment. On the other hand, Figure 9.35(b) uses a single-sided polarized coil as the secondary side, which can only capture horizontal flux at its center. The point here is that intrinsically vertical non-polarized circular topologies cannot transfer power to the single-sided polarized coil (horizontally flux-sensitive topologies) when they are centered on each other, and these need to be offset from each other to capture flux. Moreover, this power transfer is suboptimal (lower and has two narrow bands) but also means that a vehicle must be misaligned from the primary coil to transfer power. Figure 9.35(d) shows an example when the secondary uses a multicoil single-sided magnetic topology (as in Figure 34(c) or (d)) that includes two independent coils that are sensitive to both the horizontal and vertical flux components. This secondary allows high-power transfers to be achieved when centered on the primary coil. Moreover, it shows wide lateral tolerance and better coupling despite misalignments.

The Auckland team proposed an IPTS including many small power pads for RPEV, where the length of a power pad is much shorter than that of a vehicle to avoid unwanted energizing

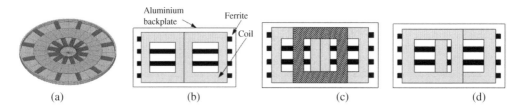

Figure 9.34 Coil structures created by the Auckland team [81] (a) a circular coil, (b) a single-sided polarized coil, (c) a single-sided polarized coil adding a coil, and (d) a single-sided polarized coil overlapping another.

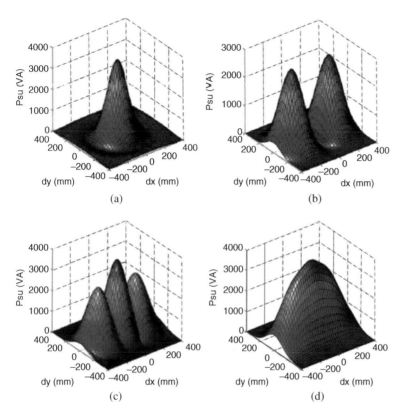

(a)   (b)

(c)   (d)

Figure 9.35 Output power by the Auckland team: (a) a circular coil, (b) a single-sided polarized coil, (c) a single-sided polarized coil adding a coil, and (d) the single-sided polarized coil of Figure 9.34(c) or (d) [83].

and loading [89], as shown in Figure 9.36. This scheme, however, requires numerous considerations such as increased control complexity and deployment in addition to the maintenance costs of the power pads in ground, which should survive from the harsh road environments, as addressed in Figure 9.3.

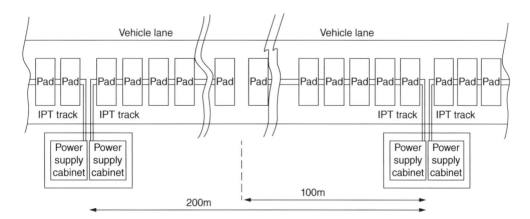

Figure 9.36 The configuration of an IPTS using many small power pads by the Auckland team [89].

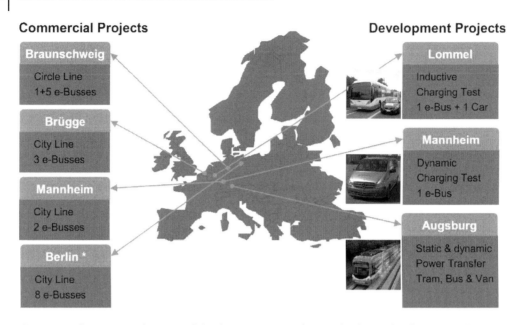

Figure 9.37 The commercialization and development projects for RPEV by the Bombardier team [96].

9.6.2 The Bombardier Research Team

Bombardier has developed several IPTS systems [90–97] for the stationary and dynamic charging of trams and buses since 2010, as shown in Figure 9.37. With the PRIMOVE project applying IPTS technologies to transportation sectors, Bombardier used an operating frequency of 20 kHz and a three-phase power system for their IPTS to obtain a higher power density without exceeding the EMF guidelines. For stationary and dynamic charging, a power transfer of 250 kW for PRIMOV trams has been achieved in Augsburg, Germany, as shown in Figure 9.38. The airgap between the primary and secondary coils is about 6 cm, and the lateral displacement of trams is as low as a few cm; hence, the design of an IPTS is much easier than that used on ordinary roadways.

Figure 9.38 The Bombardier PRIMOVE tram [96].

Figure 9.39 The coils of the golf car platform of RPEV [99]. Reproduced with permission of IEEE.

Considering large-sized vehicles such as trams, it would be wise to use the three-phase power system for an IPTS because higher power can be transferred to pick-up coils than with a single-phase power system. These merits of high-power delivery capability and increased power efficiency may offset the initial investment costs for procuring more complex and expensive HF inverters and power supply coils. It is not clear, in general, that the multiphase power systems for RPEVs [85] are superior to a single-phase power system due to its high cost and complexity. There is not much open access information about the Bombardier team's works compared to that of previous research teams. PRIMOV buses of a maximum power of 200 kW have been deployed and operated in Brunswick and Mannheim, Germany, and in Bruges, Belgium, since 2013 [97].

9.6.3 The ORNL Research Team

The Oak Ridge National Laboratory (ORNL) has been investigating the IPTS for RPEVs since 2011 [98, 99], which use many circular power supply coils and pick-up coils, as shown in Figures 9.39 and 9.40. Thus, the lateral tolerance is inherently small due to the geometrical limitation of the circular coils and because the magnetic coupling coefficient between coils fluctuates

Figure 9.40 The experimental golf car platform of RPEV [99].

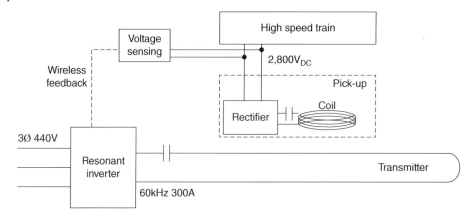

Figure 9.41 Configuration of the IPTS for a high-speed train [129].

along the power supply coils. The ORNL also used an operating frequency of around 20 kHz and achieved a maximum power transfer of 2.2 kW with an efficiency of 74%, which is limited by the 72 V lead–acid battery on the GEM vehicle.

9.6.4 The South Korea Railroad Research Institute Team

Since 2012, the Korea Railroad Research Institute (KRRI) team has been developing IPTSs for a high-speed train and achieved a maximum output power of 820 kW with a system efficiency of 83% at an airgap of 5 cm [129]. For the small pick-up size as well as cost reduction in the IPTS, an operating frequency of 60 kHz, which is three times that used for the 1G, 2G, 3G, 4G, and 5G OLEVs, was adopted while a single-phase power system was used instead of a three-phase power system due to its simple control and low-cost characteristics, as shown in Figure 9.41. To realize the 1 MW level high-frequency inverter, five 200 kW level full-bridge pulse amplitude modulation (PAM) resonant inverters using IGBT modules were connected in parallel for cost effectiveness [129].

Moreover, four pick-up sets, that is, 200 kW for each pick-up set, are connected in parallel, and the output voltage of the pick-up sets is directly controlled by resonant inverter currents through the voltage sensor and wireless feedback systems to rapidly adjust the traction inverter input voltage of 2800 V_{DC}, as shown in Figure 9.42, without any additional regulator or batteries, which would lead to large volume, weight, and cost problems on the pick-up side [129].

As mentioned in the previous section on the 6G OLEV, the large voltage stress on both the compensation capacitor banks and power supply rails becomes a significant problem of the

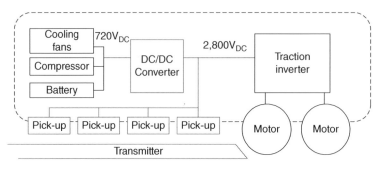

Figure 9.42 Block diagram of the high-speed train with the IPTS [129].

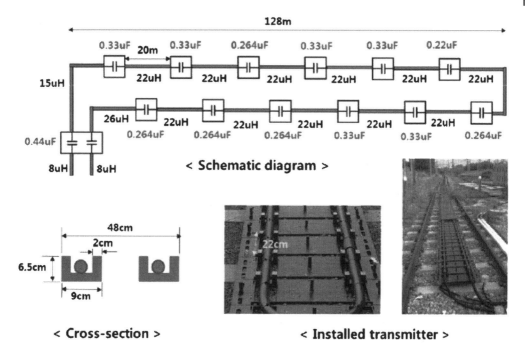

Figure 9.43 Schematic diagram and photos of the transmitter for a high-speed train [129].

high power level IPTS using a high operating frequency of more than 20 kHz. As a remedy for this problem, it is essential to use distributed compensation capacitors, which are connected in series with the transmitter and pick-up coils to effectively reduce the voltage stress between power cables, as shown in Figures 9.43 and 9.44. The structure of the power supply rail and pick-up used for the IPTS is basically the same as the 3G and 3$^+$G OLEVs because these

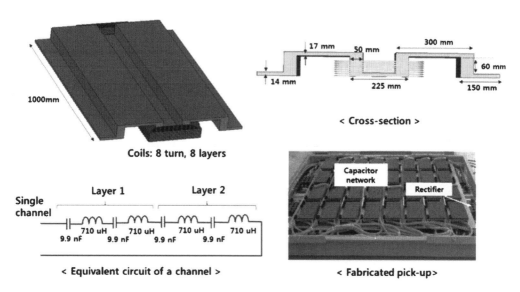

Figure 9.44 Schematic diagram and photos of the pick-up for a high-speed train [129].

Figure 9.45 View of the high-speed train with the IPTS [129].

structures can guarantee several significant advantages such as low construction cost and time, high power transfer capability, continuous power transfer along a rail, and low EMF characteristics. For the system integration test, KRRI made a 128 m-long power supply rail, as shown in Figure 9.45. From this integration test, the high-speed train using the IPTS successfully operated and accelerated to a speed of 10 km/h.

9.6.5 The Endesa Research Team

Since April 2013, Endesa has been involved in the vehicle initiative consortium for transport operation and road inductive application (VICTORIA) project and has adopted the triple charging technologies such as conventional plug-in, stationary, and dynamic charging systems based on an IPTS for RPEVs. Hence, RPEVs can be charged by a plug-in charger when parked at a bus terminal during the night. During the daytime, RPEVs can be partially charged at wireless charging bus stops and bus lanes equipped with power supply rails while moving over them to extend the vehicles' driving ranges [130]. For stationary and dynamic charging, Endesa did not adopt multiple small pads but rather a power supply rail to guarantee several advantages such as its continuous power transfer, low construction cost and time, and simple control characteristics. With a rectangular pick-up, the U-type power supply rail, which had been used for the 2G OLEVs developed by KAIST in 2011 [99], was used for an interoperability between stationary and dynamic charging, as shown in Figure 9.46. For demonstration purposes, as shown in

Figure 9.46 U-type power rail (left) and rectangular pick-up (right) used for the RPEV developed by the VICTORIA project [131]. Reproduced with permission of IEEE.

Figure 9.47 An RPEV developed by the VICTORIA project for triple charging [132].

Figure 9.47, an RPEV with a maximum power transfer of 50 kW has been deployed and operated in the 10 km bus route of the Number 16 bus in Malaga, Spain, since December 2014.

Within the total bus route of 10 km, ten-segmented power supply rails were installed along the route, totaling 300 m in length, where the eight-segmented power supply rails had an interval of 12.5 m for dynamic charging and the interval between other two-segmented power supply rails was 300 m for stationary charging, as shown in Figure 9.48. Moreover, a self-guided control system, which automatically controls the steering wheel to follow the center of a road in the same way as in the 1G OLEV, has been adopted to minimize lateral displacements for its efficient power transfer [132]. Unfortunately, there is not much open-access information about the Endesa team's works compared to that of the other research teams.

9.6.6 The INTIS Research Team

The integrated infrastructure solution (INTIS) has been investigating the IPTS for both SCEVs and RPEVs to provide engineering services and consulting from developments to field tests since 2011 [133, 134]. Now INTIS has its own 25 m-long power supply rail in a test center in Lathen, Germany, and the test center can be used to evaluate IPTSs for SCEVs and RPEVs from the component level to the completed system level.

Based on a single-phase power system, the test power supply rail of the IPTS for SCEVs and RPEVs is available for tests up to a maximum output power of 200 kW at an operating frequency of up to 35 kHz while the double U-type power supply rail, which has two U-type power rails

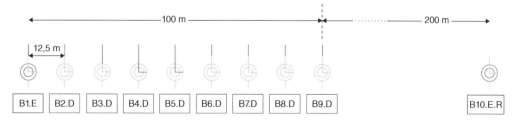

Figure 9.48 Scheme of ten-segmented power rails in Malaga, Spain [132].

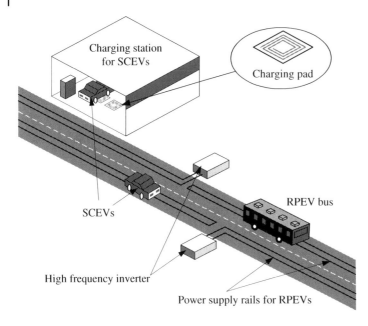

Figure 9.49 Ideal concepts for SCEVs compatible with power supply rails for RPEVs [128].

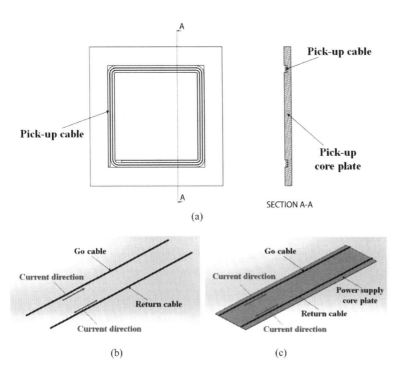

Figure 9.50 Conceptual scheme of the proposed coreless power supply rail for both RPEVs and SCEVs [128]: (a) a rectangular pick-up coil for SCEVs in accordance with SAE J2954, (b) proposed coreless power supply rail, and (c) conventional power rail used for the 3G and 3G+ OLEVs. Reproduced with permission of IEEE.

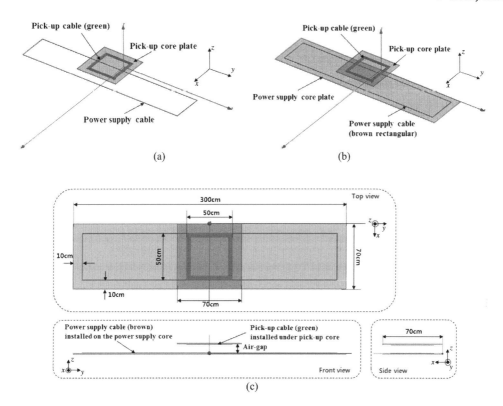

Figure 9.51 Maxwell simulation models for the proposed coreless power supply rail with a rectangular pick-up [128]: (a) bird's eye view of the proposed coreless power supply rail, (b) birds eye view of the conventional with-core power supply rail, and (c) dimensions for the simulation models. Reproduced with permission of IEEE.

in parallel, is adopted at the test center. So far, INTIS has developed two IPTSs for SCEVs and RPEVs [135], which used an operating frequency of 30 kHz for the maximum output power of 30 kW at an airgap of 10 cm.

9.7   Interoperable IPT: The Sixth-Generation (6G) OLEV

Although RPEVs are free from the battery problems and are ready for commercialization, RPEVs have not been widely used so far due to the huge initial investment cost. To cope with this problem, we may need strong motivation to build the national infrastructure for RPEVs with public consent. Therefore, one of the best deployment scenarios would be for SCEVs to be designed to be completely compatible with IPTSs for RPEVs to be wirelessly charged while moving on a road, as shown in Figure 9.49, because there is no doubt that SCEVs will be widely deployed all over the world in the near future to replace wired EV chargers due to SCEVs' convenience and safe operation. However, the design considerations of an IPTS for RPEVs are quite different from those of SCEVs because the IPTS for RPEVs would need to meet additional system requirements such as low construction cost and time, low voltage stress, large lateral tolerance, high power delivery capability, and continuous power delivery while moving on the road.

In order to manage the interoperability issue between RPEVs and SCEVs to the satisfaction of the design goals for RPEVs, the 6G OLEVs, using a new coreless power supply rail, were recently proposed [117]. As shown in Figure 9.50, the shape of the proposed coreless power

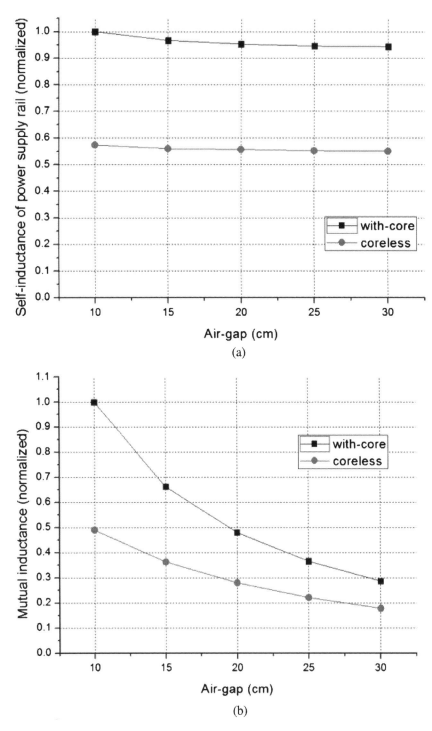

Figure 9.52 Self-inductance of a power rail with/without core plates along to an airgap [128]: (a) self-inductance and (b) mutual inductance variations. Reproduced with permission of IEEE.

supply rail is basically the same as the U- and W-type power supply rails, which were used for the 3G and 3G+ OLEVs, but there is no core plate. Therefore, the proposed coreless power supply rail can generate a uniform magnetic field along a road, and a rectangular pick-up coil determined by the SAE J2954 for SCEVs, as shown in Figure 9.50(a), is completely compatible with the power supply rail for RPEVs and continuously obtain uniform output power when moving on the road. Moreover, the construction cost and time of the proposed coreless power supply rails can be further reduced compared to conventional with-core power supply rails because core plates are totally eliminated in the proposed coreless power supply rails.

Meanwhile, the large voltage stresses on both the compensation capacitor bank and distributed power supply rail V_{l1} are a unique feature of RPEVs, which is one of the main reasons that the operating frequency is limited by around 20 kHz [99] because the voltage stress is directly proportional to its operating frequency f_s as well as its power supply current I_s as in (9.2).

In accordance with the magnetic mirror model [66], however, it is well known that the self-inductance of a power supply rail with core plates can become two times that without core plates when core plates are infinitely long and its relative permeability is infinite; hence, it is expected that the proposed coreless power supply rail will have about half of its previous voltage stress due to its reduced inductance so that the operating frequency can be increased from 20 kHz to 85 kHz to meet the SAE J2954 standard for SCEVs, with only about two times the voltage stress compared to that of 20 kHz for a given rail current. Moreover, it is also expected that the proposed coreless power supply rail can guarantee a large lateral tolerance compared to the conventional with-core power supply rails because the self-inductance variation of the pick-up coil along lateral displacements is small enough to be negligible when a coreless power supply rail is used.

In order to validate those unique characteristics of the proposed coreless power supply rail, two simulation models, which are with-core and coreless power supply rails with a rectangular pick-up for SCEVs, are proposed, as shown in Figure 9.51.

From the simulation results, it is found that the self-inductance of the proposed coreless rail is about half that of the conventional with-core power supply rail, as shown in Figure 9.52(a). At the same time, the mutual inductance of the proposed coreless power rail also becomes half that of the conventional with-core power rail, as shown in Figure 9.52(b).

Moreover, the self-inductance variation of a pick-up coil with the proposed coreless power supply rail is negligible along lateral displacements, as shown in Figure 9.53, which means that the proposed coreless power supply rail can guarantee a larger lateral tolerance of RPEVs and SCEVs compared to the conventional with-core power supply rail [128].

9.8 Conclusion

The development history of IPTS for RPEV from its advent to its current status as a state-of-the-art technology has been introduced throughout this chapter. The size, weight, efficiency, airgap, lateral tolerance, EMF, and cost of the IPTS have been substantially improved during a century, so RPEV are becoming viable solutions for future transportation. The first commercialized OLEV is an especially strong candidate for the near-future widespread use of RPEVs in public transportation. IPTSs that are more economic, compact, efficient, robust, and easy to deploy and maintain will be welcome for future commercialization.

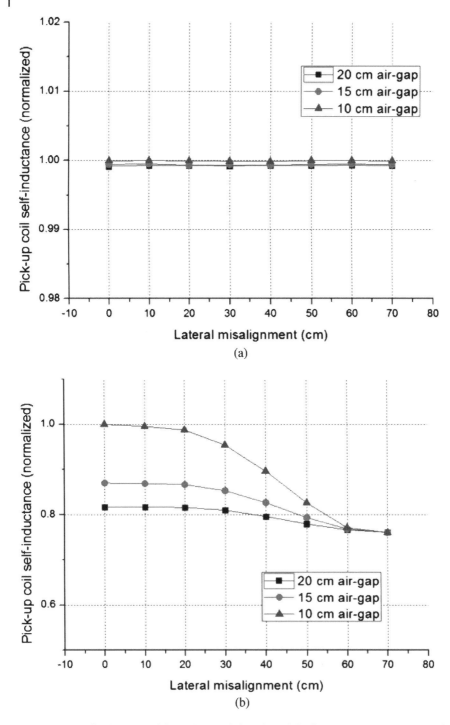

Figure 9.53 Self-inductance of the pick-up coil along lateral displacements [128]: (a) proposed coreless rail and (b) conventional with-core rail.

References

1 J.T. Salihi, P.D. Agarwal, and G.J. Spix, "Induction motor control scheme for battery-powered electric car (GM-Electrovair I)," *IEEE Trans. on Industry and General Applications*, vol. IGA-3, no. 5, pp. 463–469, September 1967.

2 J.R. Bish, and G.P. Tietmeyer, "Electric vehicle field test experience," *IEEE Trans. on Vehicular Technology*, vol. 32, no. 1, pp. 81–89, February 1983"

3 J. Dixon, I. Nakashima, E.F. Arcos, and M. Ortuzar, "Electric vehicle using a combination of ultra capacitors and ZEBRA battery," *IEEE Trans. on Industrial Electronics*, vol. 57, no. 3, pp. 943–949, March 2010.

4 C.H. Kim, M.Y. Kim, and G.W. Moon, "A modularized charge equalizer using a battery monitoring IC for series-connected Li-ion battery string in electric vehicles," *IEEE Trans. Power Electron.*, vol. 28, no. 8, pp. 3779–3787, November 2012.

5 J. Malan and M.J. Kamper, "Performance of hybrid electric vehicle using reluctance synchronous machine technology," in *IEEE Industrial Applications Conference*, 2000, pp. 1881–1887.

6 N.A. Rahim, H.W. Ping, and M. Tadjuddin, "Design of axial flux permanent magnet brushless DC motor for direct drive of electric vehicle," in *IEEE Power Electronics Society General Meeting*, 2007, pp. 1–6.

7 M. Ceraolo, A. Donato, and G. Franceschi, "A general approach to energy optimization of hybrid electric vehicles," *IEEE Trans. on Vehicular Technology*, vol. 57, no. 3, pp. 1433–1441, May 2008.

8 Y. Cheng, R. Trigui, C. Espanet, A. Bouscayrol, and S. Cui, "Specifications and designs of a PM electric variable transmission for Toyota Prius II," *IEEE Trans on Vehicular Technology*, vol. 60, no. 9, pp. 4106–4114, November 2011.

9 F.L. Mapelli, D. Tarsitano, and M. Mauri, "Plug-in hybrid electric vehicle: modeling, prototype, realization, and inverter losses reduction analysis," *IEEE Industrial Electronics*, vol. 57, no. 2, pp. 598–607, February 2010.

10 E. Tara, S. Shahidinejad, and E. Bibeau, "Battery storage sizing in a retrofitted plug in hybrid electric vehicle," *IEEE Trans. on Vehicular Technology*, vol. 59, no. 6, pp. 2786–2794, July 2010.

11 S.G. Li, S.M. Sharkh, F.C. Walsh, and C.N. Zhang, "Energy and battery management of a plug-in series hybrid electric vehicle using fuzzy logic," *IEEE Trans. on Vehicular Technology*, vol. 60, no. 8, pp. 3571–3585, October 2011.

12 P. Lombardi, M. Heuer, and Z. Styczynski, "Battery switch station as storage system in an autonomous power system: optimization issue," *IEEE Power and Energy Society General Meeting*, 2010, pp. 1–6.

13 M. Takagi. Y. Iwafune, K. Yamaji, H. Yamato, K. Okano, R. Hiwatari, and T. Ikeya, "Economic value of PV energy storage using batteries of battery-switch stations," *IEEE Trans. on Sustainable Energy*, vol. 4, no. 1, pp. 164–173, January 2013"

14 J.J. Jamian, M.W. Mustafa, Z. Muda, and M.M. Aman, "Effect of load models on battery-switching station allocation in distribution network," in *IEEE International Conference on Power and Energy*, 2012, pp. 189–193.

15 M. Takagi, Y. Iwafune, H. Yamamoto, K. Yamaji, K. Okano, R. Hiwatari, and T. Ikeya, "Energy storage of PV using batteries of battery-switch stations," in *IEEE International Symposium on Industrial Electronics (ISIE)*, 2010, pp. 3413–3419.

16 J.H. Gauss, "Can the trolley coach compete economically with the gas or diesel bus where no overhead facilities exist?," *Trans. of American Institute of Electrical Engineers*, vol. 66, no. 1, pp. 264–268, January 1947.

17 A.B. Mcmillon, "Trolley coaches replace buses," *Trans. of American Institute of Electrical Engineers*, vol. 68, no. 1, pp. 403–406, July 1949.

18 J.P. Senior, "The trolley bus," *Student's Quarterly Journal*, vol. 20, no. 77, pp. 11–16, September 1949.

19 S. Midya, D. Bormann, T. Schutte, and R. Thottappillil, "Pantograph arcing in electrified railways – mechanism and influence of various parameters – Part I: with DC traction power supply?," *IEEE Trans. on Power Delivery*, vol. 24, no. 4, pp. 1931–1939, October 2009.

20 T. Uzuka, "Faster than a speeding bullet: an overview of Japanese high-speed rail technology and electrification," *IEEE Electrification Magazine*, vol. 1, no. 1, pp. 11–20, September 2013.

21 M. Hutin and M. Leblanc, "Transformer system for electric railways," Patent US 527,857, 1894.

22 J.G. Bolger, "Supplying power to vehicles," Patent US 3,914,562, 1975.

23 J.G. Bolger, "Roadway for supplying power to vehicle and method of using the same," Patent US 4,007,817, 1977.

24 J.G. Bolger and F.A. Kirsten, "Investigation of the feasibility of a dual mode electric transportation system," Lawrence Berkeley National Laboratory Report, LBL6301, May 1977.

25 J.G. Bolger, M.I. Green, L.S. Ng, and R.I. Wallace, "Test of the performance and characteristics of a prototype inductive power coupling for electric highway systems," Lawrence Berkeley National Laboratory Report, LBL7522, p. 1, July 1978.

26 J.G. Bolger, F.A. Kirsten, and L.S. Ng, "Inductive power coupling for an electric highway system," in *Proc. IEEE 28th Vehicular Technology Conference*, 1978, pp. 137–144.

27 J.G. Bolger, L.S. Ng, D.B. Turner, and R.I. Wallace, "Testing a prototype inductive power coupling for an electric vehicle highway system," in *Proc. IEEE 29th Vehicular Technology Conference*, 1979, pp. 48–56.

28 C.E. Zell and J.G. Bolger, "Development of an engineering prototype of a road powered electric transit vehicle system," in *Proc. 32nd IEEE Vehicular Technology Conference*, 1982, pp. 435–438.

29 J.G. Bolger, "Power control system for electrically driven vehicle," Patent US 4,331,225, 1982.

30 Santa Barbara Electric Bus Project, Phase 3A-final report, Santa Barbara Research Paper, September 1983.

31 Santa Barbara Electric Bus Project, Prototype development and testing program phase 3B-final report, Santa Barbara Research Paper, Sept. 1984.

32 Santa Barbara Electric Bus Project, Test facilities development and testing program-static test report, Santa Barbara Research Paper, June 1985.

33 Santa Barbara Electric Bus Project, Prototype development and testing program phase 3C-final report, Santa Barbara Research Paper, May 1986.

34 K. Lashkari, S.E. Schladover, and E.H. Lechner, "Inductive power transfer to an electric vehicle," in *Proc. of 8th International Electric Vehicle Symposium*, 1986.

35 E.H. Lechner and S.E. Schladover, "The road powered electric vehicle – an all-electric hybrid system," in *Proc. of 8th International Electric Vehicle Symposium*, 1986.

36 J.G. Bolger and L.S. Ng, "Inductive power coupling with constant voltage output," Patent US 4,253,345, 1988.

37 S.E. Schladover, "Systems engineering of the road powered electric vehicle technology," in *Proc. of 9th International Electric Vehicle Symposium*, 1988.

38 J.G. Bolger, "Roadway power and control system for inductively coupled transportation system," Patent US 4,836,344, 1989.

39 M. Eghtesadi, "Inductive power transfer to an electric vehicle-an analytical model," in *Proc. 40th IEEE Vehicular Technology Conference*, 1990, pp. 100–104.

40 K.W. Klontz, D.M. Divan, D.W. Novotny, and R.D. Lorenz, "Contactless battery charging system," Patent US 5,157,319, 1992.

41 K.W. Klontz, D.M. Divan, D.W. Novotny, and R.D. Lorenz, "Contactless coaxial winding transformer power transfer system," Patent US 5,341,280, 1994.

42 J.G. Bolger, "Urban electric transportation systems: the role of magnetic power transfer," in *IEEE WESCON94 Conference*, 1994, pp. 41–45.

43 California PATH Program, "Road powered electric vehicle project track construction and testing program phase 3D," *California PATH Research Paper*, pp. 3–5, March 1994.

44 K.W. Klontz, D.M. Divan, D.W. Novotny, and R.D. Lorenz, "Contactless power delivery system for mining applications," *IEEE Trans. on Industry Applications*, vol. 31, no. 1, pp. 27–35, January 1995.

45 KAIST OLEV team, "Feasibility studies of On-Line Electric Vehicle (OLEV) Project," KAIST Internal Report, August 2009.

46 N.P. Suh, D.H. Cho, and Chun T. Rim, "Design of on-line electric vehicle (OLEV)," *Plenary lecture at the 2010 CIRP Design Conference*, 2010, pp. 3–8.

47 S.W. Lee, J. Huh, C.B. Park, N.S. Choi, G.H. Cho, and Chun T. Rim, "On-line electric vehicle (OLEV) using inductive power transfer system," in *IEEE Energy Conversion Congress and Exposition (ECCE)*, 2010, pp. 1598–1601.

48 J. Huh, S.W. Lee, C.B. Park, G.H. Cho, and Chun T. Rim, "High performance inductive power transfer system with narrow rail width for on-line electric vehicles," in *IEEE Energy Conversion Congress and Exposition (ECCE)*, 2010, pp. 647–651.

49 Chun T. Rim, "The difficult technologies in wireless power transfer," *Trans. of the Korean Institute of Power Electronics (KIPE)*, vol. 15, no. 6, pp. 32–39, December 2010.

50 S.W. Lee, C.B. Park, J.G. Cho, G.H. Cho, and Chun T. Rim, "Ultra slim U and W power supply and pick-up coil design for OLEV," in *Korean Institute of Power Electronics (KIPE) Annual Summer Conference*, 2010, pp. 353–354.

51 G.H. Jung, K.H. Lee, H.G. Kim, Y J. Cho, B.Y. Song, Y.D. Son, E.H. Park, J.Y. Park, J.Y. Choi, B.O. Kong, J. Huh, H.S. Son, J.G. Cho, Chun T. Rim, and S.J. Jeon, "Non-touch inductive power transfer system for OLEV," in *Korean Institute of Electrical Engineering (KIEE) Annual Summer Conference*, 2010, pp. 1054–1055.

52 C.B. Park, S.W. Lee, and Chun T. Lim, "Dynamic phasor transformation using complex Laplace transformation," in *Korean Institute of Power Electronics (KIPE) Annual Fall Conference*, 2010, pp. 46–47.

53 G.H. Jung, K.H. Lee, H.G. Kim, Y J. Cho, B.Y. Song, Y D. Son, E.H. Park, J.Y. Park, J.Y. Choi, B.O. Kong, J. Huh, H.S. Son, J.G. Cho, Chun T. Rim, and S.J. Jeon, "Power supply and pick-up system for OLEV," in *Korean Institute of Power Electronics (KIPE) Annual Summer Conference*, 2010, pp. 218–219.

54 J. Huh, W.Y. Lee, J.G. Cho, G.H. Cho, and Chun T. Rim, "A study on current source-transformer resonance inductive power transfer system," in *Korean Institute of Power Electronics Annual Summer Conference*, 2010, pp. 355–356.

55 Chun T. Rim, "Electric vehicle system," Patent WO 2010 076976, 2010.

56 Chun T. Rim, "Unified general phasor transformation for AC converters," *IEEE Trans. on Power Electron.*, vol. 26, no. 9, pp. 2465–2745, September 2011.

57 S.W. Lee, C.B. Park, J.G. Cho, G.H. Cho, and Chun T. Rim, "Ultra slim supply and pick-up coils for on-line electric vehicles (OLEV)," *Trans. of the Korean Institute of Power Electronics (KIPE)*, vol. 16, no. 3, pp. 274–282, August 2011.

58 J. Huh and Chun T. Rim, "KAIST wireless electric vehicles – OLEV," in *JSAE Annual Congress*, 2011.

59 S.W. Lee, W.Y. Lee, J. Huh, H.J. Kim, C.B. Park, G.H. Cho, and Chun T. Rim, "Active EMF cancellation method for I-type pick-up of on-line electric vehicles (OLEV)," in *IEEE Applied Power Electronics Conference and Exposition (APEC)*, 2011, pp. 1980–1983.

60 J. Huh, W.Y. Lee, G.H. Cho, B.H. Lee, and Chun T. Rim, "Characterization of novel inductive power transfer systems for on-line electric vehicles (OLEV)," in *IEEE Applied Power Electronics Conference and Exposition (APEC)*, 2011, pp. 1975–1979.

61 J. Huh, S.W. Lee, W.Y. Lee, G.H. Cho, and Chun T. Rim, "Narrow-width inductive power transfer system for on-line electrical vehicles (OLEV)," *IEEE Trans. on Power Electron.*, vol. 26, no. 12, pp. 3666–3679, December 2011.

62 S.W. Lee, C.B. Park, and Chun T. Rim, "An analysis of DQ inverter for wireless power transfer by complex Laplace-phasor transformation," in *Korean Institute of Power Electronics Annual Summer Conference*, 2011, pp. 192–193.

63 N.P. Suh, D.H. Cho, G.H. Cho, J.G. Cho, Chun T. Rim, and S.H. Jang, "Ultra slim power supply and collector device for electric vehicle," Patent KR 1010406620000, 2011.

64 S.Z. Jeon, D.H. Cho, Chun T. Rim, and G.H. Jeong, "Load-segmentation-based full bridge inverter and method for controlling same," Patent WO 2011 078424, 2011.

65 S.Z. Jeon, D.H. Cho, Chun T. Rim, and G.H. Jeong, "Load-segmentation-based 3-level inverter and method for controlling the same," Patent WO 2011 078425, 2011.

66 N.P. Suh, D.H. Cho, J.G. Cho, Chun T. Rim, J. Huh, J.H. Kim, C.S. Choi, K.H. Lee, B.Y. Song, Y.J. Cho, and C.H. Rim, "Monorail type power supply device for electric vehicle including EMF cancellation apparatus," Patent WO 2011 046374, 2011.

67 N.P. Suh, S.H. Jang, D.H. Cho, G.H. Cho, Chun T. Rim, S.W. Lee, C.B. Park, J. Huh, and H.J. Kim, "Space division multiplexed power supply and collector device," Patent WO 2011 152678, 2011.

68 N.P. Suh, S.H. Jang, D.H. Cho, G.H. Cho, Chun T. Rim, J. Huh, B.H. Lee, and Y.H. Kim, "Cross-type segment power supply," Patent WO 2011 152677, 2011.

69 J. Huh, and Chun T. Rim, "A new coil set with core for magnetic resonant systems," in *Korean Institute of Power Electronics Annual Summer Conference*, 2012, pp. 625–626.

70 N.P. Suh, S.H. Jang, D.H. Cho, G.H. Cho, Chun T. Rim, S.W. Lee, and C.B. Park, "EMI cancellation device in power supply and collector device for magnetic induction power transmission," Patent WO 2011 149263, 2012.

71 W.Y. Lee, J. Huh, S.Y. Choi, X.V. Thai, J.H. Kim, E.A. Al-Ammar, M.A. El-Kady, and Chun T. Rim, "Finite-width magnetic mirror models of mono and dual coils for wireless electric vehicles," *IEEE Trans. on Power Electronics*, vol. 28, no. 3, pp. 1413–1428, March 2013.

72 Chun T. Rim, "Trend of road powered electric vehicle technology," *Magazine of the Korean Institute of Power Electronics (KIPE)*, vol. 18, no. 4, pp. 45–51, August 2013.

73 Chun T. Rim, "The development and deployment of on-line electric vehicles (OLEV)," *IEEE Energy Conversion Congress and Exposition (ECCE)*, 2013.

74 S.Y. Choi, J. Huh, W.Y. Lee, S.W. Lee, and Chun T. Rim, "New cross-segmented power supply rails for road powered electric vehicles," *IEEE Trans. on Power Electron*, vol. 28, no. 12, pp. 5832–5841, December 2013.

75 S. Lee, B. Choi, and Chun T. Rim, "Dynamics characterization of the inductive power transfer system for on-line electric vehicles (OLEV) by Laplace phasor transform," *IEEE Trans. on Power Electron.*, vol. 28, no. 12, pp. 5902–5909, December 2013.

76 S.Y. Choi, B.W. Gu, S.Y. Jeong, and Chun T. Rim, "Ultra-slim S-type inductive power transfer system for road powered electric vehicles," in *International Electric Vehicle Technology Conference and Automotive Power Electronics in Japan (EVTeC and APE Japan)*, accepted for publication.

77 S.Y. Choi, J. Huh, W.Y. Lee, J.G. Cho, and Chun T. Rim, "Asymmetric coil sets for wireless stationary EV chargers with large lateral tolerance by dominant field analysis," *IEEE Trans. on Power Electron.*, accepted for publication.

78 S.Y. Choi, B.W. Gu, S.W. Lee, W.Y. Lee, J. Huh, and Chun T. Rim, "Generalized active EMF cancel methods for wireless electric vehicles," *IEEE Trans. on Power Electron.*, accepted for publication.

79 M. Budhia, G.A. Covic, and J.T. Boys, "Design and optimization of magnetic structures for lumped inductive power transfer systems," *IEEE Trans. on Power Electron.*, vol. 26, no. 11, pp. 3096–3108, November 2011.

80 M. Budhia, G.A. Covic, and J.T. Boys, "A new magnetic coupler for inductive power transfer electric vehicle charging systems," in *Proc. 36th Annual Conf. IEEE Ind. Electron.*, November 2010, pp. 2487–2492.

81 M. Budhia, J.T. Boys, G.A. Covic, and C.-Y. Huang, "Development of a single-sided flux magnetic coupler for electric vehicle IPT charging systems," *IEEE Trans. on Ind. Electron.*, vol. 60, no. 1, pp. 318–328, January 2013.

82 G.A. Covic, L.G. Kissin, D. Kacprzak, N. Clausen, and H. Hao, "A bipolar primary pad topology for EV stationary charging and highway power by inductive coupling," in *IEEE Energy Conversion Congress and Exposition (ECCE)*, September 2011, pp. 1832–1838.

83 M. Budhia, G.A. Covic, J.T. Boys, and C.-Y. Huang, "Development and evaluation of single sided flux couplers for contactless electric vehicle charging," in *IEEE Energy Conversion Congress and Exposition (ECCE)*, September 2011, pp. 614–621.

84 A. Zaheer, D. Kacprzak, and G.A. Covic, "A bipolar receiver pad in a lumped IPT system for electric vehicle charging applications," in *IEEE Energy Conversion Congress and Exposition (ECCE)*, September 2012, pp. 283–290.

85 G.A. Covic, J.T. Boys, M. Kissin, and H. Lu, "A three-phase inductive power transfer system for roadway power vehicles," *IEEE Trans. Ind. Electron.*, vol. 54, no. 6, pp. 3370–3378, December 2007.

86 C.S. Wang, O.H. Stielau, and G.A. Covic, "Design considerations for a contactless electric vehicle battery charger," *IEEE Trans. Ind. Electron.*, vol. 52, no. 5, pp. 1308–1314, October 2005.

87 J.T. Boys, G.A. Covic, and A.W. Green, "Stability and control of inductively coupled power transfer systems," *IEE Proc. EPA*, vol. 147, pp. 37–43, August 2002.

88 C.S. Wang, G.A. Covic, and O.H. Stielau, "Power transfer capability and bifurcation phenomena of loosely coupled inductive power transfer systems," *IEEE Trans., Industrial Electronics*, vol. 51, no. 1, pp. 148–157, 2004.

89 G.A. Covic and J.T. Boys, "Modern trends in inductive power transfer for transportation applications," *IEEE Journal of Emerging and Selected Topics in Power Electronics*, vol. 1, no. 1, pp. 28–41, March 2013.

90 J. Meins and S. Carsten, "Transferring energy to a vehicle," Patent WO 2010 000494, 2010.

91 J. Meins and K. Vollenwyder, "System and method for transferring electrical energy to a vehicle," Patent WO 2010 000495, 2010.

92 K. Vollenwyder, J. Meins, and C. Struve, "Inductively receiving electric energy for a vehicle," Patent US 0055751, 2012.

93 M. Zengerle, "Transferring electric energy to a vehicle using a system which comprises consecutive segments for energy transfer," Patent US 0217112, 2012.

94 K. Vollenwyder and J. Meins, "Producing electromagnetic fields for transferring electric energy to a vehicle," Patent US 8,544,622, 2013.

95 R. Czainski, J. Meins, and J. Whaley, "Transferring electric energy to a vehicle by induction," Patent US 0248311, 2013.

96 J. Meins, "German activities on contactless inductive power transfer," in *IEEE Energy Conversion Congress and Exposition (ECCE)*, 2013.

97 Bombardier website, http://insideevs.com/brunswick-gets-first-of-five-electric-buses-with-wireless-charging/, https://sustainablerace.com/bombardier-begins-operation-first-inductive-high-power-charging-station-primove-electric-buses/.

98 O.C. Onar, J.M. Miller, S.L. Campbell, C. Coomer, C.P. White, and L.E. Seiber, "A novel wireless power transfer for in-motion EV/PHEV charging," in *IEEE Applied Power Electronics Conference and Exposition (APEC)*, 2013, pp. 3073–3080.

99 J.M. Miller, O.C. Onar, and P.T. Jones, "ORNL developments in stationary and dynamic wireless charging," in *IEEE Energy Conversion Congress and Exposition (ECCE)*, 2013.

100 E. Fox and A. Albright, "Vehicle propelled by electricity," Patent US 281,859, 1883.

101 N.H. Kutkut, D.M. Divan, D.W. Novotny, and R.H. Marion, "Design consideration and topology selection for a 120-kW IGBT converter for EV fast charging," *IEEE Trans. on Power Electronics*, vol. 13, no. 1, pp. 238–244, January 1998.

102 C. Praisuwanna and S. Khomfoi, "A quick charger station for EVs using pulse frequency technique," in *IEEE Energy Conversion Congress and Exposition (ECCE)*, 2013, pp. 3595–3599.

103 J.D. Marus and V.L. Newhouse, "Method for charging a plug-in electric vehicle," Patent US 0234664, 2013.

104 B. Kang and G. Ceder, "Battery materials for ultrafast charging and discharging," *Nature*, vol. 458, pp. 190–193, March 2009.

105 J. Huh, W.Y. Lee, S.Y. Choi, G.H. Cho, and Chun T. Rim, "Explicit static circuit model of coupled magnetic resonance system," in *IEEE Energy Conversion Congress and Exposition (ECCE) – Asia*, May 2011, pp. 2233–2240.

106 J. Huh, W.Y. Lee, S.Y. Choi, G.H. Cho, and Chun T. Rim, "Frequency-domain circuit model and analysis of coupled magnetic resonance systems," *Journal of Power Electronics*, vol. 13, no. 2, pp. 275–286, March 2013.

107 E.S. Lee, J. Huh, X.V. Thai, S.Y. Choi, and Chun T. Rim, "Impedance transformers for compact and robust coupled magnetic resonance systems," in *IEEE Energy Conversion Congress and Exposition (ECCE)*, September 2013, pp. 2239–2244.

108 C.B. Park, S.W. Lee, and Chun T. Rim, "5m-off-long-distance inductive power transfer system using optimum shaped dipole coils," in *IEEE Energy Conversion Congress and Exposition (ECCE) – Asia*, June 2012, pp. 1137–1142

109 C.B. Park, S.W. Lee, and Chun T. Rim, "Innovative 5m-off-long-distance inductive power transfer system with optimum shaped dipole coils," *IEEE Trans. Power Electron.*, accepted for publication.

110 M. Hanazawa, N. Sakai, and T. Ohira, *"SUPRA: supply underground power to running automobiles,"* in *IEEE International Electric Vehicle Conference*, IEVC2012, Greenville, March 2012.

111 Guidelines for limiting exposure to time-varying electric and magnetic fields (up to 300 GHz), ICNIRP Guidelines, 1998.

112 Guidelines for limiting exposure to time-varying electric and magnetic fields (up to 100 kHz), ICNIRP Guidelines, 2010.

113 P.R. Bannister, "New theoretical expressions for predicting shielding effectiveness for the plane shield case," *IEEE Transactions on Electromagnetic Compatibility*, vol. EMC-10, no. 1, pp. 2–7, March 1968.

114 P. Moreno and R.G. Olsen, "A simple theory for optimizing finite width ELF magnetic field shields for minimum dependence on source orientation," *IEEE Transactions on Electromagnetic Compatibility*, vol. 39, no. 4, pp. 340–348, November 1997.

115 Y. Du, T.C. Cheng, and A.S. Farag, "Principles of power-frequency magnetic field shielding with flat sheets in a source of long conductors," *IEEE Transactions on Electromagnetic Compatibility*, vol. 38, no. 3, pp. 450–459, August 1996.

116 S.Y. Ahn, J.S. Park, and J.H. Kim, "Low frequency electromagnetic field reduction techniques for the on-line electric vehicles (OLEV)," in *IEEE International Symposium on Electromagnetic Compatibility*, 2010, pp. 625–630.

117 J.H. Kim and J.H. Kim, "Analysis of EMF noise from the receiving coil topologies for wireless power transfer," in *IEEE Asia-Pacific Symposium on Electromagnetic Compatibility*, 2012, pp. 645–648.

118 H.S. Kim and J.H. Kim, "Shielded coil structure suppressing leakage magnetic field from 100 W-class wireless power transfer system with higher efficiency," in *IEEE Microwave Workshop Series on Innovative Wireless Power Transmission Technologies, Systems and Applications*, 2012, pp. 83–86.

119 S.Y. Ahn, H.H. Park, and J.H. Kim, "Reduction of electromagnetic field (EMF) of wireless power transfer system using quadruple coil for laptop applications," in *IEEE Microwave Workshop Series on Innovative Wireless Power Transmission Technologies, Systems and Applications*, 2012, pp. 65–68.

120 S.C. Tang, S.Y.R. Hui, and H.S. Chung, "Evaluation of the shielding effects on printed circuit board transformers using ferrite plates and copper sheets," *IEEE Trans. on Power Electron.*, vol. 17, no. 6, pp. 1080–1088, November 2002.

121 X. Liu and S.Y.R. Hui, "An analysis of a double-layer electromagnetic shield for a universal contactless battery charging platform," *IEEE Power Electronics Specialists Conference (PESC)*, June 2005, pp. 1767–1772.

122 P. Wu, F. Bai, Q. Xue, and S.Y.R. Hui, "Use of frequency selective surface for suppressing radio-frequency interference from wireless charging pads," *IEEE Trans. Ind. Electron.*, accepted for publication.

123 M.L. Hiles and K.L. Griffing, "Power frequency magnetic field management using a combination of active and passive shielding technology," *IEEE Trans. on Power Delivery Electronics*, pp. 171–179, 1998.

124 C. Buccella and V. Fuina, "ELF magnetic field mitigation by active shielding," in *IEEE International Symposium on Industrial Electronics*, 2002, pp. 994–998.

125 J. Kim, J.H. Kim, and S.Y. Ahn, "Coil design and shielding methods for a magnetic resonant wireless power transfer system," *Proceedings of the IEEE*, vol. 101, no. 6, pp. 1332–1342, June 2013.

126 C.B. Park, S.W. Lee, and Chun T. Rim, "Static and dynamic analyses of three-phase rectifier with LC input filter by Laplace phasor transformation," in *Energy Conversion Congress and Exposition (ECCE)*, September 2012, pp. 1570–1577.

127 Chun T. Rim and G.H. Cho, "Phasor transformation and its application to the DC/AC analyses of frequency phase-controlled series resonant converters (SRC)," *IEEE Trans. Power Electron.*, vol. 5, no. 2, pp. 201–211, April 1990.

128 V.X. Thai, S.Y. Choi, S.Y. Jeong, and Chun T. Rim, "Coreless power supply rails compatible with both stationary and dynamic charging of electric vehicles," in 2015 *IEEE International Future Energy Electronics Conference (IEEE IFEEC 2015)*, 978-1-4799-7657-7.

129 J.H. Kim, B.S. Lee, J.H. Lee, and J.H. Baek, "Development of 1 MW inductive power transfer system for a high speed train," *IEEE Trans. on Ind. Electron.*, vol. 62, no. 10, pp. 6242–6250, October 2015.

130 Endesa web site, http://futurenergyweb.es/endesa-desarrolla-en-malaga-un-sistema-para-cargar-un -autobus-electrico-en-movimiento-y-sin-cables/?lang=en.

131 E. Mascarell, "VICTORIA: towards and intelligent e-mobility," UNPLUGGED Final Event, 2015.

132 J.A. Ruiz, "ITS systems developing in Malaga," in *2nd Congress EU Core Net Cities*, 2014.

133 INTIS web site, http://www.intis.de/intis/mobility.html.

134 Technical Article eCarTec 2014 in INTIS web site, http://www.intis.de/intis/assets/flyer_intis_testing_e.pdf, http://www.intis.de/intis/assets/ecartec2014_e.pdf.

135 Technical Information 30 kW coil system VW T5 transporter in INTIS web site, http://www.intis.de/intis/assets/ti_intis_t5_e.pdf.

10

Narrow-Width Single-Phase Power Rail (I-type)

10.1 Introduction

Due to global warming and the depletion of petroleum resources, automobile manufacturers have been developing EVs such as hybrid electric vehicles (HEVs), plug-in HEVs (PHEVs), battery EVs (BEVs), and others. Among these green cars, the BEV is one of the most promising future types of transportation. However, it is not yet welcomed into the markets by potential customers due to drawbacks such as its high price, heaviness, and the large space required by its battery. Other factors are limited lithium resources, a driving range shorter than that of a normally fueled car, a long charging time, and the frequent charging requirements.

In an effort to solve these problems, road-powered electric vehicles using IPTS have been developed [1–7]. The Partner for Advanced Transit and Highways (PATH) team in California in the USA was the first to develop a road-powered inductive power transfer electric vehicle, achieving 60% power transfer efficiency with a 7.6 cm airgap [1–3]. A research team at Auckland University, New Zealand, also proposed road-powered electric vehicles using magnetic coupled systems [4–6]. However, for the practical use of an IPTS with a road-powered EV, the airgap of an IPTS must meet the road regulation, currently 12 cm in Korea and 16 cm in Japan for example. The power transfer efficiency with a large airgap should also be sufficiently high, exceeding, for example 70%, so that the EVs can practically reduce CO_2 emissions compared to internal combustion engine vehicles. In addition, the infrastructure cost and vehicle price must be lower than other competitive vehicles.

To transfer a large amount of power, while improving the overall efficiency and reducing the voltage and current rating of the power supply, the IPTS should operate in the resonant mode [8–18]. There are several resonant modes for various IPTS applications: the parallel resonant mode [1–6, 8–11], series resonant mode [7, 13, 14], and series–parallel resonant mode [12]. It is necessary to select the appropriate resonant mode considering the input and output impedances, coil parasitic parameters, and load characteristics. The design of the resonant circuits or "resonant transformers" in the IPTS, regardless of the resonant modes, is challenging work as the dynamic and static behaviors of the IPTS, including the load and source variations, are difficult to characterize in practice [28–31]. The IPTS should also be robust to lateral displacement, which is the distance between the center of a vehicle and that of a roadway power supply a rail lane [19, 20] to ensure enough power delivery while the vehicle is operating on the road. To satisfy these requirements, OLEVs [7] have been developed and represent the most recent and advanced version of road-powered EVs.

In this chapter, a new IPTS for the OLEVs with a narrow rail width of 10 cm, a small pick-up having a large lateral displacement of about 24 cm, and a large airgap of 20 cm is explained. The IPTS was analyzed and verified by simulations and experiments, by which the maximum output

Wireless Power Transfer for Electric Vehicles and Mobile Devices, First Edition. Chun T. Rim and Chris Mi.
© 2017 John Wiley & Sons Ltd. Published 2017 by John Wiley & Sons Ltd.

power of 35 kW and the maximum efficiency of 74% at 27 kW were achieved. It is proven that the proposed IPTS, driven by a current source, is equivalently an ideal voltage source from the load side point of view [21–24].

Nomenclature

EMF	Electromagnetic field
L_p	Total inductance of a power supply rail
L_s	Total inductance of a pick-up
L_{l1}	Leakage inductance of a power supply rail
L_{l2}	Leakage inductance of a pick-up
L_m	Magnetizing inductance of a power supply rail
C_1	Compensation capacitance of a power supply rail
C_2	Compensation capacitance of a pick-up
R_L	Load resistance
R_o	Equivalent output resistance from a secondary side viewpoint to load
r_p	Internal resistance of a power supply rail
r_s	Internal resistance of a pick-up
I_S	Input current (rms)
I_{Lm}	Magnetizing current (rms)
I_o	Output current (rms)
V_{th}	Thevenin equivalent voltage (rms)
V_{Lm}	Induced voltage at the magnetizing inductance (rms)
V_o	Output voltage at the equivalent output resistance (rms)
N_1	The number of turns of two poles of a power supply rail
N_2	The number of turns of two pick-ups
n	Turn ratio N_2/N_1
ω_i	Power supply rail angular frequency
ω_s	Switching angular frequency

10.2 Narrow-Width I-Type IPTS Design

10.2.1 Previous Works on OLEVs

The concept of the OLEV, a sort of roadway wireless powered EVs, is shown for a bus in Figure 10.1. This vehicle can reduce the battery requirement by a factor of 5, which was one of the crucial obstacles in commercializing EVs.

Three versions of OLEVs have been developed since 2009 [7]. The first version, announced on February 27, 2009, is an OLEV golf car equipped with a rail 20 cm wide that includes E-type cores and a mechanically controlled pick-up that is automatically aligned to the rail. It can transfer 3 kW to the vehicle with 80% overall system efficiency through a 1 cm airgap. The second version, announced on July 14, 2009, is an OLEV bus. To increase the airgap, a flat and ultra-slim mono rail, known as a U-type IPTS, with a 140 cm width was developed. This IPTS can transfer 52 kW via ten pick-ups at 5.2 kW each for the OLEV bus through a 17 cm airgap. The maximum power transfer efficiency from the input voltage source of 60 Hz, 380 V three-phase to the regulator on the OLEV bus was 72%. The third version, announced on August 14, 2009, is an OLEV sports utility vehicle (SUV). The IPTS applied to this OLEV SUV, termed a

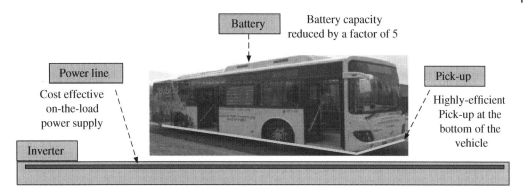

Figure 10.1 The concept of an on-line electric vehicle.

W-type IPTS, comprised a dual rail made of two mono rails put together along a roadway. To pick up more power from the W-type dual rail of the IPTS, a slim W-type multiwinding pick-up was developed to provide 15 kW per pick-up through a 20 cm airgap. The maximum power transfer efficiency was 74%. The "bone core structure" [7] was introduced as the third type of IPTS. It increases the durability of the power supply rail by applying a unique concrete pouring method. The parameters for the U-type and W-type coils are shown in Table 10.1.

The line inductance of a power supply rail for a segmentation length, which ranges for OLEVs from 3 m to 60 m, was measured as about 1 μH/m for the U-type mono rail and about 2 μH/m for the W-type dual rail. The line voltage stress due to this inductance was about 1–3 kV for a line current of 200 A; hence, a capacitor bank was serially inserted to offset the induced voltage of the inductance and high-frequency power cables with more than 5 kV insulation capability were used to withstand the high-voltage stress. This explains why the OLEV IPTS typically adopts the series resonant mode at the primary side. To nullify the high-voltage stress of the pick-up and to transfer maximum power to the vehicles, resonant capacitors were inserted serially into the pick-up as well. The line inductance and the back-electromotive force voltage, however, vary abruptly and unexpectedly as vehicles travel along the road; hence, a constant current source AC power supply [15, 16] was used as the OLEV IPTS [7].

The power supply inverter of the IPTS, operating at 60 Hz, 440 V, and with a three-voltage source, provided a power supply rail with an output current of 200 A at an operating frequency of about 20 kHz. This operating frequency of the IPTS was increased as high as necessary to transfer the required power while avoiding the creation of audible noise and minimizing the size of the IPTS, while maintaining an acceptable level of switching loss in the inverter and a rectifier

Table 10.1 Characteristics of U-type and W-type coil structures

	U-type	W-type
Rail width	140 cm	80 cm
Displacement	20 cm	15 cm
EMF	5.1 μT	5.0 μT
Airgap	17 cm	20 cm
Output power	5.2 kW/pick-up	15 kW/pick-up
Efficiency	72%	74%

for the IPTS, as well as the conduction loss of the high-frequency power cables and coils. For example, the operating frequency of the PATH team [3] was 400 Hz, but the frequencies used in more recent works [5–7] were 10 kHz, 38.4 kHz, and 20 kHz, respectively.

Question 1 Using a magnetic mirror model, verify the line inductance of a power supply rail: about 1 µH/m for the U-type mono rail and about 2 µH/m for the W-type dual rail. Appropriate assumptions can be used for dimensions.

10.2.2 Proposed I-Type Power Supply Rail of the IPTS

The IPTSs for OLEVs developed thus far require a wide rail width to provide a large airgap, resulting in a strong electromagnetic field (EMF) for pedestrians. Apart from the EMF problem, the allowable lateral displacement of the previous IPTSs was less than 20 cm, which is too small to guarantee free driving by OLEVs. To solve these problems, a new IPTS with a very narrow power supply rail width of 10 cm and alternating polarity magnetic poles along the road is explained in this chapter, as shown in Figure 10.2, where the name "I-type" stems from the front shape of the power rail, as depicted in Figure 10.2(c). Each magnetic pole consists of ferrite cores and turned cables, and the poles are connected to each other with ferrite cores. The finite element method (FEM) simulation result of the magnetic flux density B, as shown in Figure 10.3, indicates that the magnetic flux density reaches its maximum at the center of the pole, gradually decreasing as it deviates from the center and eventually reaching its minimum at the middle of each pole. The polarity of B at each pole is reversed according to the current direction of each pole; hence, the main magnetic flux is circulating through the nearest poles. A large airgap can be achieved by arranging each pole at a proper distance if necessary. Due to this alternating magnetic polarity of adjacent poles, as shown in Figure 10.2, the EMF for pedestrians around the power supply rail can be drastically reduced. This is a result of EMF cancelation by neighborhood poles with opposite polarity. Large lateral displacement can also be achieved by the wide pick-up, where a width as small as 80 cm is due to the narrow power supply rail of the proposed IPTS, as shown in Figure 10.2.

10.2.3 Proposed Power Supply Rail Design

The design parameters of the proposed IPTS include the pole size (width w_p, length l_p, height h_p, and thickness t_p), the pole distance d, the bottom plate size (width w_b and thickness t_b), and the number of turns of the power supply rail N_1, as shown in Figure 10.4. The effect of each parameter on the magnetic flux density was verified by simulations and experiments. As shown in Figure 10.2(b), the current source of each pole generates magnetic flux; therefore, the resultant magnetic flux is the superposition of each magnetic flux. For the pick-up covering two magnetic poles, as shown in Figure 10.2, the magnetic flux leaking to neighborhood poles is relatively small. A simplified magnetic circuit of the I-type IPTS, neglecting all the magnetic flux flowing to neighborhood poles, is shown in Figure 10.5. $\Re_{cp}$, $\Re_{c1}$, $\Re_{c2}$, $\Re_{air}$, $\Re_l$ are the magnetic resistances of pole height, core plate of a power rail, core plate of a pick-up, airgap between the pole and the pick-up, and airgap between two poles, respectively, whereas the magnetic leakage flux is Φ_l and the magnetic mutual flux is Φ_m. From the simplified magnetic resistance circuit, as shown in Figure 10.5(b), the magnetic mutual flux Φ_m can be calculated as follows:

$$\Phi_m = \frac{N_1 I_S}{2\Re_{cp} + \Re_{c1} + \Re_l//(\Re_{c2} + 2\Re_{air})} \cdot \frac{\Re_l}{\Re_l + (\Re_{c2} + 2\Re_{air})} \qquad (10.1)$$

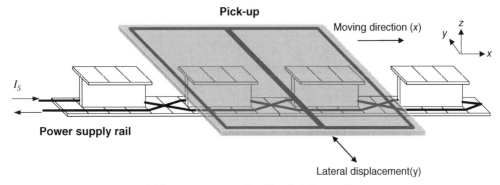

(a) I-type power supply rail and pick-up coils

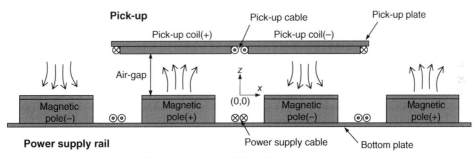

(b) Cross-section of the coils (side view)

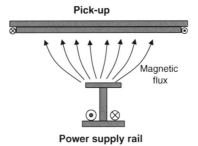

(c) Cross-section of the coils including a "I"-shape power supply rail (front view)

Figure 10.2 Proposed narrow width I-type power supply rail and pick-ups of the IPTS.

Any kth magnetic resistance $\mathfrak{R}_k$ is given by the following equation, where $\mu_0 = 4\pi \times 10^{-7}$ H/m, μ_c is a relative permeability, and A_{eff} is an effective area:

$$\mathfrak{R}_k \equiv \frac{l_k}{\mu_0 \mu_{c,k} A_{eff,k}} \tag{10.2}$$

It is very difficult to determine $\mathfrak{R}_k$ by hand because the effective area A_{eff} of the proposed I-type IPTS cannot be easily calculated; this is why we should rely on computer simulations. It can be identified from (10.1); however, the magnetic mutual flux Φ_m will increase until the pole distance d reaches an optimum value due to the increase of the parallel leakage magnetic resistance $\mathfrak{R}_l$. It will decrease again due to the increase of the magnetic resistance of the poles

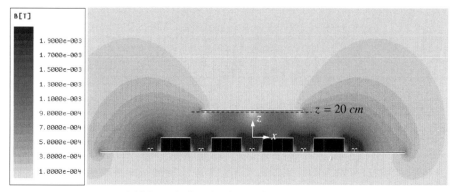

(a) Magnetic flux density profile (side view)

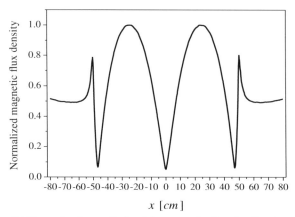

(b) Normalized magnitude of the magnetic flux density at $z = 20$ cm

Figure 10.3 Magnetic flux density distribution of the proposed I-type power supply rail with a pick-up core. The simulation was conducted by Maxwell 2D, version 12.

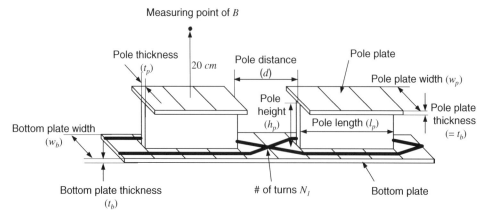

Figure 10.4 Design parameters of the proposed I-type IPTS. If the physical winding is one, the effective number of turns N_1 becomes two.

Figure 10.5 A simplified magnetic circuit of the proposed I-type IPTS neglecting all the other magnetic fluxes flowing into neighborhood poles.

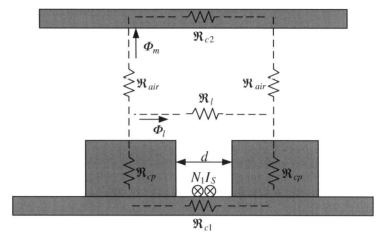

(a) The schematic diagram of magnetic resistances

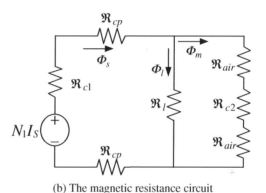

(b) The magnetic resistance circuit

and core plates as d increases. To find out the optimum distance, FEM three-dimensional simulations were performed for the range of d from 5 cm to 45 cm, as shown in Figure 10.6(a). It is found that the magnetic flux density above each pole is maximized for a pole distance of 20 cm.

The most sensitive parameter as regards the power is the number of turns of each magnetic pole, as shown in Figure 10.6(b), whereas the pole thickness and width are found to have little effect on the power transfer as long as the core of the pole is not saturated, as shown in Figure 10.6(c) and (d). For minimizing the cost of cable and ferrite core and the pick-up size and maximizing power delivery, the parameters are selected as listed in Table 10.2. The thickness of the bottom plate t_b is selected as 1 cm in order not to be saturated for maximum power delivery, and the bottom plate width w_b is selected as 10 cm as a trade-off between minimizing the power supply rail width and maximizing the power delivery. The number of turns is determined as a trade-off between the cost of the cable and the power delivery capability, and the pole length is determined also as a trade-off between minimizing the EMF for pedestrians and maximizing the power delivery. The final determination of the parameters could be possible after several trial-and-error experiments.

10.2.4   Proposed Pick-up Design

The output power of 15 kW per pick-up achieved from the previous IPTS for the OLEV SUV was a little small for an OLEV car considering the limited space at the bottom of a car to set

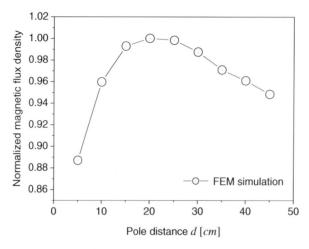

(a) FEM simulation result of the normalized magnetic flux density along the pole distance. The permeability of cores is 2000 and the power supply rail for the simulation has five poles, whose dimensions are $w_p = 7$ cm, $w_b = 10$ cm, $l_p = 30$ cm, $h_p = 10$ cm, $t_b = 1$ cm, and $t_p = 2$ cm, with one winding flowing at 200 A, 20 kHz

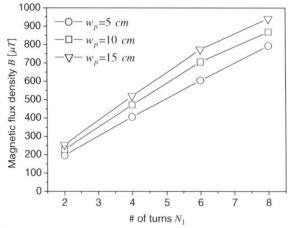

(b) Measurement results of magnetic flux density versus the number of turns

Figure 10.6 Simulation and measurement results of magnetic flux density of the power supply rail depending on design parameters. A power supply rail with ten poles was implemented for the experiment, and the measurement was conducted using two poles at the center of the rail to avoid any asymmetric results. The input current was 200 A and 20 kHz and the width w_b and the thickness t_b of the bottom plate of the power supply rail were 10 cm and 1 cm, respectively.

up pick-ups. Therefore, the output power per unit pick-up should be increased. To satisfy the power requirements of more than 20 kW [24], the proposed I-type IPTS was developed for the output power of 25 kW, 10 kW greater than the previous IPTS for the OLEV SUV. To increase the output power, pick-ups were deliberately designed based on the determined values of the parameters listed in Table 10.2. Considering the limited space at the bottom of a vehicle and lateral displacement allowance, the pick-up width and length were determined as 80 cm and 100 cm, respectively. The maximum output current is limited by the temperature rise

Figure 10.6 (*Continued*)

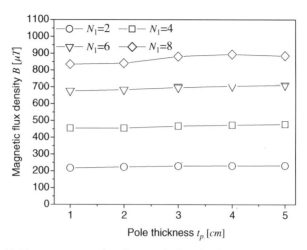

(c) Measurement results of magnetic flux density versus the pole thickness

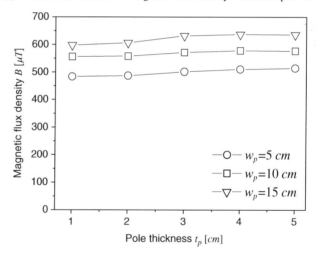

(d) Measurement results of magnetic flux density versus the pole thickness

Table 10.2 Design result of coil parameters.

	I-type
Pole distance (d)	20 cm
Pole length (l_p)	30 cm
Pole thickness (t_p)	2 cm
Pole plate width (w_p)	7 cm
Bottom plate width (w_b)	10 cm
Bottom plate thickness (t_b)	1 cm
Number of turns (N_1)	4

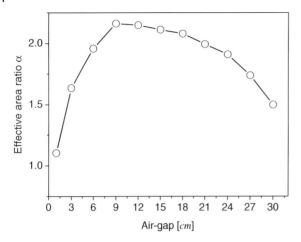

Figure 10.7 The ratio of the effective area of the proposed I-type IPTS versus the airgap, calculated by the FEM three-dimensional simulation. Due to the fringe effect of the asymmetric structures of the two coils, the effective area increases. The dimensions of the structure for simulation are given in Table 10.2.

owing to the heat generation of pick-up cables and cores; hence, the output current rating was selected as 50 A. The output open voltage rating was set to 550 V for the output power of 25 kW, considering a 50 V voltage drop due to internal resistance.

The remaining important design parameter of the pick-up is the number of turns N_2, which is determined so as to provide the output open voltage V_o as follows:

$$|V_o| = N_2 \omega_s \Phi_m = N_2 \omega_s A_{eff} |B| \tag{10.3}$$

The magnetic flux and its density of (10.3) are defined to be the peak values over a pick-up, a location that is found be the right above each pole, as shown in Figure 10.4. It is, however, not easy to find the effective area A_{eff} of (10.3) by hand; hence, it was determined using the FEM simulation. It is found that the effective area becomes about two times (α) a pole plate area A_p for the air-gap of 20 cm due to a significant fringe effect of the narrow power supply rail and the wide pick-up, as shown in Figure 10.7. Finally, the theoretical value of N_2 was determined from (10.3) as follows:

$$N_2 = \frac{|V_o|}{\omega_s A_{eff} |B|} = \frac{|V_o|}{2\pi \times f_s \, \alpha \, A_p |B|} \approx \frac{550}{2\pi \times 20k \times 2 \times 0.021 \times 0.00354} = 29.4 \tag{10.4}$$

where, the pole plate area A_p is determined from Table 10.2 by

$$A_p = w_p \times l_p = 0.07 \times 0.30 = 0.021 \tag{10.5}$$

In (10.4), f_s is the inverter switching frequency and the magnetic flux density B was determined by the FEM simulation. The number of turns of pick-ups N_2 should be adjusted in practice and finally determined as 32 after a few additional experiments.

There are many more practical considerations such as the allowable voltage and current ratings of commercially available parts, the EV battery voltage and capacity ratings, and the rectifier ratings; hence, the baseline pick-up design results should be verified and modified by experiments.

10.3 Analysis of the Fully Resonant Current-Source IPTS

10.3.1 Overall Configuration

The IPTS is composed of a power supply inverter, a power supply rail, pick-up coil(s), and full-bridge rectifier(s). Figure 10.8 shows a complete system diagram of the IPTS for the case of powering multiple vehicles and Figure 10.9 shows its schematic of an equivalent circuit for powering a vehicle case. Because the power supply rail is a controlled current source, the multiple vehicles can get electric power from the road independently of each other so far as the sum of requiring power is within the capacity of the power supply inverter. For an overload condition, some power scheduling should be adopted, though this is not explained in this chapter. The power supply inverter of the IPTS, operating from a 60 Hz, 440 V three-phase voltage source, provides a power supply rail with an output current of 200 A at an operating frequency of about 20 kHz. The switching frequency of 20 kHz was selected to transfer the required power while avoiding the creation of audible noise, minimizing the size of IPTS, and maintaining an acceptable level of switching losses in the inverter and conduction losses in the power supply rail and the pick-up due to the increase of internal AC resistance at high frequency. To nullify the high-voltage stresses and reactive powers, lumped capacitors, denoted here as C_1 and C_2 in Figure 10.9, are used. The interaction between the power supply rail and the pick-up results in an equivalent transformer with magnetizing inductance L_m. An equivalent current source is constituted by regulating the power supply rail current with the power supply inverter. For maximum power delivery, the "fully resonant scheme," where the capacitor C_2 is tuned not only to L_{l2} but also to L_m, was selected in this chapter. Figure 10.10 shows the simplified equivalent circuit of Figure 10.9.

Question 2 In Figure 10.8, what happen to the output voltage of the power supply inverter as the number of vehicles on the power supply rail increases. Replace each vehicle with an equivalent resistance and observe voltages, assuming a constant current source.

10.3.2 Current Source IPTS

The inverter used in this chapter is assumed to be operating at the switching angular frequency ω_s, which is slightly higher than the intrinsic resonant angular frequency ω_i of the power supply rail. This frequency discrepancy makes the IPTS always inductive such that a stable current switching can be guaranteed. In the case of a loosely coupled IPTS, $L_l \gg L_m$, ω_i can be

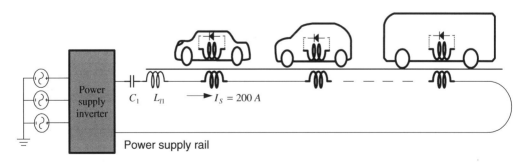

Figure 10.8 A complete system diagram, where multiple vehicles can be powered from the IPTS.

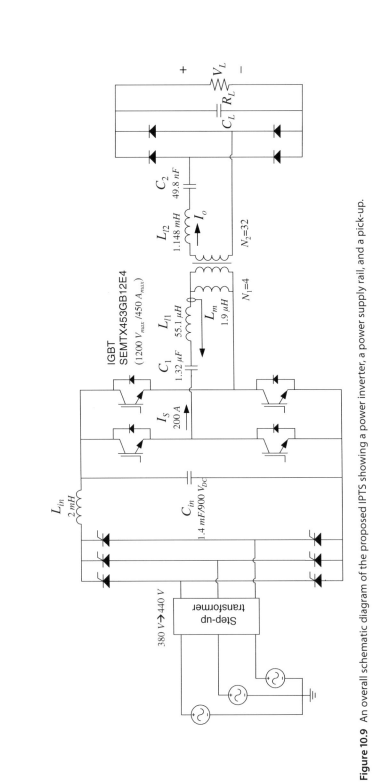

Figure 10.9 An overall schematic diagram of the proposed IPTS showing a power inverter, a power supply rail, and a pick-up.

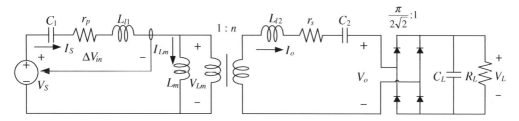

Figure 10.10 A simplified equivalent circuit of the proposed IPTS with a constant current control.

given in (10.6) below. The supply current I_S in the steady state is determined by the difference between the inverter output voltage and the magnetizing voltage ΔV_{in}, the equivalent residual input inductance ΔL determined by (10.8), and the internal resistance of the power supply rail r_p, as shown in Figure 10.10. By making the internal resistance of a power supply rail much smaller than the residual reactance by using litz wire, that is $r_p \ll (\omega_s L_{l1} - 1/\omega_s C_1)$, I_S can be determined as given in (10.7). The magnetizing voltage V_{Lm} includes the effect of the reflected impedance of the pick-up. For a constant current source IPTS, V_S should be changed in accordance with V_{Lm}. The current I_S is regulated by the inverter such that it can be regarded as a constant current source.

$$\omega_i \equiv \frac{1}{\sqrt{L_p C_1}} = \frac{1}{\sqrt{(L_{l1} + L_m) C_1}} \approx \frac{1}{\sqrt{L_{l1} C_1}} \quad (\because L_{l1} \gg L_m) \tag{10.6}$$

$$I_S = \frac{V_S - V_{Lm}}{r_p + j\left(\omega_s L_{l1} - \frac{1}{\omega_s C_1}\right)} \approx \frac{V_S - V_{Lm}}{j\left(\omega_s L_{l1} - \frac{1}{\omega_s C_1}\right)} = \frac{\Delta V_{in}}{j\omega_s \Delta L} \tag{10.7}$$

The ΔV_{in} and ΔL in (10.7) are given as follows:

$$\Delta V_{in} \equiv V_S - V_{Lm}, \ \Delta L \equiv L_{l1} - \frac{1}{\omega_s^2 C_1} \tag{10.8}$$

10.3.3 Conventional IPTS with Secondary-Only Resonance

A conventional IPTS of secondary-only resonance [13], as shown in Figure 10.11, is characterized by

$$j\omega_s L_{l2} + \frac{1}{j\omega_s C_2} = 0 \tag{10.9}$$

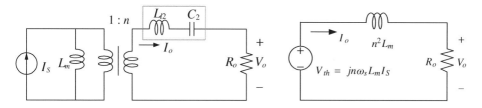

(a) An equivalent current source IPTS circuit.   (b) A simplified Thevenin equivalent circuit.

Figure 10.11 A conventional current source IPTS with secondary-only resonance and its simplified circuits.

It can be seen from the Thevenin equivalent circuit, as shown in Figure 10.11(b), that the output voltage of the conventional IPTS is substantially limited by the magnetizing inductance as follows:

$$|V_o| = \frac{n\omega_s L_m |I_S|}{\sqrt{1 + \left(\frac{n^2 \omega_s L_m}{R_o}\right)^2}} = n\omega_s L_m |I_S| \sqrt{1 - \left(\frac{n|I_o|}{|I_S|}\right)^2} \qquad (10.10)$$

In (10.10) and Figure 10.11, the output resistance R_o, as given below, is used instead of the load resistance R_L in Figure 10.10 and the full-bridge rectifier is equivalently transformed to an ideal transformer [25, 28]:

$$R_o = \frac{\pi^2}{8} R_L \qquad (10.11)$$

The output power of the conventional IPTS P_o can be obtained from (10.10) as follows:

$$P_o = \frac{|V_o|^2}{R_o} = \frac{(n\omega_s L_m |I_S|)^2}{R_o + \frac{(n^2 \omega_s L_m)^2}{R_o}} = n\omega_s L_m |I_S||I_o| \sqrt{1 - \left(\frac{n|I_o|}{|I_S|}\right)^2} \qquad (10.12)$$

The characteristics of the output voltage and output power of the conventional current source IPTS versus the output current, as described by (10.10) and (10.12), are shown in Figure 10.12, where the maximum output power is found to be too small to be used for high-power applications such as OLEV. For example, if $|I_S| = 200$ A and $n = 10$, the maximum output power is achieved at $R_o = n^2 \omega_s L_m = 23.9 \ \Omega$ and becomes only 4.8 kW.

10.3.4 Proposed Fully Resonant IPTS

To increase the output power by eliminating the voltage drop in the residual inductance in Figure 10.11(b), the fully resonant scheme, as shown in Figure 10.13, is introduced for the

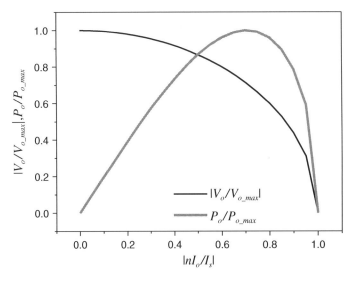

Figure 10.12 The theoretical characteristics of the output voltage and output power of the conventional current source IPTS with secondary-only resonance.

Figure 10.13 Equivalently simplified circuits of the fully resonant IPTS.

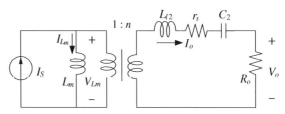

(a) A simplified circuit of the proposed IPTS: the rectifier is replaced with an equivalent resistance R_o

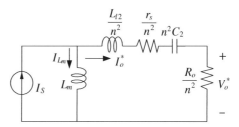

(b) A simplified circuit of the proposed IPTS eliminating the transformer, where, $I_o^* = nI_o$ and $V_o^* = V_o/n$

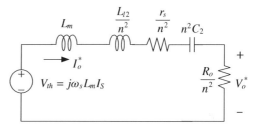

(c) A simplified Thevenin circuit of the proposed IPTS

(d) The final equivalent circuit of an ideal voltage source characteristic

proposed IPTS, where the compensation capacitor C_2 is now tuned to all inductance components as follows:

$$j\omega_s L_m + j\frac{\omega_s L_{l2}}{n^2} + \frac{1}{j\omega_s n^2 C_2} = 0 \qquad (10.13)$$

To analyze the proposed IPTS under the fully resonant condition denoted in (10.13), the circuit in Figure 10.10 is equivalently transformed from the primary side, as shown in Figure 10.13(a) to (d). The internal resistances of the power supply rail and secondary pick-up r_p and r_s, respectively, are included.

Applying the resonant condition (10.13) to the equivalent circuit in Figure 10.13(c), the output current and output voltage are determined as the two equations below in simplified forms, where $I_o{}^*$ is the reflected output current from the primary side viewpoint:

$$I_o = \frac{I_o^*}{n} = \frac{j\omega_s L_m I_S}{n\left(j\omega_s L_m + j\frac{\omega_s L_{l2}}{n^2} + \frac{1}{j\omega_s n^2 C_2}\right) + \frac{r_s + R_o}{n}} = \frac{jn\omega_s L_m I_S}{r_s + R_o} = j\frac{Q_m}{n} I_S \tag{10.14}$$

$$V_o = \frac{jn\omega_s L_m I_S R_o}{r_s + R_o} \approx jn\omega_s L_m I_S \quad for \quad r_s \ll R_o \tag{10.15}$$

In (10.14), the quality factor Q_m is defined as follows:

$$Q_m \equiv \frac{n^2 \omega_s L_m}{r_s + R_o} \tag{10.16}$$

The output voltage of the proposed IPTS V_o is nearly constant for a large output resistance; hence, it looks like an ideal voltage source, as shown in Figure 10.13(d).

From (10.15), the output power P_o is determined as follows:

$$P_o = \frac{|V_o|^2}{R_o} = \frac{(n\omega_s L_m |I_S|)^2 R_o}{(r_s + R_o)^2} \tag{10.17}$$

The output power P_o is proportional to the square of the input current $|I_S|$ and can be drastically increased for $R_o = r_s$ in theory; however, it is limited by the maximum output current I_o determined by the cable current rating and the voltage rating of the compensation capacitance of the pick-up C_2.

From Figure 10.13(b), the voltage of the magnetizing inductance V_{Lm} can be determined, considering the effect of all the secondary impedances, as follows:

$$V_{Lm} = I_S \left\{ j\omega_s L_m // \left(j\omega_s \frac{L_{l2}}{n^2} + \frac{1}{j\omega_s n^2 C_2} + \frac{r_s + R_o}{n^2} \right) \right\} = j\omega_s L_m I_S (1 - jQ_m) \tag{10.18}$$

From (10.18), the magnetizing current I_{Lm} is given as follows:

$$I_{Lm} = \frac{V_{Lm}}{j\omega_s L_m} = I_S (1 - jQ_m) \tag{10.19}$$

$$|I_{Lm}| = |I_S| \sqrt{1 + Q_m^2} \tag{10.20}$$

In (10.20), the magnetizing current I_{Lm} is drastically increased for a high output power, or a small output resistance R_o, which results in a large Q_m. The cores of the IPTS are saturated for a large magnetizing current, by which the output power P_o is also limited. The influence of core saturation on the output power and voltage is verified by the experiments in this chapter.

From Figure 10.13(b), the relationship of the currents of the proposed IPTS is found as follows:

$$I_S = I_{Lm} + I_o^* \quad or \quad I_{Lm} = I_S - I_o^* = I_S + (-I_o^*) \tag{10.21}$$

The phasor vector plots for (10.21) together with (10.14), (10.15), and (10.19) are shown in Figure 10.14, which is useful to see the phase relationship with respect to I_S. It is noted that the output current always leads to the input current by $\pi/2$ and that the magnitude of the output current may be much greater than that of the input current, which are quite unique characteristics of the proposed IPTS as well as the well-known Tesla coils. On the other hand, the input and output currents or voltages of a conventional transformer are determined by a turn ratio

Figure 10.14 Phasor vector plots of the fully resonant IPTS.

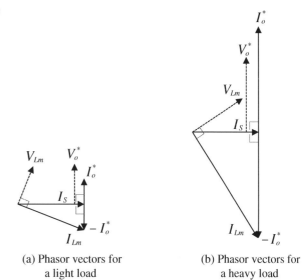

(a) Phasor vectors for
a light load

(b) Phasor vectors for
a heavy load

with little phase difference to each other. Due to the phase difference of $\pi/2$ between I_S and I_o, the EMF cancelations for I_S and I_o of the IPTS should be separated from each other. This fact has been applied to the design of pick-ups with an active EMF cancelation [27]. It is also identified from Figure 14(a) and (b) that the reflected output current $I_o{}^*$ and the magnetizing current I_{Lm} could be much larger than the input current I_S for the heavy-load case. Assuming that $n = 1$, for simplicity, the output current I_o becomes the same as I_{Lm} for a large I_o and irrelevant to I_S.

Question 3 (1) From (10.14) and (10.20), find the direct relationship between I_o and I_{Lm} without I_S. (2) Show that I_o is directly proportional to I_{Lm} for a small-load resistance, that is, large Q_m. This means that the core of IPTS may be saturated for a large output current. (3) Propose an idea to protect the IPTS from the large output current by utilizing the saturation characteristic of (2). Is there any other solution to limit output current?

10.4   Example Design and Experimental Verification

The proposed IPTS including an I-type narrow-width power supply rail and its pick-ups for experiments is shown in Figure 10.15. The number of turns of the power supply rail and the pick-up coils were selected as 4 and 32, respectively, and the airgap was increased to 20 cm to provide more space between the OLEV and the road. The input current was selected as 200 A to deliver enough power to the OLEV and the switching frequency was chosen as 20 kHz. These parameters were tentatively selected according to the proposed design; however, they could be finalized after a few experiments for the implemented OLEV systems.

The pick-up is comprised of two coils, and their induced AC voltages are rectified independently and then serially summed to provide he load with a high DC voltage. To nullify the high-voltage stress of the pick-up, the two coils were divided into four subcoils with 8 turns each, and the subcoils were serially connected through compensation capacitors. The inductances of the power supply rail and the pick-up, L_1 and L_2, respectively, were measured by a precision LCR

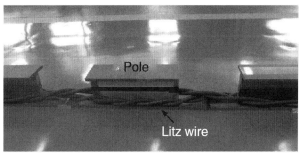

Figure 10.15 The proposed power supply rail with a narrow rail width and pick-ups for the experiments.

(a) The power supply rail for experiments

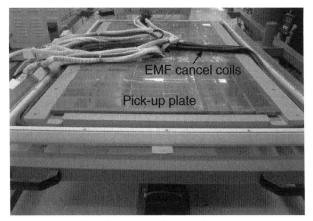

(b) The pick-ups for experiments

meter, which were 57 μH and 1.27 mH, respectively. The mutual inductance was measured as 1.9 μH. From the inductances, the compensation capacitance was determined as 49.8 nF for the proposed fully resonant scheme in (10.13). EMF cancelation coils were attached to the pick-up coils to reduce the magnetic field generated by the coil currents, as depicted in Figure 10.15(b). The major parameters of the IPTS for the experiments are listed in Table 10.3.

Figure 10.16 shows the measured waveforms of the input current, the output current, and the output voltage when the IPTS is operating in the fully resonant scheme. The output current leads to the input current by nearly $\pi/2$, and the output current and the output voltage are nearly in phase. This is in good agreement with the phasor diagram shown in Figure 10.14.

Table 10.3 Experimental conditions.

Parameter	Value	Parameter	Value
L_1	57 μH	N_2	32
L_2	1.27 mH	Airgap	20 cm
L_m	1.9 μH	Input current	200 A
C_1	1.32 μF	Switching frequency	20 kHz (tuned)
C_2	49.8 nF	Pick-up size	80 cm × 100 cm
N_1	4	load	Resistive

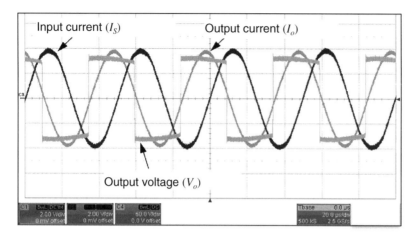

Figure 10.16 Waveforms of the input current, output current, and output voltage of the fully resonant IPTS. The input and the output current were measured at the power supply rail and the pick-up coils, respectively, and the output voltage was measured at the input of the full-bridge rectifier.

10.4.1 Output Voltage

The output voltage of the IPTS should be kept constant in theory regardless of the output current of the proposed fully resonant mode, as denoted in (10.15). The output voltage and power of the two compensation schemes, that is, the secondary-only resonance scheme and the fully resonant scheme, are compared to each other as functions of the output current, as shown in Figure 10.17. In the secondary-only resonance scheme, an abrupt voltage drop appears as I_o increases. In contrast, the output voltage of the fully resonant scheme decreases linearly as I_o increases.

The experiment results in Figure 10.18 show that no abrupt voltage drop appears until $I_o = 70$ A, but the output voltage is drastically reduced when the output current exceeds 70 A for the fully resonant scheme. It was found from the experiments that there are two categories of voltage drops. The first is a linear voltage drop and the second is an abrupt voltage drop, as shown in Figure 10.18. By tuning the switching frequency to the fully resonant frequency, as shown in Figure 10.19, it is noted that the abrupt voltage drop stems from the resonant frequency variation of the IPTS for an excessive value of I_o, at which point the pick-up cores are partially saturated.

The experiment result in Figure 10.19 shows that the optimum switching frequency increases by about 4% when the output current increases from 70 A to 100 A. The variable switching frequency scheme, however, is not applicable to solve the abrupt voltage drop problem because the inverter should drive several OLEVs with different load conditions on the road. To solve this problem, pick-ups should be designed for a higher output voltage to reduce the output current for a given power, or mechanically or electrically controlled variable inductors or capacitors should be used to compensate for the variation of inductance in the pick-up coils, which are left for further work.

10.4.2 Output Power and Efficiency

There are many system parameters involved in the output power and efficiency; therefore, experimental verifications are preferred in this chapter. To evaluate the output power and

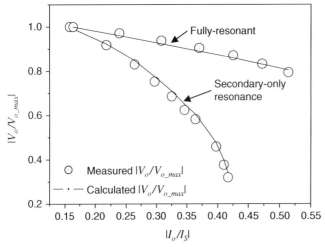

(a) The output voltage versus the current ratio I_o/I_S

Figure 10.17 Comparisons of the output voltage and the output power for the two compensation schemes: the secondary-only resonance and the fully resonant schemes.

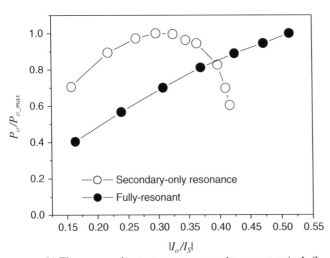

(b) The measured output power versus the current ratio I_o/I_S

efficiency of the proposed IPTS, the input power was measured at the input transformer and the output power was measured at the load resistor, as shown in Figure 10.9. The maximum output power was measured when the pick-up was correctly aligned to the center of the pole and the IPTS was under the optimum switching frequency to eliminate the effect of the partial leakage inductance variation of the pick-up coils for an excessive output current, which was 35 kW with an airgap of 20 cm.

The maximum power transfer efficiency obtained was 74% at an output of 27 kW, that is, $V_o = 408$ V and $I_o = 66.2$ A, as shown in Figure 10.20. The efficiency includes the inverter power loss, conduction losses in the power supply rail and pick-ups, core losses, and rectifier power losses. There are certain fixed power dissipations in the inverter and the power supply rail because the input current is constant regardless of the output current. In a light-load condition, the inverter power loss and the power supply rail conduction loss take a large portion of the overall input power; hence, the efficiency increases as the output power increases. As the output power increases, the fraction of these power losses is reduced and the efficiency eventually decreases.

Figure 10.18 Experimental results of the output voltage drop due to internal resistances and partial core saturations for an excessive output current. The abrupt voltage drop was alleviated when the switching frequency was appropriately changed from 19.79 kHz to 19.85 kHz to tune to the variant resonant frequency.

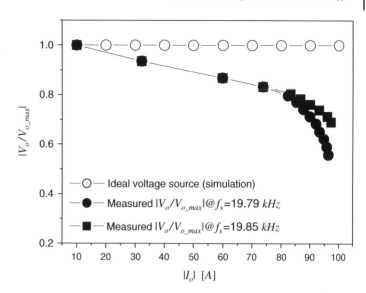

10.4.3   Spatial Power Variations

The output power variation, when the pick-up was misaligned with the center of the pole, was measured. As the pick-up moved away from the center of the pole along with the road (x-direction), the output power was reduced and reached nearly 1 kW at the cross-point of the poles, as shown in Figure 10.21. In contrast, the output power exceeded 20 kW when the pick-up was laterally moved (y-direction) within 20 cm and becomes half at 24 cm from the center of the power supply rail, as shown in Figure 10.22. The output power variation for the vehicle moving direction should be improved by other means or minimized by appropriate methods such as multiple uses of the pick-ups and power supply rails. To solve this problem is also left for further work.

Figure 10.19 Experimental results for the relationship between the abrupt output voltage change and switching frequencies. The abrupt output voltage variation for an excessive output current, here about 70 A, was improved by increasing the switching frequencies by about 4% greater than its original frequency.

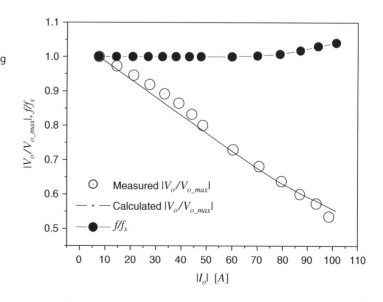

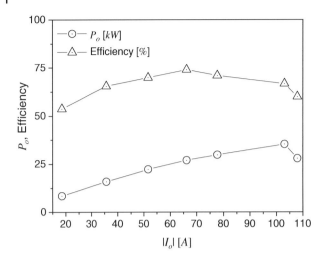

Figure 10.20 Experimental results of the output power and efficiency.

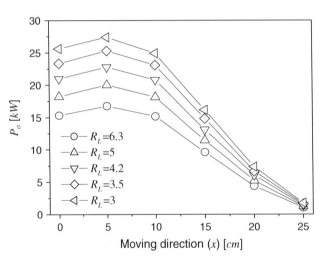

Figure 10.21 Experimental results of the output power variation versus moving direction (x).

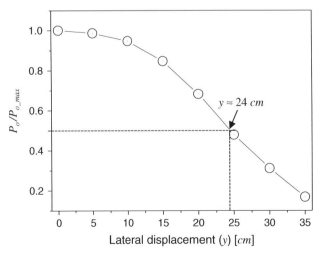

Figure 10.22 Experimental results of the normalized output power variation versus the lateral displacement (y).

Figure 10.23 Experimental results of the EMF for a pedestrian beside a roadway.

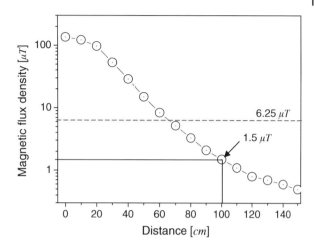

10.4.4 EMF

A magnetic field is generated whenever electrical devices are operating in general, but it should be lower than the allowable EMF level for all frequency ranges. The EMF around the power supply rail and pick-up of the proposed IPTS was measured, as shown in Figure 10.23. The EMF generated from the power supply rail was reduced considerably due to the alternating polarities of the I-type poles, as predicted in the previous sections. It was as low as 1.5 μT at a distance of 1 m from the center of the power supply rail lane, which was approximately 20 μT for previous versions of OLEV IPTSs. This EMF level meets the ICNIRP Guidelines, 6.25 μT at 20 kHz.

10.5 Conclusion

A new IPTS with an extremely narrow width of a power supply rail of 10 cm, a large airgap of 20 cm, and a low infrastructure cost is presented in this chapter and verified by simulations and experiments. The proposed fully resonant current-source IPTS has ideal voltage source characteristics and shows quite good properties, such as a larger lateral displacement, lower EMF, and higher output power than previous IPTSs for OLEV [7], which are essential for its practical use. An optimum I-type power supply rail was designed, characterized, and verified by simulations and experiments, showing that it works well in practice. This IPTS improved the airgap from 17 cm in previous IPTSs for OLEVs [7] to 20 cm, reduced the width of the power supply rail from 80 cm to 10 cm, minimized the EMF level below 6.25 μT, and achieved the maximum output power up to 35 kW and overall power efficiency to 74% at 27 kW.

Problems

10.1 You are stepping up the load voltage to double the present value by increasing the number of turns of the secondary winding of the proposed I-type IPTS of Figure 10.9 for a given volume of the pick-up. You will keep the other parameters and performances unchanged if possible, which include all the parameters of the power supply rail, source current, load power, and operating frequency.
 (a) What are the parameters inevitably changed due to this load voltage doubling?
 (b) What will become the efficiency and load-quality factor?

10.2 **(a)** How about doubling the source current instead of the load voltage of 10.1? Discuss the merits and demerits of the two approaches.

(b) Can you find a general conclusion or only a specific conclusion from the comparison of the two approaches?

References

1 J.G. Bolger, F.A. Kirsten, and LS. Ng, "Inductive power coupling for an electric highway system," in *Proc. IEEE 28th Vehicular Technology Conference*, 1978, vol. 28, pp. 137–144.

2 C.E. Zell, and J.G. Bolger, "Development of an engineering prototype of a road powered electric transit vehicle system," in *Proc. 32nd IEEE Vehicular Technology Conference*, 1982, vol. 32, pp. 435–438.

3 M. Eghtesadi, "Inductive power transfer to an electric vehicle-an analytical model," in *Proc. 40th IEEE Vehicular Technology Conference*, 1990, pp. 100–104.

4 A.W. Green, and J.T. Boys, "10 kHz inductively coupled power transfer concept and control," in *Proc. 5th Int. Conf. IEE Power Electron and Variable-Speed Drivers*, 1994, pp. 694–699.

5 G.A.J. Elliott, J.T. Boys, and A.W. Green, "Magnetically coupled systems for power transfer to electric vehicles," in *Proc. Int. Conf. Power Electron. Drive Syst.*, 1995, pp. 797–801.

6 G.A. Covic, J.T. Boys, M.L.G. Kissin, and H G. Lu, "A three-phase inductive power transfer system for road powered vehicles," *IEEE Trans. on Ind. Electron.*, vol. 54, pp. 3370–3378, December 2007.

7 S.W. Lee, J. Huh, C.B. Park, N.S. Choi, G.H. Cho, and C.T. Rim, "On-line electric vehicle using inductive power transfer system," in *IEEE Energy Conversion Congress and Exposition (ECCE)*, 2010, pp. 1598–1601.

8 C.S. Wang, O.H. Stielau, and G.A. Covic, "Design consideration for a contactless electric vehicle battery charger," *IEEE Trans. on Ind. Electron.*, vol. 52, pp. 1308–1314, October 2005.

9 P. Si, and A.P. Hu, "Analysis of DC inductance used in ICPT power pick-ups for maximum power transfer," in *Proc. IEEE/PES Transmission and Distribution Conference and Exhibition*, pp. 1–6, 2005.

10 J.T. Boys, and G.A. Covic, "DC analysis technique for inductive power transfer pick-ups," *IEEE Power Electronics Letters*, vol. 1, pp. 51–53, June 2003.

11 M.L.G. Kissin, C.Y Huang, G.A. Covic, and J.T. Boys, "Detection of the tuned point of a fixed-frequency LCL resonant power supply," *IEEE Trans. on Power Electron.*, vol. 24, pp. 1140–1143, April 2009.

12 P. Nagatsuka, N. Ehara, Y. Kaneko, S. Abe, and T. Yasuda, "Compact contactless power transfer system for electric vehicles," in *International Power Electronics Conference (IPEC)*, 2010, pp. 807–813.

13 G.B. Joung, and B.H. Cho, "An energy transmission system for an artificial heart using leakage inductance compensation of transcutaneous transfer," *IEEE Trans. on Power Electron.*, vol. 13, pp. 1013–1022, November 1998.

14 S. Valtchev, B. Borges, K. Brandisky, and J.B. Klaassens, "Resonant contactless energy transfer with improved efficiency," *IEEE Trans. on Power Electron.*, vol. 24, pp. 685–699, March 2009.

15 M. Borage, S. Tiwari, and S. Kotaiah, "Analysis and design of an LCL-T resonant converter as a constant-current power supply," *IEEE Trans. on Ind. Electronics*, vol. 52, pp. 1547–1554, December 2005.

16 B. Mangesh, T. Sunil, and K. Swarna, "LCL-T resonant converter with clamp diodes: a novel constant-current power supply with inherent constant-voltage limit," *IEEE Trans. on Ind. Electron.*, vol. 54, pp. 741–746, April 2007.

17 B.L. Cannon, J.F. Hoburg, D.D. Stancil, and S.C. Goldstein, "Magnetic resonance coupling as a potential means for wireless power transfer to multiple small receivers," *IEEE Trans. on Power Electron.*, vol. 24, pp. 1819–1825, July 2009.

18 D.L. O'Sullivan, M.G. Egan, and M.J. Willers, "A family of single-stage resonant AC/DC converters with PFC," *IEEE Trans. on Power Electron.*, vol. 24, pp. 398–408, February 2009.

19 G.A.J. Elliott, S. Raabe, G.A. Covic, and J.T. Boys, "Multiphase pickups for large lateral tolerance contactless power-transfer systems," *IEEE Trans. on Ind. Electron.*, vol. 57, pp. 1590–1598, May 2010.

20 M.L.G. Kissin, J.T. Boys, and G.A. Covic, "Interphase mutual inductance in polyphase inductive power transfer systems," *IEEE Trans. on Ind. Electron.*, vol. 56, pp. 2393–2400, July 2009.

21 J. Huh, S.W. Lee, C.B. Park, G.H. Cho, and C.T. Rim, "High performance inductive power transfer system with narrow rail width for on-line electric vehicles," in *IEEE Energy Conversion Congress and Exposition (ECCE)*, 2010, pp. 1598–1601.

22 C.S. Wang, G.A. Covic, and O.H. Stielau, "Power transfer capability and bifurcation phenomena," *IEEE Trans. on Ind. Electron.*, vol. 51, pp. 148–157, February 2004.

23 H.H. Wu, G.A. Covic, J.T. Boys, and D.J. Robertson, "A series-tuned inductive-power transfer pickup with a controllable AC-voltage output," *IEEE Trans. on Power Electron.*, vol. 26, pp. 98–109, November 2011.

24 J. Huh, B.H. Lee, W.Y. Lee, G.H. Cho, and C.T. Rim, "Characterization of novel inductive power transfer systems for on-line electric vehicles," in *IEEE Applied Power Electronics Conference and Exposition (APEC)*, 2011, pp. 1975–1979.

25 R.L. Steigerwald, "A comparison of half-bridge resonant converter topologies," *IEEE Trans. on Power Electron.*, vol. 3, pp. 174–182, April 1998.

26 C.H. Moon, H.W. Park, and S.Y. Lee, "A design method for minimum-inductance planar magnetic-resonance-imaging gradient coils considering the pole-piece effect," *Measurement Science and Technology*, vol. 10, pp. 136–141. August 1999.

27 S.W. Lee, W.Y. Lee, J. Huh, H.J. Kim, C.B. Park, G.H. Cho, and C.T. Rim, "Active EMF cancellation method for I-type pick-up of on-line electric vehicles," in *IEEE Applied Power Electronics Conference and Exposition (APEC)*, 2011, pp. 1980–1983.

28 C.T. Rim and G.H. Cho, "Phasor transformation and its application to the DC/AC analyses of frequency phase-controlled series resonant converters (SRC)," *IEEE Trans. on Power Electron.*, vol. 5, pp. 201–211, April 1990.

29 C.T. Rim, D.Y. Hu, and G.H. Cho, "Transformers as equivalent circuits for switches: general proofs and D-Q transformation-based analyses," *IEEE Trans. on Ind. Applic.*, vol. 26, no. 4, pp. 777–785, July/August 1990.

30 C.T. Rim, "Unified general phasor transformation for AC converters," *IEEE Trans. on Power Electron.*, vol. 26, no. 9, pp. 2465–2475, September 2011.

31 S. Lee, B. Choi, and C.T. Rim, "Dynamics characterization of the inductive power transfer system for on-line electric vehicles by Laplace phasor transform," *IEEE Trans. on Power Electron.*, vol. 28, no. 12, pp. 5902–5909, December 2013.

11

Narrow-Width Dual-Phase Power Rail (I-type)

11.1 Introduction

Electric vehicles (EVs) draw attention as a promising substitution for internal combustion engine vehicles. There are a few problems in EVs such as frequent fast charging and deep discharging of batteries, which degrade the lifespan of the expensive onboard batteries [1, 2]. Stationary wireless EV chargers can solve inconvenient and dangerous charging problems [3–9]; however, the limited cruising range and lifespan degradation of batteries are remaining issues.

To solve these issues of EVs, road-powered electric vehicles (RPEV) [10–23], whose power is wirelessly supplied from power supply rails buried under the roadway, were proposed as a stepping-stone to the popularization of pure EVs. To narrow the power supply rail, to improve the construction convenience, and to secure economic feasibility, the I-type power supply rail, whose width is just 10 cm, was proposed [18]. The magnetic poles of U- and W-type rails are arranged on the vertical plane of the driving direction of the vehicles, so the induced voltage of the pick-up is almost independent with respect to the position of the vehicle along the driving direction. The magnetic poles of I-type rails are alternately placed along the driving direction to reduce the width of the rail and to enlarge the airgap, as shown in Figure 11.1. Due to this alternating magnetic pole arrangement, the induced voltage of a pick-up varies depending on the positions. It goes to near 0 V at the specific positions of the pick-up on the power supply rail.

In this chapter, an inductive power transfer system (IPTS) with a dq-power supply rail based on our patent [24] is proposed, as shown in Figure 11.2, to make the induced output voltage of the pick-up as even as possible with respect to pick-up positions along the driving direction. The dq-power supply rail is composed of d-winding and q-winding, which are magnetized by two independently controlled currents to balance power supply rail currents, respectively. For the compact multiphase winding in the power supply rail, an integrated winding method with shared core is also explained. Theoretical analyses results are suggested and verified by experiments in the following sections.

11.2 Design of the Proposed dq-Power Supply Rail

In this section, the characteristics of a conventional I-type IPTS are reviewed and its drawbacks are studied to propose the dq I-type power supply rail. For a fair comparison, the same pick-ups are used for both the conventional I-type power supply rail and the proposed dq-power supply rail. The series–series resonant compensation scheme is adopted, as shown in Figure 11.2, and only static behaviors of the fundamental components of the IPTS are investigated throughout this chapter. Therefore, the variables in this chapter are static phasor, which is a complex number.

Wireless Power Transfer for Electric Vehicles and Mobile Devices, First Edition. Chun T. Rim and Chris Mi.
© 2017 John Wiley & Sons Ltd. Published 2017 by John Wiley & Sons Ltd.

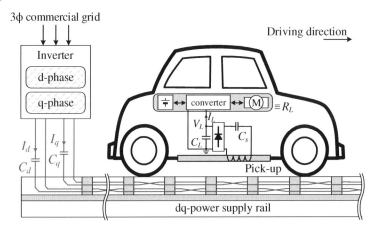

Figure 11.1 Overall system of the proposed dq I-type IPTS.

11.2.1 Spatial Induced Voltage Variation of the Conventional I-Type IPTS

The induced voltage of the pick-up V_o with respect to the positions along the driving direction of the conventional I-type IPTS is shown in Figure 11.1, where the induced voltage is proportional to the magnetic coupling between the power supply rail and pick-up. Therefore, V_o is spatially periodic; it nears zero at the positions of zero magnetic coupling and the output power P_o, which is proportional to the square of V_o, can also be zero at these positions [18]. If the vehicle is stopped at an arbitrary position on the I-type power supply rail, the vehicle may not take any power from the power supply rail.

From experimental observations [18], the spatial distribution of the induced pick-up voltage $V_o(x)$ can be approximately described, using sinusoidal form, as follows:

$$V_o(x) = \left| V_a \cos \frac{2\pi x}{l_0} \right| \tag{11.1}$$

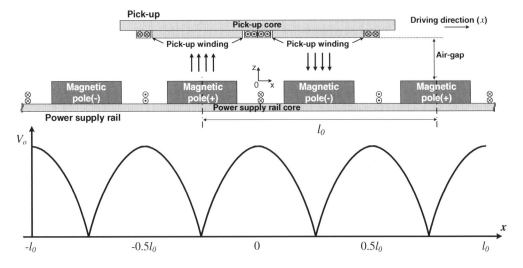

Figure 11.2 Structure of the conventional I-type IPTS and its induced voltage of the pick-up V_o w.r.t. the position of the pick-up on the power supply rail.

where x, V_a, and l_0 are displacement of the pick-up, spatial amplitude of the induced voltage, and spatial periodic length of the I-type power supply rail, respectively. The analytical expression for $V_o(x)$ is too complex to obtain at the moment.

Question 1 Justify (11.1) by using any theory or simulation if inevitable. You should reflect on the air-gap change of the model you are going to propose. Use any approximation, which is better than finding an exact but very complicated model.

11.2.2   Design and Analysis of the Proposed dq-Power Supply Rail

To compensate for the valley parts of $V_o(x)$ and to make it even, another magnetic pole that can magnetize the pick-up should be added between the magnetic poles of the conventional I-type power supply rail shown in Figure 11.1. Put simply, another power supply rail that is spatially shifted ($l_0/4$) so that its magnetic poles are located between the magnetic poles of an existing power supply rail can be added in parallel, as shown in Figure 11.3. The spatial shifting length is determined so that the peak points of the induced voltage for the added rail is placed at the valley points of the existing power supply rail. The existing power supply rail and added-power supply rail are called the d-power supply rail and q-power supply rail, respectively. With this configuration, zero-induced voltage points can be eliminated by combining the induced voltage from each power supply rail. One major advantage of the I-type power supply rail is a much lower construction cost owing to the narrow width of the power supply rail compared to the U- or W-type power supply rail, but the wider rail width of the two separated power supply rails weakens this advantage. The two separated rails should be narrowed by integrating the power supply rails into one.

To integrate the d- and q-power supply rails, the wire wound around magnetic poles of the q-power supply rail should pass through the magnetic poles of the d-power supply rail. To create

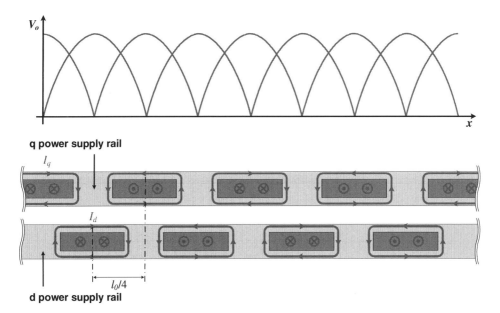

Figure 11.3 A valley compensation method using two separated-power supply rails of d- and q-power supply rails in parallel.

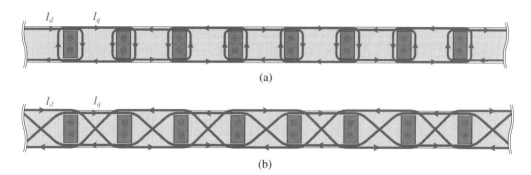

Figure 11.4 Integrated winding method of the proposed dq-power supply rail: (a) pole windings around spilt magnetic poles for integrated winding and its winding direction and (b) practical winding method on a long power supply rail with one wire for each phase.

space for the wires to pass through, each magnetic pole in Figure 11.3 should be divided into two parts, as shown in Figure 11.4(a). Using this space between the divided magnetic poles and overlapped winding, as shown in Figure 11.4 (a), two separated windings can be integrated into a single power supply rail. The winding configuration of Figure 11.4(a), where each magnetic pole is wound separately, cannot be adopted for a long power supply rail due to the high winding complexity. The separated windings of each magnetic pole can be wound by a single wire similarly to the I-type power supply rail of [18], as shown in Figure 11.4(b).

The proposed dq I-type power supply rail and pick-up are shown in Figure 11.5, where Φ_d and Φ_q are interlinked magnetic fluxes from the d- and q-power supply rails to the pick-up, respectively. The induced pick-up voltage from each interlinked flux, V_d and V_q, can be calculated as follows:

$$V_d = j\omega_o \Phi_d N_2 \tag{11.2}$$
$$V_q = j\omega_o \Phi_q N_2 \tag{11.3}$$

where N_2 and ω_o are the number of turns of the pick-up and the operating angular frequency of the IPTS, respectively. The combined magnitude of the induced voltage of the pick-up V_{0dq} is as follows:

$$V_{0dq} \equiv |V_d + V_q| = |j\omega_o N_2(\Phi_d + \Phi_q)| = \omega_o N_2|\Phi_d + \Phi_q| \tag{11.4}$$

Spatial distribution of the interlinked magnetic flux $\Phi_d(x)$ can be approximately described, using a sinusoidal form like (11.1), as follows:

$$|\Phi_d(x)| = \left|\Phi_0 \cos\left(\frac{2\pi x}{l_0}\right)\right| \tag{11.5}$$

where Φ_0 is the amplitude of the spatial function. Considering spatial shift of the q-power supply rail, $\Phi_q(x)$ is similarly obtained as follows:

$$|\Phi_q(x)| = \left|\Phi_0 \sin\left(\frac{2\pi x}{l_0}\right)\right| \tag{11.6}$$

Considering the phase component, assuming zero phase for the d-power supply rail, the spatial flux becomes

$$\Phi_d = \Phi_0 e^{j0} \cos\left(\frac{2\pi x}{l_0}\right) \tag{11.7}$$

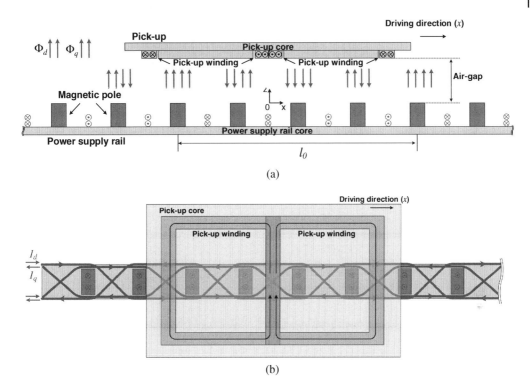

Figure 11.5 Proposed dq I-type power supply rail and pick-up: (a) side view of the proposed IPTS and (b) top view of the proposed IPTS with the winding configuration of the power supply rail and pick-up.

Assuming that the q-power supply rail is driven by a current source with an arbitrary phase difference θ, Φ_q becomes

$$\Phi_q = \Phi_0 e^{j\theta} \sin\left(\frac{2\pi x}{l_0}\right) \tag{11.8}$$

Substituting (11.7) and (11.8) into (11.4) results in the following:

$$V_{0dq} = \omega_o N_2 \Phi_0 \sqrt{1 + \sin(4\pi x / l_0)} \cos\theta \tag{11.9}$$

It can be seen from (11.9) that V_{0dq} is a function of x and θ. If the $\cos\theta$ term of (11.9) becomes zero, the dependency of V_{0dq} to x can be completely eliminated. Therefore, V_{0dq} for $\theta = 90°$ becomes

$$V_{0dq}|_{\theta=90°} = \omega_o N_2 \Phi_0 \tag{11.10}$$

As long as the phase difference between Φ_d and Φ_q is maintained as 90°, V_{0dq} is independent from the pick-up's displacement along the driving direction. To maintain the phase difference of the magnetic fluxes, the phase difference between I_d and I_q, which are the driving currents of each power supply rail, should be maintained. From Ampere's law, each phase of Φ_d and Φ_q is always identical to I_d and I_q, respectively.

11.2.3 Comparison with I-Type Power Supply Rail

Considering that a vehicle stops at an arbitrary position on the power supply rail, the spatial average power represents power receptibility of the vehicle. The output power of the pick-up is proportional to the square of the pick-up induced voltage [18]. The pick-up induced voltage distribution of an I-type IPTS is as follows:

$$V_{oI}(x) = \left| \omega_o N_2 \Phi_0 \cos \frac{2\pi x}{l_0} \right| \tag{11.11}$$

From (11.11), the spatial average pick-up induced voltage of an I-type IPTS $\bar{V}_{oI}$ is as follows:

$$\bar{V}_{oI} = \frac{1}{l_0} \int_0^{l_0} V_{oI}(x) \mathrm{d}x = \frac{2}{\pi} \omega_o N_2 \Phi_0 \tag{11.12}$$

From (11.10), the spatial distribution of V_{0dq} is as follows:

$$V_{0dq}(x) = \omega_o N_2 \Phi_{0dq} \tag{11.13}$$

The spatial average pick-up induced voltage of an I-type IPTS, $\bar{V}_{0dq}$, is as follows:

$$\bar{V}_{0dq} = \frac{1}{l_0} \int_0^{l_0} V_{0dq}(x) \mathrm{d}x = \omega_o N_2 \Phi_{0dq} \tag{11.14}$$

From (11.12) and (11.14), the interlinkage flux Φ_0 of the proposed dq I-type IPTS can be reduced for the same spatial average pick-up induced voltage. By lowering Φ_{0dq}, the volume of ferrite material can be reduced in terms of maintaining peak magnetic flux density in the ferrite cores of the power supply rail and pick-up. Because Φ_0 is proportional to current of the power supply rail, the current can also be lowered. Assuming $\bar{V}_{0dq}$ is the same as $\bar{V}_{oI}$ with the same current density of the wire in each power supply rail, the wire thickness of the power supply rail of the proposed dq I-type can be $2/\pi$ of the I-type IPTS. Two-phase configuration of the proposed dq I-type IPTS makes the winding length of that twice that of the I-type, that is, $4/\pi$ times the amount of wire of the I-type. Even though the proposed dq I-type IPTS needs 27% more wire (wire), the proposed dq I-type IPTS does not have spatially periodic zero powering positions of the I-type IPTS in expense of hardware complexity.

11.3 Circuit Design of the Proposed IPTS

In order to provide the direct and quadrature currents for the d- and q-power supply rails, respectively, appropriate control and power circuits are designed and verified by simulation in this section.

11.3.1 Power Circuit of the Proposed IPTS

Figure 11.6 shows the circuit diagram of the proposed dq I-type IPTS. Each phase of the power supply rail is driven by a dedicated inverter. The amplitudes of the output currents of the inverters I_d and I_q are controlled by a current amplitude controller so that the driving currents of the power supply rails are consistently controlled against load change and variation of the magnetizing inductance between the power supply rail and pick-up. I_{qsen} and I_{dsen} are the sensed currents of I_q and I_d by current sensors, respectively. The output current amplitude of the inverter can be regulated by controlling the DC link voltage of the inverter. The current amplitude controller

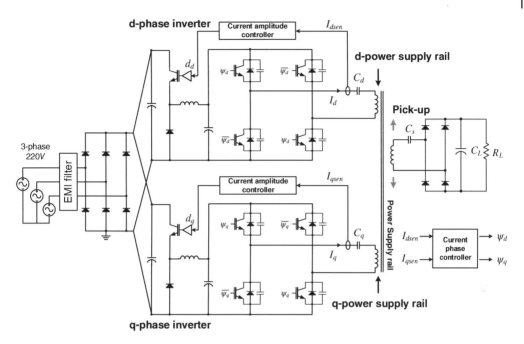

Figure 11.6 Circuit diagram of the proposed dq-type IPTS.

controls the dependent voltage source that generates the DC link voltage of the inverter so that the output current is regulated to a designated amplitude depending on the applied external control voltage. The dependent voltage source is composed of a buck converter adopting the conventional voltage mode PWM controller.

11.3.2   Phase Shift Analysis of the Power Circuit

An equivalent circuit model of the proposed dq I-type IPTS is shown in Figure 11.7(a), where V_d (V_q), C_d(C_q), L_{ld}(L_{lq}), L_{md}(L_{mq}), L_{ls}, C_s, $R_L{}'$, and n are d(q) phase inverter output voltage, d(q-phase primary resonant capacitor, leakage inductance of the d(q)-power supply rail, magnetizing inductance between the d(q)-power supply rail and pick-up, secondary leakage inductance, secondary series capacitor, referred resistance of the full-bridge rectifier of load resistance R_L, and the turn ratio of primary and secondary windings, respectively.

The combined induced voltage of the pick-up V_o can be derived as follows:

$$V_o = V_{od} + V_{oq} = n(j\omega L_{md}I_d + j\omega L_{mq}I_q) \tag{11.15}$$

Assuming $I_d \perp I_q$,

$$I_q = jI_d \tag{11.16}$$

Substituting (11.16) into (11.15),

$$V_o = j\omega n(L_{md}I_d + jL_{mq}I_d) \tag{11.17}$$

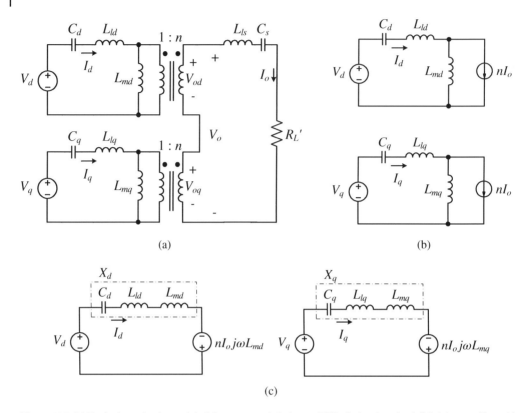

Figure 11.7 (a) Equivalent circuit model of the proposed dq I-type IPTS eliminating the full-bridge rectifier with referred load resistance R_L'. (b) Simplified circuit model of (a), eliminating the transformer. (c) Simplified circuit model of (b), using the Thevenin equivalent circuit.

With the series resonant condition of pick-up, the series resonant tank of L_{ls} and C_s can be omitted in the derivation [18]. The output current of R_L' is as follows:

$$I_o = \frac{V_o}{R_L'} = \frac{j\omega n(L_{md} + jL_{mq})I_d}{R_L'} \tag{11.18}$$

From Figure 11.7(c), I_d and I_q are derived as follows:

$$I_d = \frac{V_d + nI_o j\omega L_{md}}{jX_d} \tag{11.19a}$$

$$I_q = \frac{V_q + nI_o j\omega L_{mq}}{jX_q} \tag{11.19b}$$

Substituting (11.18) into (11.19), V_d and V_q are derived as follows:

$$V_d = I_d \left\{ 1 - \omega^2 C_d(L_{ld} + L_{md}) + \frac{\omega^3 n^2 L_{md}^2 C_d}{R_L'} + j\frac{\omega^3 n^2 L_{md} L_{mq} C_d}{R_L'} \right\} \bigg/ \omega C_d \tag{11.20a}$$

$$V_q = I_q \left\{ 1 - \omega^2 C_q(L_{lq} + L_{mq}) + \frac{\omega^3 n^2 L_{mq}^2 C_q}{R_L'} + j\frac{\omega^3 n^2 L_{mq} L_{md} C_q}{R_L'} \right\} \bigg/ \omega C_q \tag{11.20b}$$

Figure 11.8 $\Delta\theta$ w.r.t. pick-up displacement x assuming the 90° phase difference between I_d and I_q.

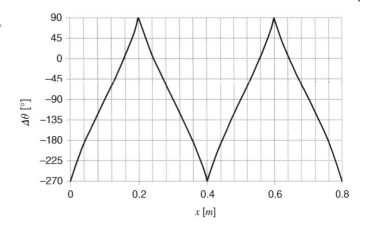

The phase difference between V_d and V_q $\Delta\theta$ is as follows:

$$\Delta\theta = \angle V_d - \angle V_d = \tan^{-1}\left(\frac{\frac{\omega^3 n^2 L_{md} L_{mq} C_d}{R_L'}}{1 - \omega^2 C_d(L_{ld} + L_{md}) + \frac{\omega^3 n^2 L_{md}^2 C_d}{R_L'}}\right)$$

$$- \left\{90 + \tan^{-1}\left(\frac{\frac{\omega^3 n^2 L_{mq} L_{md} C_q}{R_L'}}{1 - \omega^2 C_q(L_{lq} + L_{mq}) + \frac{\omega^3 n^2 L_{mq}^2 C_q}{R_L'}}\right)\right\} \tag{11.21}$$

As identified from (11.5) and (11.6), the magnetizing inductances L_{md} and L_{mq} are a function of pick-up displacement x in the equivalent circuit model:

$$L_{md} = L_{m0}\left|\sin\frac{2\pi x}{l_0}\right| \tag{11.22a}$$

$$L_{mq} = L_{m0}\left|\cos\frac{2\pi x}{l_0}\right| \tag{11.22b}$$

where L_{m0} is the maximum magnetizing inductance.

Using (11.21) and (11.22), $\Delta\theta$ can be plotted, as shown in Figure 11.8. As identified in (11.9), the relative phase difference between I_d and I_q is important to achieve the spatial invariant characteristic of the induced pick-up voltage. To guarantee the 90° phase difference between I_d and I_q, the phase $\Delta\theta$ should be controlled, as shown Figure 11.8, depending on the displacement x of the pick-up. To eliminate the phase shift dependency, a current phase controller is needed in order to maintain the phase difference between I_d and I_q regardless of x.

11.3.3 Control Circuits of the Proposed IPTS

A schematic of the current amplitude controller is shown in Figure 11.9. The same controller is used for both phases. Only the controller for the d-phase is described herein. I_{dsen} is the sensed secondary current of the isolated current transducer using the Hall effect. R_1 converts I_{dsen} into voltage. To sense the amplitude of I_d, a peak detector trances the envelope of this voltage. R_2 and R_3 are used as a voltage divider in order to make the divided voltage meet the proper input range of the error amp of the PWM controller. D_1 and D_2 are inserted to cancel out the forward

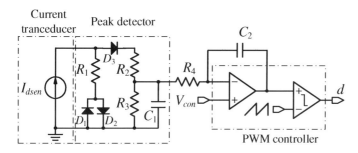

Figure 11.9 Current amplitude controller of the inverter.

drop voltage of D_3. V_{con} is the external control voltage of the PWM controller to be adjusted. Assuming that the gain of the error amplifier is infinite and R_2 and R_3 are much larger than R_1, the controlled amplitude of I_{dsen} is as follows:

$$I_{dsen} = \frac{R_2 + R_3}{R_1 R_3} V_{con} \tag{11.23}$$

Considering the sensing gain of the current transducer A_s, the controlled I_d is as follows:

$$I_d = \frac{I_{dsen}}{A_s}. \tag{11.24}$$

The current phase controller of the proposed dq I-type IPTS is shown in Figure 11.10. A clock generator generates ψ_d that is the predefined operating frequency of the ITPS. I_{dsen} is converted to voltage and ϕ_d, whose phase is the same as I_{dsen}, is generated by a comparator. To make a 90° phase difference between I_d and I_q, the phase shifter delays ϕ_d for 90° using an active integrator and a comparator. The comparator in the phase shifter generates a 90° phase-shifted pulse of ϕ_d, ϕ_{d90}. The phase difference between ϕ_{d90} and ϕ_q is compared by a phase-locked loop (PLL), and a frequency divider using T flip-flop guarantees the duty cycle of the q-phase inverter switching frequency ψ_q to 50%. The phase of ψ_q is controlled by the feedback loop of PLL so that the phase difference between ϕ_{d90} and ϕ_q becomes zero; that is, the phase difference between I_d and I_q becomes 90°. A timing diagram of the current phase controller is shown in Figure 11.11.

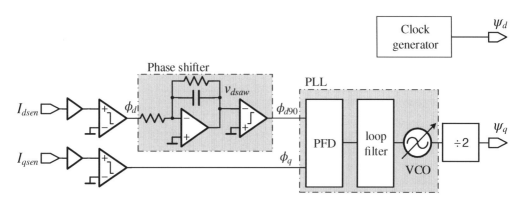

Figure 11.10 Current phase controller of the inverter of the proposed IPTS.

Figure 11.11 Timing diagram of the current phase controller.

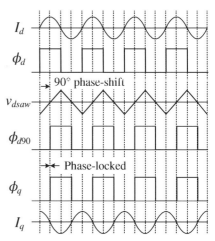

11.3.4 Simulation Verification of the Design

To verify the spatial characteristics of the proposed IPTS with respect to various pick-up displacements, transient simulation was performed using the magnetic transient simulator of ANSOFT Maxwell v.14.0, as shown in Figure 11.12. A bundle of transient simulation results for various pick-up displacements along the driving direction x is plotted as a three-dimensional plot with a color map. Because the structure of the proposed IPTS is spatially periodic, the simulation result of half of the spatial period $l_0/2$ can represent all of the spatial characteristics of the IPTS. As discussed in previous sections, V_o does not have deep valley parts depending on the displacements of the pick-up, but is nearly constant.

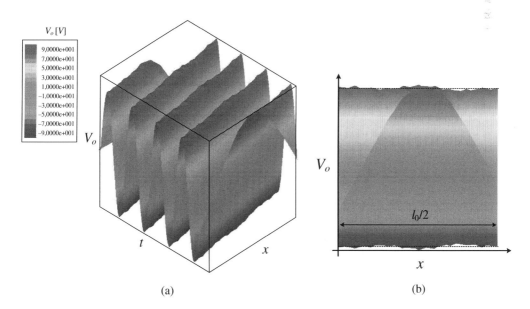

Figure 11.12 Transient FEA simulation result of the proposed IPTS: (a) simulated induced voltage bundle of the pick-up for various pick-up positions x and (b) side view of (a).

11.4 Example Design and Experimental Verifications

Figure 11.13 shows the proposed dq I-type IPTS including the dq-power supply rail, pick-up, and the inverter for experimental verification. The number of turns of each phase of the power supply rail and the pick-up are 4 and 40, respectively, and the airgap between the top of the magnetic poles of the power supply rail and pick-up core is 15 cm. As a pilot test bed to verify the spatial characteristic of the proposed IPTS, the input current of each phase of the power supply rail is scaled down to 20 A_{rms} and the operating frequency is 20 kHz, just as with the I-type rail. The pick-up winding is composed of two separated windings, which are connected in series so that the induced voltage of each pick-up is summed constructively. Part of the series resonant capacitor is installed between the windings to relieve voltage stress between the wires [18].

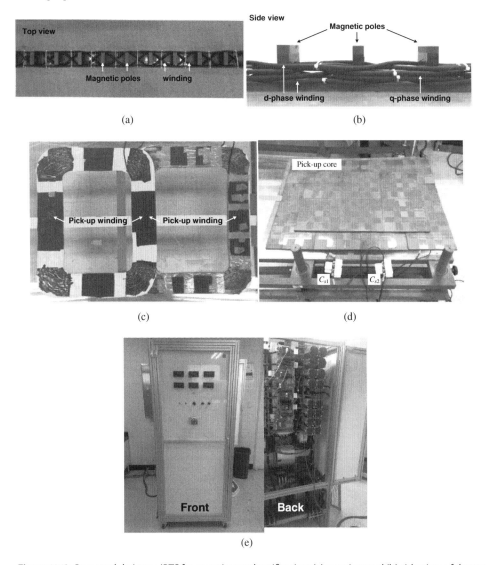

Figure 11.13 Proposed dq I-type IPTS for experimental verification: (a) top view and (b) side view of the proposed dq-power supply rail, (c) pick-up windings under the pick-up core, (d) implemented pick-up on a jig, and (e) implemented inverter for the proposed dq-IPTS.

Table 11.1 Parameters of the proposed IPTS

Parameters	Values	Parameters	Values
L_{ld}, L_{lq}	105 μH	Operating frequency	20 kHz
L_{md}, L_{mq}*	2.4 μH	N_{1d}, N_{1q}	4
C_d, C_q	660 nF	N_2	40
L_{ls}	1.89 mH	Pick-up size	90 cm × 120 cm
C_s	33.3 nF	l_0	80 cm
I_d, I_q	20 A_{rms}	Airgap	15 cm

*Measured magnetizing inductance where the pick-up is positioned at maximum output voltage.

The parameters of the proposed IPTS are summarized in Table 11.1. Because the magnetizing inductance of the proposed IPTS varies spatially, L_{md} and L_{mq} refer to the magnetizing inductance where the pick-up is positioned at the maximum output voltage for each phase of the power supply rail.

Figure 11.14 shows the waveforms of the currents of power supply rail, the detected phases of the power supply rail currents, the input phases of the PLL, and VCO control voltage of the PLL. The amplitude of the current of each phase is well regulated as the designated level by the current amplitude controller. From Figure 11.14(b), the PLL of the proposed current phase

Figure 11.14 Waveforms of the proposed IPTS: (a) currents of dq-power supply rail and its detected-phase and (b) input signals of PLL and VCO control voltage.

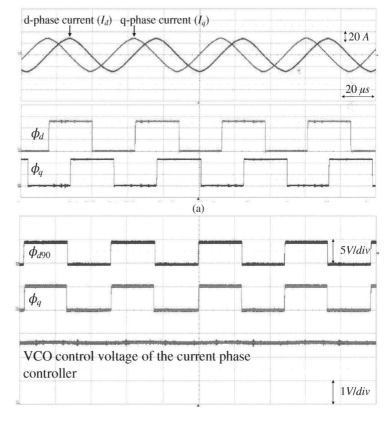

d-phase current (I_d) q-phase current (I_q)

↕20 A

20 μs

ϕ_d

ϕ_q

(a)

ϕ_{d90}

5V/div

ϕ_q

VCO control voltage of the current phase controller

1V/div

(b)

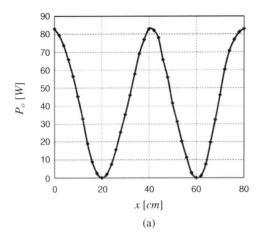

(a)

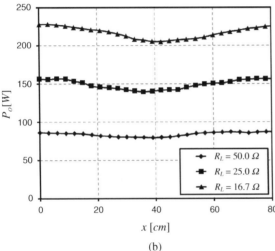

(b)

Figure 11.15 Measured output power of the proposed IPTS w.r.t. various displacements of the pick-up and load conditions. (a) Spatial output power variation when the d-phase of the power supply rail only is turned on with a load resistance of 50 Ω. (b) Spatial output power variation when both phases are turned on for various loads.

controller makes ϕ_{d90} and ϕ_q locked so that the phase difference between ϕ_{d90} and ϕ_q becomes zero. The VCO control voltage is also stably settled by a feedback loop in the steady state.

The measured spatial output power variation of the proposed dq I-type IPTS is shown in Figure 11.15. When only the d-phase of the power supply rail is turned on, the spatial output power characteristic is identical to the I-type one. There are two valleys where the output power nears zero in one spatial period l_0 of the power supply rail. The measured spatial output power characteristic with various load resistors is shown in Figure 11.15(b). Different from the characteristic of the I-type, the deep valley points do not exist, and the vehicles can be powered on the proposed dq-power supply rail with any displacements along the driving direction. The slight variation in Figure 11.15(b) comes mostly from the non-sinusoidal spatial distribution of magnetic fluxes and non-ideal control currents.

11.5 Conclusion

The main problem of the I-type IPTS, whose spatial power variation characteristic along the vehicle moving direction is 100%, has been reduced to 11%. Because the operating nature of the proposed IPTS is the same as that of the I-type, its advantages, such as narrower width, larger airgap, higher output power, and lower construction cost, are preserved. The proposed dq I-type power supply rail, composed of two phases of windings, was driven by the separated current sources whose amplitude and relative phase were controlled by the proposed current amplitude and phase controller embedded in the inverter.

Question 2 Compare the merits and demerits of the single I-type IPTS of the previous chapter and the DQ I-type IPTS of this chapter for a given dimension, power, and efficiency of IPTS, assuming that the fluctuated output power is well regulated and then provided for an on-board battery of the EV.

Problems

11.1 Extend the proposed dual IPTS to a three-phase IPTS, where an evenly distributed three-phase power supply rail and a corresponding balanced inverter are equipped.
    (a) Show that the induced output voltage also becomes constant for the same direction move of an EV.
    (b) Discuss the merits and demerits of the three-phase IPTS compared to the dual-phase IPTS in terms of winding complexity and effective area of core (i.e., the magnetic coupling factor between the power supply rail and pick-up).

11.2 Another possible solution to the fluctuated induced output voltage of the single-phase power supply rail is to adopt dual pick-up for the moving direction of an EV. Discuss the merits and demerits of the dual-phase power supply rail compared to the dual pick-up for the moving direction of an EV.

References

1 C.-S. Wang, O.H. Stielau, and G.A. Covic, "Design considerations for a contactless electric vehicle battery charger," *IEEE Trans.on Ind. Electron.*, vol. 52, no. 5, pp. 1308–1314, October 2005.

2 M.G. Egan, D.L. O'Sullivan, J.G. Hayes, M.J. Willers, and C.P. Henze, "Power factor corrected single stage inductive charger for electric vehicle batteries," *IEEE Trans. on Ind. Electron.*, vol. 54, no. 2, pp. 1217–1226, April 2007.

3 J. Sallan, J.L. Villa, A. Llombart, and J.F. Sanz, "Optimal design of ICPT systems applied to electric vehicle battery charge," *IEEE Trans. on Ind. Electron.*, vol. 56, no. 6, pp. 2140–2149, June 2009.

4 J.L. Villa, J. Sallan, J.F. Sanz Osorio, and A. Llombart, "High-misalignment tolerant compensation topology for ICPT systems," *IEEE Trans. on Ind. Electron.*, vol. 59, no. 2, pp. 945–951, February 2012.

5 M. Budhia, J.T. Boys, G.A. Covic, and C,Y Huang, "Development of a single-sided flux magnetic coupler for electric vehicle IPT charging systems," *IEEE Trans. on Ind. Electron.*, vol. 60, no. 1, pp. 318–328, January 2013.

6 R. Chen, C. Zheng, Z.U. Zahid, E. Faraci, W. Yu, J.-S. Lai, M. Senesky, D. Anderson, and G. Lisi, "Analysis and parameters optimization of a contactless IPT system for EV charger," in *IEEE APEC*, March 2014, pp. 1654–1661.

7 S.Y. Choi, J. Huh, W.Y. Lee, and C.T. Rim, "Asymmetric coil sets for wireless stationary EV chargers with large lateral tolerance by dominant field analysis," *IEEE Trans. on Power Electron.*, vol. 29, no. 12, pp. 6406–6420, December 2014.

8 J.G. Bolger, F.A. Kirsten, and L.S. Ng, "Inductive power coupling for an electric highway system," in *Proc. IEEE 28th Veh. Technol. Conf.*, March 1978, vol. 28, pp. 137–144.

9 C.E. Zell and J.G. Bolger, "Development of an engineering prototype of a road powered electric transit vehicle system," in *Proc. 32nd IEEE Veh. Technol. Conf.*, May 1982, vol. 32, pp. 435–438.

10 M. Eghtesadi, "Inductive power transfer to an electric vehicle-analytical model," in *Proc. 40th IEEE Veh. Technol. Conf.*, May 1990, pp. 100–104.

11 G.A. Covic, J.T. Boys, M.L.G. Kissin, and H.G. Lu, "A three phase inductive power transfer system for road powered vehicles," *IEEE Trans. on Ind. Electron.*, vol. 54, no. 6, pp. 3370–3378, December 2007.

12 G. Elliott, S. Raabe, G.A. Covic, and J.T. Boys, "Multiphase pickups for large lateral tolerance contactless power-transfer systems," *IEEE Trans. on Ind. Electron.*, vol. 57, no. 5, pp. 1590–1598, May 2010.

13 S.W. Lee, J. Huh, C.B. Park, N.S. Choi, G.H. Cho, and C.T. Rim, "On-line electric vehicle using inductive power transfer system," in *IEEE ECCE*, September 2010, pp. 1598–1601.

14 J. Huh, S.W. Lee, C.B. Park, G.H. Cho, and C.T. Rim, "High performance inductive power transfer system with narrow rail width for on-line electric vehicles," in *IEEE ECCE*, 2010, pp. 1598–1601.

15 M.L.G. Kissin, G.A. Covic, and J.T. Boys, "Steady-state flat-pickup loading effects in polyphase inductive power transfer systems," *IEEE Trans. on Ind. Electron.*, vol. 58, no. 6, pp. 2274–2282, June 2011.

16 J. Huh, S.W. Lee, W.Y. Lee, G.H. Cho, and C.T. Rim, "Narrow-width inductive power transfer system for online electrical vehicles," *IEEE Trans. on Power Electron.*, vol. 26, no. 12, pp. 3666–3679, December 2011.

17 S. Chopra and P. Bauer, "Driving range extension of EV with on-road contactless power transfer – a case study," *IEEE Trans. on Ind. Electron.*, vol. 60, no. 1, pp. 329–338, January 2013.

18 G.A. Covic and J.T. Boys, "Modern trends in inductive power transfer for transportation applications," *IEEE J. of Emerging and Selected Topics in Power Electron.*, vol. 1, no. 1, pp. 28–41, March 2013.

19 J. Shin, S. Shin, Y. Kim, S. Ahn, S. Lee, G. Jung, S.-J. Jeon, and D.g-H. Cho, "Design and implementation of shaped magnetic-resonance-based wireless power transfer system for road powered moving electric vehicles," *IEEE Trans. on Ind. Electron.*, vol. 61, no. 3, pp. 1179–1192, March 2014.

20 A. Zaheer, G.A. Covic, and D. Kacprzak, "A bipolar pad in a 10-kHz 300-W distributed IPT system for AGV applications," *IEEE Trans. on Ind. Electron.*, vol. 61, no. 7, pp. 3288–3301, July 2014.

21 M. Budhia, G. Covic, J. Boys, and M. Kissin, "Inductive power transfer system primary track topologies," PCT WO 2011/145953, November 24, 2011.

22 N. Suh, S. Chang, G. Cho, D. Cho, C.T. Rim, J. Huh, S. Lee, H. Kim, and C. Park, "Space-division multiple power feeding and collecting apparatus," PCT WO 2011/152678, December 8, 2011.

23 C. Park, S. Lee, S.-Y. Jeong, G.H. Cho, and C.T. Rim, "Uniform power I-type inductive power transfer system with DQ-power supply rails for on-line electric vehicles," *IEEE Trans. on Power Electron.*, vol. 30, no. 11, pp. 6446–6455, November 2015.

12

Ultra-Slim Power Rail (S-type)

12.1 Introduction

Among various EVs such as pure battery EVs (BEVs), hybrid EVs (HEVs), and plug-in hybrid EVs (PHEVs), road-powered EVs (RPEVs) have merits in solving battery-related problems [1–24]. In fact, RPEVs inherently need no battery for their traction because the required power is directly obtained from a power supply rail under the road while they are moving on it. Therefore, RPEVs are free from battery problems.

In this chapter, a new ultra-slim S-type power supply module with a 4 cm width is proposed to further reduce the construction time and cost for commercialization. By adopting the module concept for the power supply rail, the cable connection between power supply rails is no longer needed during its deployment. The proposed S-type power supply rail is optimized in terms of its core thickness and verified by FEA simulations and experiments, compared with the previous I-type power supply rail of the 4G OLEV. Experimental results for a prototype power supply module set are shown for the delivered power, efficiency, and lateral tolerance.

12.2 Ultra-Slim S-type Power Supply Rail Design

In general, the IPTS includes two subsystems [23]: one is a roadway subsystem to transfer power and the other is an on-board subsystem to receive power from the roadway subsystem. As shown in Figure 12.1, the roadway subsystem consists of a high-frequency inverter with its own utility rectifier, capacitor banks C_s, and power supply rails, while the on-board subsystem is composed of a pick-up coil set, capacitor banks C_o, a high-frequency rectifier, and a load resistor R_L, which may be replaced with a battery.

12.2.1 Overall Configuration of the Proposed Power Supply Rail and Pick-up

Through 4G developments of the OLEV, a significant improvement in lateral tolerance as well as a large airgap, high power efficiency, lower construction cost, and time reduction has been achieved. The construction cost and time of the power supply rail, however, should be further reduced for better commercialization because the construction cost of the power supply rail is critical for deploying the RPEV and longer construction time results in more traffic jams and extra deployment costs.

In order to mitigate these problems, the ultra-slim S-type power supply rail for RPEV, where the name "S-type" stems from the front shape of the power supply rail, as shown in Figure 12.2(c), is explained in this chapter. The S-type power supply rail has a width of only 4 cm

Wireless Power Transfer for Electric Vehicles and Mobile Devices, First Edition. Chun T. Rim and Chris Mi.
© 2017 John Wiley & Sons Ltd. Published 2017 by John Wiley & Sons Ltd.

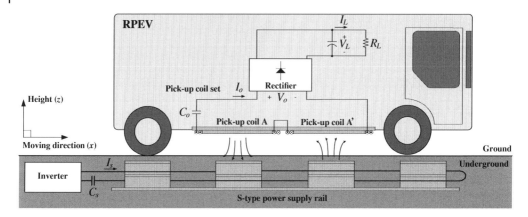

Figure 12.1 The configuration of the IPTS for an RPEV, adopting the proposed ultra-slim S-type power supply rail, where voltages and currents are rms values in the steady state.

by virtue of the S-shape configuration, which marks a decrease from the I-type module width of 10 cm. It is noteworthy that the S-type pick-up, adopted for a monorail system [27], has an S-shaped core plate, which provides higher power density and lower cost than other pick-ups such as E-, Z-, and U-types. The S-type pick-up has a similar configuration with the proposed S-type power supply rail, but the S-type pick-up moves along a pair of fixed cable while the proposed S-type power supply rail has a fixed cable firmly tied to the S-type core plate. Therefore, the details of the configuration and operating principle of the S-type pick-up and S-type power supply rail are quite different from each other even though the same S-shape core is used.

Each magnetic pole consists of ferrite core plates and power cables, and adjacent magnetic poles are connected by bottom core plates, as shown in Figure 12.2(a). The EMF for pedestrians around the power supply rail can be significantly reduced due to the opposite magnetic polarity of adjacent poles, as shown in Figure 12.2(b). The EMF generated from the I-type power supply rail and a flat-type pick-up coil set, which is basically the same structure as the proposed S-type one, was as low as 1.5 μT at a distance of 1 m from the power supply rail [17]. From this result, it is estimated that the S-type one has lower EMF than the I-type one because the width of the S-type one is narrower than that of I-type one, which will be well below the ICNIRP guideline of 27 μT. For a further EMF reduction, twisted pair cables between two magnetic poles can be used, which is left for further work. Moreover, a large lateral displacement d_{lat} is obtained [20] due to the small width w_t of the S-type power supply rail for a given pick-up width w_p, as follows:

$$d_{lat} \cong \frac{w_p}{2} - \frac{w_t}{2} \tag{12.1}$$

The proposed S-type power supply rail adopts a module concept, as shown in Figure 12.3(a), and makes it easier to fold itself by virtue of flexible thin power cables. Therefore, no power cable connection is required after being deployed, as shown in Figure 12.3(b). In general, a module may include an arbitrary number of magnetic poles, though only two poles are illustrated in Figure 12.3. For clarity, the deployment procedures of power supply rails are simplified as follows:

(a) To dig up a roadway and make enough space for the installation of power supply rails (only a 4 cm width is needed for the S-type power supply modules).

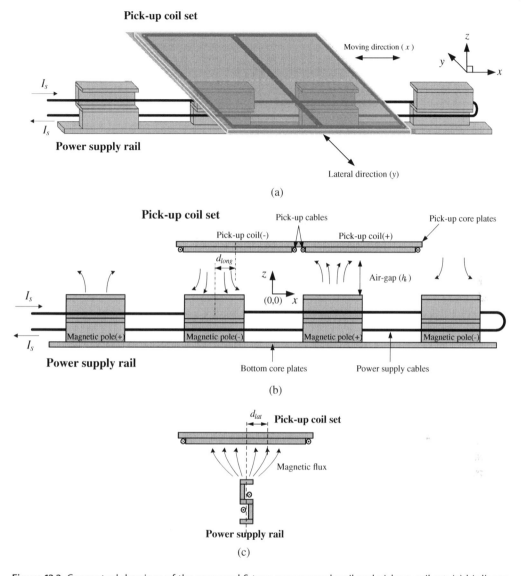

Figure 12.2 Conceptual drawings of the proposed S-type power supply rail and pick-up coil set: (a) bird's eye view, (b) side view, and (c) front view.

(b) To build infrastructures for its installation, and put power supply rails into the installation space. In addition, power cable connections between the power supply rails are needed (for the S-type modules, the cable connections are no longer needed because the modules were fabricated at a manufacturing factory with a cable connection and the modules are folded into each other and moved to a construction site).

(c) To pave up the roadway with concrete and asphalt.

In a word, the S-type one leads to a lower construction cost and deployment time because the impact of deploying the S-type power supply modules on the road surface is minimal and the change to existing road operation conditions is also greatly reduced.

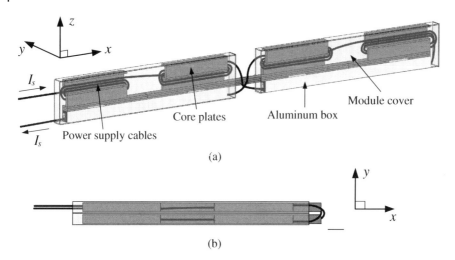

(a)

(b)

Figure 12.3 Configuration of the proposed ultra-slim S-type power supply modules including two magnetic poles: (a) bird's eye view for two unfolded modules and (b) top view of a folded module.

As a result, the proposed S-type power supply rail itself has three firm merits as follows:

(a) Substantial reduction in construction cost and time for commercialization of RPEV.
(b) Larger lateral tolerance of an IPTS.
(c) Further reduction for the EMF generated from a power supply rail for pedestrians.

12.2.2 Dimensions of the Proposed S-Type Power Supply Rail

The design parameters of the proposed S-type power supply rail include the pole widths w_t, w_m, w_b, pole thicknesses t_t, t_{up}, t_m, t_{low}, t_b, pole distance d_p, pole heights h_{up}, h_{low}, pole length l_p, airgap h_a, and the number of primary turns N_1, as shown in Figure 12.4. The pole widths are all selected as 3 cm, considering the target as 4 cm width of the power supply module and the thickness of a module cover as 0.4 cm. All the pole widths of the proposed S-type power supply rail are determined from the target output power, from which the ampere-turns $N_1 I_1$ of the power supply rail and $N_2 I_2$ of the pick-up coil set are determined. The best choice among combinations of N_1 and I_1 is selected in this chapter as 17 turns and 47 amperes to minimize the

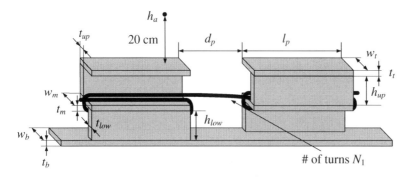

Figure 12.4 Definition of dimensions of the proposed ultra-slim S-type power supply rail.

thickness of the power supply rail for a given 800 ampere-turns. In order to meet this require-ment, the thickness of the core plates and the diameter of power cables are determined as 1 cm. Thus, the top cover width becomes 3 cm due to two layers of coil winding and one layer of core plate. The pole thicknesses will be determined by FEA simulation results in the subsequent sec-tion. The pole distance d_p of 20 cm, airgap h_a of 20 cm, and pole length l_p of 30 cm, which were optimally determined by the simulations and experiments for the I-type one [17], are reused for the proposed S-type power supply rail in this chapter for comparison. In addition, the ampere-turns N_1I_1 and N_2I_2 for the power supply rail and pick-up coil set are selected as 800 A-turn and 1500 A-turn, respectively, which were also used for the I-type power supply rail to achieve the output power of 25 kW [17].

12.2.3   Winding Methods of Power Cables

In order to reduce the conduction loss as well as the construction cost and time of power sup-ply rails, the winding methods of power cables should be carefully considered. There are two winding methods suggested in this chapter for the proposed S-type power supply rail: one is the multipole winding method and the other is the single-pole winding method, as shown in Fig-ure 12.5. The multipole method is to wound the cables for multiple poles in accordance with the concept shown in Figures 12.1 and 12.2; however, the single-pole method is to wind the cables for each pole, which is apparently different from the other concept.

For the multipole winding method, the total length of the power cables l_{multi} of a module is determined as follows:

$$l_{multi} = (mN_1)l_p + 2l_{side} + (m-1)N_1d_p \tag{12.2a}$$

$$\because l_{side} = \frac{\pi}{2} \sum_{i=odd}^{\frac{N_1-1}{2}} \left\{ t_m + \left(2\left[\frac{i+1}{2}\right] - 1\right)D \right\} \tag{12.2b}$$

where l_{side}, D, and m are the length of bended power cables at the end of each pole, the diameter of power cables, and the number of magnetic poles for each power supply module, respectively, assuming that no gap exists between the power cables and N_1 is the number of power supply side cable turns for only two magnetic poles with an odd number to connect to the next power supply module. The middle of the diameter of the power cables as well as the thickness of the

Figure 12.5 Proposed power cable winding methods for the S-type power supply rail: (a) multipole winding method and (b) single-pole winding method.

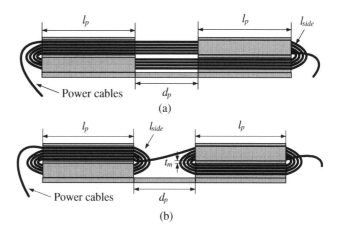

core plates are used to calculate l_{side}. Moreover, the floor function, for example, as in (12.2b), [5.9] = 5, is essential to consider as each layer of power cables can contain two power cables; for example, l_{side} should be the same when N_1 becomes "3 and 5" or "7 and 9."

During fabrication, one of the drawbacks of this method is that it is hard to wind power cables straightly due to the deflection of power cables caused by gravity. As a remedy to the problem, the single-pole winding method is considered in this chapter, where the total length of power cables l_{single} is determined as follows:

$$l_{single} = (mN_1)l_p + 2ml_{side} + (m-1)d_p \tag{12.3a}$$
$$\Delta l = l_{multi} - l_{single} = (m-1)\{(N_1-1)d_p - 2l_{side}\} \tag{12.3b}$$

For comparison purposes, the difference between the cable lengths is calculated from (12.2) and (12.3a), as given in (12.3b). Thus l_{multi} becomes larger than l_{single} for the proposed design, as shown in Figure 12.6, where l_{single} is about 20% shorter than l_{multi}. However, the reverse is also true, that is, $l_{multi} < l_{single}$ for a small d_p and large l_{side} in the case where $m > 1$ and $N_1 > 1$. It is found from (12.2b) that l_{side} increases as t_m or D increases. Therefore, it is not straightforward to choose the winding method, which should be determined case by case in practice.

Considering voltage stresses between the two winding methods, the single-pole winding method is much better than the multipole winding method because the voltage stress of the single-pole winding method between power cables is about a half that of the multipole method. In addition, a capacitor bank can be easily inserted between magnetic poles to compensate for its voltage stress in the single-pole case whereas in the multipole case this is not due to the lack of space between the magnetic poles.

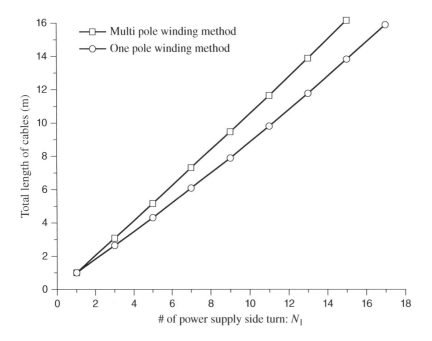

Figure 12.6 Comparison of a total cable length for the multipole winding method with that for the single-pole winding method under the proposed design. This tendency may change for different design conditions.

12.2.4 Optimum Core Size for Preventing Partial Saturation

The proposed S-type IPTS adopts the fully resonant mode [16–24] and the output power significantly drops when the resonant frequency of the pick-up deviates from the switching frequency. In order to prevent variation in the resonant frequency due to partial saturation of ferrite core plates, which is caused by the lack of ferrite core plates under the designed N_1I_1 and N_2I_2, FEA simulations using the dominant field analysis (DoFA) method [20], as also described in Chapter 15, were conducted to determine the minimum thickness of ferrite core plates. With the DoFA, the total magnetic flux density $\boldsymbol{B_t}$, represented as a vector in phasor form, can be determined not by complex AC current sources but by conventional DC current sources. The magnetic flux is induced from the two different sinusoid current sources of the power supply rail and pick-up coil set, which are quadrature in phase when the IPTS is in a fully resonant condition. The $\boldsymbol{B_t}$ can be determined from the magnetic flux densities $\boldsymbol{B_1}$ and $\boldsymbol{B_2}$ of the power supply and pick-up sides, respectively, which are in complex vector form by the superposition theorem in the Cartesian coordinate under the no core saturation condition as follows:

$$\boldsymbol{B_t} = \boldsymbol{B_1} + \boldsymbol{B_2} \tag{12.4a}$$

$$|\boldsymbol{B_t}|^2 = |\boldsymbol{B_1} + \boldsymbol{B_2}|^2 = |\boldsymbol{B_1}|^2 + |\boldsymbol{B_2}|^2 \tag{12.4b}$$

$$B_t \equiv |\boldsymbol{B_t}| = \sqrt{|\boldsymbol{B_1}|^2 + |\boldsymbol{B_2}|^2} = \sqrt{B_1^2 + B_2^2} \tag{12.4c}$$

For the S-type IPTS, the ferrite core plates of PL-5 by Samhwa Electronics were used and their saturated magnetic flux density B_{sat} was selected as 0.39 T at 100 °C instead of 0.5 T at 25 °C, considering the heat dissipations from the conduction, hysteresis, and eddy current losses of each coil set. According to the DoFA method, each of the maximum magnetic flux densities B_{1_max} and B_{2_max}, which are obtained from the conditions of $N_1I_1 = 800$ A turns, $N_2I_2 = 0$ A turns and $N_1I_1 = 0$ A turns, $N_2I_2 = 1500$ A turns, respectively, should be smaller than the reference magnetic flux density B_{ref} of 0.276 T obtained from $B_{sat}/\sqrt{2}$. From the FEA simulation results, it is found that the maximum magnetic flux density occurs at the bottom core plates and pick-up core plates as long as the pick-up coil set covers the magnetic poles symmetrically at the center. The minimum thickness of the bottom core plates t_b for the prevention of partial saturation was found to be 3 cm, as shown in Figure 12.7, where B_t is smaller than B_{sat} of 0.39 T. In addition, other parameters t_t, t_{up}, t_m, and t_{low} are determined as 1 cm and are not shown here.

To minimize the use of ferrite core plates, the bottom core plates of the S-type power supply rail are optimized by the FEA simulation, where additional core plates are inserted under the bottom core plates at $t_b = 2$ cm. As shown in Figure 12.8, each of the B_{1_max} and B_{2_max} is

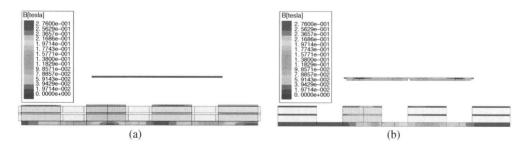

Figure 12.7 FEA simulation results for the power supply and pick-up core plates using the DoFA method: (a) $N_1I_1 = 800$ A-turn, $N_2I_2 = 0$ A-turn and (b) $N_1I_1 = 0$ A-turn, $N_2I_2 = 1500$ A-turn, where $t_b = 3$ cm.

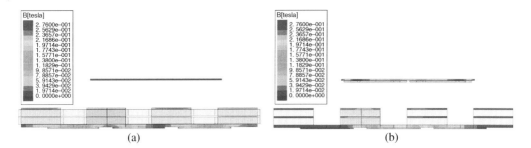

(a)   (b)

Figure 12.8 FEA simulation results for the power supply and pick-up core plates using the DoFA method by adding core plates under the bottom core plates: (a) $N_1 I_1 = 800$ A-turn, $N_2 I_2 = 0$ A-turn and (b) $N_1 I_1 = 0$ A-turn, $N_2 I_2 = 1500$ A-turn, where $t_b = 2$ cm.

under the B_{ref}, when the additional core plates with length of 40 cm, thickness of 1 cm, and width of 3 cm are inserted. By the optimization, more than 7% of the bottom core plates can be decreased, which reduces the ferrite core cost.

12.2.5 Top Cover Width and Magnetizing Inductance

For the design of the S-type power supply rail, the effect of the top cover width on the magnetizing inductance L_m should be investigated because the output power P_o is directly proportional to the square of L_m. By neglecting the magnetic resistance of ferrite core plates, where the relative permeability μ_r is typically over 2000, the total magnetic resistance between the power supply rail and pick-up coil set can be approximately determined as follows:

$$\Re \cong \frac{2h_a}{\mu_o A_{eff}} \tag{12.5}$$

Then L_m can be determined as follows:

$$L_m \cong \frac{N_1^2}{\Re} \cong \frac{N_1^2 \mu_o A_{eff}}{2h_a} = \frac{N_1^2 \mu_o \alpha w_t l_p}{2h_a} \tag{12.6}$$

where A_{eff} is defined as

$$A_{eff} \equiv \alpha w_t l_p \tag{12.7}$$

From (12.6), it is found that L_m is directly proportional to the effective area A_{eff} when other parameters are fixed; hence, it is worthwhile to know the variation of the effective area ratio α with respect to the top cover width w_t because w_t increases construction and deployment costs of the power supply rail. Since α cannot be calculated by hand at the moment, FEA simulations or experiments are crucial for obtaining α [17]. From the FEA simulation results for the proposed S-type power supply rail, α is found to decrease as the top cover width increases, as shown in Figure 12.9. The reason for this decrease of α is due to relatively less fringe effect for a larger top cover area. Comparing the proposed S-type rail with the previous I-type one, where the top cover widths were 3 cm and 7 cm, respectively, the effective area ratio of the S-type one is two times that of the I-type one. A smaller top cover width leads to a larger effective area ratio due to the fringe effect [20], as follows:

$$|V_{th}| = n\omega_s L_m |I_s| \cong \frac{n\omega_s N_1^2}{\Re} |I_s| = \frac{\omega_s N_1 N_2 \mu_o A_{eff}}{2h_a} |I_s| = \frac{\omega_s N_1 N_2 \mu_o \alpha w_t l_p}{2h_a} |I_s| \quad \because n = \frac{N_2}{N_1} \tag{12.8}$$

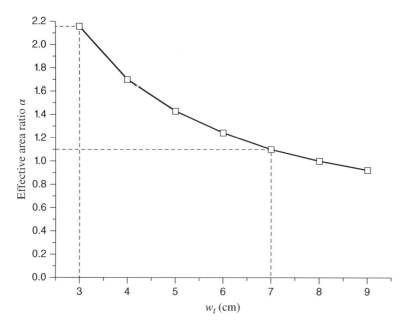

Figure 12.9 FEA simulation results of the effective area ratio versus the top cover width when h_a is 20 cm at the center.

From (12.8), it is found that the open-circuit voltage can be determined from the effective area ratio α and cover width w_t when other parameters are fixed; therefore, the open-circuit voltage for the top cover width of 3 cm of the S-type rail becomes about 84% of that for the top cover width of 7 cm of the I-type rail. This means that the open-circuit voltage decreases merely by 16% though the top cover width decreases by 2.3 times.

12.2.6 Power Loss Analysis

Even though the proposed S-type power supply rail may have high efficiency with the pick-up coil set, the effect of the power losses on the IPTS for an RPEV should be further investigated for stable operation because the IPTS deals with a high power of several tens of kilowatts and leads to several kilowatts of power loss, which are converted into heat dissipation.

The power losses can be classified as conduction losses in the power supply rail P_{con1} and pick-up coil set P_{con2}, respectively, core loss P_{core}, and full-bridge rectifier power loss P_{rec} as follows:

$$P_{loss} \equiv P_{con1} + P_{con2} + P_{core} + P_{rec} \tag{12.9a}$$
$$P_{con1} = |I_s|^2 r_1 \tag{12.9b}$$
$$P_{con2} = |I_o|^2 r_2 \tag{12.9c}$$
$$P_{rec} \cong 2|V_F||I_o| \tag{12.9d}$$
$$P_{core} = |I_o|^2 R_{oh} = P_{loss} - (P_{con1} + P_{con2} + P_{rec}) \tag{12.9e}$$

where V_F is the forward voltage of the IGBT of the rectifier.

The series hysteresis loss term R_{oh} is defined [20] as follows:

$$R_{oh} = \frac{(n\omega_s L_m)^2}{R_h} \tag{12.10a}$$

$$r_{eq} = R_{oh} + r_2 \tag{12.10b}$$

where the R_h is an equivalent hysteresis resistance [20] in the same way as a conventional transformer, as shown in Figure 12.10, and the P_{core} includes eddy current loss and hysteresis loss of the ferrite core plates and surrounding conductive materials of the IPTS.

As shown in Figure 12.10, the equivalent circuit of the IPTS including hysteresis loss resistance R_h can be represented [20], and the open-circuit output voltage can be (12.8), assuming that $R_h \gg \omega_s L_m$, as follows:

$$V_{th} = nI_s(R_h // j\omega_s L_m) = nI_s \frac{j\omega_s L_m R_h}{R_h + j\omega_s L_m} = nI_s \frac{j\omega_s L_m R_h}{R_h \left(1 + \frac{j\omega_s L_m}{R_h}\right)} \cong jn\omega_s L_m I_s + nI_s \frac{(\omega_s L_m)^2}{R_h}$$

$$\cong jn\omega_s L_m I_s \quad for \quad R_h \gg \omega_s L_m \because V_{th1} \equiv jn\omega_s L_m I_s, \quad V_{th2} \equiv nI_s \frac{(\omega_s L_m)^2}{R_h} \tag{12.11}$$

In addition, as shown in Figures 12.10(c) and (d), the Thevenin equivalent resistance r_{eq} from (12.10b) can be determined by the fully resonant condition [20] as follows:

$$jn^2\omega_s L_m + j\omega_s L_{l2} + \frac{1}{j\omega_s C_2} = 0 \tag{12.12}$$

The output resistance R_o can be defined [25, 26] as follows:

$$R_o \equiv \frac{V_o}{I_o} = \frac{8}{\pi^2} \frac{V_L}{I_L} = \frac{8}{\pi^2} R_L \tag{12.13}$$

In general, the maximum load power P_L for stable operation of the IPTS is limited by an output current I_o because the internal resistance of the pick-up coil r_2 is so high that the current leads to a temperature rise from (12.9c), which results in inferior characteristics of core plates and power cables, namely, the demagnetization of core plates and increase of electric resistance. Although the core loss is higher than the conduction loss in the pick-up coil set for the IPTS for RPEV, it does not significantly contribute to the temperature rise because the heat dissipation of the core loss is spread out to a large volume of ferrite cores whereas that of the conduction loss is concentrated in the center of the pick-up coil set, where four coil layers are partially merged.

12.3 Example Design and Experimental Verifications

The proposed S-type power supply rail and a pick-up coil set for experiments were fabricated, which allows a large effective area ratio as well as a large lateral tolerance for a given pick-up, as shown in Figure 12.11. The single-pole winding method for the power supply rail was selected due to a lower construction cost and easier fabrication compared to the multipole winding method. All the experiment parameters are the same as the simulation parameters listed in Table 12.1, where the number of the power supply side turns N_1 is determined to be 17 to enable the modules to fold themselves with flexible thin power cables. The internal resistances of the power supply rail r_1 and pick-up coil set r_2 include the equivalent series resistances (ESRs) of the capacitor banks for each side.

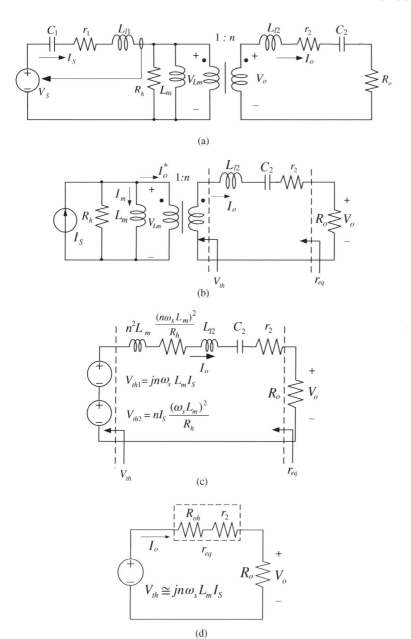

Figure 12.10 The equivalent circuits of the proposed IPTS: (a) an equivalent circuit including hysteresis loss resistance R_h, (b) simplified circuit, assuming constant current control, (c) a more simplified circuit at the output side, and (d) approximated final circuit.

The number of the pick-up side turns N_2 was determined to be 32, where 16 turns of power cables are allocated for each magnetic pole of the pick-up coil set and the 16-turn coils include two layers of 8-turn coils, as shown in Figure 12.11(c).

Due to the large inductance of the pick-up coil set, it is a challenge to compensate for its large voltage stress of about 10 kV by using a capacitor bank and may cause an isolation breakdown

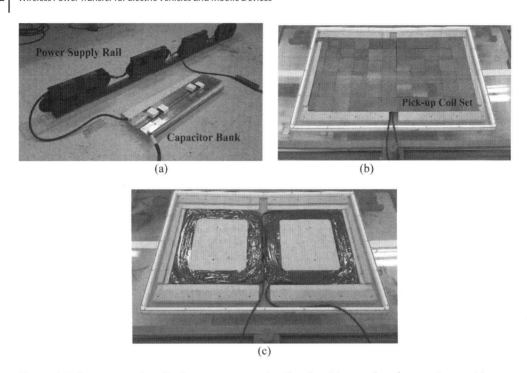

Figure 12.11 A prototype ultra-slim S-type power supply rail and a pick-up coil set for experiments: (a) power supply rail, (b) pick-up coil set with the pick-up core plates, and (c) pick-up coil set without the pick-up core plates.

Table 12.1 Measured parameters of the proposed S-type power supply rail

Parameter	Value	Parameter	Value
I_s	47 A	w_t	3 cm
f_s	20 kHz	w_m	3 cm
N_1	17 turns	w_b	3 cm
N_2	32 turns	t_t	1 cm
L_1	341 µH	t_{up}	1 cm
L_2	1.15 mH	t_m	1 cm
L_m	28.4 µH	t_{low}	1 cm
C_1	0.19 µF	t_b	2 cm
C_2	55 nF	l_p	30 cm
h_{up}	5 cm	d_p	20 cm
h_{low}	5 cm	r_1	0.08 Ω
h_a	20 cm	r_2	0.20 Ω
Pick-up size	100 cm × 80 cm (x,y)	Load	Resistive

between the pick-up layers. In order to solve this problem, a capacitor bank was divided into several capacitor banks, which are serially connected to each layer to enhance the isolation between the pick-up layers and to mitigate the voltage stress for each capacitor bank.

In addition, the proposed S-type power supply rail adopts a constant current controlled inverter [16–24], where the inverter operates at the switching frequency f_s, slightly higher than the resonant frequency f_r of the power supply rail, to always guarantee zero voltage switching (ZVS) operation.

12.3.1   Effective Area and Load Power

In order to verify the design of the proposed S-type rail, the effective area α along with the airgap h_a, the load power P_L, and the efficiency excluding an inverter were measured. Figure 12.12 shows that the effective area ratio reaches its peak value at an airgap of 14 cm and decreases as the airgap increases. As confirmed above, the measured effective area ratio at the airgap of 20 cm is as large as 3.5, which is about two times that of the I-type case [17]. As identified from (12.8), the open-circuit voltage is proportional to α and decreases by about 16% only for the proposed S-type, which may result in a decrease of load power of 29%.

As shown in Figure 12.13, the load power and efficiency along with the different resistive loads were measured at an airgap of 20 cm, where the efficiency is defined from the input of the power supply rail to the DC output of a rectifier. The efficiency was maintained at around 90% for the wide load range, and the maximum efficiency was found to be 91% at 9.5 kW, whereas the maximum load power was measured as 22 kW.

In cases where the efficiency takes account of the inverter power loss, the efficiency is significantly reduced by around 70% because the inverter used for the IPTS was designed for a higher power level IPTS of around 100 kW; therefore, the efficiency greatly improves with an adjusted inverter for the present application.

Figure 12.12 Measured effective area ratio versus airgap at the center.

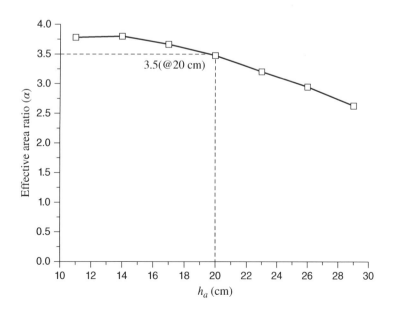

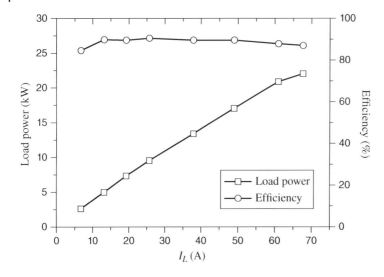

Figure 12.13 Measured load power and efficiency.

12.3.2 Power Loss and Thermal Integrity

In order to validate the thermal integrity of the IPTS, when the design output current I_o of 50 A flows into the pick-up side, the temperatures of the power supply rail and pick-up coil set, which are the highest temperatures in the IPTS, were measured as 35 °C and 50 °C, respectively, as shown in Figure 12.14, where the IPTS is safe and stable for the design condition even though an additional cooling system may be needed for the higher required power.

As mentioned earlier, the highest temperature in the IPTS lies in the center of the pick-up coil set due to the concentrated conduction loss, where four pick-up coil layers are located, as shown in Figure 12.14(b). All the power losses from (12.9a) to (12.9e) are calculated as listed in Table 12.2, and it was found that the sum of the P_{core} and P_{con2} accounts for more than 83% of the total power and the P_{core} was larger than the P_{con2}, as discussed above.

In order to obtain R_{oh}, the V_L along with various I_L were measured, as shown in Figure 12.15, and the r_{eq} can be obtained as follows:

$$r_{eq} \equiv \left| \frac{\Delta V_L}{\Delta I_L} \right| \quad for \quad V_{th} = constant \tag{12.14}$$

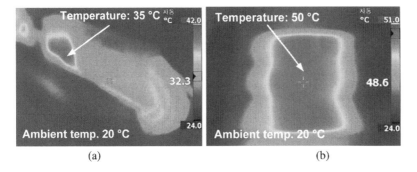

Figure 12.14 Measured highest temperatures of the S-type power supply rail and pick-up coil set after 50 min of operation the in thermal equilibrium state: (a) power supply rail and (b) pick-up coil set with the pick-up core plates.

Table 12.2 Calculated power losses of the proposed S-type IPTS

Parameter	Value
P_{loss}	2.00 kW
P_{con1}	0.18 kW
P_{con2}	0.50 kW
P_{rec}	0.15 kW
P_{core}	1.17 kW

From Figure 12.15, the r_{eq} was obtained as 0.651 Ω so that the R_{oh} can be determined as 0.45 Ω with the measured r_2 of 0.2 Ω, which has a good agreement with the extracted R_{oh} of 0.47 Ω obtained as follows:

$$R_{oh} = \frac{P_{core}}{|I_o|^2} = \frac{P_{loss} - (P_{con1} + P_{con2} + P_{rec})}{|I_o|^2} \tag{12.15}$$

12.3.3 Lateral and Longitudinal Tolerances

In order to see the spatial power variations, the load powers along lateral d_{lat} and longitudinal d_{long} displacements were measured for three different airgaps of 12 cm, 16 cm, and 20 cm; the road regulations of Korea and Japan [17] are currently 12 cm and 16 cm, respectively. As shown in Figure 12.16, it is found that the lateral tolerance increases at a higher airgap because the inductance of the pick-up coil set becomes relatively insensitive to lateral displacement as the airgap increases. In addition, the lateral tolerance of 30 cm, where the load power reached the −3 dB point having half of its maximum load power, was achieved at an airgap of 20 cm, which is 6 cm larger than the I-type one of 24 cm [17]. This lateral tolerance nearly corresponds

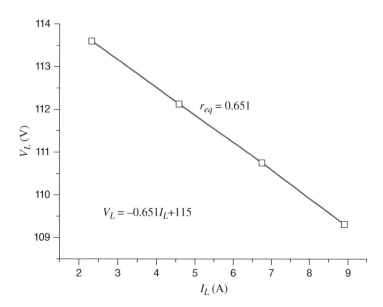

Figure 12.15 Measured equivalent resistance.

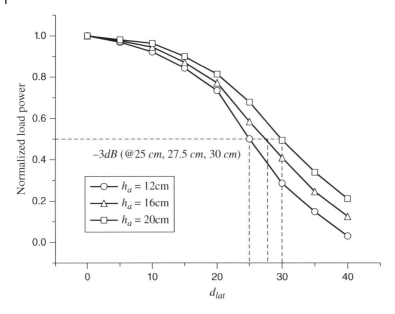

Figure 12.16 Measured normalized load power along the lateral displacement for different airgaps, where $I_s = 15$ A and $R_L = 50$ Ω.

to the theoretical value 28.5 cm of (12.1), where the pick-up width w_p is as small as 60 cm and the power supply rail width w_t is 3 cm.

Similar to the lateral tolerance case, the longitudinal tolerance increases for the higher airgap, but there is not much difference between the different airgaps, as shown in Figure 12.17. This longitudinal tolerance also nearly corresponds to the theoretical value of 15 cm [20], which is about half of a magnetic pole d_p of the power supply rail.

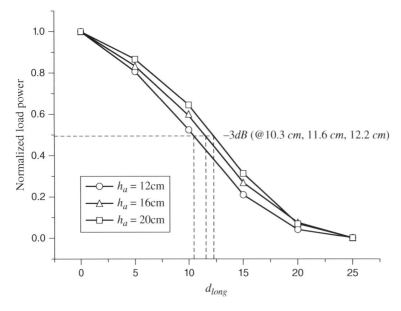

Figure 12.17 Measured normalized load power for the longitudinal direction at different airgaps, where $I_s = 15$ A and $R_L = 50$ Ω.

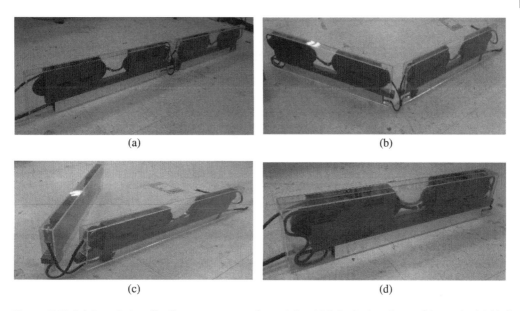

Figure 12.18 Fabricated ultra-slim S-type power supply modules: (a) fully deployed case, (b) one-third folded case, (c) two-thirds folded case, and (d) completely folded case.

12.4   Fabrication of the Flexible S-Type Power Supply Module

The proposed S-type power supply module, which has a width of only 4 cm and needs no cable connections after being deployed for the construction cost and time reduction, was fabricated as shown in Figure 12.18. The power supply module includes the power supply rail, transparent module cover, and aluminum box for capacitor banks and better heat transfer to the ground. As shown in Figure 12.19, the capacitor bank should be smaller than the aluminum box into which it is to be inserted and electrically isolated from the aluminum box because aluminum is a conductive material. In addition to this, the aluminum box also plays a role in canceling leakage magnetic fluxes flowing under the power supply rail. Moreover, each module, which has two magnetic poles, is serially connected to the capacitor bank installed in an adjacent module in order to mitigate high voltage stress for the capacitor bank due to the large self-inductance of the power supply rail. In general, thinner power supply cables for the S-type power supply rail are a more suitable choice with respect to the construction cost of a power supply rail. This is because a space usability for cable winding can be maximized with thinner power cables for

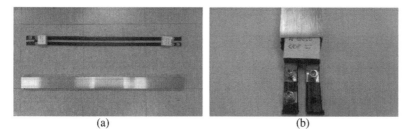

Figure 12.19 Aluminum box and capacitor banks for the S-type power supply modules: (a) separated case and (b) inserted case.

Table 12.3 Characteristics of I-type and S-type power supply rail structures.

	I-type	S-type
Rail width	10 cm	4 cm
Lateral tolerance	24 cm	30 cm
Airgap	20 cm	20 cm
Output power	27 kW/pick-up	22 kW/pick-up
Efficiency	74% at 27 kW	71% at 22 kW

a given size and ampere-turn of the power supply rail. However, it should be noted that there are several demerits to using many turns for the power supply rail, such as higher conduction loss due to the proximity effect and worse heat-transfer ability to ambient due to the high space usability for cable winding. Therefore, in this chapter, I_s of 50 A and 16 turns having a diameter of 0.9 cm was finally determined for the given 800 A-turn.

With the flexible thin power cables having a diameter of 0.9 cm, it is possible to conduct cable connection works for power supply modules not at the construction site but at a module manufacture facility to avoid long construction times, which leads to traffic jams and additional construction costs.

For commercialization, power supply modules should be robust to high humidity as well as repetitive external mechanical impacts at least for 10 years [18]. As a remedy, the S-type power supply module can be filled with epoxy to reinforce the S-type power supply rail and to protect the module from high humidity while the module cover endures external mechanical impacts.

As a conclusion, the characteristics of the I-type and S-type power supply rails are summarized, as shown in Table 12.3. Except for the rail width reduction and lateral tolerance increase, the output power and efficiency of the S-type one are slightly inferior to that of the I-type one, assuming that the pick-up size and ampere-turn of the power supply rail are the same as each other.

Question 1 In fabricating the proposed S-type power rail, as shown in Figure 12.18, the joint part of each module should be flexible but robust to electrical stress and humidity. What is your detail design of the joint part?

Question 2 As the inductance of each module may be large due to a lumped coil structure, the voltage stress could be large for a large power transfer. Calculate the voltage stress of a module, which should be compensated by the capacitor bank. This voltage stress, however, should be insulated in order not to cause voltage breakdown on the core surface.

12.5 Conclusion

The ultra-slim S-type power supply rail, which has a width of only 4 cm, for RPEVs has been fully verified throughout this chapter. In order to further reduce the construction cost of roadway infrastructure, which accounts for more than 80% of the total deployment cost for the commercialization of RPEVs, an explanation has been given of foldable power supply modules that are easy to carry and minimize construction time, where flexible power cables connect each foldable power supply module such that no connectors are needed during deployment. The decrease in the open output voltage for the S-type one is merely 17%, even though the top cover width has been decreased from the previous 7 cm (I-type) to the proposed 3 cm (S-type). By virtue of the ultra-slim shape, a large lateral displacement of 30 cm at the airgap of 20 cm was experimentally obtained, which is 6 cm larger than that of the I-type power supply rail. The S-type one has a lower EMF than the I-type one because the width of the S-type one is narrower than that of I-type one. The maximum efficiency excluding the inverter was 91% at 9.5 kW, whereas the maximum pick-up power was 22 kW.

Problems

12.1 Compare the merits and demerits of the proposed S-type IPTS and the I-type IPTS of the previous chapters for a given power and efficiency of IPTS.

12.2 The S-type IPTS has a heatsink problem like all other IPTSs in general. The molded structure with an aluminum plate can transfer heat generated from the power cable, core, and capacitor. For additional heatsink, multiple metal rods can be attached to the S-type power supply rail. Calculate the length and distance between each rod to dissipate the generated heat to the soil. Adopt appropriate assumptions to deal with this problem and consider the hottest summer and the coldest winter seasons.

References

1 J.G. Bolger, "Urban electric transportation systems: the role of magnetic power transfer," *IEEE Idea/Microelectronics Conference (WESCON94)*, 1994, pp. 41–45.

2 California PATH Program, "Road powered electric vehicle project track construction and testing program phase 3D," California PATH Research Paper, March 1994.

3 W. Zhang, S.C. Wong, C.K. Tse, and Q.H. Chen, "An optimized track length in roadway inductive power transfer systems," *IEEE Journal of Emerging and Selected Topics in Power Electronics*, vol. 2, no. 3, pp. 598–608, Sep. 2014.

4 M. Budhia, J.T. Boys, G.A. Covic, and C.-Y. Huang, "Development of a single-sided flux magnetic coupler for electric vehicle IPT charging systems," *IEEE Transactions on Industrial Electronics*, vol. 60, no. 1, pp. 318–328, January 2013.

5 G.A. Covic, L.G. Kissin, D. Kacprzak, N. Clausen, and H. Hao, "A bipolar primary pad topology for EV stationary charging and highway power by inductive coupling," in *IEEE Energy Conversion Congress and Exposition (ECCE)*, September 2011, pp. 1832–1838.

6 G.A. Covic and J.T. Boys, "Modern trends in inductive power transfer for transportation applications," *IEEE Journal of Emerging and Selected Topics in Power Electronics*, vol. 1, no. 1, pp. 28–41, March 2013.

7 G.R. Jagendra, L. Chun, G.A. Covic, and J.T. Boys, "Detection of EVs on IPT highways," *IEEE Journal of Emerging and Selected Topics in Power Electronics*, vol. 2, no. 3, pp. 584–597, September 2014.

8 L. Chun, G.R. Jagendra, J.T. Boys, and G.A. Covic, "Double-coupled systems for IPT roadway applications," *IEEE Journal of Emerging and Selected Topics in Power Electronics*, vol. 3, no. 1, pp. 37–49, March 2015.

9 J. Meins, "German activities on contactless inductive power transfer," in *IEEE Energy Conversion Congress and Exposition (ECCE)*, 2013.

10 O.C. Onar, J.M. Miller, S.L. Campbell, C. Coomer, C.P. White, and L.E. Seiber, "A novel wireless power transfer for in-motion EV/PHEV charging," in *IEEE Applied Power Electronics Conference and Exposition (APEC)*, 2013, pp. 3073–3080.

11 J.M. Miller, O.C. Onar, and P.T. Jones, "ORNL developments in stationary and dynamic wireless charging," in *IEEE Energy Conversion Congress and Exposition (ECCE)*, 2013.

12 Y. Yamauchi, K. Throngnumchai, S. Komiyama, and T. Kai, "Resonant circuit design to enhance the performance of a dynamic wireless charging system," in *International Electric Vehicle Technology Conference (EVTeC)*, May 2014, no. 20144038.

13 Y. Naruse and K. Throngnumchai, "Study on charging performance of a dynamic wireless charging system using twisted transmitter coils," in *International Electric Vehicle Technology Conference (EVTeC)*, May 2014, no. 20144037.

14 K. Throngnumchai, A. Hanamura, Y. Naruse, and K. Takeda, "Design and evaluation of a wireless power transfer system with road embedded transmitter coils for dynamic charging of electric vehicles," in *Electric Vehicle Symposium and Exhibition (EVS27)*, November 2013, pp. 1–10.

15 M. Mochizuki, Y. Okiyoneda, T. Sato, and K. Yamamoto, "2 kW WPT system prototyping for moving electric vehicle," in *International Electric Vehicle Technology Conference (EVTeC)*, May 2014, no. 20144029.

16 S.W. Lee, J. Huh, C.B. Park, N.S. Choi, G.H. Cho, and Chun T. Rim, "On-line electric vehicle (OLEV) using inductive power transfer system," in *IEEE Energy Conversion Congress and Exposition (ECCE)*, 2010, pp. 1598–1601.

17 J. Huh, S.W. Lee, W.Y. Lee, G.H. Cho, and Chun T. Rim, "Narrow-width inductive power transfer system for on-line electrical vehicles (OLEV)," *IEEE Transactions on Power Electronics*, vol. 26, no. 12, pp. 3666–3679, December 2011.

18 S.Y. Choi, J. Huh, W.Y. Lee, S.W. Lee, and Chun T. Rim, "New cross-segmented power supply rails for road powered electric vehicles," *IEEE Transactions on Power Electronics*, vol. 28, no. 12, pp. 5832–5841, December 2013.

19 S. Lee, B. Choi, and Chun T. Rim, "Dynamics characterization of the inductive power transfer system for on-line lectric vehicles (OLEV) by Laplace phasor transform," *IEEE Transactions on Power Electronics*, vol. 28, no. 12, pp. 5902–5909, December 2013.

20 S.Y. Choi, J. Huh, W.Y. Lee, J.G. Cho, and Chun T. Rim, "Asymmetric coil sets for wireless stationary EV chargers with large lateral tolerance by dominant field analysis," *IEEE Transactions on Power Electronics*, vol. 29, no. 12, pp. 6406–6419, December 2014.

21 J. Shin, S. Shin, Y. Kim, S. Ahn, S. Lee, G. Jung, S. Jeon, and D. Cho, "Design and implementation of shaped magnetic-resonance-based wireless power transfer system for road powered moving electric vehicles," *IEEE Transactions on Industrial Electronics*, vol. 61, no. 3, pp. 1179–1192, March 2014.

22 S.Y. Choi, B.W. Gu, S.W. Lee, W.Y. Lee, J. Huh, and Chun T. Rim, "Generalized active EMF cancel methods for wireless electric vehicles," *IEEE Transactions on Power Electronics*, vol. 29, no. 11, pp. 5770–5783, November 2014.

23 S.Y. Choi, B.W. Gu, S.Y. Jeong, and Chun T. Rim "Advances in wireless power transfer systems for road powered electric vehicles," *IEEE Journal of Emerging and Selected Topics in Power Electronics*, doi: 10.1109/JESTPE.2014.2343674.

24 S.Y. Choi, S.Y. Jeong, E.S. Lee, B.W. Gu, S.W. Lee, and Chun T. Rim "Generalized models on self-decoupled dual pick-up coils for large lateral tolerance," *IEEE Transactions on Power Electronics*, submitted.

25 R.L. Steigerwald, "A comparison of half-bridge resonant converter topologies," *IEEE Transactions on Power Electronics*, vol. 3, no. 2, pp. 174–182, April 1998.

26 Chun T. Rim and G.H. Cho, "Phasor transformation and its application to the DC/AC analyses of frequency phase-controlled series resonant converters (SRC)," *IEEE Transactions on Power Electronics*, vol. 5, no. 2, pp. 201–211, April 1990.

27 G.A.J. Elliott, G.A. Covic, D. Kacprzak, and J.T. Boys, "A new concept: asymmetrical pick-ups for inductively coupled power transfer monorail systems," *IEEE Transactions on Magnetics*, vol. 42, no. 10, pp. 3389–3391, 2006.

13

Controller Design of Dynamic Chargers

13.1 Introduction

The depletion of petroleum and global warming make the electric vehicle (EV) the most promising means of transport, and various types of EVs, such as hybrid electric vehicle (HEV), plug-in HEV (PHEV), and pure battery EV (BEV), have been developed. The high price, large size, and relatively short distance per charge of the battery in the EVs, however, make it difficult to commercialize them. To alleviate the battery problems, the road-powered EVs (RPEVs), which are based on the inductive power transfer system (IPTS) have been developed [1–11]. Recently, the on-line electric vehicle (OLEV), one of the most advanced RPEVs so far, has been developed and successfully deployed at various public sites [5–7]. Its concept is shown in Figure 13.1 and the overall schematic of the IPTS is shown in Figure 13.2.

To improve the power transfer efficiency, the "fully resonated current source IPTS" was proposed for the OLEV [9–11]. The IPTS constantly controls the primary coil current i_1, and the magnetizing inductance L_m and leakage inductance L_{l2} of the pick-up are resonated with the secondary capacitor C_2. By doing so, the output voltage of the IPTS is nearly constant for various load conditions. The pick-up current, however, highly fluctuates because of the abrupt in-rush movement of the vehicle and the rapid load current change. The pick-up undergoes severe voltage stress and breakdown may occur in the capacitors during this transient state. One of the worst-case scenarios is that the LC resonant tank of the pick-up, that is, i_2, is maximally energized by the inverter whereas the output voltage is zero; this is a very probable situation in practice when a fast OLEV rushes into the primary coil. In this case, i_2 can be largely fluctuated and its maximum value should be clearly identified in order to design the ratings of the C_2. Hence, not only the static analysis but also the large signal dynamic analysis on i_2 is quite essential for various load conditions and circuit parameters.

To find an appropriate dynamic model, the recently proposed Laplace phasor transform theory [19], which was developed for the dynamic analysis of phasored circuits [12–18], has been adopted in this chapter. The conventional DQ transformations [20,21] could be used to analyze AC systems if the system order is less than three, but it is nearly impossible to obtain the dynamics of the IPTS by the traditional techniques because the system order is very high, as identified from Figure 13.2. On the other hand, the linear time-invariant dynamic model for a very high order AC system can be easily obtained by the unified general phasor transform [18,19], and the analysis of the phasor transformed circuit is straightforward because conventional circuit analysis techniques such as Kirchhoff's voltage and current laws, Thevenin's theorem, and Norton's theorem can be applied to the phasored circuits.

In this chapter, the Laplace phasor transform [19] is first applied to the practical real system and the large signal dynamic model for the pick-up current of the OLEV is fully developed.

Wireless Power Transfer for Electric Vehicles and Mobile Devices, First Edition. Chun T. Rim and Chris Mi.
© 2017 John Wiley & Sons Ltd. Published 2017 by John Wiley & Sons Ltd.

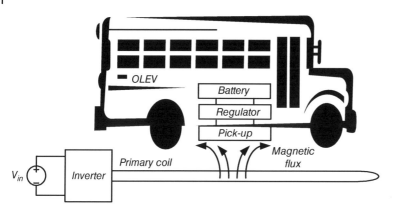

Figure 13.1 The concept of the OLEV system.

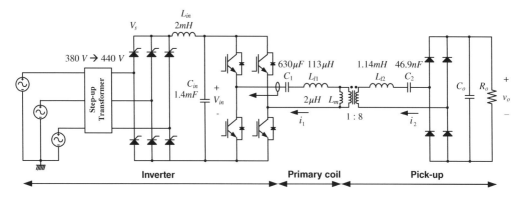

Figure 13.2 The overall schematic of the IPTS for the OLEV. The on-board regulator and battery is replaced by R_o.

13.2 Large Signal Dynamic Model for the OLEV IPTS

13.2.1 Operation Principle of the OLEV

The key operation concept of the IPTS for OLEV is to control the primary coil current i_1 as a constant current source, where the C_2 is resonated with $n^2 L_m + L_{l2}$ to maximize power transfer, as shown in Figure 13.3. Under this condition, the output voltage V_o in the steady state is equal to $n\omega_s L_m I_1$ [11], where the ω_s is the switching angular frequency and I_1 is the magnitude of the primary coil current i_1 in the steady state. Because the primary current I_1 is constant regardless of the load condition, the output voltage of the IPTS of an OLEV is kept constant;

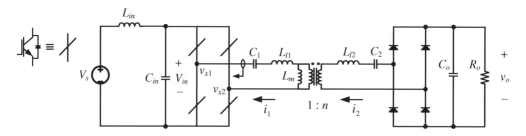

Figure 13.3 Simplified overall schematic of the IPTS for the OLEV.

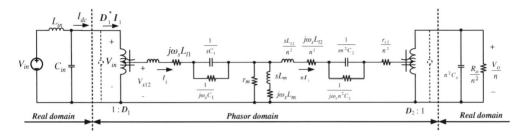

Figure 13.4 The overall Laplace phasor transformed equivalent circuit of the IPTS from the source side, removing the transformer. The bold letters indicate the complex variables in the Laplace domain. The D_1 and D_2 indicate the complex turn ratios of the complex transformers for the inverter and diode rectifier, respectively.

hence, this IPTS can effectively supply wireless power to multiple OLEVs on the primary coil simultaneously.

In addition, the intrinsic resonant frequency ω_i of the $L_m + L_{l1}$ and the C_1 is set to a slightly lower frequency than the inverter switching frequency ω_s so that the IPTS can be always inductive in the primary coil, which results in the zero voltage switching (ZVS) of the inverter [11]. Because the pick-up and primary coil of the IPTS is loosely coupled, the L_m is much lower than the L_{l1}. Therefore, the ω_i does not vary largely in accordance with the number of OLEVs on the primary rail and the air-gap between the pick-up and primary coil; hence, the stable control of the i_1 is possible at the fixed switching frequency.

13.2.2   Large Signal Dynamic Analysis of the Pick-up Current

To develop the large signal model of the pick-up current i_2, the Laplace phasor transform [19], which converts the rotatory AC domain circuit to the stationary phasor domain circuit in the frequency domain, is adopted. The overall phasor transformed circuit of the IPTS for the OLEV is shown in Figure 13.4. The dotted small circles connect the real domain and the phasor domain, and is called the "real part operator" [18]. In Figure 13.4, the complex transformer turn ratio D_1, that is, $D_1 e^{j\theta_1}$ where D_1 is the voltage conversion ratio and θ_1 is the phase difference between the source voltage V_{in} and V_{x12}, is defined as

$$V_{x12} = V_{in}D_1, \quad I_{dc} = Re\left(I_1 D_1^*\right) \tag{13.1}$$

where the detailed transform process is explained in [18] and [19]. In this process, the real part operator $Re()$ regards the Laplacian "s" as a real number, and was named the "pseudo real Laplacian" [19]. For example, $Re(sL + R + j\omega L) = sL + R$ is valid for this pseudo real Laplacian. In Figure 13.4, the r_m indicates the side effects of the coils of the IPTS, which include the iron loss and eddy current loss of the coils. The r_{s2} includes the copper loss of the pick-up coil and the dynamic resistance of the bridge diodes. Because the inverter primary coil current i_1 is constantly controlled by the inverter, it becomes a constant current source, as shown in Figure 13.5(a). In general, the system dynamics of the IPTS is much slower than the switching period of the inverter; hence, the absolute value of Laplacian "s" can be artificially much smaller than the ω_s as follows [14]:

$$|s| \ll \omega_s \tag{13.2}$$

Therefore, the impedance of the pick-up coil Z_2 can be derived using (13.2) as

$$Z_2 = \frac{sL_{l2}}{n^2} + \frac{j\omega_s L_{l2}}{n^2} + \frac{1}{sn^2 C_2 + j\omega_s n^2 C_2} = \frac{sL_{l2}}{n^2} + \frac{j\omega_s L_{l2}}{n^2} + \frac{1}{j\omega_s n^2 C_2} \cdot \frac{1}{1 + \dfrac{s}{j\omega_s}} \tag{13.3}$$

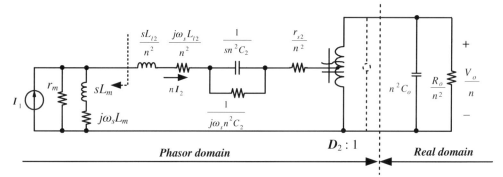

(a) Equivalent circuit of the IPTS assuming a constant current source

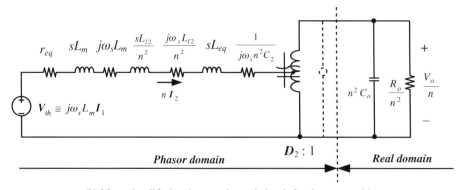

(b) More simplified and approximated circuit for the source side

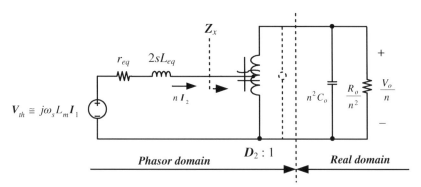

(c) The most simplified circuit in the resonant condition

Figure 13.5 Dynamic equivalent circuits of the IPTS in the complex Laplace domain.

By the condition (13.2), it can be approximately as follows:

$$Z_2 \cong \frac{sL_{l2}}{n^2} + \frac{j\omega_s L_{l2}}{n^2} + \frac{1}{j\omega_s n^2 C_2}\left(1 - \frac{s}{j\omega_s}\right) = \frac{sL_{l2}}{n^2} + \frac{j\omega_s L_{l2}}{n^2} + \frac{1}{j\omega_s n^2 C_2} + \frac{s}{\omega_s^2 n^2 C_2} \quad (13.4)$$

In (13.4), the last term can be expressed as follows:

$$\frac{s}{\omega_s^2 n^2 C_2} = sL_{eq} \quad (13.5)$$

where

$$L_{eq} = \left(L_m + \frac{L_{l2}}{n^2} \right) \cdot \left(\frac{\omega_{r2}}{\omega_s} \right)^2, \quad \omega_{r2} = \frac{1}{\sqrt{(n^2 L_m + L_{l2})C_2}} \tag{13.6}$$

To apply Thevenin's theorem to the dotted arrow region in Figure 13.5(a), the open-circuit equivalent impedance is obtained as follows:

$$\boldsymbol{Z}_{th} = (sL_m + j\omega_s L_m)//r_m = \frac{sL_m + j\omega_s L_m}{1 + \dfrac{sL_m + j\omega_s L_m}{r_m}} \tag{13.7}$$

Because the r_m is much larger than the $|j\omega_s L_m|$ or $|sL_m + j\omega_s L_m|$, (13.7) can be approximately represented as

$$\frac{sL_m + j\omega_s L_m}{1 + \dfrac{sL_m + j\omega_s L_m}{r_m}} \cong (sL_m + j\omega_s L_m)\left(1 - \frac{sL_m + j\omega_s L_m}{r_m} \right) \tag{13.8}$$

By applying the condition (13.2) to (13.8), it can be simplified as follows:

$$\boldsymbol{Z}_{th} \cong sL_m + j\omega_s L_m + \frac{\omega_s^2 L_m^2}{r_m} \tag{13.9}$$

In addition, the Thevenin voltage source is defined as

$$\boldsymbol{V}_{th} = \{(sL_m + j\omega_s L_m)//r_m\}I_1 \tag{13.10}$$

Using (13.7) to (13.9), it can be approximately expressed as follows:

$$\boldsymbol{V}_{th} \cong \left(sL_m + j\omega_s L_m + \frac{\omega_s^2 L_m^2}{r_m} \right)I_1 \tag{13.11}$$

Because $|sL_m + j\omega_s L_m|$ is much larger than $\omega_s^2 L_m^2/r_m$ so far as $r_m \gg \omega_s L_m$ and $|sL_m|$ is much smaller than $|j\omega_s L_m|$, the Thevenin voltage source of (13.11) can be simplified as follows:

$$\boldsymbol{V}_{th} \cong j\omega_s L_m I_1 \tag{13.12}$$

In addition, the equivalent resistance r_{eq} is defined as follows:

$$r_{eq} = \frac{\omega_s^2 L_m^2}{r_m} + \frac{r_{s2}}{n^2} \tag{13.13}$$

Therefore, the first approximated equivalent circuit is obtained from Figure 13.5(a) as shown in Figure 13.5(b).

If the resonant angular frequency ω_{r2} in (13.6) is equal to the switching frequency ω_s, the equivalent inductance L_{eq} is equal to $L_m + L_{l2}/n^2$ and the most simplified equivalent circuit can be obtained, as shown in Figure 13.5(c). In this figure, the dynamic model of i_2 can be derived if the $\boldsymbol{Z}_x$ is known. To obtain the $\boldsymbol{Z}_x$, the phasor transformed circuit of the diode rectifier and the output impedance Z_o is as shown in Figure 13.6. Because the diode rectifier is a sort of current-controlled rectifier (CSR), the phase of the test current $\boldsymbol{I}_x$ determines the phase of the complex

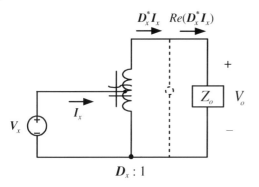

Figure 13.6 The phasor transformed circuit of the diode rectifier and the output impedance.

transformer turn ratio D_x, as follows [14]:

$$I_x = I_x e^{j\theta_x} \rightarrow D_x = D_x e^{j\theta_x}, \quad D_x = \frac{2\sqrt{2}}{\pi} \tag{13.14}$$

Therefore, the test voltage V_x is determined as

$$V_x = V_o D_x, \tag{13.15}$$

and the Z_x is determined as follows:

$$Z_x = \frac{V_x}{I_x} = \frac{V_o D_x}{I_x} = \frac{Re\left(I_x D_x^*\right) Z_o \cdot D_x}{I_x} \tag{13.16}$$

$$where\ Z_o = \frac{R_o}{n^2(1 + sC_o R_o)}$$

From (13.14), $I_x D_x^*$ can be derived as

$$I_x D_x^* = I_x e^{j\theta_x} \cdot D_x e^{-j\theta_x} = I_x D_x = Re\left(I_x D_x^*\right) \tag{13.17}$$

Finally, the Z_x is determined as

$$Z_x = \frac{I_x D_x^* Z_o \cdot D_x}{I_x} = D_x^2 Z_o \tag{13.18}$$

The final equivalent circuit can be derived, as shown in Figure 13.7, and the phasor I_2 can be obtained as

$$I_2 = jG_2 \cdot \frac{1 + sC_o R_o}{\dfrac{s^2}{\omega_2^2} + \dfrac{s}{Q_2 \omega_2} + 1} I_1 \tag{13.19}$$

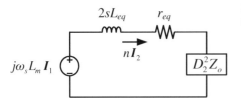

Figure 13.7 The final simplified phasor transformed circuit of the IPTS.

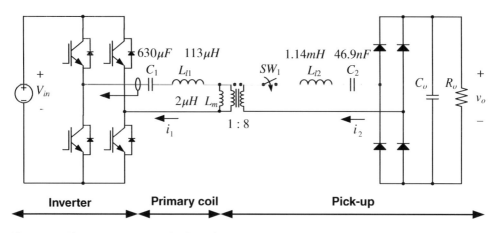

Figure 13.8 The experiment setup for the pick-up current dynamics study.

where

$$G_2 = \frac{n\omega_s L_m}{D_2^2 R_o + n^2 r_{eq}}, \quad \omega_2 = \frac{1}{n}\sqrt{\frac{D_2^2 R_o + n^2 r_{eq}}{2 L_{eq} C_o R_o}}$$

$$\tag{13.20}$$

$$Q_2 = \frac{\sqrt{(D_2^2 R_o + n^2 r_{eq})(2 L_{eq} C_o R_o)}}{n(2 L_{eq} + C_o R_o r_{eq})}, \quad \zeta_2 = \frac{1}{2 Q_2} = \frac{n(2 L_{eq} + C_o R_o r_{eq})}{2\sqrt{(D_2^2 R_o + n^2 r_{eq})(2 L_{eq} C_o R_o)}}$$

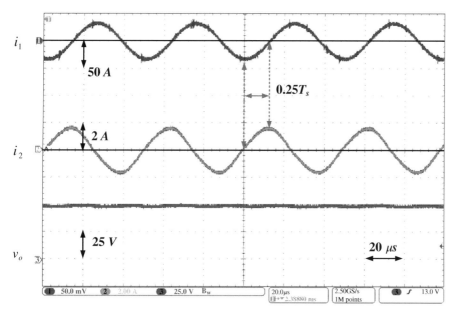

Figure 13.9 The steady-state waveforms for the primary coil current i_1, pick-up current i_2, and the output voltage v_o. The T_s is the period of the switching frequency ω_s.

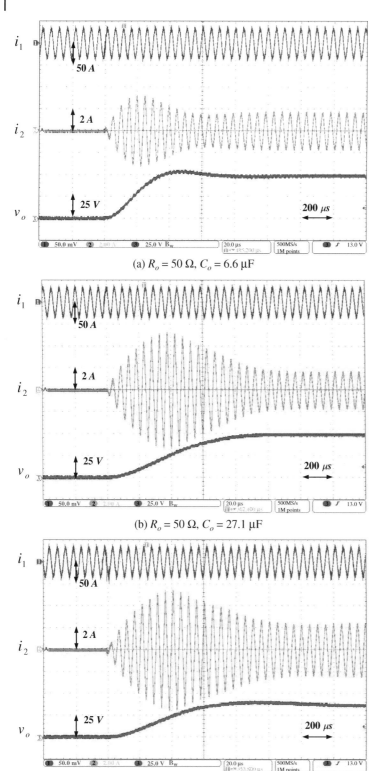

Figure 13.10 The measured step response of the pick-up current i_2 and the out voltage v_o for various load conditions.

(a) $R_o = 50\ \Omega,\ C_o = 6.6\ \mu F$

(b) $R_o = 50\ \Omega,\ C_o = 27.1\ \mu F$

(c) $R_o = 25\ \Omega,\ C_o = 27.1\ \mu F$

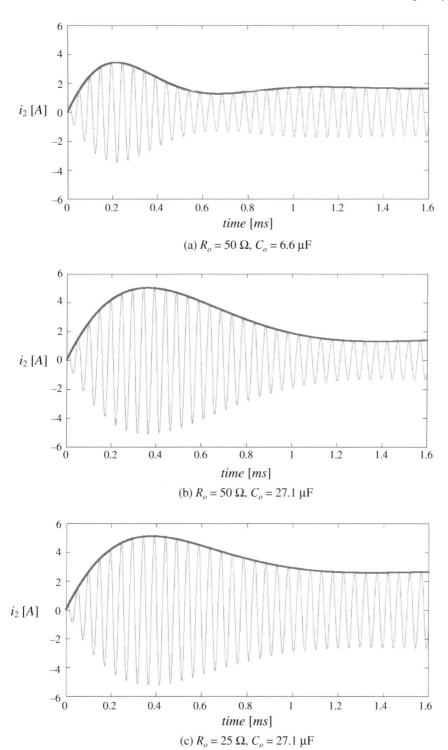

(a) $R_o = 50\ \Omega$, $C_o = 6.6\ \mu F$

(b) $R_o = 50\ \Omega$, $C_o = 27.1\ \mu F$

(c) $R_o = 25\ \Omega$, $C_o = 27.1\ \mu F$

Figure 13.11 The simulated step response (AC waveform) and the step response of (13.19) (envelope waveform) for the pick-up current i_2.

13.3 Example Design and Experimental Verifications

To experiment for the pick-up current dynamics, the switch SW_1 was inserted in the pick-up coil, as shown in Figure 13.8. The inverter constantly controls the primary coil current i_1, and the resonant frequency of the primary coil was set to slightly lower than the switching frequency of 20 kHz of the inverter. To check the resonant conditions of the pick-up coil, the steady-state waveforms were measured, as shown in Figure 13.9. As expected in (13.19), the phase difference between the i_1 and i_2 was about $\pi/2$, as shown in Figure 13.9. This means that the impedance of the $L_m + L_{l2}/n^2$ is successfully canceled by the impedance of the $n^2 C_2$.

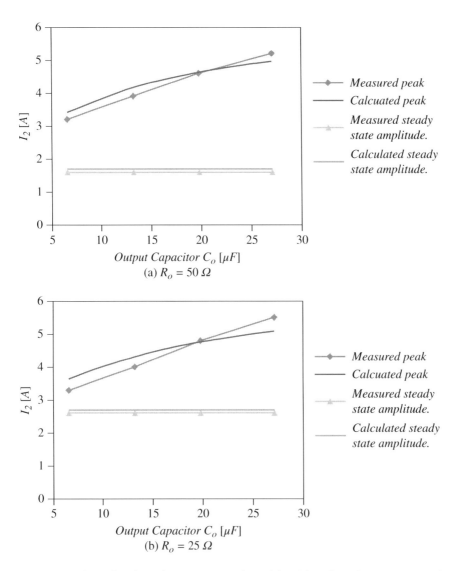

Figure 13.12 The peak and steady-state current values of the pick-up for various output capacitors.

Figure 13.13 The measured peak current of the pick-up current for various output resistances.

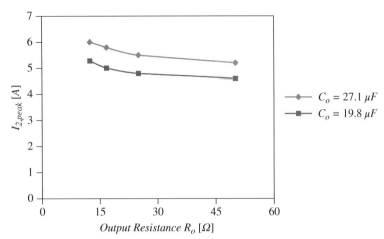

Initially, the SW_1 was open and the i_1 was set to 40 A$_{peak}$. After the i_1 was stabilized to its steady-state value, the SW_1 was turned on and the dynamics of the i_2 was observed for various load conditions, as shown in Figure 13.10. For comparison, the simulation results and the step response of (13.19) were shown in Figure 13.11. The step response of (13.19) is well matched with the simulated pick-up current by PSIM. In that process, the r_{eq} was found by comparing the experimental waveforms, and is about 0.15 Ω. The comparison plot for the maximum and the steady-state current of the i_2 is shown in Figure 13.12.

The experimental results and the step response of (13.19) show that the maximum value of the pick-up current i_2 varyies with the load conditions. The maximum value increases with a larger output capacitance for a given output resistance. The remarkable characteristic of the IPTS from the results is that the maximum values of the pick-up current for different output resistances are not significantly changed for a given output capacitor, as shown in Figure 13.13. This feature is quite advantageous for the robust IPTS development because the rated voltage of the C_{l2} does not largely change by the output resistance.

13.4 Conclusions

The dynamic model for the IPTS of the OLEV has been fully developed and verified by simulations and experiments in this chapter. By applying the Laplace phasor transform first to the very high order IPTS, a very simple second-order equivalent circuit could be obtained. With the help of this novel dynamic model, the maximum pick-up current during the transient state was successfully identified and found to be relatively unchanged for various load resistances.

Problems

13.1 What happen to the circuit of Figure 13.8 if the diode rectifier is replaced with an active rectifier with a duty cycle control?

13.2 Canthe circuit of Figure 13.2 be analyzed using any conventional dynamic phasor except the proposed one? Set up equations and try to solve them, as shown in this chapter.

References

1 J.G. Bolger, F.A. Kirsten, and L.S. Ng, "Inductive power coupling for an electric highway system," in *Proc. IEEE 28th Veh. Technol. Conference*, March 1978, vol. 28, pp. 137–144.

2 A.W. Green and J.T. Boys, "10 kHz inductively coupled power transfer concept and control," in *Proc. 5th Int. Conf. IEE Power Electron Variable-Speed Drivers*, October 1994, pp. 694–699.

3 G.A.J. Elliott, J.T. Boys, and A.W. Green, "Magnetically coupled systems for power transfer to electric vehicles," in *Proc. Int. Conf. Power Electron. Drive Syst.*, February 1995, pp. 797–801.

4 G.A. Covic, J.T. Boys, M.L.G. Kissin, and H.G. Lu, "A three-phase inductive power transfer system for road powered vehicles," *IEEE Trans. on Ind. Electron.*, vol. 54, no. 6, pp. 3370–3378, December 2007.

5 S.W. Lee, J. Huh, C.B. Park, N.S. Choi, G.H. Cho, and C.T. Rim, "On-line electric vehicle using inductive power transfer system," in *IEEE Energy Conversion Congress and Exposition*, September 2010, pp. 1598–1601.

6 N.P. Suh, D.H. Cho, and C.T. Rim, "Design of on-line electric vehicle (OLEV)," *Plenary Lecture at the 2010 CIRP Design Conference*, 2010.

7 J. Huh and C.T. Rim, "KAIST wireless electric vehicles – OLEV," in *JSAE Annual Congress*, 2011, invited paper.

8 S.W. Lee, W.Y. Lee, J. Huh, C.B. Park, and C.T. Rim, "Active EMF cancellation method for I-type pick-up of on-line electric vehicles," in *IEEE Applied Power Electronics Conference and Exposition (APEC)*, 2011, pp. 1980–1983.

9 J. Huh, B.H. Lee, W.Y. Lee, G.H. Cho, and C.T. Rim, "Characterization of novel inductive power transfer systems for on-line electric vehicles," in *IEEE Applied Power Electronics Conference and Exposition (APEC).*, March 2011, pp. 1975–1979.

10 J. Huh, S.W. Lee, C.B. Park, G.H. Cho, and C.T. Rim, "High performance inductive power transfer system with narrow rail width for on-line electric vehicles," in *IEEE Energy Conversion Congress and Exposition*, 2010, pp. 1598–1601.

11 J. Huh, S.W. Lee, W.Y. Lee, G.H. Cho, and C.T. Rim, "Narrow-width inductive power transfer system for online electrical vehicles," *IEEE Trans. Power Electron.*, vol. 26, no. 12, pp. 3666–3679, December 2011.

12 C.T. Rim, "Analysis of linear switching systems using circuit transformations," PhD Dissertation, KAIST, Seoul, February 1990.

13 C.T. Rim, D.Y. Hu, and G.H. Cho, "Transformers as equivalent circuits for switches: general proofs and D-Q transformation-based analyses," *IEEE Trans. Ind. Applic.*, vol. 26, no. 4, pp. 777–785, July/August 1990.

14 C.T. Rim and G.H. Cho, "Phasor transformation and its application to the DC/AC analyses of frequency phase-controlled series resonant converters (SRC)," *IEEE Trans. Power Electron.*, vol. 5, pp. 201–211, April 1990.

15 C.T. Rim, "A complement of imperfect phasor transformation," in *Korea Power Electronics Conference*, Seoul, 1999, pp. 159–163.

16 C.T. Rim, N.S. Choi, G.C. Cho, and G.H. Cho, "A complete DC and AC analysis of three-phase controlled-current PWM rectifier using circuit D-Q transformation," *IEEE Trans. Power Electron.*, vol. 9, no. 4, pp. 390–396, July 1994.

17 S.B. Han, N.S. Choi, C.T. Rim, and G.H. Cho, "Modeling and analysis of static and dynamic characteristics for buck type three-phase PWM rectifier by circuit DQ transformation," *IEEE Trans. on Power Electron.*, vol. 13, no. 2, pp. 323–336, March 1998.

18 C.T. Rim, "Unified general phasor transformation for AC converters," *IEEE Trans. on Power Electron.*, vol. 26, no. 9, pp. 2465–2475, September 2011.

19 C. Park, S. Lee, G.H. Cho, and C.T. Rim, "Static and dynamic analyses of three-phase rectifier with LC input filter by Laplace phasor transformation," in *IEEE Energy Conversion Congress and Exposition*, 2012, pp. 1570–1577.

20 D.C. White and H.H. Woodson, *Electromechanical Energy Conversion*, John Wiley & Sons, New York, 1959.

21 K.D.T. Ngo, "Low frequency characterization of PWM converter," *IEEE Trans. on Power Electron.*, vol. PE-1, pp. 223–230, October 1986.

14

Compensation Circuit

14.1 Introduction

Inductive power transfer systems (IPTSs) are becoming widely used in electric vehicles, consumer electronics, medical devices, lightings, factory automations, defense systems, and remote sensors [1–26]. Compensation circuits are very often introduced in an IPTS to improve the input power factor and to increase the load power by canceling out reactive powers generated from loosely coupled coils of the IPTS, as shown in Figure 14.1(a). The compensation schemes of an inductively coupled coil set can be classified by the number of compensation capacitors and inductors, the configuration of compensation circuits, and the type of sources. One of the most common compensation schemes is to use two capacitors only, which is regarded as the basic compensation scheme of an IPTS in this chapter and is the most economical way of compensation. As shown in Figure 14.1, the combination of all possible series/parallel (S/P) compensation circuits and voltage/current (V/I) sources results in eight possible compensation schemes [27–52]: current source-type primary series–secondary series (I-SS), current source-type primary series–secondary parallel (I-SP), current source-type primary parallel–secondary series (I-PS), current source-type primary parallel–secondary-parallel (I-PP), voltage source-type SS (V-SS), voltage source-type SP (V-SP), voltage source-type PS (V-PS), and voltage source-type PP (V-PP). For simplicity, the diode rectifier circuit is equivalently replaced with a load resistor R_L, as shown in Figure 14.1 [27–46, 62, 63]. I-SS, I-SP, V-PS, and V-PP are often excluded from consideration because the series circuit of a current source or the parallel circuit of a voltage source is a trivial case for circuit analysis; however, they should be included in the basic compensation schemes because they are being used, for example, I-SS [47–51] and I-SP [52]. The current source in Figure 14.1 could be a controlled current source implemented by the feedback of an inverter [47–52]. More complex compensation schemes such as LCL [53–55] and LCC [56], which have a current source characteristic, are excluded from consideration in this chapter.

Various topological evaluations have been conducted to compare the advantages and disadvantages of each compensation scheme [27–46]; however, no complete comparative analyses on important design drivers are available, and only part of the analyses are provided as follows:

1. SS, SP, PS, and PP were compared to each other in terms of the input power factor at the secondary resonance angular frequency, where the primary compensation capacitance of SS for the unity power factor was found to be independent of both the load and magnetic coupling coefficient k [27–33].
2. V-SS and V-SP were compared to each other and V-SP was found to be superior to V-SS because V-SP has load- and k-independent output voltage characteristics, maintaining a unity input power factor [17, 34–36].

Wireless Power Transfer for Electric Vehicles and Mobile Devices, First Edition. Chun T. Rim and Chris Mi.
© 2017 John Wiley & Sons Ltd. Published 2017 by John Wiley & Sons Ltd.

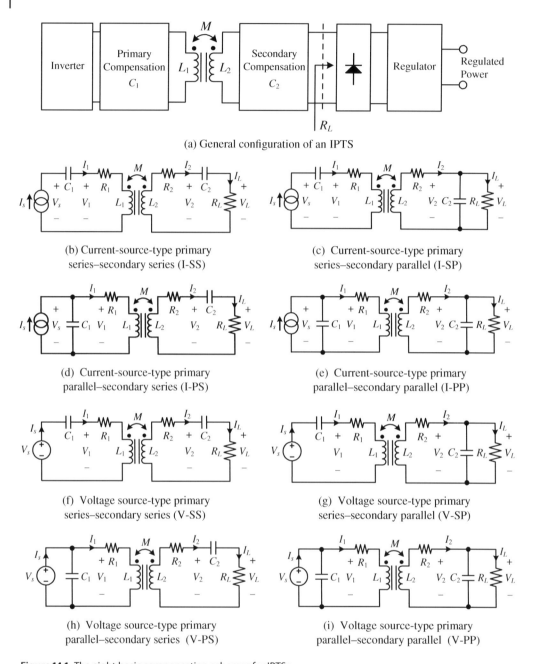

(a) General configuration of an IPTS

(b) Current-source-type primary
series–secondary series (I-SS)

(c) Current-source-type primary
series–secondary parallel (I-SP)

(d) Current-source-type primary
parallel–secondary series (I-PS)

(e) Current-source-type primary
parallel–secondary parallel (I-PP)

(f) Voltage source-type primary
series–secondary series (V-SS)

(g) Voltage source-type primary
series–secondary parallel (V-SP)

(h) Voltage source-type primary
parallel–secondary series (V-PS)

(i) Voltage source-type primary
parallel–secondary parallel (V-PP)

Figure 14.1 The eight basic compensation schemes for IPTSs.

3. V-SS and V-SP were evaluated in terms of the efficiency, considering power losses in the equivalent series resistances (ESRs) of an inductively coupled coil set, and load-independent output characteristics, where V-SS was found to have a load-independent output current characteristic and V-SP was found to have a load-independent output voltage characteristic, respectively [37–42].

4. SS and PS as well as SP and PP were compared to each other in terms of the voltage and current ratings of the inverter, where SS and SP were found to be adequate for reducing the voltage rating of the power supply and PS and PP are found to be adequate for reducing the current rating of the inverter [30, 43, 44].
5. V-SS, V-SP, V-PS, and V-PP were evaluated in terms of the allowance of no coupling ($k = 0$), where V-SS and V-SP were found to be unsafe for the power supply when $k = 0$ [45].
6. SS, SP, PS, and PP were analyzed in terms of the required copper mass for fabricating an inductively coupled coil set, where SS and SP were found to be economical, especially for high-power applications [46].

In spite of several valuable results and insights above, it is still unclear how to select the most appropriate compensation scheme for each application. There is a need for a collective and comparative work focusing on major design parameters.

In this chapter, the eight basic compensation schemes are compared to each other in terms of the following five criteria collected from the literature of [27] to [46], for a general and comprehensive evaluation of IPTSs. By appropriate selection of the source angular frequency and two compensation capacitances, the following operational characteristics of the IPTSs could become ideal:

1. The maximum efficiency
2. The maximum load power transfer
3. Load-independent output voltage or output current
4. k-independent compensation
5. Allowance of no magnetic coupling ($k = 0$)

It is identified in this chapter that only I-SS and I-SP can satisfy all of the five criteria simultaneously among the eight compensation schemes; hence, the equivalence and duality between I-SS and I-SP are further explored in the aspect of efficiency, power transfer characteristics, and voltage and current ratings of reactive components. Design guidelines for both I-SS and I-SP are proposed and experimentally verified by a 200 W prototype of air coils at the source frequency of 100 kHz.

14.2 Comparative Evaluations of the Eight Basic Compensation Schemes

In this section, the eight basic compensation schemes are comparatively evaluated under the proposed five criteria. Throughout this chapter, it is assumed that all the circuit parameters are linear and that the IPTS is operating in the steady state and includes no switching harmonics. Therefore, all of the circuit variables are represented in a phasor form; for example, I_s and V_s are the phasors of the fundamental frequency component of the source current and voltage, respectively. All the sources and capacitors are assumed to be ideal; hence, they include no ESR. In this chapter, an inductively coupled coil set is modeled as the T-model, as shown in Figure 14.2, where the mutual inductance is M and the self-inductances of the primary coil and secondary coil are L_1 and L_2, respectively. The magnetic coupling coefficient k is then determined as follows:

$$k = \frac{M}{\sqrt{L_1 L_2}} \tag{14.1}$$

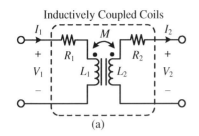

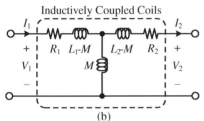

Figure 14.2 Equivalent T-model of an inductively coupled coil set: (a) model for an inductively coupled coil set and (b) equivalent T-model.

The ESRs of the inductively coupled coils set, R_1 and R_2, represent the equivalent AC resistances of the coils including conduction loss and eddy current loss in windings caused by the skin effect and proximity effect [49,61]. A generalized circuit diagram, using the T-model, is drawn for the eight basic compensation schemes, as shown in Figure 14.3, where the source represents either a current-type source or a voltage-type source and the compensation capacitors C_1 and C_2 are either series-compensated or parallel-compensated. Z_s, Z_1, Z_{m1}, Z_{m2}, Z_2, and Z_L are the impedances seen from the left to the right of the source, primary, medium-left, medium-right, secondary, and load ports, respectively.

The detailed analysis process is omitted in this chapter due to the page limit and duplicate description of equations. Instead, a unified analysis for the generalized circuit diagram of Figure 14.3 is provided and the analytical results are summarized in Table 14.1.

14.2.1 The Maximum Efficiency Condition

The efficiency of the proposed generalized IPTS, as shown in Figure 14.3, can be determined by considering power losses in R_1 and R_2. An optimum angular frequency for the maximum efficiency can be derived from the efficiency equation. In Figure 14.3, P_s, P_1, P_{m1}, P_{m2}, P_2, and P_L represent real powers at the source, primary, medium-left, medium-right, secondary, and

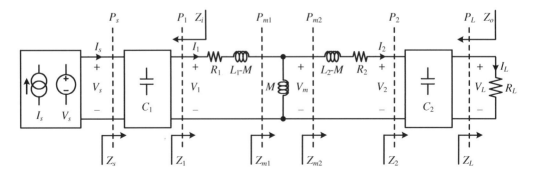

Figure 14.3 A general unified circuit diagram of an IPTS for the eight basic compensation schemes.

Table 14.1 Characteristics of the eight basic comensation schemes of the five criteria

Criteria	I-SS	I-SP	I-PS	I-PF
Maximum efficiency condition ($\omega_{\eta,m}$)	$\dfrac{\omega_2}{\sqrt{1-1/\{2(Q_2//Q_L)^2\}}} \approx \omega_2{}^a$	$\dfrac{\omega_2\sqrt{1+1/(Q_2 Q_L)}}{\sqrt[4]{1+k^2 Q_1/Q_2}} \approx \omega_2{}^b$	$\dfrac{\omega_2}{\sqrt{1-1/\{2(Q_2//Q_L)^2\}}} \approx \omega_2{}^a$	$\dfrac{\omega_2\sqrt{1+1/(Q_2 Q_L)}}{\sqrt[4]{1+k^2 Q_1/Q_2}} \approx \omega_2{}^b$
Maximum load power condition ($\omega_{P_L,m}$)	$\dfrac{\omega_2}{\sqrt{1-1/2Q_L^2}} \approx \omega_2{}^c$	ω_2	$\dfrac{\omega_1}{\sqrt{1+k^4 Q_L^2}} = \omega_2$	$\dfrac{\omega_3}{\sqrt{1+k^4 Q_1^2/(1-k^2)^2}} = \omega_2$
R_L-independent output condition (ω_{V_L} or ω_{I_L})	ω_{V_L}: ω_2 ω_{I_L}: Not exist	ω_{V_L}: Not exist ω_{I_L}: ω_2	ω_{V_L}: ω_5 ω_{I_L}: ω_1	ω_{V_L}: ω_3 ω_{I_L}: ω_5
k-independent compensation	Yes	Yes[b]	No	No
k = 0 allowance	Allowed	Allowed	Not allowed	Not allowed

Criteria	V-SS	V-SP	V-PS	V-PP
Maximum efficiency condition ($\omega_{\eta,m}$)	$\dfrac{\omega_2}{\sqrt{1-1/\{2(Q_2//Q_L)^2\}}} \approx \omega_2{}^a$	$\dfrac{\omega_2\sqrt{1+1/(Q_2 Q_L)}}{\sqrt[4]{1+k^2 Q_1/Q_2}} \approx \omega_2{}^b$	$\dfrac{\omega_2}{\sqrt{1-1/\{2(Q_2//Q_L)^2\}}} \approx \omega_2{}^a$	$\dfrac{\omega_2\sqrt{1+1/(Q_2 Q_L)}}{\sqrt[4]{1+k^2 Q_1/Q_2}} \approx \omega_2{}^b$
Maximum load power condition ($\omega_{P_L,m}$)	$\omega_1 = \omega_2$	$\omega_3 = \omega_2$	ω_4	$\omega_4\sqrt{1-(1-k^2)/(2Q_L^2)} \approx \omega_4{}^c$
R_L-independent output condition (ω_{V_L} or ω_{I_L})	ω_{V_L}: ω_5 ω_{I_L}: ω_1	ω_{V_L}: ω_3 ω_{I_L}: ω_5	ω_{V_L}: ω_4 ω_{I_L}: Not exist	ω_{V_L}: Not exist ω_{I_L}: ω_4
k-independent compensation	Yes	No	No	No
k = 0 allowance	Not allowed	Not allowed	Allowed	Allowed

[a] $1 \ll Q_2$ and $1 \ll Q_L$.
[b] $1 \ll Q_1$, $1 \ll Q_2$, $1 \ll Q_L$, and $k \ll 1$.
[c] $1 \ll Q_L$.

load ports, respectively, and are defined as follows:

$$P_s = Re(V_s I_s^*), P_1 = Re(V_1 I_1^*), P_{m1} = Re(V_m I_1^*), P_{m2} = Re(V_m I_2^*),$$
$$P_2 = Re(V_2 I_2^*), P_L = Re(V_L I_L^*) \tag{14.2}$$

The efficiency of the IPTS becomes as follows:

$$\eta \equiv \frac{P_L}{P_s} = \frac{P_1}{P_s} \cdot \frac{P_{m1}}{P_1} \cdot \frac{P_{m2}}{P_{m1}} \cdot \frac{P_2}{P_{m2}} \cdot \frac{P_L}{P_2} \tag{14.3}$$

As the internal resistances of sources and ESRs of the compensation capacitors are assumed to be negligible, the real part of Z_i becomes zero, which results in $P_s = P_1$ and $P_2 = P_L$. The real power at the medium stage is zero; hence, $P_{m1} = P_{m2}$. Then (14.3) can be simplified as follows:

$$\eta = \frac{P_{m1}}{P_1} \cdot \frac{P_2}{P_{m2}} = \frac{Re(Z_{m1})}{R_1 + Re(Z_{m1})} \cdot \frac{Re(Z_2)}{R_2 + Re(Z_2)} \tag{14.4}$$

where Z_{m1} is found from Figure 14.3 as

$$Z_{m1} \equiv j\omega M // \{j\omega(L_2 - M) + R_2 + Z_2\} \tag{14.5}$$

In (14.4), the efficiency of each stage is determined from an equivalent circuit consisting of two resistors connected in series.

As shown in Figure 14.4, Z_2 is either a series impedance Z_{2S} or a parallel impedance Z_{2P}, depending on the secondary compensation circuit, as follows:

$$Z_{2S} = R_L + \frac{1}{j\omega C_2} \tag{14.6a}$$

$$Z_{2P} = R_L // \frac{1}{j\omega C_2} \tag{14.6b}$$

Applying (14.6a) and (14.6b) into (14.4) and (14.5) results in the efficiency η_S of the secondary series compensation schemes (I-SS, I-PS, V-SS, and V-PS) and the efficiency η_P of the secondary

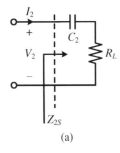

(a)

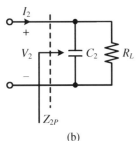

(b)

Figure 14.4 Z_2 for the secondary compensation circuits: (a) series compensation circuit and (b) parallel compensation circuit.

parallel compensation schemes (I-SP, I-PP, V-SP, and V-PP), respectively, which are obtained as follows:

$$\eta_S = \frac{Q_1(Q_2//Q_{LS})k^2\omega_n^2}{1 + Q_1(Q_2//Q_{LS})k^2\omega_n^2 + (Q_2//Q_{LS})^2(\omega_n - 1/\omega_n)^2} \cdot \frac{1}{1 + Q_{LS}/Q_2} \tag{14.7a}$$

$$\eta_P = k^2\omega_n^2 Q_1 \cdot \frac{a_1\omega_n^4 + a_2\omega_n^2 + a_3}{a_4\omega_n^6 + a_5\omega_n^4 + a_6\omega_n^2 + a_7} \cdot \frac{1}{1 + 1/(Q_2 Q_{LP}) + Q_{LP}\omega_n^2/Q_2} \tag{14.7b}$$

where the normalized angular frequency ω_n, secondary resonant angular frequency ω_2, the intrinsic quality factors of coils Q_1, Q_2, and the load quality factors Q_{LS}, Q_{LP} are defined as follows:

$$\omega_n \equiv \frac{\omega}{\omega_2}, \omega_2 \equiv \frac{1}{\sqrt{L_2 C_2}} \tag{14.8a}$$

$$Q_1 \equiv \left.\frac{\omega L_1}{R_1}\right|_{\omega=\omega_2}, Q_2 \equiv \left.\frac{\omega L_2}{R_2}\right|_{\omega=\omega_2}$$

$$Q_{LS} \equiv \left.\frac{\omega L_2}{R_L}\right|_{\omega=\omega_2}, Q_{LP} \equiv \left.\frac{R_L}{\omega L_2}\right|_{\omega=\omega_2} \tag{14.8b}$$

The coefficients in (14.7) are given as follows:

$$a_1 = Q_{LP}^4/Q_2^2 \tag{14.9a}$$

$$a_2 = Q_{LP}^2(Q_{LP} + 2/Q_2) \tag{14.9b}$$

$$a_3 = (Q_{LP} + 1/Q_2) \tag{14.9c}$$

$$a_4 = (1 + k^2 Q_1/Q_2)Q_{LP}^4 \tag{14.9d}$$

$$a_5 = k^2 Q_1 a_2 + Q_{LP}^2 \left(2 - 2Q_{LP}^2 + Q_{LP}^2/Q_2^2\right) \tag{14.9e}$$

$$a_6 = k^2 Q_1 a_3 + 2Q_{LP}^2(Q_{LP} + 1/Q_2)/Q_2 + \left(1 - Q_{LP}^2\right)^2 \tag{14.9f}$$

$$a_7 = (Q_{LP} + 1/Q_2)^2 \tag{14.9g}$$

Note that the efficiencies η_S and η_P in (14.7) are valid when the ESRs of the compensation capacitors are not significantly large and that the losses in the ESRs of the coupled coils only are considered. By taking derivatives of (14.7a) and (14.7b) with respect to ω_n, respectively, the optimum angular frequencies $\omega_{\eta_S,m}$ and $\omega_{\eta_P,m}$ that maximize the efficiencies η_S and η_P, respectively, can be found as follows:

$$\left.\frac{\partial \eta_S}{\partial \omega_n}\right|_{\omega_n=\omega_{\eta_S,m}/\omega_2} = 0 \Rightarrow \omega_{\eta_S,m} = \frac{\omega_2}{\sqrt{1 - 1/\{2(Q_2//Q_{LS})^2\}}} \tag{14.10a}$$

$$\left.\frac{\partial \eta_P}{\partial \omega_n}\right|_{\omega_n=\omega_{\eta_P,m}/\omega_2} = 0 \Rightarrow \omega_{\eta_P,m} = \omega_2 \sqrt{\frac{1 + 1/(Q_2 Q_{LP})}{\sqrt{1 + k^2 Q_1/Q_2}}} \tag{14.10b}$$

In (14.10), the fact that the efficiencies of (14.7) become zeros for a zero ω_n and increase as ω_n increases is used, which means that the efficiencies reach their maximum (not minimum) under the conditions of (14.10). A similar expression for (14.10) is available in [37]; however, the underlying process for obtaining (14.10b) is shown in this chapter.

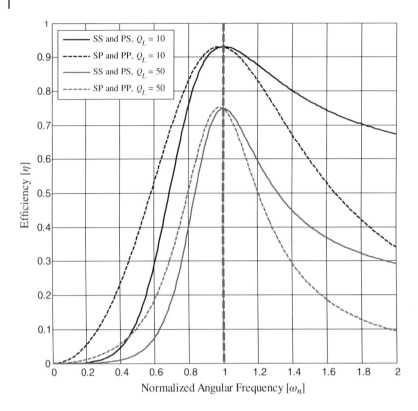

Figure 14.5 Efficiency η versus the normalized angular frequency ω_n when $k = 0.3$ and $Q_1 = Q_2 = 150$.

Assuming that $1 \ll Q_L$, $1 \ll Q_1$, $1 \ll Q_2$, and $k \ll 1$, which are quite common in many IPTS designs, (14.10) can be approximated as follows:

$$\omega_{\eta S,m} \cong \omega_{\eta P,m} \cong \omega_2 \tag{14.11}$$

Equation (14.11) indicates that the maximum efficiency for all of the eight basic compensation schemes is approximately achieved at the secondary resonant angular frequency ω_2. As shown in Figure 14.5, the maximum efficiency of any basic compensation scheme is obtained in the vicinity of ω_2 regardless of Q_L.

14.2.2 The Maximum Load Power Transfer Condition ($R_1 = R_2 = 0$)

The optimum angular frequency $\omega_{P_L,m}$ for maximizing the load power P_L is derived from the unified IPTS model of Figure 14.3 in this section. Two example cases for I-SS and V-SS are analyzed in detail in this chapter, and similar analyses can be applied for the remaining six basic compensation schemes. In this section, the ESRs of the inductively coupled coils are neglected for simplicity; otherwise, these calculations would be too complex to handle by hand.

Load powers for any current source type and voltage source type, as shown in Figure 14.3, can be determined as follows:

$$P_L \equiv Re(Z_s)|I_s|^2 \quad \text{(for the current source type)} \tag{14.12a}$$

$$P_L = Re(Y_s)|V_s|^2 \equiv Re\left(\frac{1}{Z_s}\right)|V_s|^2 \quad \text{(for the voltage source type)} \tag{14.12b}$$

where $|I_s|$ and $|V_s|$ are given values of the current-type source and voltage-type source, respectively, and Z_s is the input impedance seen from the source side, as described below:

$$Z_s - \frac{\omega_n^2 \omega_2 k^2 L_1}{Q_{LS}(\omega_n - 1/\omega_n)^2 + 1/Q_{LS}} + j\omega_n \omega_2 L_1 \left\{ 1 - \frac{\omega_1^2}{\omega_n^2 \omega_2^2} \left| \frac{k^2 \left(1 - \omega_n^2 \right)}{\left(\omega_n - 1/\omega_n \right)^2 + 1/Q_{LS}^2} \right. \right\}$$

$$\text{(for I-SS and V-SS)} \qquad (14.13)$$

where the primary resonant angular frequency ω_1 is defined by

$$\omega_1 \equiv \frac{1}{\sqrt{L_1 C_1}} \qquad (14.14)$$

Applying (14.13) to (14.12a) and taking its derivative with respect to ω_n, the maximum load power point for I-SS can be found as follows:

$$\left. \frac{\partial P_L}{\partial \omega_n} \right|_{\omega_n = \omega_{P_L,m}/\omega_2} = 0 \quad \Rightarrow \quad \omega_{P_L,m} = \frac{\omega_2}{\sqrt{1 - 1/\left(2Q_{LS}^2 \right)}} \text{ (for I-SS)} \qquad (14.15a)$$

With the assumption of $Q_{LS} \gg 1$, which is commonly acceptable in most IPTS designs, (14.15a) can be approximated as follows:

$$\omega_{P_L,m} \cong \omega_2 \text{ (for I-SS)} \qquad (14.15b)$$

Applying (14.13) to (14.12b) and taking its derivative with respect to ω_n, an optimum angular frequency for the maximum load power transfer of V-SS can be determined. It is found, however, that the outcome equation is a third-order polynomial and its solutions tend to be too complicated. A practical way of solving this problem is to set the operating angular frequency to an appropriate one, for example, $\omega \equiv \omega_2$, which intuitively minimizes the impedance of a secondary compensation circuit and which was also used in similar works [27–29] without detailed explanation. To find the condition for the maximum load power transfer of V-SS, the derivative of (14.12b) with respect to ω_1, assuming $\omega = \omega_2$, is sought in this chapter as follows:

$$\left. \frac{\partial P_L}{\partial \omega_1} \right|_{\omega_n = 1} = 0 \quad \Rightarrow \quad \omega_1 = \omega_2 = \omega_{P_L,m} \text{ (for V-SS)} \qquad (14.16a)$$

$$\because Z_s(\omega_n, \omega_1)\big|_{\omega_n = 1} = \omega_2 Q_{LS} k^2 L_1 + j\omega_2 L_1 \left(1 - \frac{\omega_1^2}{\omega_2^2} \right) \qquad (14.16b)$$

The optimum angular frequencies for the remaining six basic compensation schemes are found in a similar way and are summarized in Table 14.1, where the optimum angular frequencies for V-SP and V-PS are defined as follows:

$$\omega_3 \equiv \frac{1}{\sqrt{L_1 C_1 (1 - k^2)}} \qquad \text{(for V-SP)} \qquad (14.17a)$$

$$\omega_4 \equiv \frac{1}{\sqrt{L_2 C_2 (1 - k^2)}} \qquad \text{(for V-PS)} \qquad (14.17b)$$

Similar expressions for the optimum angular frequencies including (14.15) to (14.17) are also found for specific cases [27, 28, 32, 57]; however, a general unified approach is first shown in this chapter.

14.2.3 Load-Independent Output Voltage or Output Current Characteristic ($R_1 = R_2 = 0$)

A constant output voltage V_L or a constant output current I_L, regardless of the value of the load resistance, is one of the most useful characteristics of an IPTS to achieve load voltage regulation or overload protection. To obtain this load-independent output characteristic, regulation circuits at the secondary side with feedback signal channels linked to the primary side have been conventionally used [17,18,20,39,58,59]. However, the load-independent output characteristic could also be achieved from the inherent characteristics of some compensation schemes by a proper choice of the source angular frequency. Even though this load-independent characteristic may not completely exclude the use of regulation circuits in some applications, it provides a relatively small load voltage variation or an additional short-circuit protection capability for an IPTS.

In this section, the load-independent V_L and I_L characteristics for I-SS and V-SS are explored and the angular frequency for the load-independent V_L or I_L is derived in case it exists. The ESRs of the inductively coupled coils are neglected again for simplicity of analyses, which makes it possible to handle the computations by hand.

In order to identify the output characteristic of an IPTS, the general unified circuit of Figure 14.3 is redrawn, simplifying all the circuits to a Thevenin or a Norton equivalent circuit from the load side, as shown in Figure 14.6. An IPTS may have a Thevenin or a Norton equivalent circuit only in case either I_N or V_L becomes infinity. When $Z_o = 0$, V_L becomes the same as the Thevenin voltage V_T, which is independent of R_L because the IPTS is assumed to be linear. In other words, an IPTS has the load-independent V_L characteristic if $Z_o = 0$. For a similar reason, $Z_o = \infty$ is required for the load-independent I_L ($= I_N$) characteristic of an IPTS. Even if an IPTS has the load-independent characteristic, the V_T or I_N may not be fixed but instead may vary for airgaps and misalignments; this is why a regulator, as shown in Figure 14.1(a), is required for certain applications.

The V_T, I_N, and Z_o of I-SS and V-SS can be derived from Figures 14.3 and 14.4(a) as follows:

$$V_T = j\omega M I_s \qquad \text{(for I-SS)} \tag{14.18a}$$

$$I_N = \frac{\omega^2 M C_2 I_s}{\omega^2 L_2 C_2 - 1} \qquad \text{(for I-SS)} \tag{14.18b}$$

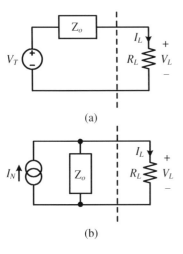

(a)

(b)

Figure 14.6 Simplified circuits for the unified general compensation schemes seen from the load side: (a) equivalent Thevenin circuit and (b) equivalent Norton circuit.

$$Z_o = j\frac{\omega^2 L_2 C_2 - 1}{\omega C_2} \qquad \text{(for I-SS)} \tag{14.18c}$$

$$V_T = \frac{\omega^2 M C_1 V_s}{\omega^2 L_1 C_1 - 1} \qquad \text{(for V-SS)} \tag{14.19a}$$

$$I_N = j\frac{\omega^3 M C_1 C_2 V_s}{\omega^4 M^2 C_1 C_2 (1 - 1/k^2) + \omega^2 (L_1 C_1 + L_2 C_2) - 1} \qquad \text{(for V-SS)} \tag{14.19b}$$

$$Z_o = j\frac{-\omega^4 M^2 C_1 C_2 (1 - 1/k^2) - \omega^2 (L_1 C_1 + L_2 C_2) + 1}{\omega C_2 (\omega^2 L_1 C_1 - 1)} \qquad \text{(for V-SS)} \tag{14.19c}$$

From (14.18c) and (14.19c), it is noted that the Z_o of I-SS is different from that of V-SS; thus, their angular frequencies for the load-independent V_L and I_L characteristics are different from each other.

First, the load-independent V_L and I_L characteristics are examined for the I-SS case. It is identified from (14.18c) that the Z_o becomes zero when the angular frequency ω_{V_L} for the load-independent V_L is as follows:

$$Z_o|_{\omega=\omega_{V_L}} = 0 \quad \Rightarrow \quad \omega_{V_L} = \omega_2 \text{ (for I-SS)} \tag{14.20}$$

Under this condition of (14.20), I_N becomes infinity, as identified from (14.18b), that is, the Norton equivalent circuit does not exist for the I-SS case. In other words, only the Thevenin equivalent circuit is valid for the I-SS case, where the load-independent V_L characteristic is obtained.

Second, the load-independent V_L and I_L characteristics are examined for the V-SS case. It is identified from (14.19c) that Z_o becomes zero when the angular frequency ω_{V_L} for the load-independent V_L is as follows:

$$Z_o|_{\omega=\omega_{V_L}} = 0$$
$$\Rightarrow \quad \omega_{V_L} = \sqrt{\frac{\omega_1^2 + \omega_2^2 \pm \sqrt{(\omega_1^2 + \omega_2^2)^2 - 4\omega_1^2 \omega_2^2 (1-k^2)}}{2(1-k^2)}} \equiv \omega_5 \qquad \text{(for V-SS)} \tag{14.21}$$

Under this condition of (14.21), the Thevenin voltage of (14.19a), which has a pole at ω_1, does not go to infinity; therefore, V-SS has the load-independent V_L characteristic. Note that $\omega_1 \neq \omega_5$ for any ω_2 and k so far as $\omega_1 > 0$. It is also identified from (14.19c) that Z_o becomes infinity when the angular frequency ω_{I_L} for the load-independent I_L is as follows:

$$Z_o|_{\omega=\omega_{I_L}} = \infty \quad \Rightarrow \quad \omega_{I_L} = \omega_1 \qquad \text{(for V-SS)} \tag{14.22}$$

Under this condition of (14.22), the Norton current of (14.19b) becomes a finite value as follows:

$$I_N = j\frac{-\omega_1 C_1 V_s}{\omega_1^2 C_1 (L_1 - M) - 1} \qquad \text{(for V-SS)} \tag{14.23}$$

Different from the I-SS case, V-SS has both load-independent V_L and I_L characteristics for different angular frequencies, ω_{V_L} and ω_{I_L}, respectively, which was also identified as a result of comparing V-SS with V-SP [37, 39].

Similar analyses can be applied for the remaining six basic compensation schemes and the results are tabulated in Table 14.1. It can be seen from Table 14.1 that V-SS and I-PS as well as V-SP and I-PP have the same ω_{V_L} and ω_{I_L} because their circuit configurations when removing the source are identical, as is the case with Z_o.

14.2.4 The *k*-Independent Compensation Characteristic

For cases in which the angular frequencies for the optimum characteristics discussed in the previous sections are magnetic coupling coefficient *k*-dependent, a frequency control scheme is needed that adjusts the source angular frequency to track the required angular frequency [17,27,28,41,42]. This control scheme adds to the cost and complexity of an IPTS, which can be a burden for many applications. Thus, *k*-independency is a highly desirable characteristic for a low-cost and highly reliable IPTS, and can be achieved by a proper choice of compensation schemes. In this section, the angular frequencies for the maximum efficiency, maximum load power, and load-independent output are revisited in terms of the *k*-independency.

As identified from (14.10a), the angular frequency for the maximum efficiency of η_S is completely independent of *k*; however, that of η_P of (14.10b) depends on *k*. Fortunately, this *k*-dependency is drastically mitigated when $k \ll 1$ and is not a serious problem in practice. It can be said that only the four schemes of I-SS, I-PS, V-SS, and V-PS have complete *k*-independency.

As identified from (14.15), the angular frequency for the maximum load power of I-SS is completely independent of *k*, and this is also true for I-SP and V-SS, as shown in Table 14.1. As also identified from Table 14.1, the angular frequencies for the maximum load power of I-PS, I-PP, V-SP, V-PS, and V-PP are not independent of *k*.

As identified from Table 14.1, at least one of the *k*-independent angular frequencies for the load-independent output characteristic exists only for the schemes of I-SS, I-SP, I-PS, and V-SS. Note that ω_1 and ω_2 are *k*-independent; however, ω_3, ω_4, and ω_5 are a function of *k*, as identified from (14.21).

From the discussions above, it can be said that I-SS and V-SS are *k*-independent in terms of the three criteria mentioned.

14.2.5 Allowance of No Magnetic Coupling ($R_1 = R_2 = 0$)

Depending on the compensation scheme and source angular frequency, excessive voltage or current stresses on circuit components can be induced when magnetic coupling is very weak or completely removed ($k = 0$). The absence of the magnetic coupling occurs when a secondary coil is seriously misaligned or totally dislocated from a primary coil, which is frequently seen in IPTSs. Hence, the safe operation of an IPTS, even in the absence of the magnetic coupling, should be guaranteed by a proper selection of compensation schemes.

The allowance of no magnetic coupling is evaluated in detail for I-SS and V-SS. Similar analyses can be applied for the remaining six basic compensation schemes, and the results are tabulated in Table 14.1. The ESRs of the inductively coupled coils are neglected again due to their slight impact on the analytical results. Note that I-SS and V-SS have the same input impedance Z_s of (14.13) seen from the source. Confining discussions to the optimum operating conditions for the three, maximum efficiency, maximum load power, and load-independent output, the behavior of the input impedance Z_s is quite simple. For the I-SS case, the source angular frequency should be ω_2 to meet the three criteria. For the V-SS case, the source angular frequency could be either ω_1 or ω_2. Applying these different frequency conditions to (14.13a) results in the following:

$$Z_s|_{k=0,\omega_n \cong 1} \cong j\omega_2 L_1 \left(1 - \frac{\omega_1^2}{\omega_2^2} \right) \qquad \text{(for I-SS)} \tag{14.24a}$$

$$Z_s|_{k=0,\omega_n=1,\omega_1=\omega_2} = 0 \qquad \text{(for V-SS)} \tag{14.24b}$$

For allowance of $k = 0$, the magnitude of Z_s should not be infinite for I-SS and zero for V-SS. From (14.24), it is seen that I-SS allows zero magnetic coupling while V-SS does not. Allowances of no magnetic coupling for the remaining six compensation schemes are tabulated in Table 14.1, and it is noteworthy that only I-SS, I-SP, V-PS, and V-PP allow $k = 0$.

From the comparative evaluations of the eight basic compensation schemes under the five criteria, it is identified that only I-SS satisfies the criteria simultaneously, which has first been identified in this chapter. Except for the weak k-dependency on maximum efficiency, I-SP is also a good candidate. Though not analyzed in detail, the reason why the well-known LCL [53–55] and LCC [56] have fairly good characteristics is that their primary side is equivalent to a current source, which makes them resemble the proposed I-SS or I-SP.

14.3 Equivalence and Duality of I-SS and I-SP

Based on the discussions in the previous section, comparisons of the eight basic compensation schemes are given in this section. The efficiency, load power, and component ratings of the eight basic compensation schemes are compared with each other when they are operating at the secondary resonance angular frequency ω_2, which is the angular frequency for the maximum efficiency. For a fair comparison, it is assumed that all of the eight basic compensation schemes have the same operating conditions of I_s, V_s, L_1, L_2, R_1, R_2, k, C_1, C_2, as well as the quality factor $Q_{LS} = Q_{LP} = Q_L$. The only difference is the value of the load resistance R_L, which should be different, as determined from (14.8b), to give the same quality factor. Throughout this section, the power, voltage, and currents are all represented in per unit for convenient comparisons between different compensation schemes; they are divided by the predefined base quantities. The base power P_b, voltage $|V_b|$, and current $|I_b|$ are defined as follows:

$$P_{b,I} \equiv \omega_2 L_1 |I_s|^2, |V_{b,I}| \equiv \omega_2 \sqrt{L_1 L_2}|I_s|, |I_{b,I}| \equiv \sqrt{L_1/L_2}|I_s| \quad \text{(for current-source type)}$$

$$(14.25a)$$

$$P_{b,V} = |V_s|^2/(\omega_2 L_1), |V_{b,V}| = \sqrt{L_2/L_1}|V_s|, |I_{b,V}| = |V_s|/(\omega_2 \sqrt{L_1 L_2})$$

$$\text{(for voltage - source type)} \quad (14.25b)$$

The detailed analyses are given for I-SS and I-SP only, in Figure 14.7, due to the duplicate description of equations and page limit. The comparison results are tabulated in Table 14.2.

14.3.1 Equivalence in the Efficiency

It is found from Table 14.1 that I-SS and I-SP have nearly the same optimum angular frequency $\omega_{\eta,m} = \omega_{\eta S,m} \cong \omega_{\eta p,m} \cong \omega_2$ for the maximum efficiency $\eta_m = \eta_{S,m} = \eta_{P,m}$ when $1 \ll Q_L, 1 \ll Q_1$,

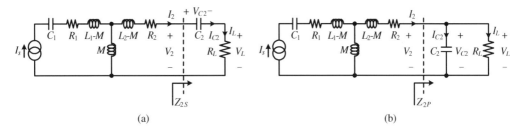

Figure 14.7 T-model-based equivalent circuits for I-SS and I-SP: (a) I-SS scheme and (b) I-SP scheme.

Table 14.2 Equivalence and duality of the eight basic compensation schemes at the secondary resonance frequency

Parameters	I-SS	I-SP	I-PS	I-PP
$\dfrac{\eta}{\eta_r}$	1	$\dfrac{Q_L+Q_2}{Q_L+Q_2+2}\cong 1^a$	1	$\dfrac{Q_L+Q_2}{Q_L+Q_2+2}\cong 1^a$
P_L (in per unit)	k^2Q_L	k^2Q_L	$k^2Q_L+\dfrac{1}{k^2Q_L}\cong k^2Q_L{}^a$	$k^2Q_L+\dfrac{(1-k^2)^2}{k^2Q_L}\cong k^2Q_L{}^a$
$\lvert V_2\rvert\cong\lvert V_{C_2}\rvert$ (in per unit)	kQ_L	kQ_L	$\dfrac{\sqrt{1+k^4Q_L^2}}{k}\cong kQ_L{}^a$	$\dfrac{\sqrt{(1-k^2)^2+k^4Q_L^2}}{k}\cong kQ_L{}^a$
$\lvert I_2\rvert\cong\lvert I_{C_2}\rvert$ (in per unit)	kQ_L	kQ_L	$\dfrac{\sqrt{1+k^4Q_L^2}}{k}\cong kQ_L{}^a$	$\dfrac{\sqrt{(1-k^2)^2+k^4Q_L^2}}{k}\cong kQ_L{}^a$
$\lvert V_L\rvert$ (in per unit)	k	kQ_L	$\dfrac{\sqrt{1+k^4Q_L^2}}{kQ_L}\cong k^a$	$\dfrac{\sqrt{(1-k^2)^2+k^4Q_L^2}}{k}\cong kQ_L{}^a$
$\lvert I_L\rvert$ (in per unit)	kQ_L	k	$\dfrac{\sqrt{1+k^4Q_L^2}}{k}\cong kQ_L{}^a$	$\dfrac{\sqrt{(1-k^2)^2+k^4Q_L^2}}{kQ_L}\cong k^a$

Parameters	V-SS	V-SP	V-PS	V-PP
$\dfrac{\eta}{\eta_r}$	1	$\dfrac{Q_L+Q_2}{Q_L+Q_2+2}\cong 1^a$	1	$\dfrac{Q_L+Q_2}{Q_L+Q_2+2}\cong 1^a$
P_L (in per unit)	$\dfrac{1}{k^2Q_L}$	$\dfrac{1}{k^2Q_L}$	$\dfrac{k^2Q_L}{1+k^4Q_L^2}\cong\dfrac{1}{k^2Q_L}{}^a$	$\dfrac{k^2Q_L}{(1-k)^2+k^4Q_L^2}\cong\dfrac{1}{k^2Q_L}{}^a$
$\lvert V_2\rvert\cong\lvert V_{C_2}\rvert$ (in per unit)	$\dfrac{1}{k}$	$\dfrac{1}{k}$	$\dfrac{kQ_L}{\sqrt{1+k^4Q_L^2}}\cong\dfrac{1}{k}{}^a$	$\dfrac{kQ_L}{\sqrt{(1-k^2)^2+k^4Q_L^2}}\cong\dfrac{1}{k}{}^a$
$\lvert I_2\rvert\cong\lvert I_{C_2}\rvert$ (in per unit)	$\dfrac{1}{k}$	$\dfrac{1}{k}$	$\dfrac{kQ_L}{\sqrt{1+k^4Q_L^2}}\cong\dfrac{1}{k}{}^a$	$\dfrac{kQ_L}{\sqrt{(1-k^2)^2+k^4Q_L^2}}\cong\dfrac{1}{k}{}^a$
$\lvert V_L\rvert$ (in per unit)	$\dfrac{1}{kQ_L}$	$\dfrac{1}{k}$	$\dfrac{k}{\sqrt{1+k^4Q_L^2}}\cong\dfrac{1}{kQ_L}{}^a$	$\dfrac{kQ_L}{\sqrt{(1-k^2)^2+k^4Q_L^2}}\cong\dfrac{1}{k}{}^a$
$\lvert I_L\rvert$ (in per unit)	$\dfrac{1}{k}$	$\dfrac{1}{kQ_L}$	$\dfrac{kQ_L}{\sqrt{1+k^4Q_L^2}}\cong\dfrac{1}{k}{}^a$	$\dfrac{k}{\sqrt{(1-k^2)^2+k^4Q_L^2}}\cong\dfrac{1}{kQ_L}{}^a$

$^a 1\ll Q_L$.

$1\ll Q_2$, and $k\ll 1$. Here, the quality factors of Q_L, Q_1, and Q_2 are assumed to be large, which is acceptable for most IPTSs.

As identified from (14.4), the efficiency of I-SS becomes the same as that of I-SP when $Z_{2S}=Z_{2P}=Z_2$, which are determined as follows:

$$Z_{2S}\big|_{\omega=\omega_2}=R_{LS}+\frac{1}{j\omega_2 C_2} \tag{14.26a}$$

$$Z_{2P}\big|_{\omega=\omega_2}=R_{LP}\,//\,\frac{1}{j\omega_2 C_2}=\frac{R_{LP}}{1+Q_{LP}^2}+\frac{1}{j\omega_2 C_2}\cdot\frac{Q_{LP}^2}{1+Q_{LP}^2} \tag{14.26b}$$

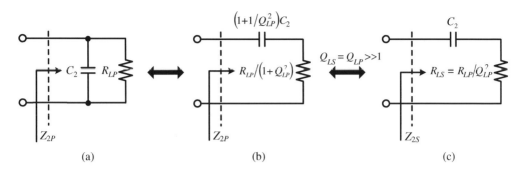

Figure 14.8 Equivalence of Z_{2S} and Z_{2P} under the conditions of a large quality factor Q_L and $Q_{LS} = Q_{LP}$: (a) parallel circuit, (b) equivalent series circuit for (a), and (c) approximated series circuit for (b).

where Q_{LS} and Q_{LP} are defined in (14.8b). Under the assumption of $Q_L \gg 1$, (14.26b) can be approximated as follows:

$$Z_{2P}|_{\omega=\omega_2} \cong \frac{R_{LP}}{Q_{LP}^2} + \frac{1}{j\omega_2 C_2} = R_{LS} + \frac{1}{j\omega_2 C_2} = Z_{2S}|_{\omega=\omega_2} \tag{14.27}$$

In (14.27), the relation of $R_{LP}/Q_{LP} = R_{LS}Q_{LS}$ is used, which is deduced from (14.8b) under the condition of $Q_{LS} = Q_{LP}$. The equivalence of Z_{2S} and Z_{2P} for a large quality factor Q_L is shown in Figure 14.8.

As shown in Figure 14.9, there is little difference in efficiency between I-SS and I-SP when Q_L is larger than 50.

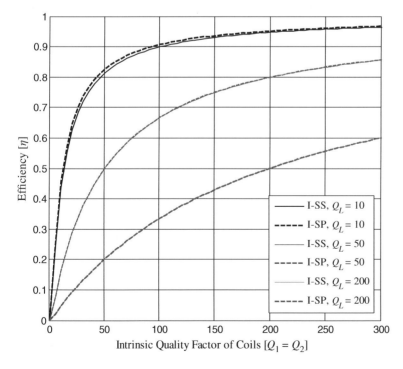

Figure 14.9 Efficiencies η versus quality factor Q_1 for I-SS and I-SP when $\omega = \omega_2$ and $k = 0.3$.

In Table 14.2, the efficiencies of the eight basic compensation schemes at ω_2 are listed. They are normalized to the reference efficiency η_r, which is defined as the efficiency of I-SS at ω_2 given from (14.7a) as follows:

$$\eta_r \equiv \eta_S|_{\omega=\omega_2} = \frac{Q_1(Q_2//Q_{LS})k^2}{1 + Q_1(Q_2//Q_{LS})k^2} \cdot \frac{1}{1 + Q_{LS}/Q_2} \tag{14.28}$$

It can be seen from Table 14.2 that all of the eight basic compensation schemes have the same efficiencies due to the equivalence of Z_{2S} and Z_{2P}, as described in (14.27).

14.3.2 Equivalence in Load Power P_L and Duality in Output Characteristics

In a similar fashion, the load power, output voltage, and output current of I-SS are compared with those of I-SP to verify the equivalence and duality of I-SS and I-SP. For I-SS, the per unit load power P_{LP}, output voltage V_{LS}, and output current I_{LS} are derived at ω_2, using the circuit model of Figure 14.7 with zero ESRs, as follows:

$$\begin{aligned} P_{LS}|_{\omega=\omega_2} &= k^2 Q_{LS} \quad &\text{(in per unit)} & \tag{14.29a} \\ |V_{LS}||_{\omega=\omega_2} &= k \quad &\text{(in per unit)} & \tag{14.29b} \\ |I_{LS}||_{\omega=\omega_2} &= kQ_{LS} \quad &\text{(in per unit)} & \tag{14.29c} \end{aligned}$$

In a similar way, the per unit ratings of I-SP are also derived as follows:

$$\begin{aligned} P_{LP}|_{\omega=\omega_2} &= k^2 Q_{LP} \quad &\text{(in per unit)} & \tag{14.30a} \\ |V_{LP}||_{\omega=\omega_2} &= kQ_{LP} \quad &\text{(in per unit)} & \tag{14.30b} \\ |I_{LP}||_{\omega=\omega_2} &= k \quad &\text{(in per unit)} & \tag{14.30c} \end{aligned}$$

Note that no constraints on the quality factor Q_L and k are enforced for (14.29) and (14.30), unlike in the previous section.

Applying the condition of $Q_{LS} = Q_{LP}$ to (14.29) and (14.30), the equivalence of I-SS and I-SP in load power is identified from (14.29a) and (14.30a). From (14.29b), (14.30b) and (14.29c), (14.30c), the duality of I-SS and I-SP in output voltage and output current is verified. I-SS has Q_L times larger I_{LS} than that of I-SP, whereas I-SP has Q_L times larger V_{LP} than that of I-SS. This is why I-SS is used for the low V_L to high I_L applications, and I-SP is used for the high V_L to low I_L applications, while their power ratings are the same.

Similar analyses can be applied for the remaining six basic compensation schemes and the results are tabulated in Table 14.2. For I-PS, I-PP, V-SS, and V-SP, the primary resonance angular frequency ω_1 is set for the maximum load power transfer, as described in Table 14.1. It can be found from Table 14.2 that the equivalence and duality holds for not only I-SS and I-SP but also for the secondary-series and secondary-parallel compensation schemes. This is because the secondary-series compensation is not distinguished from the secondary-parallel compensation under the equivalence of Z_{2S} and Z_{2P}, as described in (14.27).

14.3.3 Equivalence in Components Stresses

The voltage and current stresses on the components of I-SS and I-SP, which are crucial design parameters, are evaluated in this section. The ESRs R_1 and R_2 of coils are neglected in the following analyses due to their minor impact on the component stresses. The per unit voltage and current ratings on the secondary coil and the secondary compensation capacitor of I-SS at $\omega = \omega_2$ can be derived from Figure 14.7 as follows:

$$|V_2||_{\omega=\omega_2} = k\sqrt{1+Q_{LS}^2} \cong kQ_{LS} \quad \text{(when } Q_{LS} \gg 1, \text{ in per unit)} \tag{14.31a}$$

$$|V_{C2}||_{\omega=\omega_2} = kQ_{LS} \quad \text{(in per unit)} \tag{14.31b}$$

$$|I_2||_{\omega=\omega_2} = kQ_{LS} \quad \text{(in per unit)} \tag{14.31c}$$

$$|I_{C2}||_{\omega=\omega_2} = |I_2||_{\omega=\omega_2} = kQ_{LS} \quad \text{(in per unit)} \tag{14.31d}$$

Similarly, those of I-SP are also derived from Figure 14.7 as follows:

$$|V_2||_{\omega=\omega_2} = kQ_{LP} \quad \text{(in per unit)} \tag{14.32a}$$

$$|V_{C2}||_{\omega=\omega_2} = |V_2||_{\omega=\omega_2} = kQ_{LP} \quad \text{(in per unit)} \tag{14.32b}$$

$$|I_2||_{\omega=\omega_2} = k\sqrt{1+Q_{LP}^2} \cong kQ_{LP} \quad \text{(when } Q_{LP} \gg 1, \text{ in per unit)} \tag{14.32c}$$

$$|I_{C2}||_{\omega=\omega_2} = kQ_{LP} \quad \text{(in per unit).} \tag{14.32d}$$

Applying the condition of $Q_{LS} = Q_{LP}$ to (14.31) and (14.32), it was shown that I-SS and I-SP have nearly the same voltage and current ratings of the components on the secondary side. In other words, the compensation scheme does not affect the component stresses and can be appropriately selected for the correct purpose.

The per unit voltage and current stresses of the remaining six basic compensation schemes are tabulated in Table 14.2 by applying similar analyses to those of the previous subsection. Similar to the equivalence in load power from the previous subsection, this shows that the equivalence in the component rating is applied for the secondary-series and secondary-parallel compensation schemes.

Question 1 Show a design example based on the procedure in Figure 14.10.

From the comparative studies of the eight basic compensation schemes in terms of the efficiency, load power, and component ratings, it can be found that the secondary-series and secondary-parallel compensation schemes have equivalently the same efficiency, load power, and component ratings when the load quality factor Q_L is high. Therefore, I-SS and I-SP operating at the secondary resonance angular frequency ω_2, which meet all of the five criteria in the previous section, also have the equivalence and duality characteristics.

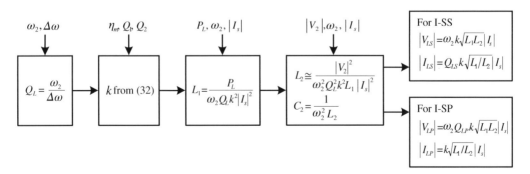

Figure 14.10 Proposed design guideline based on the analyses results for I-SS and I-SP.

14.4 A Design Guideline for I-SS and I-SP

For practical applications of I-SS or I-SP with the proposed characteristics, a design guideline based on the analyses in previous sections is suggested. The input parameters of the proposed design procedure are maximum efficiency η_m, load power P_L, frequency tolerance $\Delta\omega$, secondary resonant angular frequency ω_2, intrinsic quality factors Q_1, Q_2, secondary coil voltage rating $|V_2|$, and source current $|I_s|$. These input parameters should be determined first, considering cost, volume, mass budgets, and operating conditions. The throughput and output parameters of the design procedure are the load quality factor Q_L, magnetic coupling coefficient k, self-inductances of coils L_1 and L_2, respectively, secondary compensation capacitance C_2, and load voltage and current $|V_L|$ and $|I_L|$, respectively. The design procedure, as summarized in Figure 14.10, is composed of five steps as below.

1. Determine the load quality factor Q_L for a given frequency tolerance $\Delta\omega$ as follows [60]:

$$Q_L = \frac{\omega_2}{\Delta\omega} \qquad (14.33)$$

   For a large frequency tolerance, Q_L should be minimized. However, this low Q_L may result in low load power and high switching harmonics.

2. Calculate the magnetic coupling coefficient k for a given maximum efficiency $\eta_m \cong \eta_S \cong \eta_P$. The k is also affected by Q_1 and Q_2. The expression for k is derived from Table 14.2 as follows:

$$k = \sqrt{\frac{\eta_m}{Q_1(Q_2//Q_L)\{Q_2/(Q_2+Q_L)-\eta_m\}}} \qquad (14.34)$$

3. Calculate the primary self-inductance L_1 from (14.27a) by using the Q_L of step 1 and k of step 2, as follows:

$$L_1 = \frac{P_L}{\omega_2 Q_L k^2 |I_s|^2} \qquad (14.35)$$

4. Calculate the secondary-side LC parameters L_2 and C_2 from (14.31a) and (14.8a), respectively, by using Q_L, L_1, and k of previous steps for given ω_2, $|V_2|$, and $|I_s|$, as follows:

$$L_2 \cong \frac{|V_2|^2}{\omega_2^2 Q_L^2 k^2 L_1 |I_s|^2} \qquad (14.36a)$$

$$C_2 = \frac{1}{\omega_2^2 L_2} \qquad (14.36b)$$

5. Select a compensation scheme between I-SS and I-SP, considering the desired load voltage and current from (14.29) and (14.30). For applications requiring a high output current or constant output voltage, I-SS is appropriate, whereas I-SP is recommended for applications requiring a high output voltage or constant output current.

14.5 Example Design and Experimental Verifications

The proposed analyses and design principles were verified with an experimental kit for I-SS and I-SP, as shown in Figure 14.11.

As shown in Figure 14.12, a zero-voltage switching (ZVS) full-bridge inverter with a current control feedback loop was adopted to implement a current source. The inverter operates at

Figure 14.11 A prototype fabrication for the experiments.

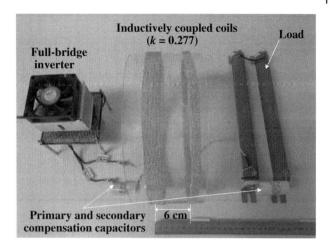

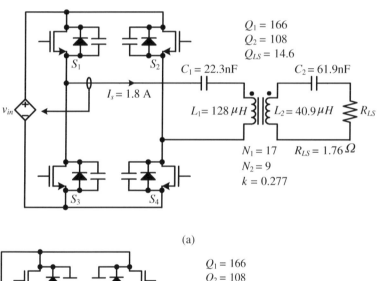

(a)

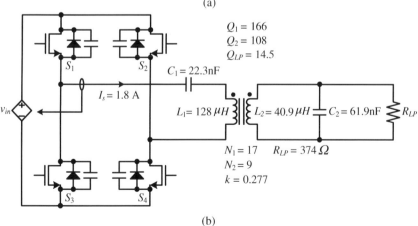

(b)

Figure 14.12 Proposed IPTS circuit schematics for the experiments: (a) full-bridge inverter with I-SS and (b) full-bridge inverter with I-SP.

Table 14.3 Experimental circuit parameters at the source frequency of 100 kHz

Circuit components	Value
IGBT power module S_1-S_4	APTGF90H60T3G (600 V_{max}, 90 A_{max})
Inductively coupled coil set ($k = 0.277$)	$N_1 = 17$ turns $L_1 = 128$ μH $R_1 = 484$ mΩ $Q_1 = 166$
	$N_2 = 9$ turns $L_2 = 40.9$ μH $R_2 = 238$ mΩ $Q_2 = 108$
Primary compensation capacitor C_1	$C_1 = 22.3$ nF
Secondary compensation capacitor C_2	$C_2 = 61.9$ nF

the source frequency of 100 kHz, which is about 5% higher than the primary side resonant frequency to guarantee the ZVS operation [47, 50, 51]. The targeted load power was 200 W and the maximum efficiency excluding the inverter was 88%; these numbers were selected for convenience, considering the availability of components. The efficiency of the inverter was 97%, which was achieved by the ZVS operation of the inverter. According to the design guideline, two inductively coupled air-coils with the magnetic coupling coefficient k of 0.277 were fabricated, where the primary and secondary self-inductances were 128 μH and 41 μH, and their number of turns were 17 and 9, respectively. The load quality factor Q_L was given as 15 for a frequency tolerance of 6.7%. All the same circuit parameters except for the load resistor R_L were used for both I-SS and I-SP, as tabulated in Table 14.3.

The efficiencies and the load powers of both I-SS and I-SP were measured w.r.t. the frequencies from 70 kHz to 130 kHz and compared with the calculation results, as shown in Figure 14.13(a) and (b), excluding the loss in the inverter, when the input current is controlled as 1.8 A. As identified from Table 14.1, the maximum efficiency and load power were obtained when $\omega = \omega_2$ for both I-SS and I-SP. The calculation results matched well with the experimental results. A little discrepancy was observed, which is mostly due to the errors in ESRs. The voltage and current stresses of the secondary coil were also measured, as shown in Figure 14.13(c) and (d). The measured results were also close to the calculation results; in particular, the measured results of I-SS were very close to those of I-SP. Therefore, the equivalences in the efficiency, the load power, and the component stresses were fully demonstrated.

14.6 Conclusion

A general and systematic comparison of the eight basic compensation schemes was conducted in this chapter. First, source frequencies were evaluated in terms of the five criteria, which are essential in most IPTSs. Throughout detailed analyses, it was found that I-SS is the only

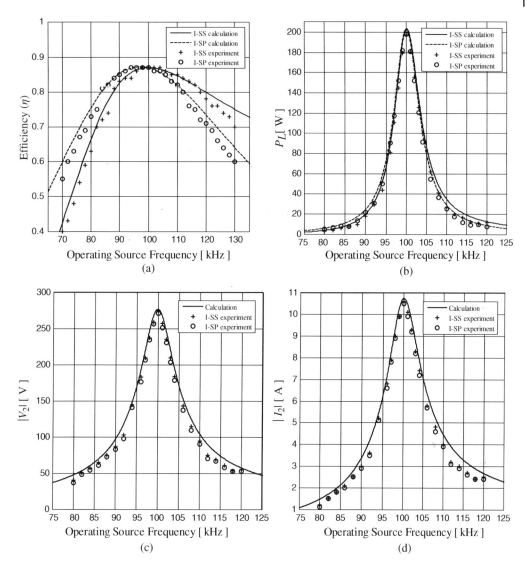

Figure 14.13 Experimental results for the equivalences of I-SS and I-SP: (a) efficiency, (b) load power, (c) voltage stress of the secondary coil, and (d) current stress of the secondary coil.

compensation scheme that meets all the five criteria successfully. That is, its frequencies for the maximum efficiency, maximum load power transfer, and load-independent output characteristic are all the same and independent of the coupling coefficient k while it can guarantee the safe operation of the inverter in the absence of the magnetic coupling. Excluding the weak k-dependency on the maximum efficiency frequency, I-SP was also found to be a suitable compensation scheme. Second, the efficiency, load power, and component ratings were evaluated for the eight basic compensation schemes at the secondary resonance frequency, which is the maximum efficiency frequency. It is shown that the secondary-series and secondary-parallel compensation schemes have the equivalence in the efficiency, load power, and component ratings provided that the load quality factor Q_L is high.

Therefore, the comparative studies in this chapter identified new aspects on the eight basic compensation schemes, which showed that I-SS and I-SP are the only compensation schemes that can satisfy all the five criteria simultaneously and have no difference in the efficiency, load power, and component ratings. Based on the analyses results, a design guideline for practical applications of I-SS and I-SP was suggested and experimentally verified by a 200 W prototype of air coils at 100 kHz.

References

1 G.A. Covic and J.T. Boys, "Inductive power transfer," *Proceedings of the IEEE*, vol. 101, no. 6, pp. 1276–1289, June 2013.

2 E. Abel and S. Third, "Contactless power transfer – an exercise in topology," *IEEE Trans. Magn.*, vol. MAG-20, no. 5, pp. 1813–1815, September-November 1984.

3 A.W. Green and J.T. Boys, "10 kHz inductively coupled power transfer-concept and control," in *Proc. 5th Int. Conf. on Power Electron. Variable-Speed Drives*, October 1994, pp. 694–699.

4 Y. Hiraga, J. Hirai, A. Kawamura, I. Ishoka, Y. Kaku, and Y. Nitta, "Decentralised control of machines with the use of inductive transmission of power and signal," in *Proc. IEEE Ind. Appl. Soc. Annual Meeting*, 1994, vol. 29, pp. 875–881.

5 A. Kawamura, K. Ishioka, and J. Hirai, "Wireless transmission of power and information through one high-frequency resonant AC link inverter for robot manipulator applications," *IEEE Trans. Ind. Applic.*, vol. 32, no. 3, pp. 503–508, May/June 1996.

6 J.M. Barnard, J.A. Ferreira, and J.D. van Wyk, "Sliding transformers for linear contactless power delivery," *IEEE Trans. on Ind. Electron.*, vol. 44, no. 6, pp. 774–779, December 1997.

7 D.A.G. Pedder, A.D. Brown, and J.A. Skinner, "A contactless electrical energy transmission system," *IEEE Trans. on Ind. Electron.*, vol. 46, no. 1, pp. 23–30, February 1999.

8 J.T. Boys, G.A. Covic, and A.W. Green, "Stability and control of inductively coupled power transfer systems," *IEE Proceedings – Electric Power Applications*, vol. 147, no. 1, pp. 37–43, 2000.

9 K.I. Woo, H.S. Park, Y.H. Choo, and K.H. Kim, "Contactless energy transmission system for linear servo motor," *IEEE Trans. Magn.*, vol. 41, no. 5, pp. 1596–1599, May 2005.

10 G.A.J. Elliott, G.A. Covic, D. Kacprzak, and J.T. Boys, "A new concept: asymmetrical pick-ups for inductively coupled power transfer monorail systems," *IEEE Trans. Magn.*, vol. 42, no. 10, pp. 3389–3391, October 2006.

11 P. Sergeant and A. Van den Bossche, "Inductive coupler for contactless power transmission," *IET Electr. Power Applic.*, vol. 2, pp. 1–7, 2008.

12 J.T. Boys and A.W. Green, "Intelligent road studs – lighting the paths of the future," *IPENZ Trans.*, vol. 24, no. 1, pp. 33–40, 1997.

13 H.H. Wu, G.A. Covic, and J.T. Boys, "An AC processing pickup for IPT systems," *IEEE Trans. of Power Electron. Soc.*, vol. 25, no. 5, pp. 1275–1284, May 2010.

14 H.H. Wu, G.A. Covic, J.T. Boys, and D. Robertson, "A series tuned AC processing pickup," *IEEE Trans. of Power Electron. Soc.*, vol. 26, no. 1, pp. 98–109, January 2011.

15 D. Robertson, A. Chu, A. Sabitov, and G.A. Covic, "High powered IPT stage lighting controller," in *Proc. IEEE Int. Symp. Ind. Electron.*, Gdansk, Poland, June 27–30, 2011, pp. 1974-1979.

16 J.E.I. James, A. Chu, A. Sabitov, D. Robertson, and G.A. Covic, "A series tuned high power IPT stage lighting controller," in *Proc. IEEE Energy Conversion Congress and Exposition*, Phoenix, AZ, USA, September 17–22, 2011, pp. 2843–2849.

17 G.B. Joung and B.H. Cho, "An energy transmission system for an artificial heart using leakage inductance compensation of transcutaneous transformer," *IEEE Trans. on Power Electron.*, vol. 13, no. 6, pp. 1013–1022, November 1998.

18 W. Guoxing, L. Wentai, M. Sivaprakasam, and G.A. Kendir, "Design and analysis of an adaptive transcutaneous power telemetry for biomedical implants," *IEEE Trans. on Circuits Syst. I, Reg. Papers*, vol. 52, no. 10, pp. 2109–2117, October 2005.

19 P. Si, A.P. Hu, J.W. Hsu, M. Chiang, Y. Wang, S. Malpas, and D. Budgett, "Wireless power supply for implantable biomedical device based on primary input voltage regulation," in *Proc. 2nd IEEE Conf. on Ind. Electron. Applic.*, 2007, pp. 235–239.

20 P. Si, A.P. Hu, S. Malpas, and D. Budgett, "A frequency control method for regulating wireless power to implantable devices," *IEEE Trans. on Biomed. Circuits Syst.*, vol. 2, no. 1, pp. 22–29, March 2008.

21 B.J. Heeres, D.W. Novotny, D.M. Divan, and R.D. Lorenz, "Contactless underwater power delivery," in *Proc. IEEE Power Electron. Specialists Conf.*, 1994, vol. 1, pp. 418–423.

22 K.W. Klontz, D.M. Divan, D.W. Novotny, and R.D. Lorenz, "Contactless power delivery system for mining applications," *IEEE Trans. on Ind. Applic.*, vol. 31, no. 1, pp. 27–35, January/February 1995.

23 B.-M. Song, R. Kratz, and S. Gurol, "Contactless inductive power pickup system for Maglev applications," in *Proc. IEEE 37th Ind. Applic. Conf.*, 2002, pp. 1586–1591.

24 J. Jia, W. Liu, and H. Wang, "Contactless power delivery system for the underground flat transit of mining," in *Proc. 6th Int. Conf. on Electr. Mach. Syst.*, Beijing, China, 2003, pp. 282–284.

25 T. Kojiya, F. Sato, H. Matsuki, and T. Sato, "Construction of non-contacting power feeding system to underwater vehicle utilizing electromagnetic induction," in *Proc. OCEANS Conf. V, Europe*, 2005, pp. 709–712.

26 H.H. Wu, M.Z. Feng, J.T. Boys, and G.A. Covic, "A wireless multi-drop IPT security camera system," in *Proc. 4th IEEE Conf. on Ind. Electron. Applic.*, Xian, China, May 25–27, 2009, pp. 70–75.

27 W. Chwei-Sen, G.A. Covic, and O.H. Stielau, "Power transfer capability and bifurcation phenomena of loosely coupled inductive power transfer systems," *IEEE Trans. on Ind. Electron.*, vol. 51, no. 1, pp. 148–157, February 2004.

28 W. Chwei-Sen, O.H. Stielau, and G.A. Covic, "Design considerations for a contactless electric vehicle battery charger," *IEEE Trans. on Ind. Electron.*, vol. 52, no. 5, pp. 1308–1314, October 2005.

29 W. Chwei-Sen, G.A. Covic, and O.H. Stielau, "General stability criterions for zero phase angle controlled loosely coupled inductive power transfer systems," in *2001 IECON Conf.*, pp. 1049–1054.

30 K. Aditya and S.S. Williamson, "Comparative study of series–series and series–parallel compensation topologies for electric vehicle charging," in *2014 ISIE Conf.*, pp. 426–430.

31 K. Aditya and S.S. Williamson, "Comparative study of series–series and series–parallel compensation topology for long track EV charging application," in *2014 ITEC Conf.*, pp. 1–5.

32 Y.-H. Chao, J.J. Shieh, C.-T. Pan, and W.-C. Shen, "A closed-form oriented compensator analysis for series–parallel loosely coupled inductive power transfer systems," in *2007 PESC Conf.*, pp. 1215–1220.

33 S. Chopra and P. Bauer, "Analysis and design considerations for a contactless power transfer system," in *2011 INTELEC Conf.*, pp. 1–6.

34 A.J. Moradewicz and M.P. Kazmierkowski, "Contactless energy transfer system with FPGA-controlled resonant converter," *IEEE Trans. Ind. Electron.*, vol. 57, no. 9, pp. 3181–3190, September 2010.

35 J. How, Q. Chen, S.-C. Wong, C.K. Tse, and X. Ruan, "Analysis and control of series/series–parallel compensated resonant converters for contactless power transfer," *IEEE Journal of Emerging and Selected Topics in Power Electronics*, vol. PP, no. 99, p. 1, 2014.

36 S.-Y. Cho, I.-O. Lee, S. Moon, G.-W. Moon, B.-C. Kim, and K.Y. Kim, "Series–series compensated wireless power transfer at two different resonant frequencies," in 2013 *ECCE Conf.*, pp. 1052–1058.

37 W. Zhang, S.-C. Wong, C.K. Tse, and Q. Chen, "Analysis and comparison of secondary series- and parallel-compensated inductive power transfer systems operating for optimal efficiency and load-independent voltage-transfer ratio," *IEEE Trans. on Power Electron.*, vol. 29, no. 6, pp. 2979–2990, June 2014.

38 W. Zhang, S.-C. Wong, and C.K. Tse, "Compensation technique for optimized efficiency and voltage controllability of IPT systems," in *2012 ISCAS Conf.*, pp. 225–228.

39 W. Zhang, S.-C. Wong, C. K. Tse, and Q. Chen, "Load-independent current output of inductive power transfer converters with optimized efficiency," in *2014 IPEC Conf.*, pp. 1425–1429.

40 W. Zhang, S.-C. Wong, C.K. Tse, and Q. Chen, "Analysis and comparison of secondary series- and parallel- compensated IPT systems," in *2013 ECCE Conf.*, pp. 2898–2903.

41 Z. Wei, W. S.-C., C.K. Tse, and C. Qianhong, "Design for efficiency optimization and voltage controllability of series–series compensated inductive power transfer systems," *IEEE Trans. on Power Electron.*, vol. 29, no. 1, pp. 191–200, January 2014.

42 X. Ren, Q. Chen, L. Cao, X. Ruan, S.-C. Wong, and C.K. Tse, "Characterization and control of self-oscillating contactless resonant converter with fixed voltage gain," in *2012 IPEMC Conf.*, pp. 1822–1827.

43 O.H. Stielau and G.A. Covic, "Design of loosely coupled inductive power transfer systems," *International Conference on Power System Technology*, 2000, pp. 85– 90.

44 H. Abe, H. Sakamoto, and K. Harada, "A noncontact charger using a resonant converter with parallel capacitor of the secondary coil," *IEEE Trans. on Ind. Applic.*, vol. 36, no. 2, pp. 444–451, March–April 2000.

45 J.L. Villa, J. Sallan, J.F. Sanz Osorio, and A. Llombart, "High-misalignment tolerant compensation topology for ICPT systems," *IEEE Trans. on Ind. Electron*, vol. 59, no. 2, pp. 945–951, February 2012.

46 J. Sallan, J.L. Villa, A. Llombart, and J.F. Sanz, "Optimal design of ICPT systems applied to electric vehicle battery charge," *IEEE Trans. on Ind. Electron.*, vol. 56, no. 6, pp. 2140–2149, June 2009.

47 J. Huh, S.W. Lee, W.Y. Lee, G.H. Cho, and Chun T. Rim, "Narrow-width inductive power transfer system for online electrical vehicles," *IEEE Trans. on Power Electron.*, vol. 26, no.12, pp. 3666–3679, Dec. 2011.

48 S.Y. Choi, B.W. Gu, J. Huh, W.Y. Lee, J.G. Cho, and Chun T. Rim, "Asymmetric coil sets for wireless stationary EV chargers with large lateral tolerance by dominant field analysis," *IEEE Trans. on Power Electron.*, vol. 29, no. 12, pp. 6406–6420, Dec. 2014.

49 C.B. Park, S.W. Lee, and Chun T. Rim, "Innovative 5 m-off-distance inductive power transfer systems with optimally shaped dipole coils," *IEEE Trans. on Power Electron.*, vol. 30, no. 2, pp. 817–827, February 2015.

50 C.B. Park, S.W. Lee, G.-H. Cho, S.-Y. Choi, and Chun T. Rim, "Two-dimensional inductive power transfer system for mobile robots using evenly displaced multiple pick-ups," *IEEE Trans. on Ind. Appl.*, vol. PP, no. 99, p. 1, June 2013.

51 S.W. Lee, B. Choi, and Chun T. Rim, "Dynamics characterization of the inductive power transfer system for on-line electric vehicles by Laplace phasor transform," *IEEE Trans. on Power Electron.*, vol. 28, no. 12, pp. 5902–5909, December 2013.

52 B.H. Choi, J.P. Cheon, J.H. Kim, and Chun T. Rim, "7 m-off-long-distance extremely loosely coupled inductive power transfer systems using dipole coils," in *2014 ECCE Conf.*, pp. 858–863.

53 M.L.G. Kissin, C.-Y. Huang, G.A. Covic, and J.T. Boys, "Detection of the tuned point of a fixed-frequency LCL resonant power supply," *IEEE Trans. on Power Electron.*, vol. 24, no. 41, pp. 1140–1143, April 2009.

54 H. Hao, G.A. Covic, and J.T. Boys, "An approximate dynamic model of LCL-T-based inductive power transfer power supplies," *IEEE Trans. on Power Electron.*, vol. 29, no. 10, pp. 5554–5567, October 2014.

55 C.Y. Huang, J.T. Boys, and G.A. Covic, "LCL pickup circulating current controller for inductive power transfer systems," *IEEE Trans. on Power Electron.*, vol. 28, no. 4, pp.2081–2093, April 2013.

56 Z. Pantic, B. Sanzhong, and S. Lukic, "ZCS LCC-compensated resonant inverter for inductive-power-transfer application," *IEEE Trans. on Ind. Electron.*, vol. 58, no. 9, pp. 3500–3510, August 2011.

57 K. Okada, K. Limura, N. Hoshi, and J. Haruna, "Comparison of two kinds of compensation schemes on inductive power transfer systems for electric vehicle," in *2012 VPPC Conf.*, 2012, pp. 766–771.

58 L.Z. Ning, R.A. Chinga, R. Tseng, and L. Jenshan, "Design and test of a high-power high-efficiency loosely coupled planar wireless power transfer system," *IEEE Trans. on Ind. Electron.*, vol. 56, no. 5, pp. 1801–1812, May 2009.

59 H.H. Wu, G.A. Covic, J.T. Boys, and D.J. Robertson, "A series-tuned inductive-power-transfer pickup with a controllable AC-voltage output," *IEEE Trans. on Power Electron.*, vol. 26, no. 1, pp. 98–109, January 2011.

60 S.Y. Choi, B.W. Gu, S.Y. Jeong, and Chun T. Rim, "Advances in wireless power transfer systems for road powered electric vehicles," *IEEE Journal of Emerging and Selected Topics in Power Electronics*, vol. PP, no. 99, 2015.

61 S.-H. Lee and R.D. Lorenz, "Development and validation of model for 95%-efficiency 220-W wireless power transfer over a 30-cm air gap," *IEEE Trans. on Ind. Applic.*, vol. 47, no. 6, pp. 2495–2504, November/December 2011.

62 Chun T. Rim and G.-H. Cho, "Phasor transformation and its application to the DC/AC analyses of frequency phase-controlled series resonant converters (SRC)," *IEEE Trans. on Power Electron.*, vol. 5, no. 2, pp. 201–211, April 1990.

63 R.L. Steigerwald, "A comparison of half-bridge resonant converter topologies," *IEEE Trans. on Power Electron.*, vol. 3, no. 2, pp. 174–182, April 1988.

15

Electromagnetic Field (EMF) Cancel

15.1 Introduction

Wireless power has been increasingly used in mobile devices [1–4], industry clean rooms [5–7], medical applications [8–10], and electrified transportations [11–46] because of its convenient, clean, robust, and safe characteristics. Widespread use of the wireless power, however, is not yet come true because it is still expensive, less power efficient, and potentially harmful for pedestrians. An electromagnetic field (EMF) is inevitably generated from any wireless power transfer systems (WPTS), which include primary coils for transmitting power and secondary coils for receiving power, but this EMF level should be well regulated for the safety of pedestrians.

A good way of reducing the EMF is to increase the magnetic coupling factor of the WPTS by appropriate design of coils so that magnetic leakage flux, which is the major source of the EMF, can be minimized. Except for the contact-type WPTS, where high magnetic coupling is achievable, this method cannot be generally applied to the WPTS with a large airgap between the primary and secondary coils where the magnetic coupling factor is very low. For the loosely coupled cases of WPTS, proactive measures to cancel out the EMF should be used because of the large leakage flux.

There are passive and active EMF cancel methods in general, where the former relies on the use of ferromagnetic materials, conductive materials, and selective surfaces for radio frequency interference [47–56] whereas the latter requires countercurrent sources [57–59]. If applicable, the passive EMF cancel method should be used because it is simple, cheap, and robust in most cases. For applications of high power or a very large airgap where a very large EMF exceeding ICNIRP Guidelines [60, 61] is generated, the active EMF cancel method cannot be used. Wireless electric vehicles (WEV) [11–46], which include wireless stationary electric vehicle charging (battery EV and PHEV) [11–30] and road-powered electric vehicles (RPEV) [31–46], are one of the typical examples using the active EMF cancel solution, where a few tens of kW wireless power can be transferred through a few tens of cm airgap. However, the passive EMF cancel solution has been found to be not effective [48, 49, 51], as verified during the development of on-line electric vehicles (OLEV) [39–46]. To find an effective and economic active EMF cancel solution is crucial for the commercialization of the OLEV, which is a viable solution of the RPEV now. An active EMF cancel design was applied to the primary coil side of the RPEV developed by the PATH team [35], where the purpose of the canceling coil is to mitigate the EMF generated from the primary coil when there is no RPEV on the road but potential pedestrians beside the road. Another active EMF cancel design was applied to the secondary coil side of the OLEV [59], where the countercurrent is dynamically controlled by adjusting resonant capacitors, sensing the EMF. However, the EMF is generated both from primary coils and secondary coils of the WPTS; therefore, it must be canceled simultaneously to mitigate the total EMF level.

Wireless Power Transfer for Electric Vehicles and Mobile Devices, First Edition. Chun T. Rim and Chris Mi.
© 2017 John Wiley & Sons Ltd. Published 2017 by John Wiley & Sons Ltd.

Furthermore, a systematic way of active EMF cancel must be secured so that the EMF can be considered from the beginning of the design of WPTS.

In this chapter, a generalized design method for canceling the total EMF of WEV, which can be extended to any WPTS, is explained. By adding an active EMF cancel coil to each primary coil and secondary coil, respectively, the EMF generated from each main coil can be independently canceled by their corresponding cancel coils. Thus, three general design methods are established and are verified by the recently developed I-type inductive power transfer systems (IPTS), which have a narrow rail width structure with alternating magnetic polarity along with a roadway.

15.2 Proposed General Active EMF Cancel Methods

In this chapter the active EMF cancel system is composed of main coil(s), cancel coil(s), and other elements such as resonant capacitors, additional EMF shields, and harnesses, as shown in Figure 15.1. The proposed EMF cancel methods can be applied to not only IPTS but also coupled magnetic resonance systems (CMRSs) because WPTS includes both IPTS and CMRS, which has been found to be just a special form of IPTS [62]. Throughout this chapter, it is assumed that the proposed active cancel system is linear time-invariant, that is, cores are unsaturated and circuit parameters are constant, and that the system is in the steady state, even though the proposed methods may hold in the transient state.

Ideally, the EMF should be canceled for any area of pedestrians at any time, regardless of load conditions; however, residual EMF after canceling is inevitable because the number of cancel coils is limited in practice and the EMF cancel position(s) does not coincide with the EMF generating points. As shown in Figure 15.1, the total EMF at pedestrians B_t, which is the residual EMF, can be described as follows:

$$B_t = B_{m1} + B_{m2} + B_c \tag{15.1}$$

where the magnetic fields B_{m1} and B_{m2} are generated from the primary main coil and secondary main coil, respectively, and the counter-magnetic field B_c is provided by one or more cancel

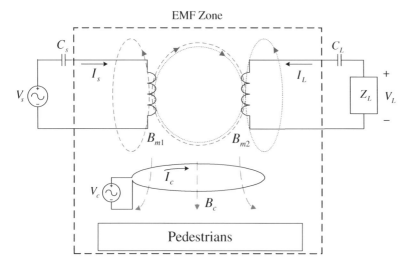

Figure 15.1 The configuration of an active EMF cancel system in general.

coil(s). Every magnetic field $\boldsymbol{B}_k$ in this chapter is represented in the Cartesian coordinate as follows:

$$\boldsymbol{B}_k \equiv B_{kx}\boldsymbol{x}_0 + B_{ky}\boldsymbol{y}_0 + B_{kz}\boldsymbol{z}_0 \tag{15.2}$$

In (15.2), the coefficients are of complex scalar form, which represent the phasor components of AC in the steady state, and $\boldsymbol{x}_0, \boldsymbol{y}_0, \boldsymbol{z}_0$ are unit vectors. From (15.1), a general requirement for the active EMF cancel can be described so that $|\boldsymbol{B}_t|$ should be lower than an EMF guide line B_{ref} over the area of pedestrians (x, y, z) by appropriate control of $\boldsymbol{B}_c$ as follows:

$$|\boldsymbol{B}_t| = |\boldsymbol{B}_t(x, y, z, \theta, I_o)| \leq B_{ref} \tag{15.3}$$

where the time dependency of the EMF is represented in the rotary reference frame as the phase difference θ between the primary and secondary coils and the load dependency is represented as I_o, respectively. The phase θ may arbitrarily change in accordance with the resonant conditions and $\boldsymbol{B}_{m2}$ is proportional to I_o in general, as shown in Figure 15.2; hence, the control of cancel coil(s) for providing appropriate counter EMF $\boldsymbol{B}_c$ would be quite challenging, especially for the single-cancel coil case. Accurate sensing of EMF and complicated control circuits are essential for this adaptive EMF cancel, as identified from the literature [59], but is not highly recommended for practical use.

A smart method to meet (15.3) without the use of the adaptive EMF cancel is proposed, as shown in Figure 15.3, which is quite effective for every space–time–load condition. The EMF can be significantly reduced by providing counter magnetic fields $\boldsymbol{B}_{c1}$ and $\boldsymbol{B}_{c2}$ generated from the primary cancel coil and secondary cancel coil, respectively, where magnetic fields $\boldsymbol{B}_{m1}$

Figure 15.2 The phasor diagram for the *x* component example in an active EMF cancel system using adaptive control of a cancel coil: (a) light load and (b) heavy load.

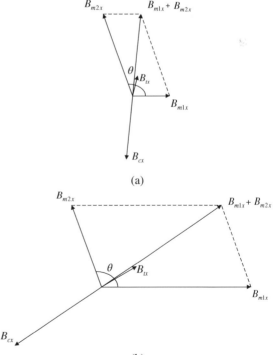

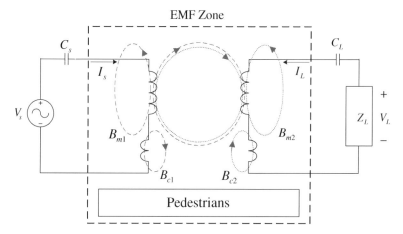

Figure 15.3 Independent self EMF cancel (ISEC) method for primary and secondary sides, fetching the cancel currents from their corresponding main coils.

and $\boldsymbol{B}_{m2}$ are generated from the primary main coil and secondary main coil, respectively, as follows:

$$\boldsymbol{B}_t = (\boldsymbol{B}_{m1} + \boldsymbol{B}_{c1}) + (\boldsymbol{B}_{m2} + \boldsymbol{B}_{c2}) \cong 0 \tag{15.4}$$

where

$$\boldsymbol{B}_{c1} \cong -\boldsymbol{B}_{m1}, \quad \boldsymbol{B}_{c2} \cong -\boldsymbol{B}_{m2} \tag{15.5}$$

Note that the current of each cancel coil is fetched from its corresponding main coil; hence, the magnetic flux of a cancel coil is in phase with that of a main coil. Therefore, the resultant EMF $\boldsymbol{B}_t$ becomes independent of both phase and load conditions, as depicted in Figure 15.4. Note that (15.4) and (15.5) hold for an arbitrary phase difference between the primary and secondary sides.

This simple and effective EMF cancel method had so far been applied only to either the primary side or secondary side. Now this method can be generally applicable to the EMF cancel for any WPTS, where a few secondary coils can be magnetically coupled with a few primary coils.

Question 1 Why is the ISEC an active cancel method? There is no sensing and control part, so it looks like a passive cancel method. Actually, it cancels current flows automatically in the passive cancel system.

The EMF of (15.4) cannot be completely canceled over a wide range of pedestrians; instead, it should be mitigated under the EMF guide line as follows:

$$|\boldsymbol{B}_t| = |(\boldsymbol{B}_{m1} + \boldsymbol{B}_{c1}) + (\boldsymbol{B}_{m2} + \boldsymbol{B}_{c2})| \equiv |\boldsymbol{B}_1 + \boldsymbol{B}_2| \le B_{ref} \tag{15.6}$$

where

$$\boldsymbol{B}_1 = \boldsymbol{B}_{m1} + \boldsymbol{B}_{c1}, \quad \boldsymbol{B}_2 = \boldsymbol{B}_{m2} + \boldsymbol{B}_{c2} \tag{15.7}$$

Figure 15.4 The phasor diagram for the *x* component example in an active EMF cancel system using the proposed ISEC design method, which is valid for not only arbitrary load but also arbitrary phase difference: (a) light load and (b) heavy load.

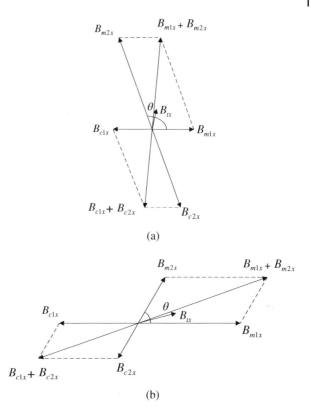

(a)

(b)

As shown in Figure 15.5, the primary current and secondary current of the resonant type WPTS are generally in quadrature insofar as each side is in full resonance [36–49,62], which holds not only for IPTS but also for CMRS. For the resonant-type WPTS, (15.6) becomes as follows [63]:

$$B_t \equiv |\boldsymbol{B}_t| = \sqrt{|\boldsymbol{B}_1|^2 + |\boldsymbol{B}_2|^2} = \sqrt{B_1^2 + B_2^2} \leq B_{ref} \quad \because B_1 \equiv |\boldsymbol{B}_1|, \ B_2 \equiv |\boldsymbol{B}_2| \tag{15.8}$$

where the following orthogonal conditions for the quadrant phasors, i.e. $\theta = \pi/2$, are used.

$$B_{1x} \perp B_{2x}, \quad B_{1y} \perp B_{2y}, \quad B_{1z} \perp B_{2z} \tag{15.9}$$

Figure 15.5 The phasor diagram for the *x* component example in an active EMF cancel system using the proposed 3DEC design method, which is in resonance.

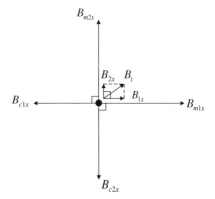

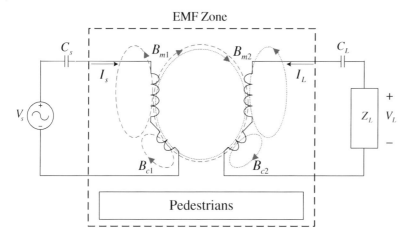

Figure 15.6 A bad design example for cancel coils, which are strongly coupled with the main magnetic linkage. By applying the proposed LFEC design method, the EMF can be effectively canceled out without any reduction in induced load voltage.

Assuming that $B_2 \leq B_1$, without the loss of generality, (15.8) can be rewritten as follows:

$$B_t = \sqrt{B_1^2 + B_2^2} \leq \sqrt{B_1^2 + B_1^2} = \sqrt{2}B_1 \leq B_{ref} \tag{15.10}$$

It is identified from (15.10) that the total EMF B_t becomes less than the EMF guide line B_{ref} if only a dominant residual EMF B_1 is 3 dB ($=\sqrt{2}$ times) less than B_{ref}. In other words, if both of the residual EMF B_1 and B_2 are 3 dB less than B_{ref}, the EMF is always mitigated under the guide line. Therefore, the EMF design of the primary side can be completely isolated from that of the secondary side. Moreover, only a dominant residual EMF needs to be mitigated under the EMF guide line. This method is named, in this chapter, as the "3dB dominant EMF cancel," which is quite useful for the EMF cancel design of WPTS, where one of the coils is either close to pedestrians or its EMF is relatively much stronger than the other. For the OLEV case, the secondary side pick-up coil generates a large EMF in most cases compared to the primary side power supply rail [43].

Question 2 In the case of out of resonance, (15.9) may not be met. What then happens to (15.10)?

EMF cancel coils may be coupled with the magnetic linkage, as shown in Figure 15.6. Under this condition, an effective total magnetic coupling between the primary and secondary sides becomes smaller because the cancel coils nullify magnetic linkage, which results in a lower induced load voltage. As shown in Figure 15.3, it is highly recommended to design the cancel coils so that the magnetic linkage should not intersect them. This design method together with previously mentioned ones will be verified by experiments.

15.3 Design Examples of Active EMF Cancels for WEV

Adopting the proposed three EMF design methods, that is ISEC, 3DEC, and LFEC, design examples for U-type and W-type IPTS of OLEV as well as a rectangular type of wireless stationary EV charger are shown in this section.

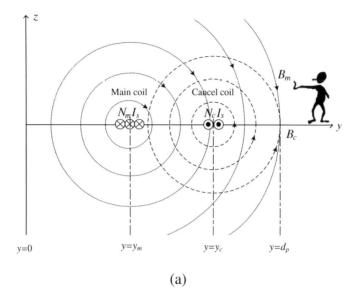

(a)

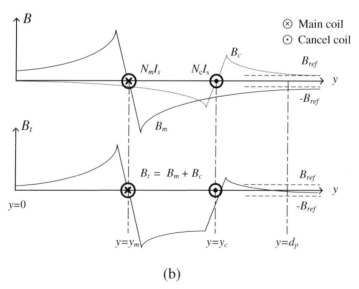

(b)

Figure 15.7 A simplified EMF cancel architecture for the infinitely long main coil at $y = y_m$ and cancel coil at $y = y_c$: (a) architecture and (b) EMF.

In practice, the calculation of EMF of WEV is neither straightforward nor analytical; hence, simulation or heuristic experiment cannot be used for EMF canceling in general. It is not intended to show any analytical design method in this chapter, but a simplified EMF cancel architecture, as shown in Figure 15.7(a), is examined to provide a good background to the idea.

In Figure 15.7(b), the EMF at the measurement point $y = d_p$ for the infinitely long coils can be easily determined from Ampere's law as follows:

$$B_t = - \left\{ \frac{\mu_o N_m I_s}{2\pi(y_p - y_m)} - \frac{\mu_o N_c I_s}{2\pi(y_p - y_c)} \right\} z_0 = \frac{\mu_o I_s N_m}{2\pi} \left(\frac{1}{y_p - y_c} \cdot \frac{N_c}{N_m} - \frac{1}{y_p - y_m} \right) z_0 \quad (15.11)$$

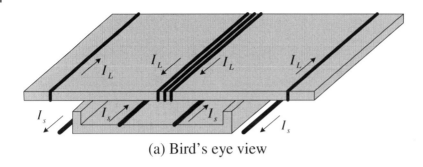

(a) Bird's eye view

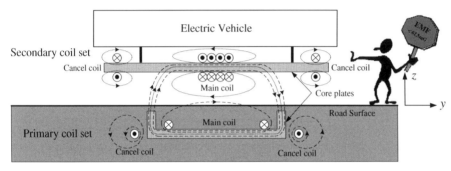

(b) Cross-section view

Figure 15.8 The configuration of U-type primary main coil and other coils for OLEV, applying the proposed three design methods.

In (15.11), the EMF can be mitigated under the EMF guide line B_{ref} by the proper selections of the turn ratio N_c/N_m, which is usually less than unity, and the position of the cancel coil y_c for a given measurement point y_p, as shown in Figure 15.7(b). In practice, the EMF is severely distorted by the cores used to focus or shield the magnetic field and the metallic body of the WEV, etc.; hence, direct use of (17.11) can never be allowed to happen, but the idea that appropriate number of turns and position of cancel coils may effectively mitigate the EMF can be used for the experimental design process.

As shown in Figure 15.8, the EMF of the U-type IPTS [40], which includes a very long U-shaped core of the primary main coil and a plate core width of the secondary main coil, can be actively canceled to meet the ICNIRP Guidelines [60] of 6.25 μT (= 62.5 mG) for 20 kHz. For the OLEV case, the primary coil set has a two-turn main cable and two separate single-turn cancel cables, whereas the secondary coil set has a multiturn main cable and multiturn cancel cables. By applying the proposed ISEC design method, the cancel coils are independently provided for each primary and secondary side, connected in series with the corresponding main coils. By applying the LFEC design method, the cancel coils are located sufficiently apart from the main magnetic linkage path, as shown in Figure 8(b). Though not shown in detail due to the page limit, the exact locations and the number of turns of cancel coils can be determined by applying the 3DEC method. In this case, the primary and secondary sides should be independently tuned by simulations or experiments so that each EMF can be less than −3 dB of B_{ref}.

The W-type IPTS [40], as shown in Figure 15.9, which has a very long W-shaped core of the primary main coil and a plate core of the secondary main coil, is now designed. Typically, the primary coil set has a one-turn cable including a no-cancel coil [39–46, 64] because the EMF generated from the primary main coil roughly diminishes as a function of the inverse

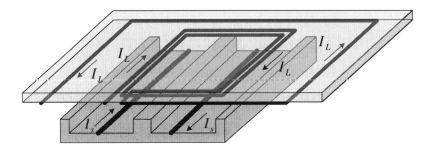

(a) Bird's eye view

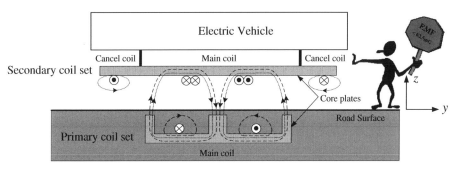

(b) Cross-section view

Figure 15.9 The configuration of the primary and secondary coils for OLEV with the proposed active EMF cancel coils.

of the square of the distance from the center of the primary coil and becomes less than the ICNIRP Guidelines of 62.5 mG for pedestrians [45]. Therefore, the cancel coil is installed with the secondary coil set only in accordance with the ISEC design method. By applying the LFEC design method, the cancel coil is sufficiently separated from the main magnetic linkage path, as shown in Figure 15.9(b). The exact location and number of turns of the cancel coil can be determined by applying the 3DEC method to the secondary coil set only, assuming that the secondary side is dominant over the primary side. A detailed procedure is also omitted here because the design of this W-type IPTS is similar to that of the I-type IPTS, to be explored in detail in the next section.

The wireless stationary EV chargers [11–30] may have either rectangular-type main coils, as shown in Figure 15.10, or circular type ones. By applying the ISEC and LFEC methods to both primary and secondary sides of the EV chargers, two cancel coils are installed, as shown in Figure 15.10. Depending on applications and operating conditions, one or more cancel coils may be omitted from consideration if the EMF level is not high and the 3DEC method can be applied accordingly. A detailed procedure is omitted again because the pick-up of this EV charger is nearly half of the subsequent I-type IPTS.

15.4 EMF Cancel Design and Analysis of the I-Type IPTS of OLEV

In this section, the EMF of I-type IPTS, which includes a very long I-shaped primary core and a set of two secondary main coils with a plate core, as shown in Figure 15.11(a), is actively

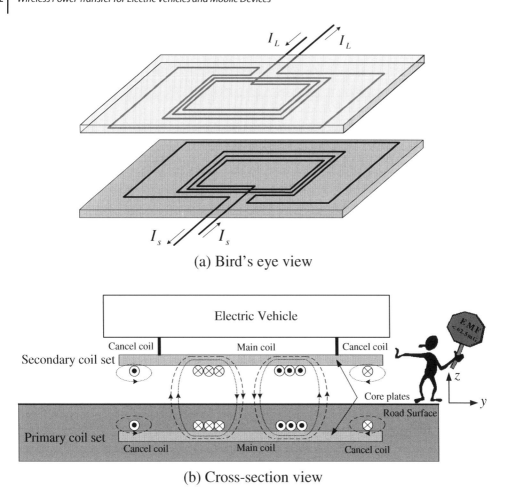

Figure 15.10 The configuration of the rectangular shaped primary and secondary coils with proposed active EMF cancel coils.

canceled using the proposed design methods. Because the polarity of magnetic poles B_1 is alternating along with the forward direction x, the secondary coils constitute a combined pick-up coil whose polarity of magnetic flux B_2 is also alternating. Due to this alternating magnetic flux distribution of the I-type core, the EMF generated from the primary side is typically very low [45]; hence, no cancel coil is required for the primary side in most cases, which is similar to the W-type IPTS, but the EMF level of the I-type one is much more enhanced. As will be shown in the subsequent section, the currents of secondary coils are not necessarily equal in phasor though their magnitudes are the same, as follows:

$$|I_L^+| = |I_L^-|, \quad \because I_L^+ \neq I_L^- \tag{15.12}$$

Applying the ISEC method to the secondary pick-up side only, the cancel coil architecture is obtained, as shown in Figure 15.11(b). According to the LFEC method, the cancel coils should be excluded from the main magnetic linkage path; hence, the cancel coils are displaced from the core, as shown in Figure 15.12(a). Furthermore, these cancel coils should be located at the upside of the core, as shown in Figure 15.12(a), which is the side view of the I-type IPTS. If it is not the case where the cancel coils are at the bottom of the core, as shown in Figure 15.12(b), the

Figure 15.11 The proposed active EMF cancel design for an I-type IPTS: (a) bird's eye view and (b) front view.

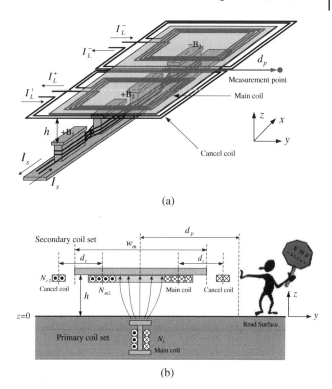

(a)

(b)

cancel coils intersect with the magnetic linkage flux, which violates the LFEC design method. Concerning the EMF of pedestrians, the forward direction (x) EMF can be ignored because normally the distance from the pick-up coil is larger than 2 m, but the side direction (y) EMF should be carefully dealt with because the distance w_p is typically only 1 m for a passenger car and 1.2 m for a bus.

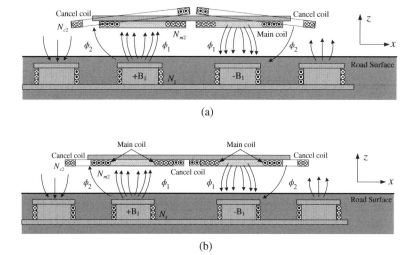

(a)

(b)

Figure 15.12 The side view of the I-type IPTS showing the importance of the proposed LFEC design method. (a) Case A: correct application of the design method. (b) Case B: incorrect application of the design method, which results in an unwanted load voltage reduction.

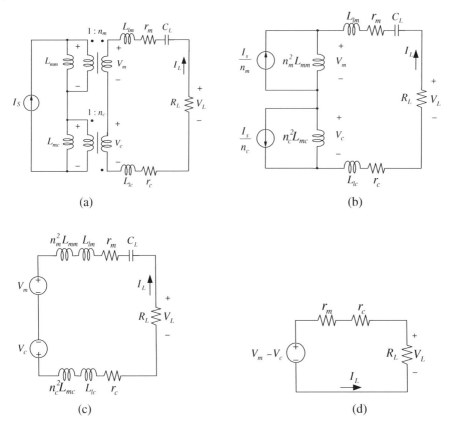

Figure 15.13 Equivalently simplified circuits of the IPTS: (a) proposed IPTS, (b) simplified circuit eliminating the transformers, (c) Thevenin equivalent circuit, and (d) final equivalent circuit in resonance.

The proposed I-type IPTS, as shown in Figures 15.11 and 15.12(a), is modeled to determine the system performances such as load voltage and power. As shown in Figure 15.13, where a secondary coil set only is modeled this time, it is assumed that the primary current I_s is controlled to be constant and that the parasitic magnetic linkage due to the cancel coils is not negligible, where L_{mm}, L_{mc} are the magnetizing inductances, L_{lm}, L_{lc} are the leakage inductances, n_m, n_c are the turn ratios, and r_m, r_c are the internal resistances of the main coil and cancel coil, respectively. The cancel coils are wound in the opposite direction to the main coils, as shown in Figure 15.13(a). The transformers can be eliminated and the magnetizing inductances and the primary current can be transferred to the secondary side, as shown in Figure 15.13(b). Applying the Thevenin theorem to the circuit of Figure 15.13(b), the induced voltages of the main coils V_m and cancel coils V_c in Figure 15.13(c) can be determined as follows:

$$V_m = j\omega_s L_{mm} I_s n_m \quad \because n_m \equiv N_{m2}/N_1 \tag{15.13a}$$
$$V_c = j\omega_s L_{mc} I_s n_c \quad \because n_c \equiv N_{c2}/N_1 \tag{15.13b}$$

The compensation capacitor C_L is tuned to cancel all the inductance components of Figure 15.13(c) as follows:

$$j\omega_s \left(n_m^2 L_{mm} + n_c^2 L_{mc} + L_{lm} + L_{lc} \right) + \frac{1}{j\omega_s C_L} = 0 \tag{15.14}$$

Applying (15.14) to Figure 15.13(c) and exploiting (15.13), the load voltage V_L can be calculated from Figure 15.13(d) as follows:

$$V_L = \frac{R_L}{r_m + r_c + R_L}(V_m - V_c) = \frac{jn_m\omega_s L_{mm} I_s R_L}{r_m + r_c + R_L}\left(1 - \frac{n_c L_{mc}}{n_m L_{mm}}\right) \tag{15.15}$$

In (15.15), the L_{mm} is little varied by the location of the EMF cancel coil, as identified from Figure 15.12, and is determined as follows:

$$L_{mm} = \frac{N_1 \phi_1}{I_s} \tag{15.16}$$

On the other hand, the L_{mc} is largely varied by the location of the EMF cancel coil. For Case A, as shown in Figure 15.12(a), where the cancel coil is out of the main magnetic linkage, the $L_{mc,A}$ is nearly zero; however, for Case B of Figure 15.12(b), the $L_{mc,B}$ is not negligible, that is,

$$L_{mc,A} \cong 0 \tag{15.17a}$$

$$L_{mc,B} = \frac{N_1(\phi_1 + \phi_2)}{I_s} > \frac{N_1 \phi_1}{I_s} = L_{mm} \tag{15.17b}$$

Comparing (15.17b) with (15.16), it can be seen that $L_{mc,B}$ is a little bit larger than L_{mm}. Bearing this fact in mind, the load voltage for Cases A and B are determined, respectively, as follows:

$$V_{L,A} \cong \frac{jn_m\omega_s L_{mm} I_s R_L}{r_m + r_c + R_L} \tag{15.18a}$$

$$V_{L,B} = \frac{jn_m\omega_s L_{mm} I_s R_L}{r_m + r_c + R_L}\left(1 - \frac{N_{c2} L_{mc}}{N_{m2} L_{mm}}\right) \tag{15.18b}$$

Note from (15.18b) that a significant voltage drop may occur when the ratio N_{c2}/N_{m2} is not small, which will be verified by experiments.

The load power for the proposed design, which is for Case A, is then obtained as follows:

$$P_L = \frac{|V_L|^2}{R_L} \cong \frac{R_L(n_m\omega_s L_{mm}|I_s|)^2}{(r_m + r_c + R_L)^2} \tag{15.19}$$

As identified from (15.18) and (15.19), the internal resistance r_c of the cancel coil should be as small as possible so as not to deteriorate the load voltage and power.

As implied by (15.12), the two sets of secondary coils, a set of which is defined as a main coil and a cancelation coil, can be connected to full-bridge output rectifier(s) in two ways, as shown in Figure 15.14. For the AC side connection case, where two pairs of the coils are directly connected in series and a rectifier only is therefore adopted, as shown in Figure 15.14(a), the AC load currents are exactly the same; however, the rectified DC load currents are the same for the DC side connection case, where each pair of coils are connected to each output rectifier and DC output voltages are summed, as shown in Figure 15.14(b). The currents are as follows:

$$I_L^+ = I_L^- \equiv I_L \text{ for the AC side connection case} \tag{15.20a}$$

$$|I_L^+| = |I_L^-|, \quad \because I_L^+ \neq I_L^- \text{ for the DC side connection case} \tag{15.20b}$$

In the case of Figure 15.14(a), the EMF generated by each cancel coil should be exactly the same, which will result in less effective EMF canceling for an unbalanced induced voltage; therefore, the DC connection case of Figure 15.14(b) is preferred in this chapter, where the total size and volume of the rectifiers is little changed from the case of Figure 15.14(a) because the output voltages have been halved.

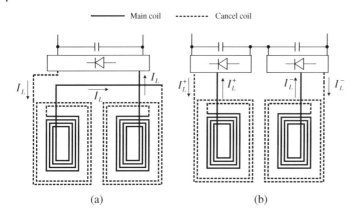

Figure 15.14 Possible connections of the secondary coils and rectifier(s): (a) AC side connection and (b) DC side connection.

15.5 Example Design and Experimental Verifications for the I-Type IPTS of OLEV

The proposed design is verified by a prototype I-type IPTS [45], where the source current is $I_s = 50$ A, the switching frequency is 20 kHz, and the number of primary coil is $N_1 = 4$. As shown in Figure 15.15, the pick-up was built for experiments, whose total size is 120 cm × 80 cm, using 1 cm-thick core plates of 10 cm × 10 cm. The baseline load current is $I_L = 30$ A, whereas the baseline number of turns of a secondary main coil and a secondary cancel coil are $N_{m2} = 16$ and $N_{c2} = 5$, and the airgap h is 20 cm, unless specified. The r_m and r_c were measured as 27.9 mΩ and 16.2 mΩ, respectively.

As shown in Figure 15.15, two cases of EMF cancel methods were tested, where the measured total voltages for the two sets of coils are listed in Table 15.1. In the case where the cancel coils were located above the pick-up core (Case A), the load voltage was merely reduced by 6.6% compared to the case of no active cancel. For Case B, however, the voltage drop was as much as 29.2%, as predicted by (15.18b). Here the voltage drop for Case B is calculated while neglecting the effects of internal resistances and ϕ_2, that is $r_m + r_c \ll R_L$ and $L_{mc} \cong L_{mm}$, respectively, as follows:

$$\%\Delta V_L \equiv \frac{V_{L,no\ cancel} - V_{L,with\ cancel}}{V_{L,no\ cancel}} \cong \frac{N_{c2}L_{mc}}{N_{m2}L_{mm}} \cong \frac{N_{c2}}{N_{m2}} \tag{15.21}$$

Figure 15.15 Two cases of the center part of lateral EMF coil location: (a) Case A and (b) Case B. (Be cautious about the dotted line box.)

Table 15.1 The effect of the location of the center part of the lateral EMF cancel coil

Total voltage	Case A (V)	Case B (V)
$|V_m|$ (Main coil voltage)	109.0	109.0
$|V_c|$ (Cancel coil voltage)	7.1	32.0
$|V_L|$ (Load voltage)	101.8	77.2
Load voltage drop	6.6%	29.2%

As can be seen from (15.21), the theoretical voltage drop is 31.3% for $N_{m2} = 16$ and $N_{c2} = 5$, which is quite close to the experimental measurement. Thus, it has been verified that the proposed LFEC design method can improve the load voltage by as much as 22.6%.

Because the magnetic field is a vector, the cancel coil should generate the magnetic field of not only the opposite direction but also the same magnitude and phase for any place for pedestrians. Therefore, it is not straightforward to apply the cancelation methods in the previous section in practice. Fortunately, it is observed from various simulations and experiments in this chapter that the horizontal direction, that is, the z-axis, is the strongest EMF area. This means that to cancel the horizontal EMF is of prime importance, neglecting other components at the moment. To find the proper location and the number of turns by (15.11) is, however, not possible because it is applicable only to the coils with no core. Hence, the EMF of the proposed I-type pick-up was directly measured to determine the parameters, as shown in Figure 15.16. For practical purposes, only the EMF values for pedestrians were measured in this chapter. Finding volumetric EMF patterns near the pick-up, considering electromagnetic interactions with the vehicle, is left for further works.

Concerning the height of the cancel coils, they must be lower than the bottom of a vehicle but higher than the airgap from the ground. Therefore, an appropriate location for them is at the same height as the main coils. Under this constraint, the optimum spacing between the main coils and the cancel coils d_c and the optimum number of turns of the cancel coil N_{c2} can be experimentally determined, as shown in Figure 15.16. The EMF at the height of 20 cm became $4\,\mu T$ for $N_{c2} = 5$ and $d_c = 22$ cm, which are the optimum design parameters that best mitigate the EMF for pedestrians. Under this optimum active cancel condition, the EMF was measured for various distances and heights from the center of a pick-up, as shown in Figure 15.17. The EMF is well within $6.25\,\mu T$ for almost everywhere, which shows that the proposed design of the active cancel coil is quite appropriate for OLEV.

It is also found that the proposed active cancel coil deteriorates the load voltage a little, as shown in Figure 15.18. The load voltage drop after adding the cancel coils was increased up to 5% as the number of turns increased; however, this small voltage drop is of no practical concern. In addition, the conduction losses of the main coil and cancel coil for the rated load current of 30 A, as estimated below,

$$P_{Lm} = I_L^2 r_m \tag{15.22a}$$
$$P_{Lc} = I_L^2 r_c \tag{15.22b}$$

are just 25.1 W and 14.6 W, respectively. Therefore, the additional loss due to the cancelcoil corresponds to merely 0.24% of the rated load power of 6 kW.

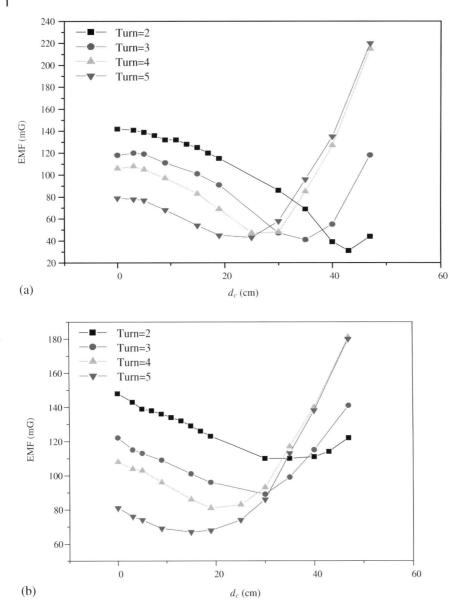

Figure 15.16 The EMF measured at 1 m for various locations and the number of turns of the cancel coil: (a) EMF at $h = 20$ cm and (b) EMF at $h = 0$ cm.

For the higher load power where the load current is larger than 30 A, the EMF should be further mitigated; hence, a few additional EMF cancel methods, which are compatible with the proposed active EMF cancel methods, have been developed in this chapter.

The EMF cancel effect was a little enhanced when the cancel coils were covered with a plate core, as shown in Figure 15.19. This can be explained by using the magnetic mirror concept [64], where the magnetic field is intensified by the core. The measured EMF was reduced from 5.8 μT to 5 μT with a 10 cm plate core, and was further reduced to 4.4 μT with a 20 cm plate core. The size of the plate core and the mitigation of EMF should be traded off.

Figure 15.17 The EMF measured for various distances and heights. (a) The measured EMF ($N_{c2} = 5$ and $h = 0$ m) for the lateral direction. (b) The measured EMF ($N_{c2} = 5$ and $d_p = 1$ m and 1.2 m) for the height direction.

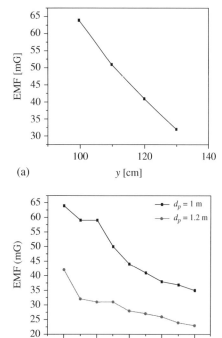

(a)

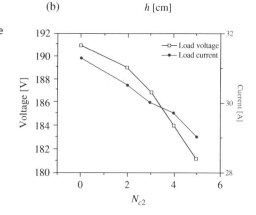

(b)

Figure 15.18 The measured load voltage drop due to the proposed active EMF cancel, which is negligible in practice.

Figure 15.19 The EMF cancel effect is enhanced by a plate core on the cancel coils: (a) 10 cm plate core and (b) 20 cm plate core.

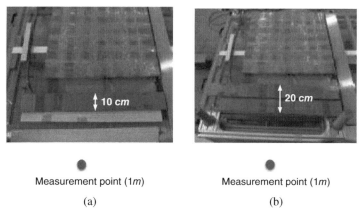

(a)   (b)

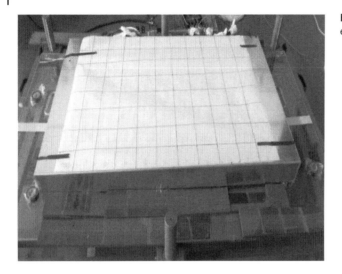

Figure 15.20 The EMF cancel is enhanced by a plate core and Al cover.

The EMF was also mitigated by an aluminum (Al) cover, which covers the main coils but does not cover the cancel coil, as shown in Figure 15.20. The effect of the gap between the Al plate and the pick-up core on the EMF was investigated, as shown in Figure 15.21. The EMF was mitigated by up to 15% using this method.

The measured EMF for possible combinations of the proposed passive cancel methods are shown in Figure 15.22, where the EMF of the case using active cancel coils, Al cover, and plate core becomes half of the case using cancel coils only. For the load condition of $I_L = 62.5$ A, where the load power was measured as 12 kW, the EMF is now as low as 4.4 μT, which is just a −3 dB value of the ICNIRP Guidelines of 6.25 μT. The design method of 3DEC has thus been verified. Note that the measured EMF so far is not the EMF from the secondary side coils only but the total EMF; hence, the 4.4 μT EMF has still a 3 dB design margin in practice because the EMF from the primary side, as can be noticed from Figure 15.21, is quite negligible for the large load power.

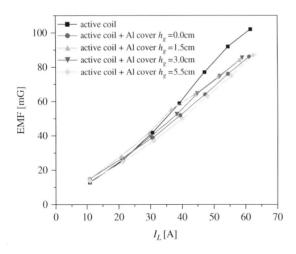

Figure 15.21 Measured EMF at 1 m for different gaps h_g of an Al cover on the pick-up core.

Figure 15.22 Measured EMF at 1 m for the four possible combinations of the EMF cancel methods.

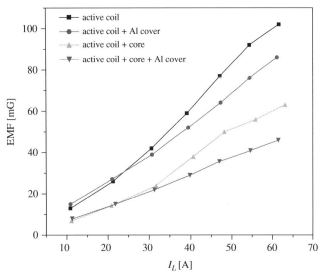

15.6   Conclusion

Three novel active EMF cancel methods, ISEC, 3DEC, and LFEC, have been applied to the design of I-type IPTS and verified by experiments. Furthermore, the proposed passive EMF cancel methods such as the Al plate cover and plate core cover have been found to be quite effective in mitigating the residual EMF. Thus, the EMF at 1 m became as low as 4.4 μT for the 62.5 A load current.

Problems

15.1   You are considering the application of the proposed three EMF cancel methods to stationary EV chargers. Identify the difference between the dynamic EV charging and the stationary EV charger in terms of the EMF cancel methods.

15.2   Can you suggest the general theory of EMF cancel by a metal plate? Quite often a metal plate is used to mitigate magnetic flux leakage and its goal is the same as the core plate. The behavior of EMF cancel is somewhat different. There is not yet a general theory. Note that Maxwell equations cannot give us analytic results but only simulation results. If you are able to determine the induced current over a metal plate for an arbitrarily given magnetic flux distribution, then you will succeed in finding the theory.

References

1 A.P. Hu and S. Hussmann, "Improved power flow control for contactless moving sensor applications," *IEEE Power Electron. Lett.*, vol. 2, no. 4, pp. 135–138, December 2004.

2 T. Hata and T. Ohmae, "Position detection method using induced voltage for battery charge on autonomous electric power supply system for vehicles," in *Proc. 8th IEEE Int. Workshop on Adv. Motion Control*, 2004, pp. 187–191.

3 B. Choi, J. Nho, H. Cha, T. Ahn, and S. Choi, "Design and implementation of low-profile contactless battery charger using planar printed circuit board windings as energy transfer device," *IEEE Trans. on Ind. Electron.*, vol. 51, no. 1, pp. 140–147, February 2004.

4 S.Y.R. Hui and W.C. Ho, "A new generation of universal contactless battery charging platform for portable consumer electronic equipment," *IEEE Trans. on Power Electron.*, vol. 20, no. 3, pp. 620–627, May 2005.

5 K.I. Woo, H.S. Park, Y.H. Choo, and K.H. Kim, "Contactless energy transmission system for linear servo motor," *IEEE Trans. on Magnetics*, vol. 41, no. 5, pp. 1596–1599, May 2005.

6 G.A.J. Elliott, G.A. Covic, D. Kacprzak, and J.T. Boys, "A new concept: asymmetrical pick-ups for inductively coupled power transfer monorail systems," *IEEE Trans. on Magnetics*, vol. 42, no. 10, pp. 3389–3391, October 2006.

7 P. Sergeant and A. Van den Bossche, "Inductive coupler for contactless power transmission," *IET Electr. Power Applic.*, vol. 2, pp. 1–7, 2008.

8 G. Wang, W. Liu, M. Sivaprakasam, and G.A. Kendir, "Design and analysis of an adaptive transcutaneous power telemetry for biomedical implants," *IEEE Trans. on Circuits Syst. I, Reg. Papers*, vol. 52, no. 10, pp. 2109–2117, October 2005.

9 P. Si, A.P. Hu, J.W. Hsu, M. Chiang, Y. Wang, S. Malpas, and D. Budgett, "Wireless power supply for implantable biomedical device based on primary input voltage regulation," in *Proc. 2nd IEEE Conf. on Ind. Electron. Applic.*, 2007, pp. 235–239.

10 P. Si, A.P. Hu, S. Malpas, and D. Budgett, "A frequency control method for regulating wireless power to implantable devices," *IEEE Trans. Biomed. Circuits Syst.*, vol. 2, no. 1, pp. 22–29, March 2008.

11 R. Laouamer, M. Brunello, J.P. Ferrieux, O. Normand, and N. Buchheit, "A multi-resonant converter for noncontact charging with electromagnetic coupling," in *Proc. IEEE Industrial Electronics Society (IECON)*, vol. 2, 1997, pp. 792–797.

12 H. Sakamoto, K. Harada, S. Washimiya, and K. Takehara, "Large air-gap coupler for inductive charger," *IEEE Trans. on Magnetics*, vol. 35, no. 5, pp. 3526–3529, September 1999.

13 J. Hirai, K. Tae-Woong, and A. Kawamura, "Study on intelligent battery charging using inductive transmission of power and information," *IEEE Trans. on Power Electron.*, vol. 15, no. 2, pp. 335–345, 2000.

14 F. Nakao, Y. Matsuo, M. Kitaoka, and H. Salcamoto, "Ferrite core couplers for inductive chargers," in *IEEE Power Conversion Conference*, 2002, pp. 850–854.

15 R. Mecke and C. Rathge, "High frequency resonant inverter for contactless energy transmission over large air gap," in *Proc. 35th Annual IEEE Power Electron. Spec. Conf.*, June 2004, vol. 3, pp. 1737–1743.

16 Y. Kamiya, Y. Daisho, F. Kuwabara, and S. Takahashi, "Development and performance evaluation of an advanced electric micro bus transportation system," in *JSAE Annual Spring Congress*, 2006, pp. 7–14.

17 M. Budhia, G.A. Covic, J.T. Boys, and C.-Y. Huang, "Development and evaluation of single sided flux couplers for contactless electric vehicle charging," in *IEEE Energy Conversion Congress and Exposition (ECCE)*, 2011, pp. 614–621.

18 M. Budhia, G. Covic, and J. Boys, "Design and optimization of circular magnetic structures for lumped inductive power transfer systems," *IEEE Trans. on Power Electron.*, vol. 26, no. 11, pp. 3096–3107, 2011.

19 S. Hasanzadeh, S. Vaez-Zadeh, and A.H. Isfahani, "Optimization of a contactless power transfer system for electric vehicles," *IEEE Trans. on Veh. Technol.*, vol. 51, no. 8, pp. 3566–3573, 2012.

20 Y. Iga, H. Omori, T. Morizane, N. Kimura, Y. Nakamura, and M. Nakaoka, "New IPT-wireless EV charger using single-ended quasi-resonant converter with power factor correction," *IEEE Renewable Energy Research and Applications (ICRERA)*, 2012, pp. 1–6.

21 H.H. Wu, A. Gilchrist, K. Sealy, and D. Bronson, "A 90 percent efficient 5 kW inductive charger for EVs," in *IEEE Energy Conversion Congress and Exposition (ECCE)*, 2012, pp. 275–282.

22 F. Sato, J. Murakami, H. Matsuki, K. Harakawa, and T. Satoh, "Stable energy transmission to moving loads utilizing new CLPS," *IEEE Trans. onMagnetics*, vol. 32, no. 5, pp. 5034–5036, September1996.

23 F. Sato, J. Murakami, T. Suzuki, H. Matsuki, S. Kikuchi, K. Harakawa, H. Osada, and K. Seki, "Contactless energy transmission to mobile loads by CLPS-test driving of an EV with starter batteries," *IEEE Trans. on Magnetics*, vol. 33, no. 5, pp. 4203–4205, September 1997.

24 C. Wang, O.H. Stielau, and G.A. Covic, "Design considerations for a contactless electric vehicle battery charger," *IEEE Trans. on Ind. Electron.*, vol. 52, no. 5, pp. 1308–1314, October 2005.

25 J. Sallan, J.L. Villa, A. Llombart, and J.F. Sanz, "Optimal design of ICPT systems applied to electric vehicle battery charge," *IEEE Trans. on Ind. Electron.*, vol. 56, no. 6, pp. 2140–2149, June 2009.

26 J.L. Villa, J. Sallan, A. Llombart, and J.F. Sanz, "Design of a high frequency inductively coupled power transfer system for electric vehicle battery charge," *Appl. Energy*, vol. 86, no. 3, pp. 355–363, 2009.

27 T. Maruyama, K. Yamamoto, S. Kitazawa, K. Kondo, and T. Kashiwagi, "A study on the design method of the light weight coils for a high power contactless power transfer systems," in *2012 15th International Conference on Electrical Machines and Systems (ICEMS)*, 2012, pp. 1–6.

28 Y. Nagatsuka, N. Ehara, Y. Kaneko, S. Abe, and T. Yasuda, "Compact contactless power transfer system for electric vehicles," in *International Power Electronics Conference (IPEC)*, 2010, pp. 807–813.

29 M. Chigira, Y. Nagatsuka, Y. Kaneko, S. Abe, T. Yasuda, and A. Suzuki, "Small-size light-weight transformer with new core structure for contactless electric vehicle power transfer system," in *IEEE Energy Conversion Congress and Exposition (ECCE)*, 2011, pp. 260–266.

30 M. Budhia, G.A. Covic, and J.T. Boys, "A new magnetic coupler for inductive power transfer electric vehicle charging systems," in *36th Annual Conference of the IEEE Industrial Electronics Society (IECON)*, 2010, pp. 2481–2486.

31 J.G. Bolger and F.A. Kirsten, "Investigation of the feasibility of a dual mode electric transportation system," Lawrence Berkeley Laboratory Report, 1977.

32 J.G. Bolger, F.A. Kirsten, and L.S. Ng, "Inductive power coupling for an electric highway system," in *Proc. IEEE 28th Vehicular Technology Conference*, 1978, pp. 137–144.

33 C.E. Zell and J.G. Bolger, "Development of an engineering prototype of a road powered electric transit vehicle system," in *Proc. 32nd IEEE Vehicular Technology Conference*, 1982, pp. 435–438.

34 M. Eghtesadi, "Inductive power transfer to an electric vehicle-an analytical model," in *Proc. 40th IEEE Vehicular Technology Conference*, 1990, pp. 100–104.

35 California PATH Program, "Road powered electric vehicle project track construction and testing program phase 3D," California PATH Research Paper, March 1994.

36 A.W. Green and J.T. Boys, "10 kHz inductively coupled power transfer concept and control," in *Proc. 5th Int. Conf. IEEE Power Electron Variable-Speed Drivers*, October 1994, pp. 694–699.

37 G.A.J. Elliott, J.T. Boys, and A.W. Green, "Magnetically coupled systems for power transfer to electric vehicles," in *Proc. Int. Conf. Power Electron. Drive System*, February 1995, pp. 797–801.

38 G.A. Covic, J.T. Boys, M.L.G. Kissin, and H.G. Lu, "A three-phase inductive power transfer system for road powered vehicles," *IEEE Trans. on Ind. Electron.*, vol. 54, no. 6, pp. 3370–3378, December 2007.

39 N.P. Suh, D.H. Cho, and Chun T. Rim, "Design of on-line electric vehicle (OLEV)," *Plenary Lecture at the 2010 CIRP Design Conference*, 2010, pp. 3–8.

40 S.W. Lee, J. Huh, C.B. Park, N.S. Choi, G.H. Cho, and Chun T. Rim, "On-line electric vehicle using inductive power transfer system," in *IEEE Energy Conversion Congress and Exposition (ECCE)*, 2010, pp. 1598–1601.

41 J. Huh and Chun T. Rim, "KAIST wireless electric vehicles – OLEV," in *JSAE Annual Congress*, 2011.

42 J. Huh, S.W. Lee, C.B. Park, G.H. Cho, and Chun T. Rim, "High performance inductive power transfer system with narrow rail width for on-line electric vehicles," in *IEEE Energy Conversion Congress and Exposition (ECCE)*, 2010, pp. 647–651.

43 S.W. Lee, W.Y. Lee, J. Huh, H.J. Kim, C.B. Park, G.H. Cho, and Chun T. Rim, "Active EMF cancellation method for I-type pick-up of on-line electric vehicles," in *IEEE Applied Power Electronics Conference and Exposition (APEC)*, 2011, pp. 1980–1983.

44 J. Huh, W.Y. Lee, G.H. Cho, B.H. Lee, and Chun T. Rim, "Characterization of novel inductive power transfer systems for on-line electric vehicles," in *IEEE Applied Power Electronics Conference and Exposition (APEC)*, 2011, pp. 1975–1979.

45 J. Huh, S.W. Lee, W.Y. Lee, G.H. Cho, and Chun T. Rim, "Narrow-width inductive power transfer system for on-line electrical vehicles," *IEEE Trans. on Power Electron.*, vol. 26, no. 12, pp. 3666–3679, December 2011.

46 S.Y. Choi, J. Huh, W.Y. Lee, S.W. Lee, and Chun T. Rim, "New cross-segmented power supply rails for road powered electric vehicles," *IEEE Trans. on Power Electron.*, vol. 28, no. 12, pp. 5832–5841, December 2013.

47 P.R. Bannister, "New theoretical expressions for predicting shielding effectiveness for the plane shield case," *IEEE Transactions on Electromagnetic Compatibility*, vol. EMC-10, issue 1, pp. 2–7, March 1968.

48 P. Moreno and R.G. Olsen, "A simple theory for optimizing finite width ELF magnetic field shields for minimum dependence on source orientation," *IEEE Transactions on Electromagnetic Compatibility*, vol. 39, no. 4, pp. 340–348, November 1997.

49 Y. Du, T.C. Cheng, and A.S. Farag, "Principles of power-frequency magnetic field shielding with flat sheets in a source of long conductors," *IEEE Transactions on Electromagnetic Compatibility*, vol. 38, no. 3, pp. 450–459, Aug. 1996.

50 S.Y. Ahn, J.S. Park, and J.H. Kim, "Low frequency electromagnetic field reduction techniques for the on-line electric vehicle (OLEV)," in *IEEE International Symposium on Electromagnetic Compatibility*, 2010, pp. 625–630.

51 J.H. Kim and J.H. Kim, "Analysis of EMF noise from the receiving coil topologies for wireless power transfer," in *IEEE Asia-Pacific Symposium on Electromagnetic Compatibility*, 2012, pp. 645–648.

52 H.S. Kim and J.H. Kim, "Shielded coil structure suppressing leakage magnetic field from 100 W-class wireless power transfer system with higher efficiency," in *IEEE Microwave Workshop Series on Innovative Wireless Power Transmission Technologies, Systems and Applications*, 2012, pp. 83–86.

53 S.Y. Ahn, H.H. Park, and J.H. Kim, "Reduction of electromagnetic field (EMF) of wireless power transfer system using quadruple coil for laptop applications," in *IEEE Microwave Workshop Series on Innovative Wireless Power Transmission Technologies, Systems and Applications*, 2012, pp. 65–68.

54 S.C. Tang, S.Y.R. Hui, and H.S. Chung, "Evaluation of the shielding effects on printed circuit board transformers using ferrite plates and copper sheets," *IEEE Trans. on Power Electron.*, vol. 17, pp. 1080–1088, November 2002.

55 X. Liu, and S.Y.R. Hui, "An analysis of a double-layer electromagnetic shield for a universal contactless battery charging platform," in *IEEE Power Electronics Specialists Conference (PESC)*, June 2005, pp. 1767–1772.

56 P. Wu, F. Bai, Q. Xue, and S.Y.R. Hui, "Use of frequency selective surface for suppressing radio-frequency interference from wireless charging pads," *IEEE Trans. on Ind. Electron.*, to be published.

57 M.L. Hiles and K.L. Griffing, "Power frequency magnetic field management using a combination of active and passive shielding technology," *IEEE Trans. on Power Delivery Electronics*, pp. 171–179, 1998.

58 C. Buccella and V. Fuina, "ELF magnetic field mitigation by active shielding," in *IEEE International Symposium on Industrial Electronics*, 2002, pp. 994–998.

59 J. Kim, J.H. Kim, and S.Y. Ahn, "Coil design and shielding methods for a magnetic resonant wireless power transfer system," *Proceedings of the IEEE*, vol. 101, no. 6, pp. 1332–1342, June 2013.

60 Guidelines for limiting exposure to time-varying electric and magnetic fields (up to 300 GHz), ICNIRP Guidelines, 1998.

61 Guidelines for limiting exposure to time-varying electric and magnetic fields (up to 100 kHz), ICNIRP Guidelines, 2010.

62 E.S. Lee, J. Huh, X.V. Thai, S.Y. Choi, and Chun T. Rim, "Impedance transformers for compact and robust coupled magnetic resonance systems," in *IEEE Energy Conversion Congress and Exposition (ECCE)*, September 2013, pp. 2239–2244.

63 S.Y. Choi, J. Huh, W.Y. Lee, J.G. Cho, and Chun T. Rim, "Asymmetric coil sets for wireless stationary EV chargers with large lateral tolerance by dominant field analysis," *IEEE Trans. on Power Electron.*, accepted.

64 W.Y. Lee, J. Huh, S.Y. Choi, X.V. Thai, J.H. Kim, E.A. Al-Ammar, M.A. El-Kady, and Chun T. Rim, "Finite-width magnetic mirror models of mono and dual coils for wireless electric vehicles," *IEEE Trans. on Power Electron.*, vol. 28, no. 3, pp. 1413–1428, March 2013.

16

Large Tolerance Design

16.1 Introduction

Electric vehicles (EVs), such as hybrid electric vehicles (HEVs), plug-in HEVs (PHEVs), and pure battery EVs (BEVs), have been developed, but they have not been widely used due to battery problems, namely, high cost, heaviness, large installation space, and frequent charging due to a low energy capacity. In order to mitigate the battery problems, RPEVs using inductive power transfer systems (IPTSs) have been proposed [1–21]. By virtue of the IPTS, the RPEV can be powered during its operation and can be free from the battery problems. In general, the IPTS adopts resonant circuits to maximize the power transfer, which is a challenging design issue due to the static and dynamic characteristics of the resonant circuits and IPTS; however, the dynamic and static behaviors of the IPTS can be characterized by using phasor transformation [22–25].

In the IPTS for RPEVs, a sufficient airgap between a power supply rail and pick-up coils, a large lateral tolerance between the centers of the power supply rail and pick-up coils, and a low electromagnetic field (EMF) for pedestrians are important design issues. In order to meet the requirements, a narrow-width I-type power supply rail was proposed [11, 12]. However, the width of the pick-up coils should be very large in order to get a large lateral tolerance, which results in a higher EMF from the pick-up coils than permitted by the ICNIRP Guidelines for pedestrians [26]. Moreover, the inductance of the pick-up coils becomes large as its width increases, which leads to a high voltage stress on compensation capacitors in a resonant condition and a narrow 3 dB bandwidth due to the higher quality factor.

As a remedy to these problems, a dual pick-up coil with active EMF cancel coils for an I-type power supply rail was suggested [1–3], where the active EMF cancel coils are wound in the opposite direction to that of the dual pick-up coil to lower the EMF for pedestrians. The dual pick-up coil [14], in which two pick-up coils are aligned along with the moving direction (y), as shown in Figure 16.1(a), has an inherently low EMF for pedestrians and a strong magnetic coupling between the power supply rail and pick-up coils compared to a single pick-up coil. Although the dual pick-up coil can mitigate the EMF level, the lateral tolerance has not yet been improved.

In order to increase the lateral tolerance, a set of laterally displaced two pick-up coils, in which two coils overlap each other, as shown in Figure 16.1(b), may be used [10]. In this case, each coil achieves its peak power when it is aligned with the center of a power supply rail, as shown in Figure 16.2(a). Therefore, it is expected that a larger lateral tolerance can be obtained if a higher load power from each coil is appropriately selected, as shown in Figure 16.2(b).

Throughout the development of the IPTS [10], however, the peak power was obtained only at the overlapped area, and in practice the power was abruptly dropped to zero out of this area,

Wireless Power Transfer for Electric Vehicles and Mobile Devices, First Edition. Chun T. Rim and Chris Mi.
© 2017 John Wiley & Sons Ltd. Published 2017 by John Wiley & Sons Ltd.

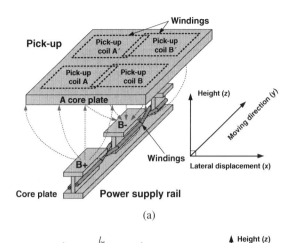

Figure 16.1 The proposed self-decoupled dual pick-up coils for a large lateral tolerance and low EMF level on an I-type power supply rail example: (a) bird's eye view and (b) front view, where the pick-up coils A—A– are coupled with the power supply rail but the pick-up coils B-B′ are not.

which unexpectedly resulted in a much smaller lateral tolerance, as shown in Figure 16.2(c). The lateral tolerance, which is defined in this chapter as the allowable lateral displacement from the center to give less than −3 dB load voltage, can be roughly estimated from Figure 16.2(b) and (c) as follows:

$$l_{t,decoupled} \cong 2l_w - l_d \quad \text{for} \quad V_{oa} \cup V_{ob} \tag{16.1a}$$

$$l_{t,coupled} \cong l_d \qquad \text{for} \quad V_{oa} \cap V_{ob} \tag{16.1b}$$

where (16.1a) is the theoretical goal to be achieved in this chapter and (16.1b) is the experimental result that has been obtained so far. One of the merits of the laterally displaced two pick-up coils, called in this chapter a "self-decoupled pick-up coil," is its low self-inductance of each coil for a given width of the pick-up. Without the self-decoupled pick-up coil, a pick-up coil with a large width should be used, resulting in high self-inductance.

In this chapter, the phenomenon of the small lateral tolerance for the two overlapped pick-up coils, as shown in Figure 16.2, is thoroughly explained and a self-decoupled dual pick-up coil is first proposed as a solution to both the large lateral tolerance and the low self-inductance. The concept of a self-decoupled coil itself is not new and was first introduced in the field of magnetic resonance imaging (MRI) [27] two decades ago. The concept was recently applied to the EV stationary charging systems [28, 29], where a self-decoupled pick-up coil was used to enlarge the lateral tolerance. However, thus far, no analytical model for the self-decoupled pick-up coil has been given. Hence, generally applicable theoretical models for the self-decoupled coil are explained in this chapter.

As an example, the self-decoupled dual pick-up coils for RPEV, as shown in Figure 16.1(a), are analyzed, simulated, and experimentally verified in this chapter. Thus, an optimum decoupling

Figure 16.2 Load voltage profile over the lateral displacement: (a) the voltage of each dual pick-up coil, (b) expected large lateral tolerance, and (c) observed small lateral tolerance.

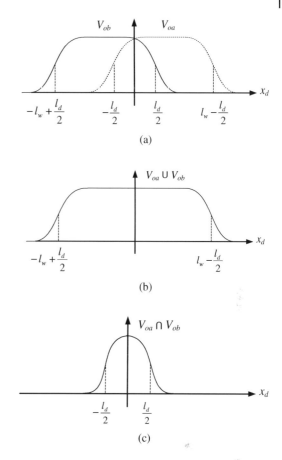

distance is systematically determined that can be generally applied to any self-decoupled coil regardless of the coil types, such as single/dual, stationary/dynamic charging, and core/coreless. Moreover, the proposed self-decoupled coils are compatible with any compensation method, including serial, parallel, and serial–parallel. By virtue of the proposed coil, not only a large lateral tolerance but also a large airgap can be achieved. In addition, a low voltage stress of compensation capacitors and a low EMF for pedestrians can be obtained without the use of active EMF cancel coils, as shown in Figure 16.1(b). The equivalent circuits and mathematical models for the self-decoupled dual pick-up coil are explained for both cases of with and without a core plate, where the finite-width magnetic mirror theory [30] is adopted to reflect the effect of a core plate on magnetic inductance.

16.2 Self-Decoupled Dual Pick-up coils with an I-Type Power Supply Rail

16.2.1 Overall Configuration of the Proposed Pick-up Coils

The IPTS for RPEV makes it possible to obtain the required power from a power supply rail while it is moving, as shown in Figure 16.3. The IPTS includes two subsystems [15]: one is the roadway subsystem to provide power for the RPEV and consists of a line frequency rectifier, a high frequency inverter, a primary capacitor bank, and a power supply rail; the other is the on-board subsystem for receiving the required power from the roadway subsystem, which includes pick-up coils, a secondary capacitor bank, a high frequency rectifier, and a regulator.

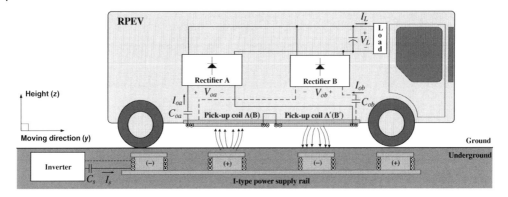

Figure 16.3 The side view of the IPTS for RPEV, adopting the proposed self-decoupled dual pick-up coil sets. Voltages and currents in the circuit are rms values in the steady state.

The dual pick-up coil A–A′ overlaps the dual pick-up coil B–B′, as shown in Figures 16.1 and 16.3, where an I-type power supply rail is displaced under the ground. Each dual pick-up coil is resonated and rectified separately and is then connected in parallel to produce the highest DC output voltage for the load, as shown in Figure 16.3. The rectified DC output voltage may be connected in series; however, the conduction loss attributing from both dual pick-up coils becomes doubled in this case.

As shown in Figure 16.4, there are eight magnetic couplings (magnetizing inductances) between the coils of the proposed IPTS:

(a) The couplings between the power supply rail and dual pick-up coil A–A′ (L_{ma})
(b) The couplings between the power supply rail and dual pick-up coil B–B′ (L_{mb})
(c) The coupling between coil A and coil B ($L_m/2$)
(d) The coupling between coil A′ and coil B′ ($L_m/2$)
(e) The coupling between coil A and coil A′
(f) The coupling between coil B and coil B′
(g) The coupling between coil A and coil B′
(h) The coupling between coil B and coil A′

For simplicity, the internal coupling cases of dual pick-up coils (e) and (f) will be regarded as a lumped coil, the cross-coupling cases (g) and (h) will not be separately modeled, and the effect of lateral displacement on the coupling changes of L_{ma}, L_{mb}, and L_m will be highly focused in this chapter.

Throughout the paper, it is assumed that there is neither tilt nor misalignment between the EV and power supply rail except for a lateral displacement. It is also assumed that each pick-up coil is located just above the center of each pole of the power supply rail in the moving direction (y), as shown in Figure 16.3. All of the circuit parameters of each dual pick-up coil are assumed to be the same and the circuit is assumed to be operating in the steady state.

16.2.2 Problems of the Non-decoupled Pick-up Coils

As the center of the overlapped pick-up coils A and B (and also A′ and B′) is well aligned with the power supply rail, that is, $x_d = 0$, the induced voltage of the pick-up coil A is exactly the same as that of the pick-up coil B (that of A′ and B′ as well); hence, there is no coupling effect between the pick-up coils A and B (A′ and B′ as well) due to the symmetrical circuit configuration. If this is not this case, as shown in Figure 16.1(b), the induced voltages and resonating currents of the pick-up coils A and B (A′ and B′ as well) are different from one other, resulting in a coupling current between the pick-up coils A and B (A′ and B′ as well).

As shown in Figure 16.3, a series–series compensation is used in this chapter as an example. Note that the proposed self-decoupled model is valid for other compensation schemes as well, which will be explained in this chapter. A circuit diagram for the proposed IPTS is suggested, as shown in Figure 16.4(a), considering the magnetic coupling between the power supply rail and each dual pick-up coil (denoted as L_{ma} and L_{mb}, respectively), where the dual pick-up coils are magnetically coupled to each other (denoted as L_m). As widely adopted in the RPEV [10–15], the source current I_s is assumed to be constant, as it is regulated by the inverter. As shown in Figure 16.4(b), the source-side circuit can be simplified as Thevenin equivalent voltage sources

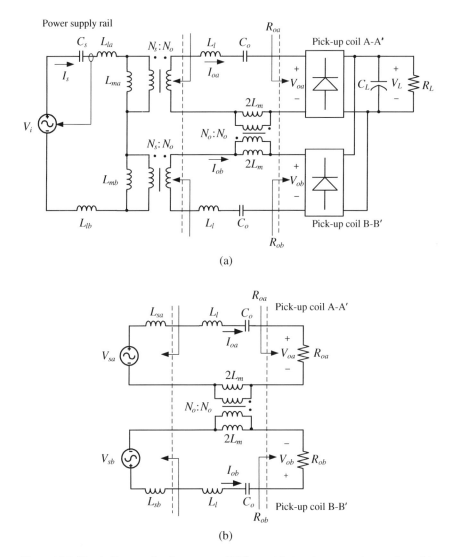

Figure 16.4 Circuit diagrams for the proposed IPTS, considering the magnetic coupling of the power supply rail and each dual pick-up coil, where the dual pick-up coils are magnetically coupled to each other: (a) an explicit transformer circuit and (b) a simplified Thevenin equivalent circuit.

and inductors [11–15] as follows:

$$V_{sa} = j\omega_s L_{ma} I_s n \tag{16.2a}$$

$$V_{sb} = j\omega_s L_{mb} I_s n \tag{16.2b}$$

$$L_{sa} = n^2 L_{ma} \tag{16.2c}$$

$$L_{sb} = n^2 L_{ma} \tag{16.2d}$$

$$\because n \equiv N_o/N_s \tag{16.2e}$$

In (16.2), V_{sa} and V_{sb} are the induced voltages, represented in phasor form, for the dual pick-up coils A–A′ and B–B′, respectively; n is the turn ratio of the power supply rail N_s and a dual pick-up coil N_o, and L_l is the leakage inductance of a dual pick-up coil; and C_o is the series compensation capacitor for pick-up side resonance. The diode rectifiers are also simplified as equivalent resistances R_{oa} and R_{ob}, respectively, considering the fundamental components of the voltages and currents of the rectifiers as follows [31, 32]:

$$R_{oa} \equiv \frac{V_{oa}}{I_{oa}} = \frac{2\sqrt{2}}{\pi} \frac{V_L}{I_{oa}} \tag{16.3a}$$

$$R_{ob} \equiv \frac{V_{ob}}{I_{ob}} = \frac{2\sqrt{2}}{\pi} \frac{V_L}{I_{ob}} \tag{16.3b}$$

As shown in Figure 16.4, the magnetizing inductance of the dual pick-up coils L_m is segmented into two $2L_m$ for each side so that each pick-up coil looks symmetrical. Also note that L_{sa} and L_{sb} as well as V_{sa} and V_{sb} are variable as L_{ma} and L_{mb} change along the lateral displacement, while L_l and L_m are nearly unaffected.

Let us now examine the behavior of the circuit of Figure 16.4(b) for both the aligned and misaligned cases of the pick-up, as shown in Figure 16.5. Depending on the lateral displacement, the induced voltages V_{sa} and V_{sb} and circulating currents I_{oa} and I_{ob} vary, which results in the variation of the equivalent resistances of (16.3). When the pick-up is exactly aligned with the power supply rail, all the circuit parameters and variables are symmetrical, as shown in Figure 16.5(a), where the equivalent resistance becomes as follows:

$$R_o \equiv \frac{V_{oa}}{I_{oa}} = \frac{8}{\pi^2} \frac{V_L}{I_L/2} = \frac{16}{\pi^2} R_L \tag{16.4}$$

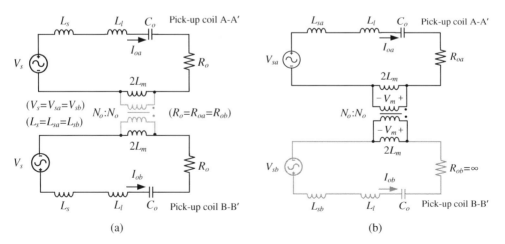

(a)   (b)

Figure 16.5 Equivalent circuits of Figure 16.4(b) when the power supply rail is (a) well-aligned ($|x_d| < l_d/2$) or (b) misaligned ($|x_d| > l_d/2$) with the pick-up.

Then no current flows around the transformer, which results in two independent circuits whose resonant frequency ω_{r0} is determined as follows:

$$\omega_{r0} = \frac{1}{\sqrt{(L_s + 2L_m + L_l)C_o}} \quad \text{for} \quad x_d = 0 \tag{16.5}$$

As shown in Figure 16.5(b), when the pick-up is misaligned so that the $V_{sb} = V_m$ condition may be eventually met, the circulating current I_{ob} becomes zero, which results in $R_{ob} = \infty$, as identified from (15.3b). Under this condition, the resonant frequency ω_{r1} is determined differently from (16.5) as follows:

$$\omega_{r1} = \frac{1}{\sqrt{(L_{sa} + L_m + L_l)C_o}} \quad \text{for} \quad |x_d| \gg l_d/2 \tag{16.6}$$

It is clearly identified from the two specific cases of (16.5) to (16.6) that the misaligned resonant frequency of a dual pick-up coil ω_{r1} may deviate from the operating frequency, which is normally tuned to the central resonant frequency ω_{r0}. In a more generalized misaligned case in which each A–A′ coil and B–B′ coil has different resonant frequency, the analysis of Figure 16.5(b) is not straightforward. It is evident, however, that the power delivery for this misaligned case is much less than the aligned case because of the number of less induced voltages as well as reduced circulating currents due to mistuned resonating circuits. In practice, the delivered power drops sharply when the resonating circuit gets out of resonance. In this way, the load voltage profile becomes very narrow for this non-decoupled pick-up case where L_m is not zero, as shown in Figure 16.2(c).

Even though the above discussions on the non-decoupled pick-up coils have been in regards to the series–series compensation example, the coupling effect mentioned above is valid for any compensation scheme, including series–parallel and parallel–parallel. This is because the misalignment of the pick-up causes the detuning of resonating circuits, regardless of the compensation scheme. Moreover, this misalignment effect is also valid for a serially connected output case, even though it is shown only for the parallel connected case of Figures 16.3 and 16.4.

A complete analysis of Figure 16.4(b) for the misaligned non-decoupled case is not provided here, though it is not impossible, because it is not very useful in practice. Instead, the case of the self-decoupled in which L_m is zero will be highly focused, which is quite useful in practice.

16.2.3 Self-Decoupled Single Pick-up Coil Set without a Core Plate

For obtaining a sufficient power regardless of the pick-up misalignment, there are generally two possible solutions: one is to use an adaptive frequency control of the inverter to ensure the pick-up is always tuned and another is to minimize the difference between ω_{r0} and ω_{r1} of (16.5) and (16.6). In practice, the former is hardly used for RPEVs because many vehicles with different resonant conditions may be on a power supply rail. The latter, however, is easy to implement if L_m is appropriately nullified. The condition $L_m = 0$ is called "self-decoupled" in this chapter, and the resonant frequencies of the pick-up become as follows:

$$\omega_{r0} = \frac{1}{\sqrt{(L_s + L_l)C_o}} \quad \text{for} \quad x_d = 0 \quad \text{and} \quad L_m = 0 \tag{16.7a}$$

$$\omega_{r1} = \frac{1}{\sqrt{(L_{sa} + L_l)C_o}} \quad \text{for} \quad |x_d| \gg l_d/2 \quad \text{and} \quad L_m = 0 \tag{16.7b}$$

$$\rightarrow \omega_r \equiv \omega_{r0} \cong \omega_{r1} = \frac{1}{\sqrt{(L_s + L_l)C_o}} \quad \text{if} \quad L_s \cong L_{sa} \quad \text{and} \quad L_m = 0 \tag{16.7c}$$

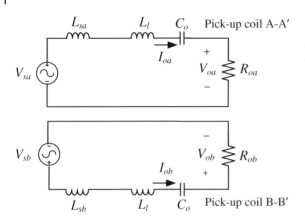

Figure 16.6 A general equivalent circuit of Figure 16.4(b) when the pick-up coils are self-decoupled from each other, that is, $L_m = 0$.

Comparing (16.7a) with (16.7b), it is found from (16.7c) that there is little frequency difference between the aligned and misaligned cases if only the $L_s \cong L_{sa}$ condition is met. This remaining condition for a large lateral tolerance will be verified by experiments in this chapter. Thus, a general equivalent circuit for the proposed self-decoupled dual pick-up coils is obtained, where two independent circuits operate regardless of misalignment, as shown in Figure 16.6.

In this chapter, the L_m is nullified by adjusting the distance x_1 between two pick-up coils, as shown in Figure 16.7. In order to develop a preliminary analytic model, only the single self-decoupled pick-up coil A–B without a core plate, instead of the dual pick-up coils A–A' and B–B' of Figure 16.1, is considered at the moment, where two co-planar rectangular windings of a width w_c overlap each other with a distance of l_d, which is similar with double-D (DD), bipolar (BP), and tripolar (TP) coils [28, 29, 33].

In order to determine an optimum distance x_1 for $L_m = 0$, the L_m is analytically obtained as follows:

$$L_m \equiv \frac{N_o \phi_m}{I_{oa}} \tag{16.8a}$$

$$\therefore \phi_m = \phi_1 + \phi_2 + \phi_3 + \phi_4 \tag{16.8b}$$

Because the superposition theorem can be applied to the coreless case (i.e., free space) of Ampere's law, the total mutual magnetic flux ϕ_m is the sum of the ϕ_1, ϕ_2, ϕ_3, and ϕ_4, which are the intersecting magnetic fluxes to coil B, induced from the left, right, bottom, and top of the pick-up coil A, respectively, as shown in Figure 16.8. The magnetic flux induced from the

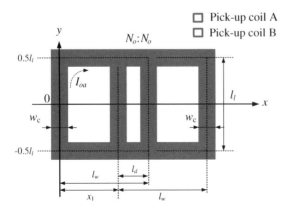

Figure 16.7 The top view of the coupled pick-up coil model without a core plate.

Figure 16.8 The intersecting magnetic flux generated from each side of coil A: (a) ϕ_1 for the left side, (b) ϕ_2 for the right side, and (c) ϕ_3 for the down side.

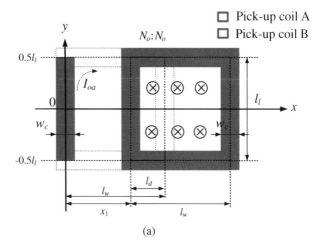

(a)

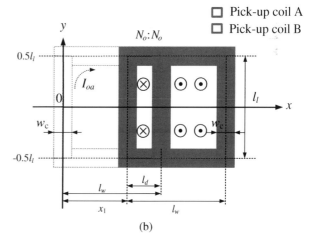

(b)

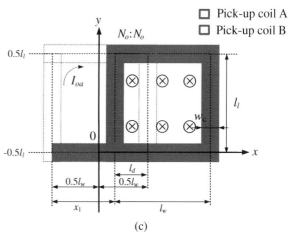

(c)

$N_o{:}N_o$

bottom side of the pick-up coil A is equal to that from the top side, so the top side case is not shown in the Figure 16.8. Then the total mutual magnetic flux becomes

$$\phi_m = \phi_1 + \phi_2 + 2\phi_3 \quad \because \phi_3 = \phi_4 \tag{16.9}$$

In order to make L_m be "0," ϕ_m should be "0" by changing the pick-up coil B x_1.

Assuming that the winding width is much smaller than the coil width, that is, $w_c \ll l_w$, the coil winding can be regarded as a thin wire; hence, ϕ_1 can be approximately obtained by using Biot–Savart's law as follows:

$$
\phi_1 \cong \int_{x_1}^{x_1 + l_w} \int_{-0.5 l_l}^{0.5 l_l} B_l(x, y) dy dx \tag{16.10a}
$$

$$
= -\frac{\mu_o N_o I_{oa}}{2\pi} \left\{ \sqrt{l_l^2 + x_1^2} + l_w - \sqrt{l_l^2 + (x_1 + l_w)^2} - l_l \cdot \ln \frac{l_w + x_1}{x_1} \right\}
$$

$$
\because B_1(x, y) = \frac{\mu_o N_o I_{oa}}{4\pi x} \left\{ \frac{y - l_l/2}{\sqrt{(y - l_l/2)^2 + x^2}} - \frac{y + l_l/2}{\sqrt{(y + l_l/2)^2 + x^2}} \right\} \tag{16.10b}
$$

In a similar way, ϕ_2 and ϕ_3 can be determined and ϕ_m is finally obtained for $0 < x_1 < l_w$ as follows:

$$
\phi_m = -\frac{2\mu_o N_o I_{oa}}{\pi} \left\{ \frac{2\sqrt{l_l^2 + x_1^2} - \sqrt{l_l^2 + (l_w + x_1)^2} - \sqrt{l_l^2 + (l_w - x_1)^2}}{2} \right.
$$

$$
\left. + \frac{l_l}{4} \cdot \ln \frac{x_1^2}{l_w^2 - x_1^2} - \frac{l_w - x_1}{2} \cdot \left(\ln \frac{l_l - 0.5 w_c}{0.5 w_c} - 2 \right) \right\} \tag{16.11}
$$

In order to verify (16.11), a three-dmensional finite element analysis (FEA) simulation for L_m was performed for $N_o I_{oa} = 1$ A-turn, $w_c = 0.5$ mm, $l_w = 50$ cm, and $l_l = 10$ cm, 50 cm, and 90 cm, respectively. As shown in Figure 16.9, the simulated L_m is quite well matched to the analyzed L_m,

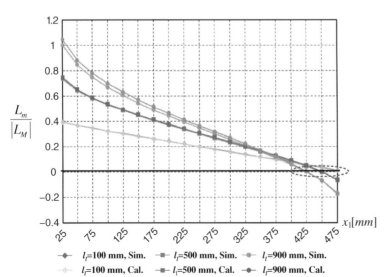

Figure 16.9 The three-dimensional FEA simulated L_m without a core plate and the calculated L_m for various lengths l_l of the pick-up coils A and B. The L_M is the maximum mutual inductance between the two pick-up coils for $l_l = 90$ cm, $l_w = 50$ cm, and $x_1 = 2.5$ cm without a core plate.

which is calculated from (16.8a) and (16.11). Moreover, an optimum self-decoupling distance x_{opt} is obtained for a given coil structure, which is found to be a little less than l_w.

When $x_1 = x_{opt}$, that is, $L_m = 0$, the resonant frequency ω_r of each pick-up coil and compensating circuit, as shown in Figure 16.6, is nearly unchanged for the lateral displacement $|x_d| < l_w - l_d/2$. Therefore, the lateral tolerance can be increased from $l_d/2$ to $l_w - l_d/2$ by using the proposed self-decoupled pick-up coils. In this way, the EMF for pedestrians is greatly mitigated when the pick-up coil A is on the power supply rail because pick-up coil B, which is close to the pedestrians, is decoupled from not only the power supply rail but also pick-up coil A, which results in little circulating current and quite low EMF generation. The detail estimation of the EMF reduction highly depends on the coil configuration and the location of a pedestrian [2].

16.2.4 Self-Decoupled Single Pick-up Coil Set with a Core Plate

In this section, discussions in the previous section will continue but with a difference in the core plate. In practice, a core plate is widely used to increase the magnetic coupling between the power supply rail and pick-up coils and to shield the RPEV from the unwanted magnetic flux leakage, as shown in Figures 16.1 and 16.3. The core plate influences not only the coupling between the power supply rail and pick-up coils but also the coupling between pick-up coils A and B, as shown in Figure 16.10. Due to the very complex interaction of the core plate, the magnetic flux induced by pick-up coil A, intersecting to pick-up coil B, is almost impossible to calculate. To solve this problem, the magnetic mirror model [24], which provides a magnetic flux density equation of closed form when a coil winding is placed on a core plate, is adopted. According to the magnetic mirror model, the magnetic flux with an infinitely large core plate and an infinite permeability is ideally two times that of the flux without the core plate. The proposed pick-up coils have a finite size core plate ($l_a \times l_b$) with a finite relative permeability of μ_r, so the multiplying effect would be a little less than two by a correction factor α. Then the total mutual magnetic flux of the with-core-plate case ϕ_{mc} is modified from the without-core-plate case ϕ_m of (16.11) as follows:

$$\phi_{mc} = 2\alpha\phi_m \tag{16.12}$$

Figure 16.10 Top view of the decoupled pick-up coil model with a core plate.

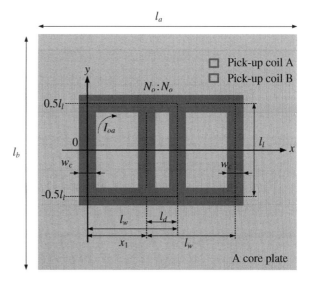

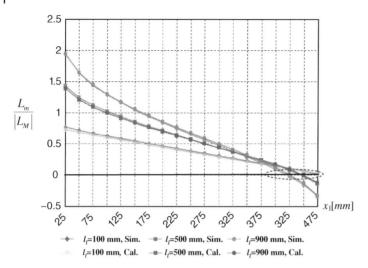

Figure 16.11 Comparison of the three-dimensional FEA simulated L_m with a core plate ($l_a = 120$ cm, $l_b = 120$ cm, $\mu_r = 2000$) with the calculated L_m with a core plate for various lengths l_l of pick-up coils A and B. The L_M is the maximum of L_m without a core plate for $l_l = 90$ cm, $l_w = 50$ cm, and $x_1 = 2.5$ cm.

In order to verify (16.12), a three-dimensional FEM simulation was done; its results are compared with (16.11), as shown in Figure 16.11, where α is obtained as 0.925, which gives the best curve fitting. As expected from (16.12), the L_m with a core plate is about two times that of the L_m without a core plate, while the self-decoupling point x_{opt} remains unchanged. The investigation on a general relationship between the core size and α is left for further work.

16.3 Example Design and Experimental Verifications

In order to verify the proposed analytic models of the self-decoupled coils, an experimental IPTS set was fabricated, as shown in Figure 16.12, where the inverter is operating at about $f_s = 20$ kHz, which is always a little bit higher than the resonant frequency f_r to guarantee the zero voltage switching (ZVS) operation of the inverter [15]. The circuit parameters are appropriately selected and measured, as listed in Table 16.1. The detailed configuration of the power supply rail specified in this chapter is the same as that of reference [12] except for the number of turns, where all the design parameters of the power supply rail are optimized by simulations and experiments.

16.3.1 The Self-Decoupled Single Pick-up Coil Set with/without a Core Plate

In order to verify the proposed analytic models for the self-decoupled single pick-up coil set with/without a core plate, a pick-up including only two pick-up coils A and B was assembled first, as shown in Figure 16.13. The number of turns of each pick-up coil is 10, that is, $N_o = 10$, and the size of the core plates is $l_a = 120$ cm and $l_b = 120$ cm. The shape of the coils is not a perfect rectangle, but a round-cornered rectangle due to the difficulty in winding. The core plate is fabricated by assembling ferrite core pieces of 10 cm × 10 cm × 1 cm.

The mutual inductance L_m between pick-up coils A and B with/without a core plate for different x_1 were measured and compared with the calculation result, as shown in Figure 16.14. As expected from the magnetic mirror model (16.12) and the previous simulation results of Figure 16.11, the mutual inductance with a core plate is nearly two times that without a core plate, and the self-decoupling distance x_{opt} to meet the condition $L_m = 0$ is unchanged. Here,

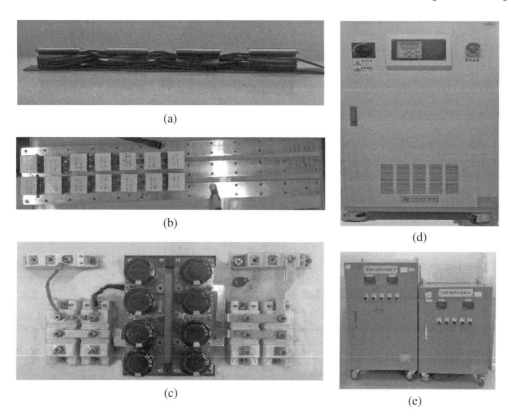

Figure 16.12 An experimental IPTS set: (a) power supply rail, (b) capacitor bank using the KEMET film capacitors, (c) rectifiers, (d) inverter, and (e) load bank.

Table 16.1 Experimental conditions of the IPTS

Parameters	Values	Parameters	Values
I_s	10 A	h_g	10 cm
N_s	10	t_{core}	1 cm
f_s	20 kHz	μ_r	2000

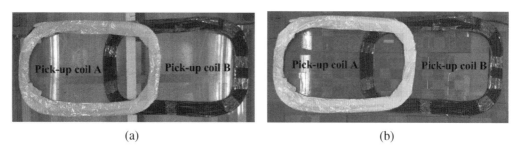

Figure 16.13 The experiment set of the self-decoupled single pick-up coils A and B: (a) without a core plate and (b) with a core plate.

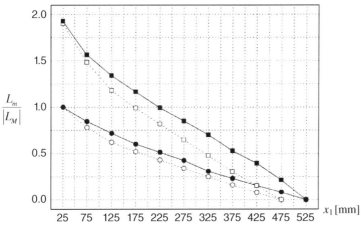

Figure 16.14 Measured mutual inductance L_m between pick-up coils A and B for various distances with/without a core plate. The L_M is the maximum L_m for $l_l = 36$ cm, $l_w = 62$ cm, and $x_1 = 2.5$ cm without a core plate.

the x_{opt} shows a slight discrepancy between the experiment results and calculation results due to the asymmetrical and rounded shape of pick-up coils A and B.

16.3.2 The Self-Decoupled Dual Pick-up Coil Set with/without a Core Plate

A self-decoupled dual pick-up coil set with/without a core plate was also assembled for experimental verification of the design, as shown in Figure 16.15. Each dual pick-up coil set includes two serially connected pick-up coils A–A′ or B–B′, where the number of turns of each pick-up coil is 10, that is, $N_o = 20$, and the size of a core plate is now $l_a = 120$ cm and $l_b = 180$ cm.

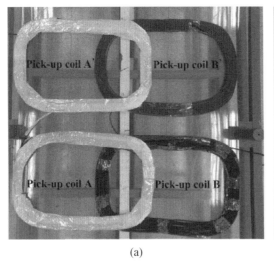
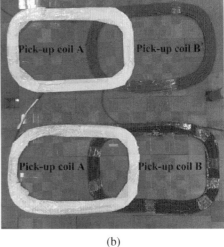

(a)   (b)

Figure 16.15 The proposed self-decoupled dual pick-up coil set for the experiments: (a) without a core plate and (b) with a core plate.

Figure 16.16 Measured mutual inductances for various positions of the pick-up B coils with/without a core plate. The L_M is the maximum mutual inductance between the two pick-up coils for $l_l = 36$ cm, $l_w = 62$ cm, and $x_1 = 2.5$ cm without a core plate.

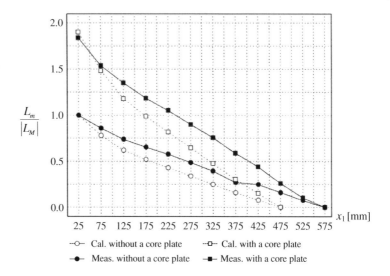

The mutual inductance L_m between the dual pick-up coil sets A–A′ and B–B′ with/without a core plate with respect to x_1 was measured and compared with the calculation result for the single pick-up coil set, as shown in Figure 16.16. Note that the calculation results of (16.11) and (16.12) are valid only for the single pick-up coil set A–B and there is no appropriate model developed for the dual pick-up case. Similar to the single pick-up coil case, the mutual inductance with a core plate is about two times that without a core plate and the optimum distance x_{opt} with/without a core plate is unchanged, as identified from Figure 16.16. However, there is an appreciable difference between the measured L_m for the dual pick-up coil set and the calculated L_m for the single pick-up coil set, which stems from the cross-coupling between the pick-up coils A′–B and A–B′.

The magnetic flux contribution of the coil A′ to the coil B is the opposite to that of the coil A to the coil B when x_1 is large; therefore, the cross-coupling increases the value of L_m, which results in the increase of x_{opt}. In this way, the self-decoupling distance x_{opt} has been increased from 52.5 cm to 57.5 cm for the dual pick-up coil set. This increase in the x_{opt} is beneficial for the enlargement of a lateral tolerance.

16.3.3 The Load Power of the Self-Decoupled Dual Pick-up Coil Set with a Core Plate

In order to confirm the effect of a self-decoupled dual pick-up coil set with a core plate, the load power and voltage were measured along with a lateral displacement x_d with respect to x_1 and quality factor Q, as shown in Figure 16.17. For simplicity, the results for only half of the lateral displacement are shown by averaging out the left- and right-side measurements. As the overlap l_d is too small, as shown in Figure 16.17(a), or it is too large, as shown in Figure 16.17(c) and (d), the lateral tolerance becomes small. When it is self-decoupled, that is, $x_1 = x_{opt} = 57.5$ cm, the lateral tolerance is as large as $l_t = 110$ cm for $Q = 5$, as identified from Figure 16.17(b). This large lateral tolerance is quite close to the theoretical estimation of (16.1a), which is $l_t \cong 2l_w - l_d = 119.5$ cm. For a large enough load power of 1.5 kW, the load resistance should be small, that is, the quality factor should be large, $Q = 60$, as shown in Figure 16.17(b). Under this condition, the lateral tolerance becomes $l_t = 90$ cm, which is about 1.5 times that of a coil width $l_w = 62$ cm.

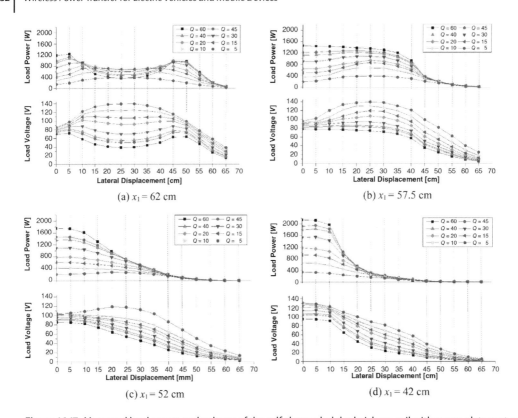

Figure 16.17 Measured load power and voltage of the self-decoupled dual pick-up coil with a core plate w.r.t. x_1.

The measured load power and voltage, as shown in Figure 16.17, reflect the power loss factors of internal coil resistances, core losses, and ESRs of capacitors, which could not be considered in the proposed model. The effects of these non-ideal circuit parameters on the lateral displacement and power efficiency are left for further works. The peak load power is normally achieved at $x_d = 0$ and becomes high as the overlap increases, as identified from Figure 16.17. The peak load power is as much as 2.2 kW, as shown in Figure 16.17(d). This is due to the increased induced voltage of V_s with the virtue of a large coupling area ($l_c \times l_l$) between the power supply rail and pick-up coils, as shown in Figure 16.18. This high power, however, cannot be utilized in practice because of the evidently large penalty of the reduced lateral tolerance.

Therefore, it can be concluded that the self-decoupled dual pick-up coil set is the optimum solution that provides both a large lateral tolerance and high load power.

16.4 Conclusions

The proposed self-decoupled dual pick-up coils and I-type power supply rail were verified to have a large lateral tolerance and high power. The proposed optimum distance between a coupled coil set was found to be so general that it is applicable to cases with/without a core plate or any compensation schemes. Moreover, the proposed model is applicable to single/dual pick-ups; hence, the self-decoupled pick-up coils can be used not only for RPEV but also for stationary charging systems. Detailed analyses for the cross-magnetic coupling effect and the magnetic mirror effect of a finite sized core plate are left for further work.

Figure 16.18 The magnetic flux linkage between the power supply rail and the dual pick-up coils A–A′ and B B′: (a) small overlapped case and (b) large overlapped case.

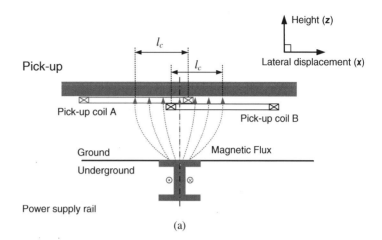

(a)

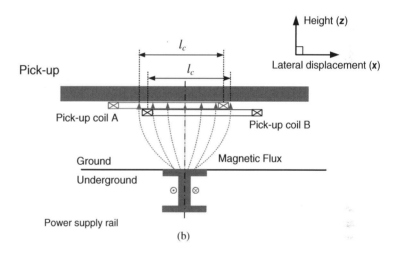

(b)

Problems

16.1 Try to find the self-decoupled distance between two identical circular coils with a diameter of D.

16.2 Repeat 16.1 for two asymmetric circular coils with diameters of D_1 and D_2, respectively.

16.3 Think about the application of the self-decoupled coils except for the proposed large tolerance design. What about the case when you need to make multiple coils on a plate without mutual inductance?

References

1 S.W. Lee, W.Y. Lee, J. Huh, H.J. Kim, C.B. Park, G.H. Cho, and Chun T. Rim, "Active EMF cancellation method for I-type pick-up of on-line electric vehicles (OLEV)," in *IEEE Applied Power Electronics Conference and Exposition (APEC)*, 2011, pp. 1980–1983.

2 S.Y. Choi, B.W. Gu, S.W. Lee, W.Y. Lee, J. Huh, and Chun T. Rim, "Generalized active EMF cancel methods for wireless electric vehicles," *IEEE Trans. on Power Electronics*, vol. 29, no. 11, pp. 5770–5783, November 2014.

3 N.P. Suh, D.H. Cho, Chun T. Rim, J.W. Kim, G.H. Jung, J. Huh, K.H. Lee, Y.D. Son, J.Y. Choi, E.H. Park, Y.J. Cho, J.C. Jang, Y.H. Kim, and H.G. Kim, "Collector device for electric vehicle with active cancellation of EMF," *KR Patent* 10-1038759, May 27, 2011.

4 J.G. Bolger, "Urban electric transportation systems: the role of magnetic power transfer," in *IEEE WESCON94 Conference*, 1994, pp. 41–45.

5 G.A. Covic and J.T. Boys, "Modern trends in inductive power transfer for transportation applications," *IEEE Journal of Emerging and Selected Topics in Power Electronics*, vol. 1, no. 1, pp. 28–41, March 2013.

6 J. Meins, "German activities on contactless inductive power transfer," in *IEEE Energy Conversion Congress and Exposition (ECCE)*, 2013.

7 O.C. Onar, J.M. Miller, S.L. Campbell, C. Coomer, C.P. White, and L.E. Seiber, "A novel wireless power transfer for in-motion EV/PHEV charging," in *IEEE Applied Power Electronics Conference and Exposition (APEC)*, 2013, pp. 3073–3080.

8 J.M. Miller, O.C. Onar, and P.T. Jones, "ORNL developments in stationary and dynamic wireless charging," in *IEEE Energy Conversion Congress and Exposition (ECCE)*, 2013.

9 N.P. Suh, D.H. Cho, and Chun T. Rim, "Design of on-line electric vehicle (OLEV)," *Plenary Lecture at the 2010 CIRP Design Conference*, 2010, pp. 3–8.

10 S.W. Lee, J. Huh, C.B. Park, N.S. Choi, G.H. Cho, and Chun T. Rim, "On-line electric vehicle (OLEV) using inductive power transfer system," in *IEEE Energy Conversion Congress and Exposition (ECCE)*, 2010, pp. 1598–1601.

11 J. Huh, S.W. Lee, C.B. Park, G.H. Cho, and Chun T. Rim, "High performance inductive power transfer system with narrow rail width for on-line electric vehicles," in *IEEE Energy Conversion Congress and Exposition (ECCE)*, 2010, pp. 647–651.

12 J. Huh, S.W. Lee, W.Y. Lee, G.H. Cho, and Chun T. Rim, "Narrow-width inductive power transfer system for on-line electrical vehicles (OLEV)," *IEEE Trans. on Power Electronics*, vol. 26, no. 12, pp. 3666–3679, December 2011.

13 S.Y. Choi, J. Huh, W.Y. Lee, S.W. Lee, and Chun T. Rim, "New cross-segmented power supply rails for road powered electric vehicles," *IEEE Trans. on Power Electronics*, vol. 28, no. 12, pp. 5832–5841, December 2013.

14 S.Y. Choi, J. Huh, W.Y. Lee, J.G. Cho, and Chun T. Rim, "Asymmetric coil sets for wireless stationary EV chargers with large lateral tolerance by dominant field analysis," *IEEE Trans. on Power Electronics*, vol. 29, no. 12, pp. 6406–6420, December 2014.

15 S.Y. Choi, B.W. Gu, S.Y. Jeong, and *Chun T. Rim* "Advances in wireless power transfer systems for road powered electric vehicles," *IEEE Journal of Emerging and Selected Topics in Power Electronics*, accepted for publication.

16 M. Budhia, G.A. Covic, and J.T. Boys, "Design and optimization of magnetic structures for lumped inductive power transfer systems," *IEEE Trans. on Power Electronics*, vol. 26, no. 11, pp. 3096–3108, November 2011.

17 G.A. Covic, J.T. Boys, M. Kissin, and H. Lu, "A three-phase inductive power transfer system for roadway power vehicles," *IEEE Trans. on Industrial Electronics*, vol. 54, no. 6, pp. 3370–3378, December 2007.

18 C.S. Wang, G.A. Covic, and O.H. Stielau, "Power transfer capability and bifurcation hhenomena of loosely coupled inductive power transfer systems," *IEEE Trans. on Industrial Electronics*, vol. 51, no. 1, pp. 148–157, 2004.

19 M. Budhia, J.T. Boys, G.A. Covic, and C.-Y. Huang, "Development of a single-sided flux magnetic coupler for electric vehicle IPT charging systems," *IEEE Trans. on Industrial Electronics*, vol. 60, no. 1, pp. 318–328, January 2013.

20 M. Budhia, G.A. Covic, and J.T. Boys, "A new magnetic coupler for inductive power transfer electric vehicle charging systems," *IEEE Trans. on Industrial Electronics*, pp. 2487–2492, November 2010.

21 G.R. Nagendra, J.T. Boys, G.A. Covic, B.S. Riar, and A. Sondhi, "Design of a double coupled IPT EV highway," in *IEEE Industrial Electronics Society, IECON 2013*, November 2013, pp. 4606–4611.

22 Chun T. Rim and G.H. Cho, "New approach to analysis of quantum rectifier-inverters," *IEEE Electronic Letters*, vol. 25, no. 25, pp. 1744–1745, December 1989.

23 Chun T. Rim, "Unified general phasor transformation for AC converters," *IEEE Trans. on Power Electronics*, vol. 26, no. 9, pp. 2465–2475, September 2011.

24 J. Huh, W.Y. Lee, S.Y. Choi, G.H. Cho, and Chun T. Rim, "Frequency-domain circuit model and analysis of coupled magnetic resonance systems," *Journal of Power Electronics*, vol. 13, no. 2, March 2013.

25 S.W. Lee, B. Choi, and Chun T. Rim, "Dynamics characterization of the inductive power transfer system for on-line electric vehicles by Laplace phasor transform," *IEEE Trans. on Power Electronics*, vol. 28, no. 12, pp. 5902–5909, December 2013.

26 Guidelines for limiting exposure to time-varying electric and magnetic fields (up to 100 kHz), ICNIRP Guidelines, 1998.

27 D. Kwiat, S. Saoub, and S. Einav, "Calculation of the mutual induction between coplanar circular surface coils in magnetic resonance imaging," *IEEE Trans. on Biomedical Engineering*, vol. 39, pp. 433–436, 1992.

28 G.A. Covic, M.L.G. Kissin, D. Kacprzak, N. Clausen, and H. Hao, "A bipolar primary pad topology for EV stationary charging and highway power by inductive coupling," in *IEEE Energy Conversion Congress and Exposition (ECCE)*, 2011, pp. 1832–1838.

29 A. Zaheer, D. Kacprzak, and G.A. Covic, "A bipolar receiver pad in a lumped IPT system for electric vehicle charging applications," in *IEEE Energy Conversion Congress and Exposition (ECCE)*, 2012, pp. 283–290.

30 W.Y. Lee, J. Huh, S.Y. Choi, X.V. Thai, J.H. Kim, E.A. Al-Ammar, M.A. El-Kady, and Chun T. Rim, "Finite-width magnetic mirror models of mono and dual coils for wireless electric vehicles," *IEEE Trans. on Power Electronics*, vol. 28, no. 3, pp. 1413–1428, March 2013.

31 R.L. Steigerwald, "A comparison of half-bridge resonant converter topologies," *IEEE Trans. on Power Electronics*, vol. 3, no. 2, pp. 174–182, April 1988.

32 Chun T. Rim and G.H. Cho, "Phasor transformation and its application to the DC/AC analyses of frequency phase-controlled series resonant converters (SRC)," *IEEE Trans. on Power Electronics*, vol. 5, no. 2, pp. 201–211, April 1990.

33 S. Kim, A. Zaheer, G. Covic, and J. Boys, "Tripolar pad for inductive power transfer systems," in *IEEE 40th Annual Conference of IEEE Industrial Electronics Society (IECON)*, Sheraton, Dallas, TX, 2014.

17

Power Rail Segmentation and Deployment

17.1 Introduction

Electric vehicles (EVs) are identified as the future transportation, but not widespread because of the relatively short driving distance for a single charge, more frequent recharging of the battery than engine cars, the lack of charging infrastructure, and the high price required for its battery. In an effort to resolve these problems, road-powered EVs (RPEVs) with an inductive power transfer system (IPTS) were proposed [1–6, 23–29, 37, 38]. The IPTS comprises an inverter, power supply rails built under roads, and pick-up(s) at the bottom of vehicles. Wireless power is transferred from the power supply rails to moving EVs by inductive couplings. In order to transfer a large power, the IPTS should operate in the resonant mode so that the overall efficiency of the IPTS can be increased [7–29]. Recently, the power and efficiency of the IPTS have reached 100 kW and 83%, respectively [23–29]. The design of the resonant circuits in the IPTS, regardless of the resonant modes, is a challenging issue due to the dynamic and static behaviors of the IPTS in practice; however, the dynamic and static behaviors of the IPTS can be characterized by using phasor transformation [30–33]. Moreover, the design of the power supply and pick-up coils can be implemented without simulations and experiments by applying improved magnetic mirror models [36]. Unlike pure EVs and hybrid EVs, RPEVs require their own roadways where power supply rails are implemented. The rails should be activated when RPEVs are on the roadway but they should be deactivated to prevent pedestrians from potentially harmful electromagnetic fields (EMF) [34] when no RPEVs are on. To meet these requirements, the power supply rail is segmented into many subpower supply rails, that is subrails. Each subrail is activated by providing a high-frequency current through a switch box from an inverter. There are several ways of controlling the segmented subrail current by the switch box, where the cable length and the number of activated subrails at times are different from each other. From a cost point of view, the cable length is of great concern because it is responsible for about 20% of the construction cost of the power supply rail [35]. In addition, for multiple driving of RPEVs, arbitrary subrails should be concurrently activated by using an inverter only.

In this chapter, new cross-segmented power supply rails (X-rail) are explained. Each subrail is connected through an autocompensation switch box, which can change the current direction of a pair of power cables. Hence, adding the current of the two pairs of power cables results in the activation mode while nullifying the current produces the silence mode. To compensate for the variable line inductance of the rail due to the change of current direction, a coupling transformer with two capacitors is introduced. The proposed rail does not need power supply cables; hence, the cable cost for the commercialization of RPEVs can be drastically reduced. Multiple driving of RPEVs is also possible by activating each subrail independently using the switch boxes. In addition, the EMF for the silence mode is drastically reduced if a twisted pair of

Wireless Power Transfer for Electric Vehicles and Mobile Devices, First Edition. Chun T. Rim and Chris Mi.
© 2017 John Wiley & Sons Ltd. Published 2017 by John Wiley & Sons Ltd.

power cables is used. Moreover, copper nets can also be used for further reduction of the EMF. The proposed X-rail was implemented for experiments and verified for practical applications.

Nomenclature

A_c Cross-sectional area of a coupling transformer
l_c Total length of a centralized switching power supply rail
l_d Total length of a distributed switching power supply rail
l_x Total length of a cross-segmented power supply rail
l_0 Length of inlet cable
l_1 Length of subrail cable
l_2 Length between two subrails
l_c Effective magnetic path length of a coupling transformer
d_{air} Airgap of a coupling transformer
n Number of sub-rails
N_{1c} Number of primary side turns for a coupling transformer
N_{2c} Number of second side turns for a coupling transformer
h_m Measuring height from the top of the pole
C_1 Compensation capacitance of an autocompensation circuit in both activation and silence modes
C_2 Compensation capacitance of an autocompensation circuit in activation mode
L_{on} Line inductance of a subrail in activation mode.
L_{off} Line inductance of a sub-rail in silence mode
L_{lc} Leakage inductance of a coupling transformer
L_{mc} Magnetizing inductance of a coupling transformer.
I_s Supply current (rms)
$f_{r,on}$ Resonant frequency of the activation mode for a branch of each auto compensation circuit
$f_{r,off}$ Resonant frequency of the silence mode for a branch of each autocompensation circuit

17.2   Cross-Segmented Power Supply Rail Design

17.2.1   Previous Works on Segmented Power Supply Rails

The power supply rails should be activated when RPEVs are on the roadway but should be deactivated when no RPEVs are on the roadway to prevent pedestrians from potentially harmful EMF [34]. To meet these requirements, the power supply rail is segmented into many subpower supply rails, that is, subrails. Each subrail is activated through a switch box from an inverter, thus providing high-frequency current.

There are two types of segmented power supply rails that have been developed for the online electric vehicles (OLEV) [23–25]: one is the centralized switching type and another is the distributed switching type. The centralized switching power supply rail consists of a few subrails, a bundle of supply cables, and a centralized switch box, as shown in Figure 17.1(a); the inverter is connected to one of several pairs of supply cables through the switch box one at a time. Assuming the inlet l_0 and the gap l_2 are much smaller than the length of a subrail l_1, the total length of cables l_c can be determined as follows:

$$l_c \approx n(n+1)l_1 \tag{17.1}$$

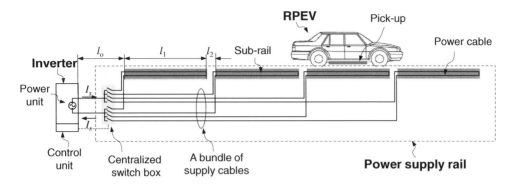

(a) Centralized switching power rail consists of a bundle of supply cables and a centralized switch box

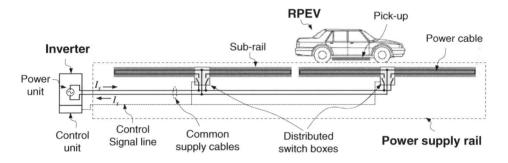

(b) Distributed switching power supply rail consists of common supply cables and multiple switch boxes

Figure 17.1 Conventional segmented power supply rails for RPEV [18–20].

where n is the number of subrails. One of the drawbacks of this method is that a subrail can only be activated by the inverter one at a time.

In the distributed switching power supply rail, as shown in Figure 17.1(b), the system is composed of a few subrails, a pair of common power supply cables, and multiple switch boxes located between two subrails, which are controlled by the control unit. Based on the assumption that the inlet l_0 and the gap l_2 are much smaller than the length of a subrail l_1, the total length of cables l_d for Figure 17.1(b) can be determined as follows:

$$l_d \approx 2(2n - 1)l_1 \tag{17.2}$$

From (17.1) and (17.2), it is found that the total length of cable for the distributed switching segmentation can be reduced compared to the centralized one, that is, $l_c \geq l_d$ for $n \geq 1$, and the reduction effect becomes significant for large n. For example, only 34.5% of cable is required for the distributed rail when $n = 10$, as shown later in Figure 17.4. The distributed switching power supply rail, however, requires common power supply cables, which increase the construction cost and conductive power loss, as shown in Figure 17.2, where the cables are protected by the fiber-reinforced polymer (FRP) pipes. With regards to the control capability of subrails, there is still the same problem as that of centralized switching.

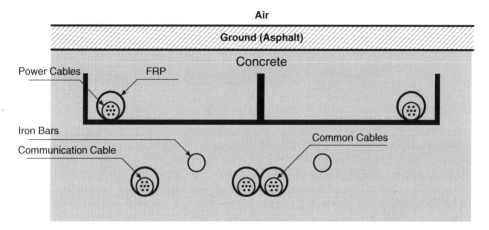

Figure 17.2 Cross-section view of the power supply rail of the distributed switching power supply rail.

Question 1 Answer for the W-type power rail of Figure 17.2. (1) Why is the FRP used except for protection? (2) What is the purpose of iron bars? (3) What is the difference between the core and side widths of the core? (4) If you can arbitrarily change the width of the core, what is the optimum width distribution, when considering magnetic saturation due to power cables?

17.2.2 Proposed Cross-Segmented Power Supply Rails (X-Rail)

As the remedy for the above-mentioned problems, a new cross-segmented power supply rail (X-rail) [37,38] is explained in this chapter. The proposed X-rail consists of segmented subrails, autocompensation switch boxes, control signal lines, and roadway harnesses, as shown in Figure 17.3. The subrail is specially made of twisted power cables, core, and copper nets, which will be described in detail in Section 7.3.

The EMF generated by the power supply cables under the subrails, as shown in Figure 17.2, should be minimized to meet ICNIRP Guidelines [34]. Furthermore, the inverter should be able to drive multiple RPEVs; however, the conventional segmentations allow only one activated subrail to be driven at a time, as shown in Figure 17.1(a) and (b).

Regarding the bundle of two half-current-rating cables as a cable, for fair comparison, the total length of the cable l_x is found in a similar way to (17.1) and (17.2), as follows:

$$l_x \approx 2n\,l_1$$
$$\approx l_d/2 \quad for \quad n \gg 1 \tag{17.3}$$

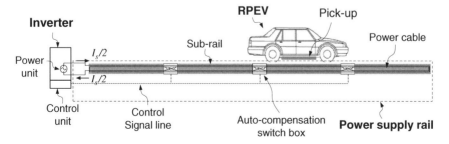

Figure 17.3 Proposed cross-segmented power supply rail consists of switch boxes and subrails without any common supply cable.

Figure 17.4 Comparison of the total length of power cables of the proposed X-rail (bottom) with that of the conventional switching rails (upper).

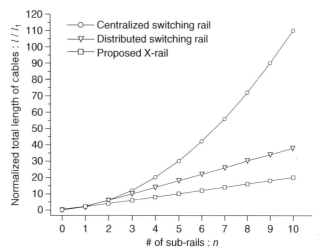

From (17.3), it is clearly identified that the cable cost of the proposed X-rail is nearly halved compared to the distributed switching segmentation of (17.2). The cable length reduction effect of the proposed scheme becomes severe as the number of segmented subrails increases, as shown in Figure 17.4.

In order to realize the X-rail, three versions of power rail types are considered. They are U-type core mono rail [23–25], W-type core dual rail [23–25], and I-type core power rail [26–29], which have been developed since 2009 as the IPTS of OLEV. The major characteristics of the three versions of the power supply rails are summarized in Table 17.1.

The detailed scheme of the W-type dual power supply rail, having two pairs of power cables, is shown in Figure 17.5, which explains how to activate a subrail. The cross-section and top views of the dual rail in activation mode is shown in Figure 17.5(a), where the current directions of the cables in a bundle are the same. Thus, an appropriate magnetic flux is generated and passes through the pick-up, whereas little magnetic flux is generated when the current directions of the cables in a bundle are opposite, as shown in Figure 17.5(b); thus it is called the silence mode. In addition, the proposed X-rail can also be realized for the U-type mono power supply rail and the I-type power supply rail, as shown in Figures 17.6 and 17.7.

The switch box of the proposed X-rail should change the current direction of a cable of the bundle by controlled switches; hence, it is implemented in this chapter using four sets of bidirectional power switches, a transformer, and four compensation capacitors, as shown in

Table 17.1 Characteristics of U-type, W-type, and I-type power supply rails.

	U-type	**W-type**	**I-type**
Rail width	140 cm	80 cm	10 cm
Lat. tolerance	20 cm	15 cm	24 cm
EMF	5.1 μT	5.0 μT	1 μT
Airgap	17 cm	20 cm	20 cm
Output power	5.2 kW/pick-up	15 kW/pick-up	25 kW/pick-up
Efficiency	72 %	74 %	80 %

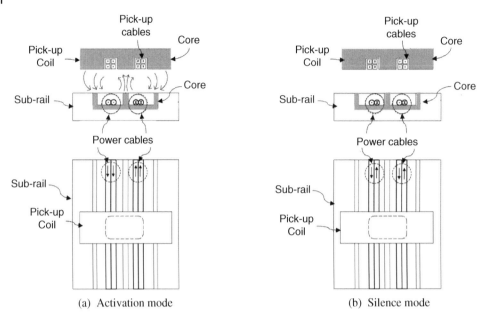

(a) Activation mode (b) Silence mode

Figure 17.5 The proposed X-segment for a dual rail on activation and silence modes: cross-section view (top) and plane view (below).

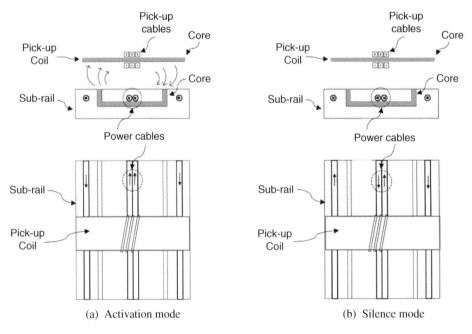

(a) Activation mode (b) Silence mode

Figure 17.6 The proposed X-segment for a mono rail on activation and silence modes: cross-section view (top) and plane view (below).

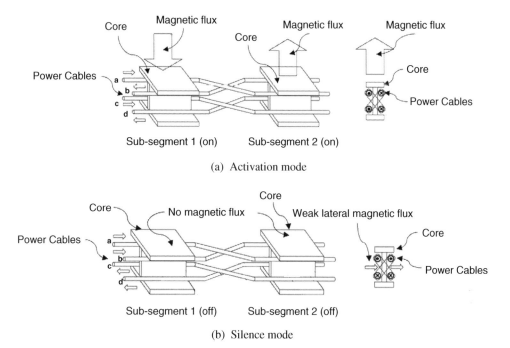

Figure 17.7 The proposed X-segment for an I-type rail on activation and silence modes: perspective view (left) and cross-section view (right).

Figure 17.8. Each switch box is controlled individually by its control signal, and the current direction of a pair of power cables is changed by the two sets of power switches; thus, the status of the next subrail is changed accordingly. To get an autocompensation capability against the inductance change of next subrails, a few capacitors and a coupling transformer are used.

The inverter for the X-rail is assumed to be operating at the switching frequency f_s, which is slightly higher than the resonant frequency f_r of the X-rail. This frequency discrepancy makes the rail always inductive so that zero voltage switching (ZVS) can be guaranteed [26, 28, 29].

The line inductance of a subrail is drastically changed by the operating modes, that is, L_{on} and L_{off} for the activation and silence modes, respectively. The resonant frequency of a

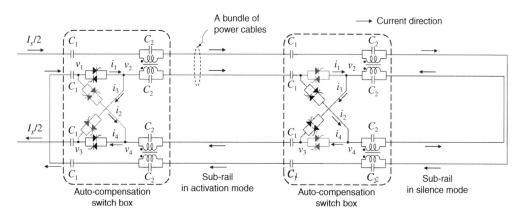

Figure 17.8 The circuit configuration and current direction of the X-rail for different switch connections.

subrail should, however, be unchanged to keep the ZVS condition for the inverter regardless of the operating mode. To solve this problem, the autocompensation box composed of two C_2 capacitors and a coupling transformer is introduced, as shown in Figure 17.8. In practice, the transformer includes finite values of magnetizing inductance L_{mc} and leakage inductance L_{lc}. It is assumed that the magnetizing impedance is high enough and the leakage impedance is low enough compared with their corresponding impedances, as shown in Figure 17.9. For the activation mode, the circuit becomes symmetrical w.r.t. the transformer; hence, the magnetizing

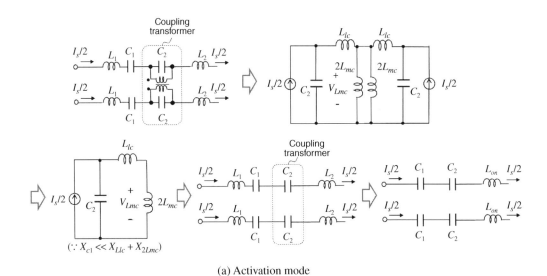

(a) Activation mode

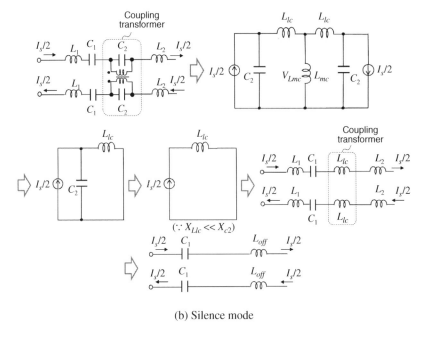

(b) Silence mode

Figure 17.9 Equivalent circuits of the autocompensation circuit for different operating modes.

inductance can be equally split into two, as shown in Figure 17.9(a). For an effective resonance, the reactance of C_2 should be much smaller than that of parallel inductances, as follows:

$$X_{C_2} \ll X_{L_{lc}} + X_{2L_{mc}} \leftrightarrow \frac{1}{\omega_s C_2} \ll \omega_s(L_{lc} + 2L_{mc}) \cong 2\omega_s L_{mc} \quad for \quad L_{lc} \ll L_{mc}$$

$$\Rightarrow \quad \frac{1}{2\omega_s^2 C_2} \ll L_{mc} \tag{17.4}$$

For the silence mode, the circuit becomes cross-symmetric w.r.t. the transformer; hence, the voltage of the magnetizing inductance is zero, as shown in Figure 17.9(b). Now the reactance of C_2 should be much larger than that of the leakage reactance as follows:

$$X_{C_2} \gg X_{L_{lc}} \leftrightarrow \frac{1}{\omega_s C_2} \gg \omega_s L_{lc} \quad \Rightarrow \quad L_{lc} \ll \frac{1}{\omega_s^2 C_2} \tag{17.5}$$

It can be summarized from (17.4) and (17.5) that the coupling transformer should be so designed that the magnetizing and leakage inductances meet the following condition:

$$L_{lc} \ll \frac{1}{\omega_s^2 C_2} \ll 2L_{mc} \tag{17.6}$$

The transformer, as shown in Figure 17.10, was designed according to conventional transformer design rules, meeting the criteria of (17.6), where the key parameters are listed in Table 17.2. Under the condition of (17.6), the transformer can be regarded as an ideal transformer. Hence, the resonant frequencies $f_{r,on}$, $f_{r,off}$ of the activation and silence modes for a branch of each autocompensation circuit of Figure 17.9(a) and (b), respectively, are determined as follows:

$$f_{r,on} = \frac{1}{2\pi\sqrt{L_{on}C_1//C_2}} \tag{17.7a}$$

$$f_{r,off} = \frac{1}{2\pi\sqrt{L_{off}C_1}} \tag{17.7b}$$

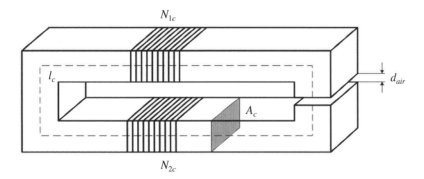

Figure 17.10 Design parameters of the proposed coupling transformer.

Table 17.2 Design result of coupling transformer parameters.

Parameter	Value
L_{mc}	220 μH
L_{lc}	5 μH
l_c	80 cm
d_{air}	0.2 cm
A_c	40 cm²
μ_s	2000
N_{1c}	10 turns
N_{2c}	10 turns
P_{max}	81 kW
I_{max}	54 A
V_{max}	1.5 kV

By appropriate selection of the value of C_2, the resonant frequency can remain unchanged as follows:

$$f_{r,on} = f_{r,off} \leftrightarrow L_{on} \frac{C_1 C_2}{C_1 + C_2} = L_{off} C_1$$

$$\therefore C_2 = C_1 \frac{L_{off}}{L_{on} - L_{off}}$$

(17.8)

For example, the inductance of the 2 m long I-type subrail in the activation mode is about 40 μH ($=2L_{on}$), but it decreases to 8 μH ($=2L_{off}$) in the silence mode. For the resonant frequency $f_r = 19$ kHz, C_1 becomes 17.5 μF. Then $C_2 = 4.4$ μF is obtained from (17.8).

17.3 Example Design and Experimental Verifications of the X-Rail

To verify the feasibility of the X-rail, details of parameters of the experimental set, as shown in Figure 17.11, are listed in Table 17.3. An experimental set composed of a prototype switch box and two I-type subrails 2 m long was implemented, as shown in Figure 17.12.

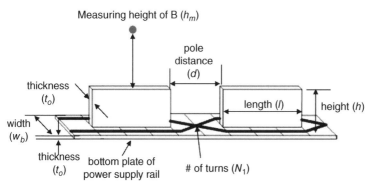

Measuring height of B (h_m)

pole distance (d)

thickness (t_o)

length (l)

height (h)

width (w_b)

thickness (t_o)

bottom plate of power supply rail

# of turns (N_1)

Figure 17.11 Parameter definitions of the I-type power supply rail.

Table 17.3 Experimental conditions for the proposed X-rail.

Parameter	Value	Parameter	Value
L_{on}	20 μH	D	20 cm
L_{off}	4 μH	L	20 cm
C_1	17.47 μF	H	10 cm
C_2	4.47 μF	t_o	1 cm
I_s	10 A	N_1	2 turns
f_s	20 kHz (tuned)	w_b	10 cm

Thyristors were used for the bidirectional switches since neither a fast switching time nor a forced turn-off is required even though they operate at a high operating frequency. In order to turn off the thyristors, the inverter is temporary turned off for a few ms to cease the current completely; then the inverter is turned on again after a transition of thyristor switching. The supply current I_s was selected as 10 A for experimental verification purposes, where the rated current for practical applications is 100–200 A [23–29]. The switching frequency f_s was chosen as 20 kHz, which has been used for the recently developed RPEV [23–29] to avoid audible acoustic noise.

The currents and voltages of the thyristors were measured in the activation and silence modes for different measuring points, as shown in Figures 17.13 and 17.14. The measurement results show that most current flows into the turn-on thyristors while a negligibly small leakage current flows into the turn-off thyristors.

As shown in Figure 17.15, the EMF without pick-up was measured on each subrail of the activation and silence modes for three different measuring heights. The twisted cables were used, as shown in Figure 17.16, for the effective reduction of the EMF. The measurement results show that the EMF of the silence mode is much smaller than that of the activation mode. The EMF decreases as the measuring height increases, as shown in Figure 17.15. However, the EMF between the two poles for the practical application where I_s = 200 A is 20 times larger than this experimental result is due to the different current levels, and the EMF at the 5 cm height, which corresponds to the ground level of Figure 17.2, is found to be higher than the ICNIRP Guidelines of 6.25 μT, even in the silence mode.

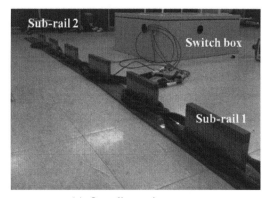

(a) Overall experiment set.

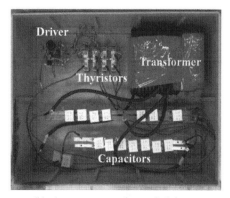

(b) Autocompensation switch box.

Figure 17.12 The experimental set of the X-rail.

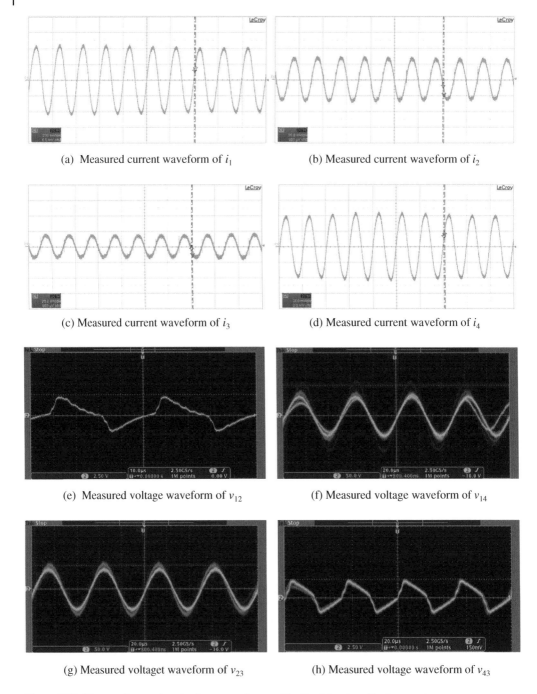

(a) Measured current waveform of i_1

(b) Measured current waveform of i_2

(c) Measured current waveform of i_3

(d) Measured current waveform of i_4

(e) Measured voltage waveform of v_{12}

(f) Measured voltage waveform of v_{14}

(g) Measured voltaget waveform of v_{23}

(h) Measured voltage waveform of v_{43}

Figure 17.13 Measured current and voltage waveforms of the thyristors in activation mode.

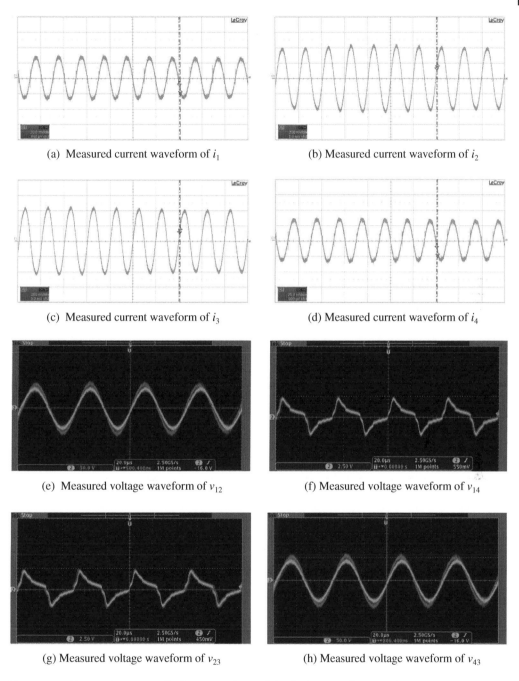

(a) Measured current waveform of i_1

(b) Measured current waveform of i_2

(c) Measured current waveform of i_3

(d) Measured current waveform of i_4

(e) Measured voltage waveform of v_{12}

(f) Measured voltage waveform of v_{14}

(g) Measured voltage waveform of v_{23}

(h) Measured voltage waveform of v_{43}

Figure 17.14 Measured current and voltage waveforms of the thyristors in silence mode.

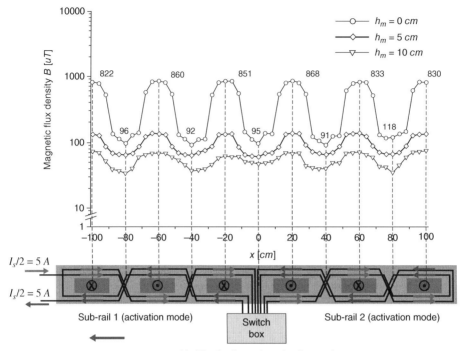

(a) All subrails are in activation mode

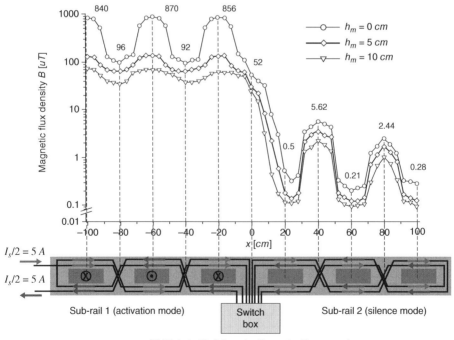

(b) Right half of the subrails are in silence mode

Figure 17.15 Measurement results of EMF on the subrails for the activation and silence modes.

Figure 17.16 Twisted cables used in the proposed X-rail for the further reduction of EMF.

In order to mitigate the high EMF of the silence mode between two poles, copper nets were used for the further reduction of the EMF, as shown in Figure 17.17. It was found that more than 10 times the reduction in the EMF level was achieved by using the proposed double copper nets. The measurement results of the EMF level for $I_s = 10$ A at $h_m = 5$ cm is less than 0.24 μT, as shown in Figure 17.18, which corresponds to 4.8 μT for $I_s = 200$ A; this is now lower than the ICNIRP Guidelines.

Question 2 Can you explain qualitatively and quantitatively, if possible, the EMF reduction effect of Figure 17.17?

For readers who are interested in the power transfer between the proposed I-type power supply rail and pick-up coil, it is recommended to look at the experiment of a similar I-type

Figure 17.17 Measurement results of EMF using copper nets for the further reduction at the cross point.

(a) No copper net

(b) Small diameter copper net

(c) Large diameter copper net

(d) Double copper nets

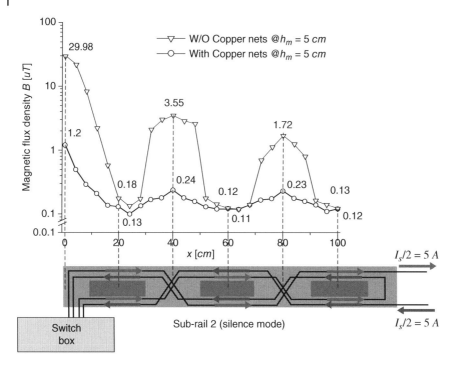

Figure 17.18 Measurement results of EMF on the subrails for the silence modes using the double copper nets for further reduction at the cross point.

[29], where the maximum output power is 35 kW at a 20 cm airgap for a 200 A supply current, and the maximum power transfer efficiency is 74%.

17.4 Conclusion

The X-rail suggested for an RPEV is quite cost-effective by halving the power cable use. This X-rail enables us to activate several RPEVs simultaneously and independently, which is a unique merit never achieved by previous segmented power rails. Moreover, the EMF for the silence mode is very low due to the proposed twisted power cables and double copper nets; hence, it is quite safe for pedestrians and other cars passing over the I-type power rail.

Problems

17.1 Compared to previous power rail segmentation methods, the proposed X-rail has a demerit of using semiconductor devices and a transformer in each subrail, as shown in Figures 17.8 and 17.10.
   (a) The semiconductor devices are relatively weak compared to passive devices and need health monitoring. What information should you collect from each subrail and how can you gather the information?
   (b) What will be the size of the transformer if the transferred power and operating frequency are both doubled?

17.2 What is the optimized length of segmentation in a power rail? Consider the length of EVs and their efficiency, which is determined by the resistance of the segmented power rail.

References

1 J.G. Bolger, F.A. Kirsten, and L.S. Ng, "Inductive power coupling for an electric highway system," in *Proc. IEEE 28th Vehicular Technology Conference*, 1978, vol. 28, pp. 137–144.

2 C.E. Zell and J.G. Bolger, "Development of an engineering prototype of a road powered electric transit vehicle system," in *Proc. 32nd IEEE Vehicular Technology Conference*, 1982, vol. 32, pp. 435–438.

3 M. Eghtesadi, "Inductive power transfer to an electric vehicle-analytical model," in *Proc. 40th IEEE Veh. Technol. Conf.*, May 1990, pp. 100–104.

4 A.W. Green and J.T. Boys, "10 kHz inductively coupled power transfer concept and control," in *Proc. 5th Int. Conf. IEE Power Electron Variable-Speed Drivers*, October 1994, pp. 694–699.

5 G.A.J. Elliott, J.T. Boys, and A.W. Green, "Magnetically coupled systems for power transfer to electric vehicles" in *Proc. Int. Conf. Power Electron. Drive Syst.*, February 1995, pp. 797–801.

6 G.A. Covic, J.T. Boys, M.L.G. Kissin, and H.G.Lu, "A three-phase inductive power transfer system for road powered vehicles," *IEEE Trans. Ind. Electron.*, vol. 54, no. 6, pp. 3370–3378, December 2007.

7 H. Matsumoto, Y. Neba, K. Ishizaka, and R. Itoh, "Model for a three-phase contactless power transfer system," *IEEE Trans. on Power Electron.*, vol. 26, pp. 2676–2687, September 2011.

8 H. Matsumoto, Y. Neba, K. Ishizaka, and R. Itoh, "Comparison of characteristics on planar contactless power transfer systems," *IEEE Trans. on Power Electron.*, vol. 27, pp. 2980–2993, June 2012.

9 H.H. Wu, G.A. Covic, J.T. Boys, and D.J. Robertson, "A series-tuned inductive-power transfer pickup with a controllable AC-voltage output," *IEEE Trans. on Power Electron.*, vol. 26, pp. 98–109, November 2011.

10 M. Budhia, G.A. Covic, and J.T. Boys, "Design and optimization of circular magnetic structures for lumped inductive power transfer systems," *IEEE Trans. on Power Electron.*, vol. 26, pp. 3096–3108, November 2011.

11 H.L. Li, A.P. Hu, and G.A. Covic, "A direct AC–AC converter for inductive power transfer systems," *IEEE Trans. on Power Electron.*, vol. 27, pp. 661–668, February 2012.

12 C.S. Wang, O.H. Stielau, and G.A. Covic, "Design consideration for a contactless electric vehicle battery charger," *IEEE Trans. on Ind. Electron.*, vol. 52, pp. 1308–1314, October 2005.

13 P. Si and A.P. Hu, "Analysis of DC inductance used in ICPT power pick-ups for maximum power transfer," in *Proc. IEEE/PES Transmission and Distribution Conference and Exhibition*, 2005, pp. 1–6.

14 J.T. Boys and G.A. Covic, "DC analysis technique for inductive power transfer pick-ups," *IEEE Power Electronics Letters*, vol. 1, pp. 51–53, June 2003.

15 M L.G. Kissin, C.Y Huang, G.A. Covic, and J.T. Boys, "Detection of the tuned point of a fixed-frequency LCL resonant power supply," *IEEE Trans. on Power Electron.*, vol. 24, pp. 1140–1143, April 2009.

16 P. Nagatsuka, N. Ehara, Y. Kaneko, S. Abe, and T. Yasuda, "Compact contactless power transfer system for electric vehicles," in *International Power Electronics Conference (IPEC)*, 2010, pp. 807–813.

17 G.B. Joung and B.H. Cho, "An energy transmission system for an artificial heart using leakage inductance compensation of transcutaneous transfer," *IEEE Trans. on Power Electron.*, vol. 13, pp. 1013–1022, November 1998.

18 S. Valtechev, B. Borges, K. Brandisky, and J.B. Klaassens, "Resonant contactless energy transfer with improved efficiency," *IEEE Trans. on Power Electron.*, vol. 24, pp. 685–699, March 2009.

19 M. Borage, S. Tiwari, and S. Kotaiah, "Analysis and design of an LCL-T resonant converter as a constant-current power supply," *IEEE Trans. on Ind. Electron.*, vol. 52, pp. 1547–1554, December 2005.

20 B. Mangesh, T. Sunil, and K. Swarna, "LCL-T resonant converter with clamp diodes: a novel constant-current power supply with inherent constant-voltage limit," *IEEE Trans. on Ind. Electron.*, vol. 54, pp. 741–746, April 2007.

21 B.L. Cannon, J.F. Hoburg, D.D. Stancil, and S.C. Goldstein, "Magnetic resonance coupling as a potential means for wireless power transfer to multiple small receivers," *IEEE Trans. on Power Electron.*, vol. 24, pp. 1819–1825, July 2009.

22 D.L. O'Sullivan, M.G. Egan, and M.J. Willers, "A family of single-stage resonant AC/DC converters with PFC," *IEEE Trans. on Power Electron.*, vol. 24, pp. 398–408, February 2009.

23 N.P. Suh, D.H. Cho, and C.T. Rim, "Design of on-line electric vehicle (OLEV)," *Plenary Lecture at the 2010 CIRP Design Conference*, 2010.

24 S.W. Lee, J. Huh, C.B. Park, N.S. Choi, G.H. Cho, and C.T. Rim, "On-line electric vehicle using inductive power transfer system," in *IEEE Energy Conversion Congress and Exposition (ECCE)*, 2010, pp. 1598–1601.

25 J. Huh and C.T. Rim, "KAIST wireless electric vehicles – OLEV," in *JSAE Annual Congress*, 2011.

26 J. Huh, S.W. Lee, C.B. Park, G.H. Cho, and C.T. Rim, "High performance inductive power transfer system with narrow rail width for on-line electric vehicles," in *IEEE Energy Conversion Congress and Exposition* (ECCE), 2010, pp. 647–651.

27 S.W. Lee, W.Y. Lee, J. Huh, H.J. Kim, C.B. Park, G.H. Cho, and C.T. Rim, "Active EMF cancellation method for I-type pick-up of on-line electric vehicles," in *IEEE Applied Power Electronics Conference and Exposition (APEC)*, 2011, pp. 1980–1983.

28 J. Huh, W.Y. Lee, G.H. Cho, B.H. Lee, and C.T. Rim, "Characterization of novel inductive power transfer systems for on-line electric vehicles," in *IEEE Applied Power Electronics Conference and Exposition (APEC)*, 2011, pp. 1975–1979.

29 J. Huh, S.W. Lee, W.Y. Lee, G.H. Cho, and C.T. Rim, "Narrow-width inductive power transfer system for on-line electrical vehicles," *IEEE Trans. on Power Electron.*, vol. 26, no. 12, pp. 3666–3679, December 2011.

30 C.T. Rim and G.H. Cho, "Phasor transformation and its application to the DC/AC analyses of frequency phase-controlled series resonant converters (SRC)," *IEEE Trans. Power Electron.*, vol. 5, no. 2, pp. 201–211, April 1990.

31 C.T. Rim, D.Y. Hu, and G.H. Cho, "Transformers as equivalent circuits for switches: general proofs and D-Q transformation-based analyses," *IEEE Trans. Ind. Applic.*, vol. 26, no. 4, pp. 777–785, July/August 1990.

32 C.T. Rim, "Unified general phasor transformation for AC converters," *IEEE Trans. Power Electron.*, vol. 26, pp. 2465–2745, September 2011.

33 S.W. Lee, C.B. Park, and C.T. Rim, "Static and dynamic analyses of three-phase rectifier with LC input filter by Laplace phasor transformation," in *IEEE ECCE 2012*, September 2012, pp. 1570–1577.

34 ICNIRP Guidelines, "International commission on non-ionizing radiation protection," 1998, www.icnirp.de/documents/emfgdl.pdf.

35 KAIST OLEV Team, "Feasibility studies of On-Line Electric Vehicle Project, KAIST Internal Report, August 2009.

36 W.Y. Lee, J. Huh. S. Choi, X.V. Thai, J.H. Kim, E.A. Al-Ammar, M.A. El-Kady, and C.T. Rim, "Finite-width magnetic mirror models of mono and dual coils for wireless electric vehicle," *IEEE Trans. on Power Electron.*, vol. 28, pp. 1413–1428, March 2013.

37 N.P. Suh, S.H. Jang, D.H. Cho, G.H. Cho, J.G. Cho, C.T. Rim, J. Huh, B.H. Lee, and Y.H. Kim, "Cross type segment power supply," Patent Application No. 1020100052341, patented.

38 N.P. Suh, S.H. Jang, D.H. Cho, G.H. Cho, J.G. Cho, C.T. Rim, J. Huh, B.H. Lee, Y.H. Kim, W.Y. Lee, and H.J. Kim, "Cross-segment feed device capable of turning on/turning off individual modules," Patent Application No. PCT/KR2011/004069, patented.

Part IV

Static Charging for Pure EVs and Plug-in Hybrid EVs

This part of the book is for stationary chargers for electric vehicles (EVs) such as pure battery EVs (BEVs) and plug-in hybrid EVs (PHEVs). This part is relatively small compared with the previous part on dynamic charging, not because this part is less important but because I have less experience and fewer papers on this issue. There are numerous other articles for specific issues not addressed here.

This part will start from an introduction to static charging and asymmetric coils for large-tolerance EV chargers are explained, which is one of the ways of enlarging lateral tolerance. DQ coils are also explained for the same purpose, then capacitive power transfer for EV chargers is explained by Chris Mi, and finally foreign object detection (FOD) will be covered. Note that FOD can also be used in dynamic charging, but it is introduced in this part because FOD is more frequently used in static charging.

Wireless Power Transfer for Electric Vehicles and Mobile Devices, First Edition. Chun T. Rim and Chris Mi.
© 2017 John Wiley & Sons Ltd. Published 2017 by John Wiley & Sons Ltd.

18

Introduction to Static Charging

18.1 The Need for Electric Vehicles (EVs) and Wireless Electric Vehicles (WEVs)

Why are most people still not daily driving an EV even though the number of EVs is rapidly growing? Actually, I do not have an EV at the moment even though I have been looking for a good candidate as a researcher on EVs. As a customer, the EV is still very expensive and inconvenient compared to existing conventional vehicles. The structure of EVs is relatively simple and, therefore, easy to maintain; however, it is very expensive and heavy because of the battery. EVs have a much shorter driving distance for a full charge compared to conventional internal combustion engine (ICE) vehicles. Of course, some extraordinary EVs such as Tesla Motors have a comparable driving distance but at a doubled or tripled price of other vehicles.

It can be said that EVs have not been widely commercialized because of battery and charging problems, as shown in Figure 18.1. It is remarkable that the battery car was commercialized in the late nineteenth century, which is about 20 years earlier than the commercialization of ICE. However, EVs were forced out of the market as ICEs became mass produced. At that time, the battery was too heavy, required a long time to recharge, had a relatively short driving distance, and was expensive. Amazingly, these problems of the battery are still true compared to current ICEs, even though they have been mitigated. The expensive price of EVs is mainly due to the battery price, and other problems of the battery also impede the commercialization of EVs. The improvement of the battery is very slow and does not follow Moor's law because it is not governed by electronics but is governed by chemistry, which took nearly a century to improve.

Another main cause of obstacles for commercialization of EVs is the charging problem, which is related to the battery but must be separated from it. In other words, even though an innovative battery will become available, still we have charging problems. For example, we should have 30C (1C corresponds to the rated power or energy of a battery for an hour) of charging capacity in order to recharge an EV in 2 minutes, enen supposing we have such a good battery. For a 50 kWh battery, we should have at least a 1.5 MW power rating charger and a power distribution facility to support it. In practice, we should recharge an EV frequently in order to avoid the battery being empty. If a cable charger is used, it is very inconvenient to plug in and plug out daily, and it is potentially dangerous to deal with the connector of the cable manually under moisture conditions. Currently, a quick charger can recharge an EV battery up to 80% within 20 minutes, which is too long a time for the customer accustomed to a 2 minute refueling. Two possible solutions to the above-mentioned problems are summarized in Figure 18.2.

Wireless Power Transfer for Electric Vehicles and Mobile Devices, First Edition. Chun T. Rim and Chris Mi.
© 2017 John Wiley & Sons Ltd. Published 2017 by John Wiley & Sons Ltd.

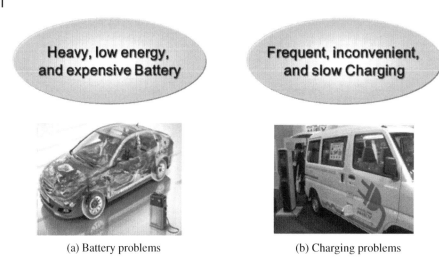

(a) Battery problems   (b) Charging problems

Figure 18.1 Major obstacles for commercialization of electric vehicles.

The first is a road-powered EV (RPEV), which is explained in a previous part of this book. An RPEV, or on-line electric vehicle (OLEV), is one of the possible solutions to the battery problems, because it does not rely on the battery but gets power directly from roadway power rails. Even though the RPEV has a battery for auxiliary purposes, the size of the battery is a few times smaller than the pure battery EV, which is not an obstacle for commercialization.

The second solution to the obstacles is stationary wireless charging, which is the main issue of this chapter of the book, as shown in Figure 18.2(b). Using static wireless charging, the inconvenient and dangerous cable charging problems can be completely solved. Of course, the slow charging problem cannot be directly solved by wireless charging, and it must be mitigated by other means such as interoperable roadway power rails, on which the static chargers can be charged. Static wireless charging can be applicable to plug-in EVs (PHEVs) and battery EVs (BEVs).

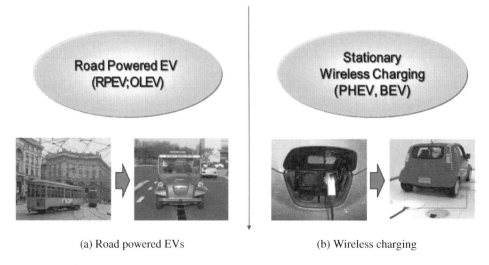

(a) Road powered EVs   (b) Wireless charging

Figure 18.2 Two possible solutions for current electric vehicles.

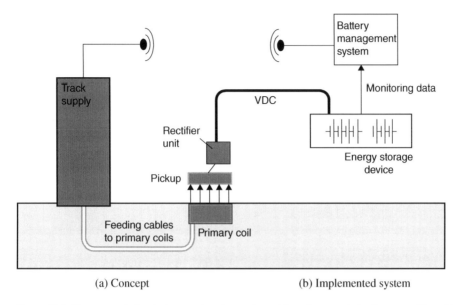

(a) Concept   (b) Implemented system

Figure 18.3 The concept of a stationary charging system at bus stops made by Germany, Wampfler Co.

I would like to call the stationary and dynamic charging EV solutions as wireless EVs (WEVs), which embrace wireless power and wireless communications for EVs. As EVs evolve to information platforms in the future, wireless communication will be essential to them, which results in ubiquitously connected cars. Wireless power will make the WEV complete.

18.2 Overview of Existing Static EV Chargers

Let me show you a few implemented examples of static chargers for EVs, as shown in Figures 18.3 to 18.7. Static wireless chargers for EV buses are shown in Figures 18.3 to 18.6. Because of the short time at each stop, the power level should be high enough to charge in such a short

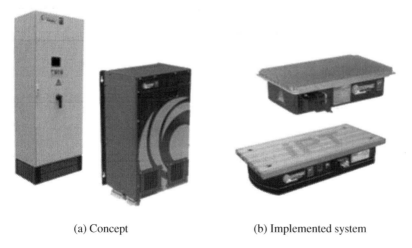

(a) Concept   (b) Implemented system

Figure 18.4 The implemented stationary charging system at bus stops made by Germany, Wampfler Co.

Figure 18.5 Stationary charging system of the 30 kW to 150 kW level at bus stops, deployed in Waseda University, Japan.

Figure 18.6 The deployed Tx coil of the stationary charging system at bus stops, deployed in Waseda University, Japan.

time. However, that for the EV passenger car, as shown in Figure 18.7, may be slowly charged; hence, the power level is as low as 3.3–6.6 kW. The system efficiency of the practically implemented static charger, which is measured from utility power to the input of an on-board battery, ranges from 85–93%, though 95–98% efficiency was reported for the experimental prototype system where the airgap is very small.

18.3 Design Issues on Static EV Chargers

The system requirements of static EV chargers may include cost, reliability, lifetime, availability, power, efficiency, tolerances (height, longitudinal, lateral), electromagnetic field (EMF), foreign object detection (FOD), robustness against mechanical shock for given source and load conditions, airgap, load change, and operating temperature. Hardware and software implementation issues such as insulation, waterproof, user interface, monitor, communications, and information are also important for the development of the system. Furthermore, standardization and regulation issues together with the EMF and FOD must be considered for commercialization, which is dealt with in the subsequent section.

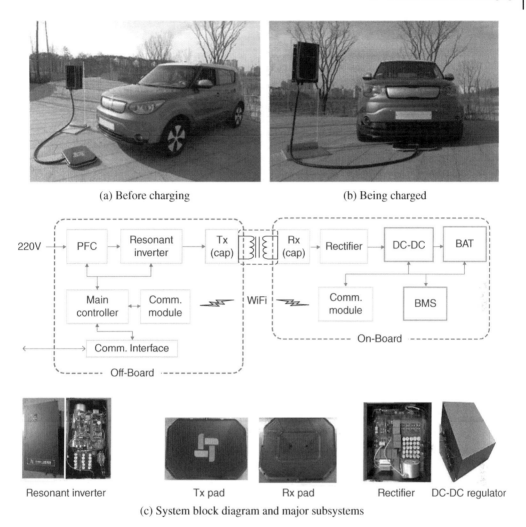

(a) Before charging

(b) Being charged

Resonant inverter   Tx pad   Rx pad   Rectifier   DC-DC regulator

(c) System block diagram and major subsystems

Figure 18.7 Stationary charging system of a 3.3 kW to 6.6 kW level for passenger cars at a parking area, deployed in Jeju Island, South Korea (Green Power Technologies, http://www.egreenpower.com/pro1_ref1.php).

Similar to dynamic charging of EVs, a static charging system has numerous design issues to meet the above-mentioned various requirements [2–4], as summarized in Table 18.1.

The underlined issues in Table 18.1 are more specific to static chargers than dynamic chargers. The most important design issues, discussed in detail in this chapter, are coil design and standardization.

18.3.1 Coil Design Issues in General

Coil design would be the most substantial part of IPT system design because the transmitting (Tx) coil and receiving (Rx) coil determine power transfer performances such as output voltage, power level, coil-to-coil efficiency, and height/lateral tolerances. The configuration of a static charger includes Tx and Rx windings and cores, as shown in Figure 18.8. The goal of coil design is to focus the magnetic field from the Tx to Rx coils so that the leakage flux is minimized to

Table 18.1 Major design issues of stationary EV chargers

- Switching frequency selection
- Coil design (coupling factor, size, tolerance, efficiency, EMF)
- Compensation circuit design (Gv, voltage/current ratings)
- Converter design (inverter, rectifier, regulator, ZVS, PF)
- Quality factor selection & resonant frequency variation
- Insulation and thermal issues
- Control of IPTS (short/open circuit, high voltage, protections)
- Communications between Tx and Rx
- Foreign object detection (metal/living objects)
- Location & airgap detections (vision/RF/inductive/capacitive)
- Budgets (cost, mass, volume, power loss, reliability, etc.)

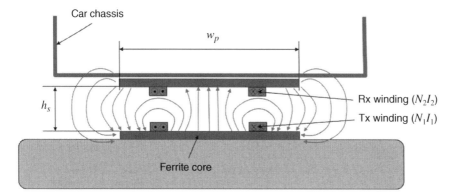

Figure 18.8 Configuration of a stationary charging system, when the Tx coil is energized and the Rx coil current is zero.

a lower electromagnetic field (EMF). In detail, to design the coil is to determine the ampere-turns (N_1, N_2, I_1, I_2) and detail dimensions of coils for a given airgap, width of coil, operating frequency, power supply voltage, delivered output power, and ambient operating temperature $(h_s, w_p, f_s, V_s, P_o, T)$.

18.3.2 Optimum Radius Design of Coils

For example, the optimum radii (r_1, r_2, r_3) of Tx and Rx windings can be analytically determined, when the coil size r_4 is given, as shown in Figure 18.9. The magnetic flux generated from Tx for zero Rx current can be calculated from the magnetic circuit of Figure 18.9(b) as follows:

$$\phi_{Tx} = \frac{N_1 I_1}{\Re_1 + \Re_2} \quad \because \Re_1 \cong \frac{h_s}{\mu_0 A_1} = \frac{h_s}{\mu_0 \pi\, r_2^2}, \quad \Re_2 \cong \frac{h_s}{\mu_0 A_2} = \frac{h_s}{\mu_0 \pi\, (r_4^2 - r_2^2)} \quad for \quad h_s \ll r_4$$

$$\cong \frac{N_1 I_1 \mu_0 \pi}{h_s \left(\dfrac{1}{r_2^2} + \dfrac{1}{r_4^2 - r_2^2} \right)} = \frac{N_1 I_1 \mu_0 \pi}{h_s \left(\dfrac{r_4^2}{r_2^2 \,(r_4^2 - r_2^2)} \right)} = \frac{N_1 I_1 \mu_0 \pi}{h_s r_4^2} x^2 (1 - x^2) \quad \because x \equiv \frac{r_2}{r_4}, r_2 = \frac{r_1 + r_3}{2}$$

$$(18.1)$$

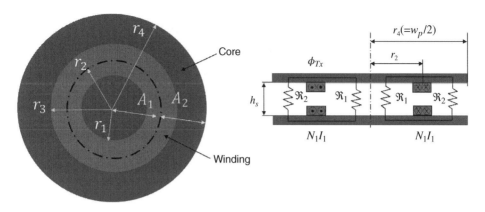

(a) Configuration of Tx and Rx coils (plane view)   (b) Magnetic flux and magnetic resistance (side view)

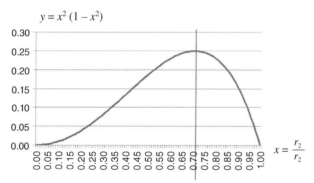

(c) Normalized magnetic flux as function of normalized radius of winding

Figure 18.9 Optimum radii of Tx/Rx windings for a given coil size, when Tx is identical with Rx.

By derivation of the magnetic flux of (18.1) w.r.t. x, the maximum point is found as follows:

$$y_m = x^2(1 - x^2)\Big|_{x_m^2 = 0.5} \quad \therefore x_m = \frac{1}{\sqrt{2}} \cong 0.707 \tag{18.2a}$$

$$\phi_{Tx,m} \cong \frac{N_1 I_1 \mu_0 \pi}{h_s r_4^2} x^2(1 - x^2)\Big|_{x_m^2 = 0.5} = \frac{N_1 I_1 \mu_0 \pi}{4 h_s r_4^2} = \frac{N_1 I_1 \mu_0 \pi}{h_s w_p^2} \quad \because h_s \ll w_p \tag{18.2b}$$

It is identified from (18.2) that the magnetic flux, that is, the induced output voltage of Rx becomes its maximum when the average radius of winding is about 71% of the coil outer radius. This optimum point corresponds to the radius where the area of inner circle is the same as that of the outer circle, as shown in Figure 18.9(a). Therefore, this design guideline to an optimum coil radius of "equal area" can be generalized to any shape of Tx and Rx coils such as rectangular and round shapes.

Question 1 Find the optimized coil size of a rectangular Tx and Rx.

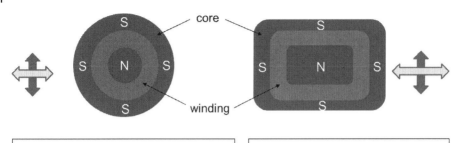

✓ Omnidirectional characteristics	✓ Symmetric characteristics
✓ Simplest structure to manufacture	✓ Max. area for a given rectangle
✓ Min. conduction loss of winding cable (Max. area for a given cable length)	✓ Suitable for rectangular-shape applications such as EV chargers and smartphone chargers
(a) Circular	(b) Rectangular

Figure 18.10 Shape of Tx and Rx coils (plane view).

18.3.3 Selection of Coil Types in Coil Design

There are a few selection issues in the design of Tx and Rx coils of static chargers. One of them is the coil shape, as shown in Figure 18.10. The merits of the two types are listed above. The arrows depict lateral tolerances of two directions: the larger the arrow length the bigger is the tolerance. "S" and "N" denote the magnetic polarity. There could be other shapes such as round shape, which is basically rectangular but its corners are round, and octagonal shape. Basically, the shape of coil is largely governed by the shape of the mother body depending on applications.

As shown in Figure 18.11, you should select the number of poles among single, double, and triple poles, which is not shown here. The lateral tolerance of a double pole could be better than

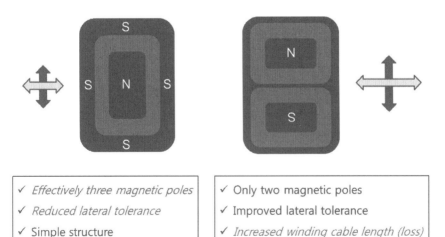

✓ *Effectively three magnetic poles*	✓ Only two magnetic poles
✓ *Reduced lateral tolerance*	✓ Improved lateral tolerance
✓ Simple structure	✓ *Increased winding cable length (loss)*
(a) Single pole	(b) Dual pole

Figure 18.11 Number of pole in Tx and Rx coils (plane view).

that of a single pole because the number of magnetic poles is only two for the double pole coil. The demerits are written in italic letters, as shown in Figure 18.11. The cable length of winding may be larger for the double-pole coil, as identified from Figure 18.11(b).

Question 2 Find the tolerances of Figure 18.11(a) and (b) for a given coil size.

What is the optimum gap between the outer winding and the outer core of Figure 18.11(b)? Is there any optimum gap that exists? The answer is "no there is no such optimum gap." In other words, the winding size could be bigger than the core, which was actually found in Auckland University team's work, as shown in Figure 18.12. The core length of Figure 18.12(b) to (d) could even be less than the inner winding size.

Figure 18.12 Circular and rectangular pads together with power distribution over x–y plane displacements [1].

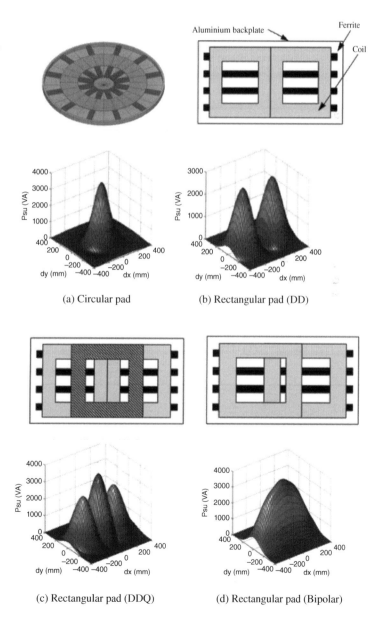

(a) Circular pad

(b) Rectangular pad (DD)

(c) Rectangular pad (DDQ)

(d) Rectangular pad (Bipolar)

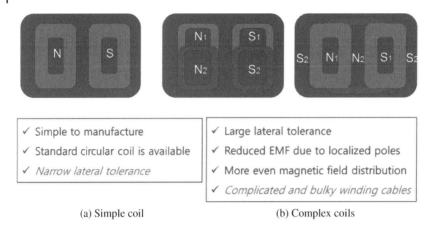

✓ Simple to manufacture	✓ Large lateral tolerance
✓ Standard circular coil is available	✓ Reduced EMF due to localized poles
✓ *Narrow lateral tolerance*	✓ More even magnetic field distribution
	✓ *Complicated and bulky winding cables*

(a) Simple coil   (b) Complex coils

Figure 18.13 Structure of Tx and Rx coils (plane view).

The coil could be of a simple structure, as shown in Figure 18.13(a); however, sometimes it becomes a combination of a few coils, as shown in Figure 18.13(b), where the left is a dual coil of dual pole and the right is a dual coil with a single coil. Compared to complex coils, the lateral tolerance of simple coils could be narrower, where multiple magnetic poles with a DQ phase are often adopted in the complex coils, as shown in Figure 18.13(b).

You should also determine the coil type depending on the magnetic field shape you want, as shown in Figure 18.14, where loop coil and dipole coil examples are given. For a given size of the Tx and Rx coils, the dipole coil has a larger lateral tolerance in the side direction but has a large EMF, which needs a good magnetic shield.

18.3.4 Selection of Core Structure in a Coil Design

The core could be dense or sparse structures, as shown in Figure 18.15. Note that the total amount of core use could remain unchanged by the average density of the core if the core

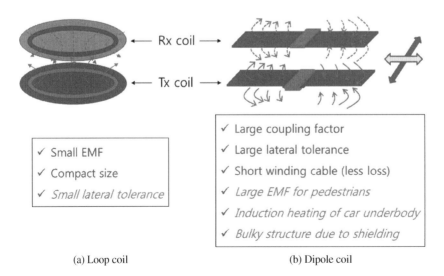

✓ Small EMF	✓ Large coupling factor
✓ Compact size	✓ Large lateral tolerance
✓ *Small lateral tolerance*	✓ Short winding cable (less loss)
	✓ *Large EMF for pedestrians*
	✓ *Induction heating of car underbody*
	✓ *Bulky structure due to shielding*

(a) Loop coil   (b) Dipole coil

Figure 18.14 Magnetic field shape of Tx and Rx coils (bird's eye view).

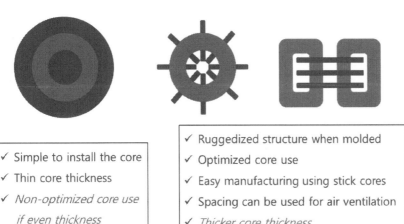

✓ Simple to install the core ✓ Thin core thickness ✓ *Non-optimized core use* *if even thickness*	✓ Ruggedized structure when molded ✓ Optimized core use ✓ Easy manufacturing using stick cores ✓ Spacing can be used for air ventilation ✓ *Thicker core thickness* ✓ *EMF leak behind stick cores*
(a) Dense core structure	(b) Sparse core structures

Figure 18.15 Core structure of Tx and Rx coils (plane view).

thickness of a dense core structure is optimized. If this is not the case, the core use of a sparse structure tends to be smaller than the dense structure. Note that the sparse structure was first proposed by myself to make a rugged structure of Tx coil in concrete [5–12].

Question 3 Explain why the sparse core structures of Figure 18.15(b) automatically become an optimized core structure, where magnetic field density inside each core is even.

Another category of core structure is the use of the core, as shown in Figure 18.16. The core is not necessarily always used, especially when the operating frequency is too high to use any core, as shown in Figure 18.16(b). Note that the coreless coil is quite often referred to as the coupled magnetic resonance system (CMRS), whereas the coil with a core is called the

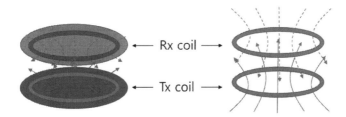

✓ High coupling factor ✓ Low magnetic flux leakage ✓ Reduced size (compact) ✓ *Increased core mass* ✓ *Core loss* ✓ *Inductance changes (h_s, d_x)*	✓ High Q due to little inductance change ✓ Large current due to no core loss ✓ *Large EMF due to magnetic flux leakage* ✓ *Large volume due to shield space* ✓ *Affected by adjacent metal objects* ✓ *Tends to operate at a few MHz (CMRS)*
(a) With core	(b) Coreless

Figure 18.16 Core use of Tx and Rx coils (bird's eye view).

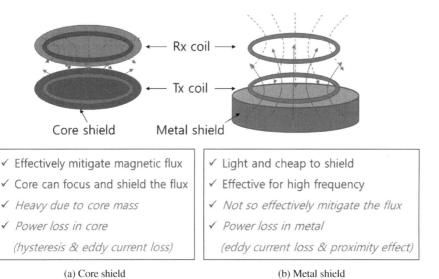

✓ Effectively mitigate magnetic flux	✓ Light and cheap to shield
✓ Core can focus and shield the flux	✓ Effective for high frequency
✓ *Heavy due to core mass*	✓ *Not so effectively mitigate the flux*
✓ *Power loss in core*	✓ *Power loss in metal*
(hysteresis & eddy current loss)	*(eddy current loss & proximity effect)*
(a) Core shield	(b) Metal shield

Figure 18.17 Magnetic field shield of Tx and Rx coils (bird's eye view).

conventional inductive power transfer system (IPTS). The people who study CMRS think that it is quite different from IPTS due to its inherent large Q and no change of primary and secondary inductances. The wonderful characteristics of CMRS, however, do not come from its resonant type but from its coreless structure, as I announced at the Special Session of Wireless Power Transfer of the 2013 IEEE ECCE Conference. As denoted in Figure 18.16, there are a lot of penalties for the coreless structure.

18.3.5 Magnetic Field Shield for Coil Design

We have two shielding methods: core shield and metal shield, as shown in Figure 18.17. Both shields are often used to enhance the magnetic shielding effect. Note again that there is no theory available for a metal shield design compared to a core shield design.

18.3.6 Misalignment Issues in the Designs of Coil, Compensation Circuit, and Controller

A very important and difficult problem in IPT design is the position tolerance issue arising from airgap change and misalignments between Tx and Rx coils, as shown in Figure 18.18. As the coils are misaligned with each other, the induced voltage drops and in general and in resonant frequencies of Tx and Rx sides usually change as well. Therefore, the position tolerance affects not only coil design but also compensation circuit, controller, and converter designs (power rating, ZVS condition, power factor, etc.).

Question 4 Discuss the impact of each position error of Figure 18.18 on the induced voltage and resonant frequency of the Rx coil.

18.3.7 Location Detection Issue

In order to position the Rx coil to be aligned with the Tx coil, often the vision system such as the rear camera is used, as shown in Figure 18.19. Of course other methods such as position detection systems using Rx or Tx coil are under development.

Figure 18.18 The position tolerance problem, arising from relative position changes of Tx and Rx coils.

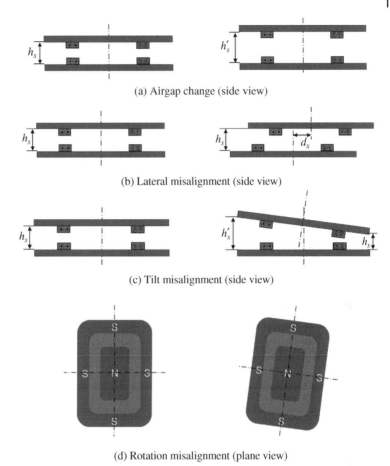

(a) Airgap change (side view)

(b) Lateral misalignment (side view)

(c) Tilt misalignment (side view)

(d) Rotation misalignment (plane view)

Navigation Monitor

Rear Camera

Figure 18.19 A vision system example used to detect the relative position between Tx and Rx coils.

Aggressive researchers are highly encouraged to develop location detection systems that operate under all weather conditions.

18.4 Standard and Regulation Issues on Static EV Chargers

Different from dynamic chargers, the standardization of static chargers, that is, wireless EV chargers, is nearly established and mostly prepared for commercialization. Regulations on EMF and others are also now well established. In this section, a few important standard and regulation issues are introduced. More standards and regulations of different nations and international organizations are summarized in a paper from Qualcomm Incorporated [13].

18.4.1 SAE J2954 Standards: Selection of Operating Frequency, Power Level, etc.

Because the switching losses of inverter and rectifier, core loss, induced voltage, power, and coil size are highly dependent on the switching frequency, the selection of switching frequency is of prime importance in the design of a static charger. Overall system efficiency, cost, temperature of the device, and system reliability are also affected by the switching frequency. Different from dynamic chargers, the line-to-line voltage of a primary power rail is not of great concern, and the switching frequency of static chargers tends to be higher.

There is a well-established frequency standard in static EV chargers, called SAE J2954, where the nominal frequency is 85 kHz and the frequency range is 81.38–90.00 kHz. The Society of Automotive Engineers (SAE) established the committee J2954 to provide safety standards on static EV wireless chargers. The SAE standard is quite active in EV chargers, as shown in Table 18.2, which includes not only wireless charging but also various technical issues on plug-in EVs.

SAE deals with the power level of static EV chargers as well. There are three power levels classed as follows:

- 3.7 kW (garage overnight charging)
- 7.7 kW (private/public parking)
- 22 kW (fast charge)

The coil sizes, maximum misalignment, EMF radiation level, and measurement method are mostly determined but have not yet released to the public at the time this book was being written.

18.4.2 ICNIRP Guidelines: EMF and E-Field

Due to potential adverse effects, though not always medically well proven, electric field (E-field), magnetic field (B-field), that is, electromagnetic field (EMF), and radio wave (RF wave) are regulated by most nations and international organizations. Potential physical effects of AC E-field, B-field, and RF wave on a human body and electronic equipment are induced current, generated heat, and chemical reaction, which is found only when the wavelength of the RF wave is shorter than that of UV light.

One of the most frequently referred regulations in WPT is the International Commission on Non-Ionizing Radiation Protection (ICNIRP) Guidelines, which have been established to limit human exposure to time-varying EMF with the aim of preventing adverse health effects. The guidelines provide reference safety restrictions on both the E-field and B-field for occupational exposure and general public exposure. Note that ICNIRP is not a regulation but a guideline;

Table 18.2 SAE standard activities for EV chargers

Document	Title – Works in Progress
J1772	SAE Electric Vehicle and Plug-in Hybrid Electric Vehicle Conductive Charge Coupler
J2836/3	Use Cases for Communication between Plug-in Vehicles and the Utility Grid for Reverse Power Flow
J2836/4	Use Cases for Diagnostic Communication for Plug-in Vehicles
J2836/5	Use Cases for Communication between Plug-in Vehicles and Their Customers
J2836/6	Use Cases for Wireless Charging Communication between Plug-in Electric Vehicles and the Utility Grid
J2847/1	Communication between Plug-in Vehicles and the Utility Grid
J2847/2	Communication between Plug-in Vehicles and Off-Board DC Chargers
J2847/3	Communication between Plug-in Vehicles and the Utility Grid for Reverse Power Flow
J2847/4	Diagnostic Communication for Plug-in Vehicles
J2847/5	Communication between Plug-in Vehicles and Their Customers
J2847/6	Wireless Charging Communication between Plug-in Electric Vehicles and the Utility Grid
J2894/2	Power Quality Requirements for Plug In Vehicle Chargers – Part 2: Test Methods
J2931/1	Digital Communications for Plug-in Electric Vehicles
J2931/4	Broadband PLC Communication for Plug-in Electric Vehicles
J2931/5	Telematics Smart Grid Communications between Customers, Plug-In Electric Vehicles (PEV), etc.
J2931/6	Digital Communication for Wireless Charging Plug-in Electric Vehicles
J2931/7	Security for Plug-in Electric Vehicle Communications
J2953	Plug-In Electric Vehicle (PEV) Interoperability with Electric Vehicle Supply Equipment (EVSE)
J2954	Wireless Charging of Electric and Plug-in Hybrid Vehicles
J2990	Hybrid and EV First and Second Responder Recommended Practice
J3009	Trapped Energy – Reporting and Extraction from Vehicle Electrical Energy Storage System

hence, it is not mandatory to meet the guideline for public use and government regulation may be weaker or stronger than the guideline.

Recently, existing "ICNIRP Guidelines 1998" has been replaced by "ICNIRP Guidelines 2010," as shown in Figure 18.20.

As summarized in Table 18.3 for the frequency range of 3 kHz to 10 MHz, which is well within most WPT applications, the human body E-field limitation for general public exposure is 83 V/m and for occupational exposure is 170 V/m, respectively. The B-field reference is 27 μT for general public exposure and 100 μT for occupational exposure as well.

When you design WPT systems operating in the frequency range of 3 kHz to 10 MHz, you need to consider the E-field and B-field references of 83 V/m and 27 μT, respectively, as denoted in bold letters of Table 18.3, not to exceed the general public exposure level. As mentioned above, you should also look at each government's regulations when you are designing a WPT system to be used in such a nation. For example, the B-field to general public exposure in South Korea should be less than 6.25 μT at 20 kHz, which has been established based on ICNIRP Guidelines 1998.

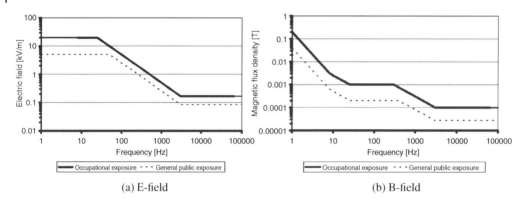

(a) E-field   (b) B-field

Figure 18.20 ICNIRP Guidelines 2010.

Table 18.3 ICNIRP Guidelines 2010 E-field and B-field references for 3 kHz to 10 MHz

Exposure Characteristic	E-Field Reference	B-Field Reference
Occupational	170 V/m	100 μT
General public	**83 V/m**	**27 μT**

It is relatively easy to meet the E-field regulations, but it is usually quite demanding to meet the B-field regulations.

18.4.3 SAR Regulation

Another regulation issue, which will not be addressed in detail here, is the specific absorption rate (SAR). SAR is the rate of radio wave absorption in watts per kilogram, which needs not to be considered under 100 kHz because of a negligible amount of SAR. Usually, SAR is considered for applications such as smartphones and medical instruments, where its radio frequency is higher than 10 MHz [14, 15]. The FCC limit for public exposure from radiation devices such as cellular telephones is 1.6 watts per kilogram (1.6 W/kg).

SAR can be derived from the fundamental electrical engineering principle, as shown in Figure 18.21.

The power dissipation can be calculated from ohmic heat generated from infinitesimal resistance as follows:

$$dP \equiv (dI)^2 dR = (|\vec{J}(\vec{x})| dydz)^2 \frac{dx}{\sigma(\vec{x})dydz} = |\vec{J}(\vec{x})|^2 \frac{dxdydz}{\sigma(\vec{x})} = |\sigma(\vec{x})\vec{E}(\vec{x})|^2 \frac{dxdydz}{\sigma(\vec{x})} = \sigma(\vec{x})|\vec{E}(\vec{x})|^2 dV$$

$$\because dV = dxdydz \tag{18.3}$$

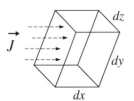

Figure 18.21 SAR calculation for an infinitesimal volume of a resistive object.

where $\sigma(\vec{x})$ is the conductivity (Siemens/m) and $\vec{E}(\vec{x})$ is the electric field (V/m) of an object under examination at a position $\vec{x}$. Note from (18.3) that the current density vector $\vec{J}(\vec{x})$ is assumed to be perpendicular to the dy–dz plane, which is true without loss of generality. From (18.3), SAR_1 at a point can be calculated as follows [15]:

$$SAR_1 \equiv \frac{dP}{dm} = \frac{dP}{dV}\frac{dV}{dm} = \frac{dP}{dV}\frac{1}{\rho(\vec{x})} = \frac{\sigma(\vec{x})|\vec{E}(\vec{x})|^2}{\rho(\vec{x})} \ [\text{W/kg}] \tag{18.4}$$

where $\rho(\vec{x})$ is the density (kg/m^3).

SAR_2 for a volume is calculated by averaging (or integrating) over a specific volume V_0 (typically a 1 gram or 10 gram area) as follows [14]:

$$SAR_2 \equiv \frac{1}{V_0} \int\limits_{object} \frac{\sigma(\vec{x})|\vec{E}(\vec{x})|^2}{\rho(\vec{x})} dV = \int\limits_{object} \frac{\sigma(\vec{x})|\vec{E}(\vec{x})|^2}{\rho(\vec{x})} dV \ [\text{W/kg}] \ (\because V_0 = 1) \tag{18.5}$$

Question 5 Equation 18.5 is valid only when the conductivity is real.

1) Extend (18.5) to the case when the conductivity is a complex number.
2) Simplify (18.5) for the case of evenly distributed density, that is, $\rho(\vec{x}) = \rho_0$.

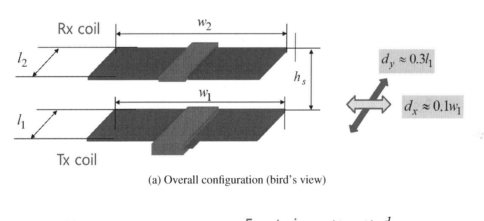

(a) Overall configuration (bird's view)

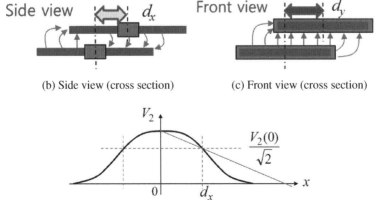

(b) Side view (cross section)   (c) Front view (cross section)

(d) Induced voltage at Rx coil and its 3dB point

Figure 18.22 Lateral tolerances of two identical Tx and Rx dipole coils.

18.5 Conclusion

The numbers of static wireless chargers are booming now, and lots of R&D works have been done, which results in standardization of IPT. Like other emerging technologies and markets, there will be no end of the story in an actively competing world. Instead, the stories will end because of lack of interest from the public. In other words, there is a lot of space left for newcomers to this world of static wireless chargers.

After a decade, we will definitely have been using smartphones that have not even been developed now. Who knows the future of wireless EV chargers? The coil types, compensation types, power levels, control/communication methods, even operating frequency may change. At least new wireless EV chargers will be studied and developed that are not necessarily a current standard, which is a consensus of best technologies in the past.

Problems

18.1 Find the lateral tolerances of identical Tx and Rx dipole coils under the conditions of $h_s \ll l_1, h_s \ll w_1$, as shown in Figure 18.22. You can use the magnetic resistance concept to calculate magnetic flux linkage. Note that the values given in Figure 18.22(a) are rough estimates and that you can find out exact numbers for them.

18.2 Find the lateral tolerances of asymmetric Tx and Rx dipole coils under the conditions of $h_s \ll l_1, h_s \ll w_1$, where the widths and lengths of Tx and Rx are different from each other.

References

1 G.A. Covic and J.T. Boys, "Modern trends in inductive power transfer for transportation applications," *IEEE Journal of Emerging and Selected Topics in Power Electronics*, vol. 1, no. 1, pp. 28–41, March 2013.

2 S.Y. Choi, S.Y. Jeong, E.S. Lee, B.W. Gu, S.W. Lee, and C.T. Rim, "Generalized models on self-decoupled dual pick-up coils for a large lateral tolerance," *IEEE Trans. on Power Electronics*, vol. 30, no. 11, pp. 6434–6445, November 2015.

3 Y.H. Son, B.H. Choi, E.S. Lee, G.C. Lim, G.-H. Cho, and C.T. Rim, "General unified analyses of two-capacitor inductive power transfer systems: equivalence of current-source SS and SP compensations," *IEEE Trans. on Power Electronics*, vol. 30, no. 11, pp. 6030–6045, November 2015.

4 Y.H. Son, B.H. Choi, G.-H. Cho, and C.T. Rim, "Gyrator-based analysis of resonant circuits in inductive power transfer systems," *IEEE Trans. on Power Electronics*, vol. 31, no. 10, pp. 6824–6843, October 2016.

5 C.T. Rim *et al.*, "Load-segmentation-based full bridge inverter and method for controlling same," US Patent Application 13/518,213, 2009.

6 C.T. Rim *et al.*, "Ultra slim power supply device and power acquisition device for electric vehicle," US Patent Application 13/262,879, 2010.

7 C.T. Rim *et al.*, "Power supply device, power acquisition device and safety system for electromagnetic induction-powered electric vehicle," US Patent Application 13/202,753, 2010.

8 C.T. Rim *et al.*, "Power supply apparatus for on-line electric vehicle, method for forming same and magnetic field cancelation apparatus," US Patent Application 13/501,691, 2010.

9 C.T. Rim *et al.*, "Modular electric-vehicle electricity supply device and electrical wire arrangement method," US Patent Application 13/510,218, 2010.

10 C.T. Rim *et al.*, "Electric vehicle systems," 10-2008-0135426, South Korea, 2010.

11 C.T. Rim *et al.*, "Power supply system and method for electric vehicle," 10-0944113, South Korea, 2010.

12 C.T. Rim *et al.*, "Method and device for designing a current supply and collection device for a transportation system using an electric vehicle," US Patent Application 13/810,066, 2011.

13 K.A. Grajski, R. Tseng, and C. Wheatley, "Loosely-coupled wireless power transfer: physics, circuits, standards," in *Microwave Workshop Series on Innovative Wireless Power Transmission: Technologies, Systems, and Applications (IMWS), 2012 IEEE MTT-S International*, 2012, pp. 9–14.

14 P. Bernardi, M. Cavagnaro, S. Pisa, and E. Piuzzi, "Specific absorption rate and temperature elevation in a subject exposed in the far-field of radio-frequency sources operating in the 10-900-MHz range," *IEEE Transactions on Biomedical Engineering*, vol. 50, no. 3, pp. 295–304, March 2003.

15 N. Firoozy and M. Shirazi, "Planar inverted-F antenna (PIFA) design dissection for cellular communication application." *Journal of Electromagnetic Analysis and Applications*, vol. 3, no. 10, p. 6, 2011, doi: 10.4236/jemaa.2011.310064.

19

Asymmetric Coils for Large Tolerance EV Chargers

19.1 Introduction

Battery charging systems play an important role in the commercialization of EVs. Conductive-type EV chargers, using AC or DC connectors, have been commercialized, but they are not widely welcome by the public because of their heavy and inconvenient charging cables. In order to resolve this problem, wireless EV chargers using an inductive power transfer system (IPTS), which typically includes a power supply coil set (or primary coil set) and a pick-up coil set (or secondary coil set), have been developed [1–39]. Among them, a circular loop coil type [1–11] or a rectangular loop coil type [12–17] is widely used because of its compact structure and low electromagnetic field (EMF) for pedestrians. However, the magnetic coupling between the coil sets rapidly degrades as the pick-up coil set is misaligned from the center of the power supply coil set as a penalty of the low EMF. Thus, the tolerance of horizontal displacement is typically only 10 cm [2, 4, 8, 12, 16], which is too narrow for an ordinary car driver to fit in. In principle, the tolerance can be enlarged by increasing the diameter of the coils, but this idea cannot be applicable in practice because of the limited space in the bottom of a car and increased EMF for pedestrians.

On the other hand, road-powered electrical vehicles (RPEVs) also need IPTSs of large lateral tolerance since they were introduced in the 1990s by the PATH team [18–21]. Various dynamic EV charging systems for RPEVs have been developed in a few institutes, including Bombardier and Auckland University [22–24]. Due to the inevitable deviation from the center of the driving path, the lateral tolerance for an RPEV should be larger than that of the stationary wireless EV charger. However, the lateral tolerance achieved by the PATH team is only about 10 cm [18], which is one of the reasons why the RPEV was not commercialized. Recently, significant improvement in lateral tolerance as well as high power efficiency, a large airgap, and cost reduction were achieved by the on-line electrical vehicles (OLEVs) [25–32]. A larger lateral displacement of 23 cm was achieved by using a double-sided pick-up coil [26], which consists of a rectangular core plate and a vertically wound coil. The pick-up coils for RPEVs have a rectangular shaped coil instead of a circular shaped one so it can maximize induced voltage when it sweeps along a longitudinal shaped primary coil [26–31]. The lateral tolerances achieved so far from the RPEVs including OLEVs, however, are not enough for the wireless stationary EV charging applications, which require a lateral tolerance of at least 30 cm.

A few double-sided coils have been developed for the stationary EV chargers [33–35], which have the same configuration of a power supply coil as that of a pick-up coil. Therefore, the size of a pick-up coil set is relatively large for the car but that of a power supply coil set is relatively small. The lateral tolerances are enhanced to 23 cm [33, 34] and 28 cm [35], respectively. However, the

Wireless Power Transfer for Electric Vehicles and Mobile Devices, First Edition. Chun T. Rim and Chris Mi.
© 2017 John Wiley & Sons Ltd. Published 2017 by John Wiley & Sons Ltd.

problem of self-heating by magnetic flux generated from the pick-up coil set and high EMF around the coil sets remained unresolved.

In this chapter, a new coil set of asymmetry configuration for a very large lateral tolerance, relative to existing systems, is proposed for applications to wireless stationary EV chargers. The width of a pick-up coil set is much smaller than that of a power supply coil set so that a larger lateral tolerance as well as a lower EMF for a given space in a car could be achieved. Thus, the proposed coil set satisfies the ICNIRP Guidelines [41] of the EMF. DofA is explained for the simulation of resonating coils of IPTS. The local saturation of ferrite core was effectively modeled and experimentally verified. A prototype IPTS including the proposed coil set was verified by both simulations and experiments.

Nomenclature

A_{eff}	Effective area that mutual flux passes through
d_{lat}	Lateral tolerance of pick-up coil displacement
d_{long}	Longitudinal tolerance of pick-up coil displacement
D_{lat}	Lateral distance from the end of pick-up to a passenger
W_c	Lateral available space at the bottom of a car
L_c	Longitudinal available space at the bottom of a car
α	Effective area ratio of a pick-up coil
β	Ratio of lateral displacement to a half of the width of power supply cover core plate
γ	Ratio of lateral displacement to a half of the length of power supply cover core plate
n	Turn ratio of power supply and pick-up coil sets
N_1	Number of turns of power supply coil set
N_2	Number of turns of pick-up coil set
w_1	Width of power supply bottom core plate
w_2	Width of pick-up roof core plate
w_{p1}	Width of power supply core pillar
w_{p2}	Width of pick-up core pillar
w_{c1}	Width of power supply cover core plate
w_{c2}	Width of pick-up cover core plate
w_{t1}	Tip width of power supply cover core plate
w_{t2}	Tip width of pick-up cover core plate
l_1	Length of power supply bottom core plate
l_2	Length of pick-up roof core plate
l_d	Distance between two power supply cover core plates
l_{p1}	Length of power supply core pillar
l_{p2}	Length of pick-up core pillar
l_{c1}	Length of power supply cover core plate
l_{c2}	Length of pick-up cover core plate
l_{t1}	Tip length of power supply cover core plate
l_{t2}	Tip length of pick-up cover core plate
t_1	Thickness of power supply bottom core plate
t_2	Thickness of pick-up roof core plate
t_{c1}	Thickness of power supply cover core plate
t_{c2}	Thickness of pick-up cover core plate
h	Airgap between power supply and pick-up coil sets
h_1	Height of power supply core pillar

h_2 Height of pick-up core pillar
L_m Magnetizing inductance of power supply coil set
L_{l1} Leakage inductance of power supply coil set
L_{l2} Leakage inductance of pick-up coil set
C_1 Compensation capacitance of power coil set
C_2 Compensation capacitance of pick-up coil set
R_L Load resistance
R_h Hysteresis resistance
I_s Power supply coil current (rms)
I_o Pick-up coil current (rms).
f_s Switching frequency
B_t Total magnetic flux density induced from power supply coil set and pick-up coil set
B_1 Magnetic flux density induced from power supply coil set
B_2 Magnetic flux density induced from pick-up coil set

19.2 Design of Proposed IPTS for EV Chargers

19.2.1 Design Considerations of the Proposed IPTS

The proposed IPTS is composed of a power supply coil set, a pick-up coil set, an inverter with its rectifier, resonant capacitors, and a load resistor, as shown in Figure 19.1. A coil set is composed of two power cables, two core pillars, two cover core plates, and a bottom core plate for a power supply coil set (or a roof core plate for a pick-up coil set), as shown in Figures 19.2 and 19.3. The two power cables of each coil set constitute a magnetic path, and the current direction of a power cable is the opposite of another one so that they can strengthen the magnetic flux of each other, as shown in Figure 19.2. Each coil has a magnetic cover, as shown in Figures 19.2 and 19.3, which increases the effective magnetic plate area so that magnetic resistance can be consequently lowered [31].

The coil sets should be designed so that they have a large enough lateral tolerance with an appropriately low EMF and insensitivity to airgap displacement, maintaining no core saturation

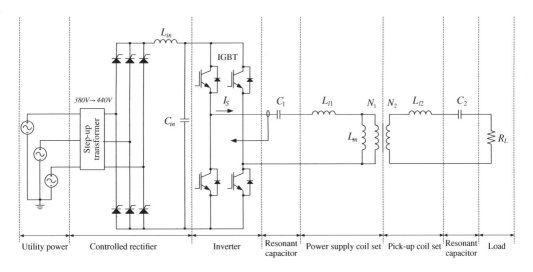

Figure 19.1 Overall circuit configuration of the proposed IPTS.

Pick-up coil set

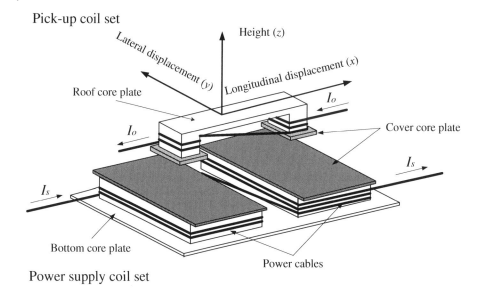

Figure 19.2 Bird's eye view of the power supply coil set and pick-up coil set.

with high power efficiency. A list of requirements for a practical stationary EV charger to fulfill these characteristics is suggested in Table 19.1, considering accommodation in an ordinary passenger car, whose available bottom space ($W_c \times L_c$) is assumed to be 200 cm in width and 100 cm long. The nominal airgap, typically as low as 10 cm for conventional EV chargers [2–4, 7, 14, 17],

Figure 19.3 Design parameters of the proposed IPTS.

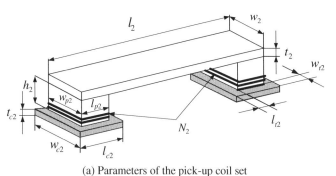

(a) Parameters of the pick-up coil set

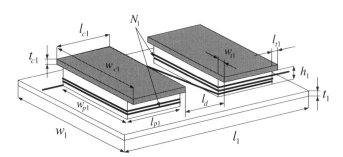

(b) Parameters of the power supply coil set

Table 19.1 Requirements of IPTS for a practical wireless stationary EV charger

Requirements	Proposed specifications	Remarks
Airgap (z axis)	15 cm	
Longitudinal tolerance (x-axis)	±20 cm	
Lateral tolerance (y-axis)	±40 cm	
Height tolerance (z-axis)	±5 cm	
EMF ($y = 100$ cm)	6 µT	ICNIRP Guidelines: 6.25 µT
Pick-up coil set	Must be small and light	No regulation
Power supply coil set	No limitations in size and weight	No regulation

has been increased to 15 cm in this chapter for better flexibility of a height change of a car with the virtue of the proposed coil configuration. This airgap requirement is quite challenging for the conventional EV chargers. The airgap tolerance is accordingly increased to ±5 cm because the proposed configuration of the coil sets is relatively insensitive to height change.

As shown in Figures 19.2 and 19.3, the width of a pick-up coil set is much smaller than that of a power supply coil; hence, the magnetic coupling is widely maintained for lateral displacement, as shown in Figure 19.4. Furthermore, the fringe effect becomes large for this asymmetric configuration of coils [31]. As shown in Figures 19.4 and 19.5, the allowable displacement of a pick-up coil is between the territory of a power supply coil set and a pick-up coil set; therefore, the approximated lateral and longitudinal tolerances are found as follows:

$$\frac{w_{c1}}{2} - \frac{w_{c2}}{2} \leq d_{lat} \leq \frac{w_{c1}}{2} + \frac{w_{c2}}{2} \tag{19.1}$$

$$\frac{l_{c1}}{2} - \frac{l_{c2}}{2} \leq d_{long} \leq \frac{l_{c1}}{2} + \frac{l_{c2}}{2} \tag{19.2}$$

The detailed verification of the lateral and longitudinal tolerances will be made in the subsequent sections. As identified from (19.1) and (19.2), the tolerances can be as large as possible with a large size power supply coil set and a small size pick-up coil set. This is why the proposed asymmetric coil structure, where the size of a pick-up coil set is much smaller than that of a power supply coil set, has very large tolerances. The tolerances for this asymmetric coil configuration become insensitive to the pick-up size as the ratio of two coil sizes increases, and can be further simplified as follows:

$$d_{lat} \cong \frac{w_{c1}}{2} \quad \text{for} \quad w_{c2} \ll w_{c1} \tag{19.3}$$

$$d_{long} \cong \frac{l_{c1}}{2} \quad \text{for} \quad l_{c2} \ll l_{c1} \tag{19.4}$$

The large size of the power supply coil set, however, leads to high voltage stress due to its high inductance. Therefore, a mechanical stopper, as shown in Figure 19.6(a), can be used to mitigate the longitudinal tolerance requirement, which results in the small size of the power supply coil set and its low inductance. In this way, the longitudinal displacement can be completely eliminated regardless of the wheel diameter, that is, $d_{long} = 0$; however, the requirement of it is given to be 20 cm in this chapter to show the versatile capability of the proposed configuration.

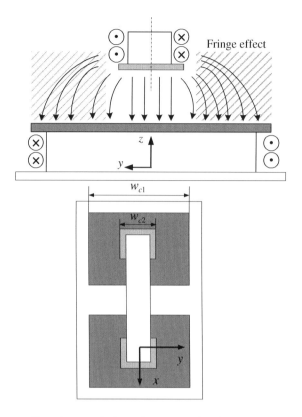

(a) Magnetic coupling characteristic at the center of a pick-up coil: front view (left) and plane view (right)

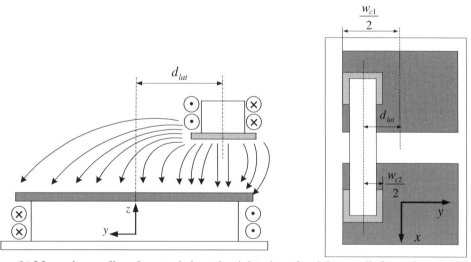

(b) Magnetic coupling characteristic at the right edge of a pick-up coil: front view (left) and plane view (right)

Figure 19.4 Proposed asymmetric configuration of the IPTS showing uniform magnetic coupling characteristics over lateral displacement.

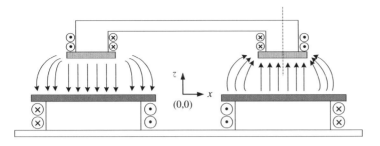

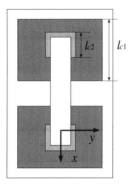

(a) Magnetic coupling characteristic at the center of a pick-up coil: side view (up) and plane view (down)

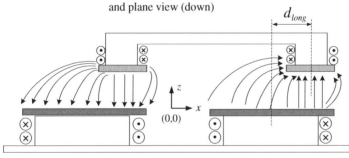

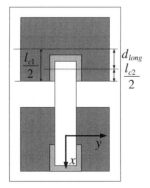

(b) Magnetic coupling characteristic at the right edge of a pick-up coil: side view (up) and plane view (down)

Figure 19.5 Proposed asymmetric configuration of the IPTS showing uniform magnetic coupling characteristics over longitudinal displacement.

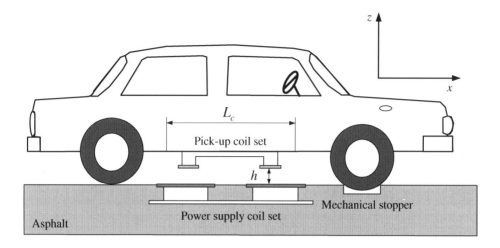

(a) The side view, where a mechanical stopper is used to eliminate longitudinal misalignment.

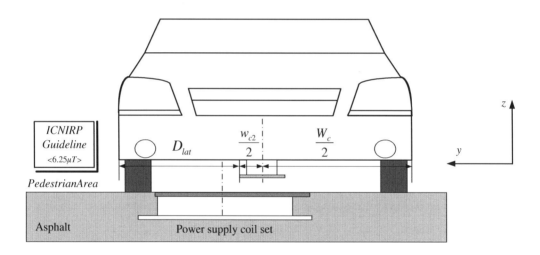

(b) The front view, showing EMF concern for an unattended misalignment.

Figure 19.6 Deployment of the proposed IPTS for a passenger car.

The EMF is drastically reduced by the proposed asymmetric coil sets, because the coil sets have alternating magnetic flux through adjacent poles to cancel each other [30, 31]. The EMF of an IPTS is generated from both the power supply and pick-up coil sets, where the EMF of a pick-up coil set is much larger than that of a power supply coil set for a heavy load power [29]. Therefore, the EMF consideration for a pick-up set is quite important, which is mainly determined by the ampere·turns of the pick-up coil $I_o N_2$ and the distance D_{lat} for a pedestrian determined as follows:

$$D_{lat} = \frac{W_c - w_{c2}}{2} \left(\cong \frac{W_c}{2} \quad \text{for} \quad w_{c2} \ll W_c \right) \tag{19.5}$$

As identified from (19.5), D_{lat} is maximized for a small size pick-up, which results in minimized EMF. Because a magnetic circuit is established for the longitudinal direction, magnetic flux leakage is significantly mitigated in this way.

19.2.2 Dimensions of Power Supply Coil and Pick-up Coil Sets

The design of the coil sets is mainly focused on the dimensions, number of turns, and allowable current levels of them in this chapter, whereas other conventional issues are less addressed. The design is primarily verified by the Ansoft Maxwell 3D (ver. 14) finite element analysis (FEA) simulations and then verified by experiments for the designed coil sets in the next section.

First of all, the width and length of power supply coil w_{c1} and l_{c1} determined from (19.3) and (19.4), respectively, are as follows:

$$w_{c1} \cong 2d_{lat} \tag{19.6}$$
$$l_{c1} \cong 2d_{long} \tag{19.7}$$

From Table 19.1, d_{lat} and d_{long} are 40 cm and 20 cm, respectively; therefore, w_{c1} and l_{c1} become 80 cm and 40 cm, respectively. As identified from (19.3) and (19.4), (19.6) and (19.7) are valid only when $w_{c2} \ll w_{c1}$ and $l_{c2} \ll l_{c1}$, respectively; hence, w_{c2} and l_{c2} should be much smaller than 80 cm and 40 cm, respectively. In this chapter, w_{c2} and l_{c2} are selected as 30 cm and 20 cm, respectively, considering implementation difficulties of the pick-up size, which can be further reduced for a compact design if required. In this chapter, the optimum design of a pick-up size is not deeply considered, which is left for further work.

The distance between two power supply cover core plates l_d should not be too large in order to reduce the lengths of the coil sets; however, it should not be too small in order to avoid large self-circulation of magnetic flux of the power supply coils. Therefore, an optimum distance can be found by an appropriated trade-off for given coil sizes and airgap, which is determined as 20 cm in this chapter, considering similar physical constraints of a previous work [31]. Thus l_d can be further reduced for a lower airgap because of stronger magnetic coupling in this case. Other dimensions of the coil sets are less sensitive to the system performance so far as they are not concerned with core saturation. For example, l_1, l_2, w_1, and w_2 are appropriately selected to fit in the above parameters. The heights of coils h_1 and h_2 are also not critical to the system performances and are determined as 10 cm, considering cable harness in this chapter; however, they can be drastically reduced to 1 cm, for example, as desired. Other miscellaneous parameters such as l_{p1}, l_{p2}, w_{p1}, w_{p2}, l_{t1}, l_{t2}, w_{t1}, and w_{t2} are accordingly chosen to accommodate previously determined dimensions.

The selected design parameters including t_1, t_{c1}, t_2, and t_{c2}, which will be determined by simulations, are summarized in Table 19.2. The switching frequency is selected as 20 kHz for the coherence of previous related works [25–32] for comparison purposes; however, it can be further increased for reducing coil sizes as long as core loss is not severe.

19.2.3 Dominant Field Analysis (DoFA) Method for Magnetic Flux Simulation in Complex Vector Domain

The total magnetic flux density $\boldsymbol{B_t}$, represented as a vector of phasor form, cannot be determined by conventional DC current simulations because it is induced from both the current of the power supply coil set and the pick-up coil set, which are quadrature in phase with each other as long as in the fully resonant condition [28–31]. The compensation capacitors C_1 and C_2 are added in series with the power supply and pick-up coil sets, respectively, and the C_2 completely

Table 19.2 Design parameters of the proposed IPTS (nominal)

Parameter	Value	Parameter	Value
I_s	30 A	l_1	110 cm
f_s	20 kHz	l_2	70 cm
n	5	l_{p1}	30 cm
N_1	3 turns	l_{p2}	10 cm
N_2	15 turns	l_{c1}	40 cm
w_1	90 cm	l_{c2}	20 cm
w_2	20 cm	l_{t1}	5 cm
w_{p1}	70 cm	l_{t2}	5 cm
w_{p2}	20 cm	t_1	2 cm
w_{c1}	80 cm	t_2	3 cm
w_{c2}	30 cm	t_{c1}	2 cm
w_{t1}	5 cm	t_{c2}	3 cm
w_{t2}	5 cm	h_1	10 cm
l_d	30 cm	h_2	10 cm

offsets not only the secondary leakage inductance L_{l2} but also the magnetizing inductance L_m. Though it is possible to analyze it in the time domain by inserting two sinusoid currents in quadrature, it is not recommended in practice because it consumes much simulation time and, moreover, it is hardly possible to present the time-varying three-dimensional vector results in the two-dimensional domain. Instead, the magnitude of $\boldsymbol{B}_t$, that is, $|\boldsymbol{B}_t|$ is examined, which can easily be calculated using the proposed new method in this chapter, which uses conventional DC magnetic flux densities for the power supply coil set and the pick-up coil set. With an unsaturated core, $\boldsymbol{B}_t$ can be determined from $\boldsymbol{B}_1$ and $\boldsymbol{B}_2$ of complex vector form by applying the superposition theorem to the Cartesian coordinate as follows:

$$\boldsymbol{B}_t = \boldsymbol{B}_1 + \boldsymbol{B}_2 \tag{19.8}$$

where

$$\boldsymbol{B}_t \equiv B_{tx}\boldsymbol{x}_0 + B_{ty}\boldsymbol{y}_0 + B_{tz}\boldsymbol{z}_0 \tag{19.9a}$$

$$\boldsymbol{B}_1 \equiv B_{1x}\boldsymbol{x}_0 + B_{1y}\boldsymbol{y}_0 + B_{1z}\boldsymbol{z}_0 \tag{19.9b}$$

$$\boldsymbol{B}_2 \equiv B_{2x}\boldsymbol{x}_0 + B_{2y}\boldsymbol{y}_0 + B_{2z}\boldsymbol{z}_0 \tag{19.9c}$$

In (19.9), the coefficients are of a complex scalar form and $\boldsymbol{x}_0$, $\boldsymbol{y}_0$, $\boldsymbol{z}_0$ are unit vectors. The magnitude of $\boldsymbol{B}_t$ is then found as follows:

$$
\begin{aligned}
|\boldsymbol{B}_t|^2 &= |\boldsymbol{B}_1 + \boldsymbol{B}_2|^2 = |B_{1x} + B_{2x}|^2 + |B_{1y} + B_{2y}|^2 + |B_{1z} + B_{2z}|^2 \\
&= |B_{1x}|^2 + |B_{2x}|^2 + |B_{1y}|^2 + |B_{2y}|^2 + |B_{1z}|^2 + |B_{2z}|^2 \\
&= |B_{1x}|^2 + |B_{1y}|^2 + |B_{1z}|^2 + |B_{2x}|^2 + |B_{2y}|^2 + |B_{2z}|^2 \\
&= |\boldsymbol{B}_1|^2 + |\boldsymbol{B}_2|^2
\end{aligned}
\tag{19.10}
$$

where the following orthogonal conditions for the quadrant phasor are used:

$$B_{1x} \perp B_{2x}, \quad B_{1y} \perp B_{2y}, \quad B_{1z} \perp B_{2z} \tag{19.11}$$

It is noteworthy from (19.10) that the magnetic flux analysis for any resonant IPTS, so far as its filters are tuned to resonant frequency and contain no saturated core, can be done by conventional DC magnetic flux analyses for power supply coils and pick-up coils. Furthermore, this principle can be generally applicable to a coupled magnetic resonant system (CMRS), where each coil has a quadrature phase against adjacent coils [40]. It is also identified from (19.11) that there is no way of reducing EMF generated from one coil by using another coil that has a quadrature phase. Therefore, EMF canceling must be performed independently for each coil.

From (19.10), the magnitude can be determined as follows:

$$B_t \equiv |\boldsymbol{B}_t| = \sqrt{|\boldsymbol{B}_1|^2 + |\boldsymbol{B}_2|^2} = \sqrt{B_1^2 + B_2^2} \quad \because B_1 \equiv |\boldsymbol{B}_1|, B_2 \equiv |\boldsymbol{B}_2| \tag{19.12}$$

For the case that the magnetic flux density induced from a coil is larger than that of another coil, (19.12) can be approximated as follows:

$$B_t = \sqrt{B_1^2 + B_2^2} = \sqrt{B_1^2 \left(1 + B_2^2/B_1^2\right)} \cong B_1 \left(1 + 0.5 B_2^2/B_1^2\right) \quad \text{for} \quad B_2 \ll B_1 \tag{19.13}$$

For example, if $B_2/B_1 = 0.5$, then B_t becomes nearly the same as B_1, that is, $B_t \cong 1.12 B_1 (\cong B_1)$. This means that the design of the power supply coil set and that of the pick-up coil set can be separated from each other if the magnetic flux contribution of one to another is less than half. In other words, each coil can be designed independently considering dominant magnetic flux, so far as it does not saturate any core. Therefore, it is worthy to have a name for this method, the *dominant field analysis* (DoFA), which is applicable to a time-varying magnetic field of vector form as well as an electric field as long as the two fields are orthogonal phasors of each other.

With the DoFA, it is possible to analyze the magnetic flux density of one coil set in the case when the other coil set is inactive, that is, the current is zero. By the FEA simulation, as shown in Figure 19.7, the magnetic flux density distribution for inside and around the core can be determined for the current of the power supply coil set only, where the current of the pick-up coil set is zero. The analysis for the pick-up core will be done in the subsequent section. It is identified from Figure 19.7 that the proposed design provides strong magnetic coupling between two coil sets and rapidly decreasing EMF characteristics for pedestrians.

Question 1 What happens to (19.12) and (19.13) if (19.11) is no longer valid?

19.2.4 Lateral and Longitudinal Tolerances

The lateral and longitudinal tolerances can be defined as the marginal displacements where magnetic flux density in the pick-up core decreases to 3 dB as the pick-up moves from the center position. As verified by the FEA simulation, the magnetic flux density in the pick-up core along the lateral and longitudinal displacements reaches 3 dB, as shown in Figure 19.8, roughly at half of w_{c1} or l_{c1}, as predicted in (19.3) and (19.4). Furthermore, the tolerances are insensitive to airgap variation, which was not easily achievable by previous works. Therefore, it is expected that half of the maximum output power of the proposed IPTS is guaranteed within d_{lat} and d_{long}.

If desired, the lateral and longitudinal tolerances of the proposed coil set can be further improved by widening the power supply coil since there is no considerable constraint on the size of the power supply, which is buried under the ground.

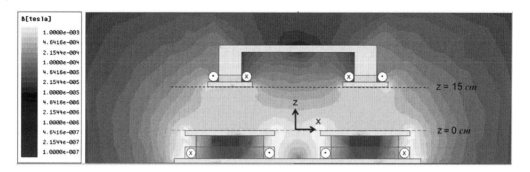

(a) Magnetic flux density profile (side view)

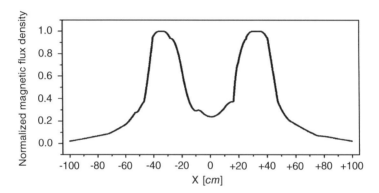

(b) Normalized magnetic flux density at z = 15 cm

Figure 19.7 Simulated magnetic flux density distribution of the proposed IPTS, indicating strong magnetic coupling between coils.

19.2.5 Mitigation of Airgap Variation Effects

As discussed in the previous section, the proposed coil sets were designed to maximize the effective area of a core plate A_{eff} [31] by the fringe effect, as shown in Figure 19.4, for increasing mutual magnetic flux. Thus, the output voltage of the proposed IPTS can be less sensitive to airgap variations. By neglecting the magnetic resistance of the ferrite core whose relative permeability μ_r is over 2000, the total magnetic resistance of the coil sets can be approximated to that of the airgap as follows:

$$\mathfrak{R} \approx \frac{2h}{\mu_o A_{eff}}, \tag{19.14}$$

where A_{eff} can be defined as a function of the pick-up size [31], as shown by

$$A_{eff} \equiv \alpha l_{c2} w_{c2} \tag{19.15}$$

Then the magnetizing inductance can be calculated as follows:

$$L_m \cong \frac{N_1^{\,2}}{\mathfrak{R}} \cong \frac{N_1^{\,2} \mu_o \alpha l_{c2} w_{c2}}{2h} \tag{19.16}$$

Figure 19.8 FEA simulation results of magnetic flux density in the pick-up core for the parameters in Table 19.2.

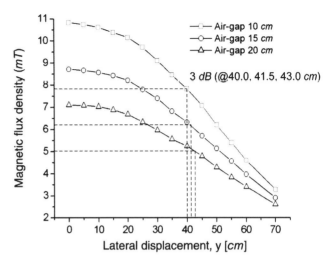

(a) Magnetic flux density versus lateral displacement

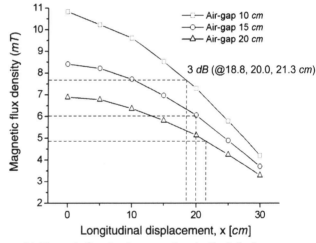

(b) Magnetic flux density versus longitudinal displacement

From the FEA simulations for the proposed coil sets of the parameters in Table 19.2, the magnetizing inductance is found to be decreasing as the airgap increases, as shown in Figure 19.9(a); however, the effective area ratio α, which is calculated from (19.16) for the obtained L_m from the simulation, is increasing as the airgap increases up to 20 cm, as shown in Figure 19.9(b). This means that the proposed IPTS is relatively insensitive to airgap variations where it is less than 20 cm. Different from (19.16), where L_m is apparently proportional to the inverse of h, the magnetic linkage is roughly proportional to the inverse of $\sqrt{h}$ due to the fringe effect, as identified from Figure 19.9.

The output voltage variation due to the airgap change still remains, even though it has been mitigated, but it can be effectively managed by a regulator attached to the load, which is not shown in Figure 19.1.

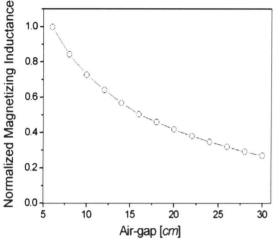

(a) Normalized magnetizing inductance versus airgap at the center

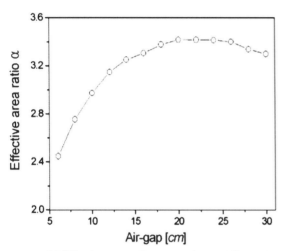

(b) Effective area ratio versus airgap at the center

Figure 19.9 FEA simulation results of normalized magnetizing inductance and effective area ratio for the parameters in Table 19.2.

19.2.6 Prevention of Partial Saturation

The proposed IPTS adopts a fully resonant mode [31], where the output power drastically decreases [29, 31] if the resonant frequency of the pick-up deviates from the switching frequency. In order to prevent this variation of the resonant frequency due to partial saturation of the ferrite core, which is caused by excessive source current I_s or output current I_o, a large enough thickness of core should be used. As the current increases, the magnetic flux density inside the ferrite core is increased until the saturation value of the ferrite core, 0.3 T, is reached. Partial saturation occurs locally where the magnetic flux is concentrated, as shown in Figure 19.10, where the pick-up coil set only is shown for different output current levels and the source current of the power supply coil set is made to be zero. In order to avoid the partial saturation, it is enough to examine this pick-up coil set case according to the proposed principle discussed in Section 19.2.3. The power supply coil set is relatively large and its core plates can

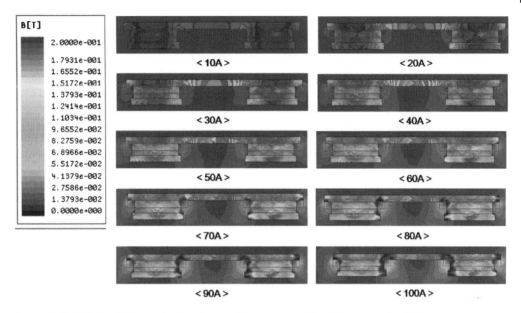

Figure 19.10 FEA simulation results showing partial saturation of the pick-up core for different output currents I_o, where $t_{c2} = 3$ cm and $t_2 = 3$ cm.

be as thick as needed, with the magnetic flux due to itself kept relatively unvaried; hence, the main focus in this chapter is on the pick-up coil set design.

It has been identified from the FEA simulations for the parameters in Table 19.2 that the power supply core begins to be partially saturated, though not shown here, when the source current exceeds 100 A for $N_1 = 3$. In the pick-up core, which is of great concern, partial saturation starts when the output current exceeds roughly 60 A for $N_1 = 15$, as shown in Figure 19.10. The number of turns can be changed within these ampere-turns. Based on these simulations, though all of them are not shown in the figure, the optimum thicknesses of the power supply and pick-up core plates were determined as $t_1 = t_{c1} = 2$ cm and $t_2 = t_{c2} = 3$ cm, respectively.

19.2.7 A New Hysteresis Loss Model

The proposed power supply coil set is driven by a constant current controlled inverter [28–31], where the inverter is operating at the switching angular frequency $\omega_s = 2\pi f_s$, which is slightly higher than the resonant angular frequency of the power supply coil. Because of this frequency discrepancy, the impedance of IPTS is always inductive and thus the zero current switching (ZCS) operation of the inverter, as shown in Figure 19.1, is always guaranteed, where a small snubber capacitor across each switch is not shown for simplicity. The partial core saturation discussed in the previous section can be modeled as an equivalent hysteresis resistance R_h, as shown in Figure 19.11, in the same way of a conventional transformer case.

The supply current I_S in the steady state can be determined of the phasor form, neglecting the effects of the resistance of the power supply coil r_1 and R_h at the moment, as follows:

$$I_S = \frac{V_S - V_{Lm}}{r_1 + j\left(\omega_s L_{l1} - \dfrac{1}{\omega_s C_1}\right)} \cong \frac{V_S - V_{Lm}}{j\left(\omega_s L_{l1} - \dfrac{1}{\omega_s C_1}\right)} = \frac{\Delta V_{in}}{j\omega_s \Delta L} \quad \because r_1 \ll \omega_s \Delta L \qquad (19.17)$$

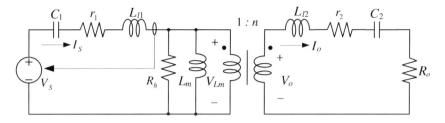

(a) An equivalent circuit including hysteresis loss resistance R_h

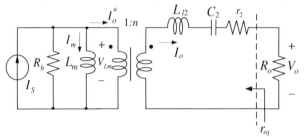

(b) Simplified circuit, assuming constant current control

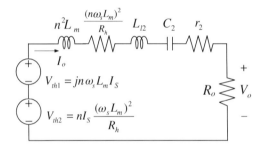

(c) More simplified circuit at the output side

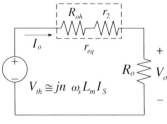

(d) Approximated final circuit

Figure 19.11 The equivalent circuit of the proposed IPTS under constant current control, including hysteresis loss and conduction loss.

where

$$\Delta L \equiv L_{l1} - \frac{1}{\omega_s^2 C_1}, \quad \Delta V_{in} \equiv V_S - V_{Lm} \tag{19.18}$$

For a constant $|\Delta V_{in}|$, as identified from (19.17), V_S should be changed in accordance with V_{Lm}, which is changed by the output current I_o as well as the variation of L_m, for a constant current source IPTS. The current I_S is appropriately regulated by the inverter such that it can be regarded as a constant current source. From Figure 19.11, the open-circuit voltage can be determined, assuming that $R_h \gg \omega_s L_m$, as a function of the input current I_S, magnetizing inductance L_m, switching angular frequency ω_s, and turn ratio $n = N_2/N_1$, as follows:

$$V_{th} = nI_S(R_h//j\omega_s L_m) = nI_S \frac{j\omega_s L_m R_h}{R_h + j\omega_s L_m} = nI_S \frac{j\omega_s L_m R_h}{R_h\left(1 + \frac{j\omega_s L_m}{R_h}\right)} \cong jn\omega_s L_m I_S \left(1 \quad \frac{j\omega_s L_m}{R_h}\right)$$

$$= jn\omega_s L_m I_S + nI_S \frac{(\omega_s L_m)^2}{R_h} \quad \because V_{th1} \equiv jn\omega_s L_m I_S, \quad V_{th2} \equiv nI_S \frac{(\omega_s L_m)^2}{R_h} \tag{19.19}$$

As shown in Figure 19.11(d), the contribution of V_{th2} to the output voltage or current compared to that of V_{th1} in (19.19) is negligible so far as $\omega_s L_m/R_h \ll 1$, because the square of it is quite small as follows:

$$|V_{th}| = n\omega_s L_m I_S \sqrt{1 + \left(\frac{\omega_s L_m}{R_h}\right)^2} \cong n\omega_s L_m I_S \left[1 + \frac{1}{2}\left(\frac{\omega_s L_m}{R_h}\right)^2\right]$$

$$\cong n\omega_s L_m I_S = |V_{th1}| \quad \text{for} \quad R_h \gg \omega_s L_m \tag{19.20}$$

In addition, the following Thevenin equivalent resistance from the output side can be obtained as shown in Figure 11(c) and (d).

$$Z_{th} = n^2(R_h//j\omega_s L_m) + j\omega_s L_{l2} + \frac{1}{j\omega_s C_2} + r_2 \cong jn^2\omega_s L_m + \frac{(n\omega_s L_m)^2}{R_h} + j\omega_s L_{l2} + \frac{1}{j\omega_s C_2} + r_2$$

$$= R_{oh} + r_2 \quad \because R_{oh} \equiv \frac{(n\omega_s L_m)^2}{R_h}, \quad jn^2\omega_s L_m + j\omega_s L_{l2} + \frac{1}{j\omega_s C_2} = 0 \tag{19.21}$$

From Figure 19.11(b), the power loss P_h of the equivalent core loss resistance R_h can also be determined as follows:

$$P_h = \frac{|V_{Lm}|^2}{R_h} \cong \frac{|\omega_s L_m(I_S - I_o^*)|^2}{R_h} = \frac{(\omega_s L_m)^2}{R_h}(|I_S|^2 + |nI_o|^2) = \frac{(\omega_s L_m|I_S|)^2}{R_h}\left(1 + \frac{n^2|I_o|^2}{|I_S|^2}\right)$$

$$\approx \frac{(n\omega_s L_m|I_o|)^2}{R_h} \equiv |I_o|^2 R_{oh} \quad \text{for} \quad \frac{n^2|I_o|^2}{|I_S|^2} \gg 1 \tag{19.22}$$

The quadrature relationship between phasors $I_S \perp I_o$, which is the result of resonance [31], as well as a heavy load condition were used in (19.22), for which the hysteresis loss becomes dominant. It is identified from (19.21) and (19.22) that the hysteresis loss can be represented as a series resistance at the output side because the power loss is proportional to the square of the output current. Therefore, the equivalent resistance r_{eq} of Figure 19.11(d) is found to be the sum of the hysteresis term and conduction loss term as follows:

$$r_{eq} = R_{oh} + r_2 \tag{19.23}$$

As identified from (19.21) to (19.23), the higher the hysteresis loss, the smaller the hysteresis resistance, which results in the larger equivalent series resistance at the output side. The feasibility of this hysteresis loss model will be verified in the subsequent section by experiments.

Question 2 In Figure 19.11, it is assumed that an equivalent current source is driven by a voltage source, as shown in Figure 19.11(a). (1) It could be true in the steady state, but is it still valid for the transient state? (2) How about the case where the Laplace transfer function of the current is known as a function of the voltage source? Is it a current source or a voltage source circuit?

19.3 Example Design and Experimental Verifications

The proposed coil sets for experiments, allowing large lateral and longitudinal displacements as well as robustness to airgap displacement, is shown in Figure 19.12. The height and weight of the pick-up coil set are 16 cm and 54 kg, respectively, which seem to be higher and heavier than other pick-ups. However, the thickness and weight of the pick-up coil set significantly depend on the output power level of the pick-up coil set. Through the optimal design process of the pick-up coil set, which not dealt with here, it is expected that the height and weight of the pick-up set can be drastically reduced for the same output power by using super core plates. All the experimental parameters are the same as the simulation parameters listed in Table 19.2. In order to verify the design of the proposed IPTS, the open-circuit output voltage along the airgap, lateral displacement, and longitudinal displacement were measured.

19.3.1 Open-Circuit Output Voltage

Figure 19.13 shows the open-circuit output voltage along the airgap, which corresponds to the theoretical value of (19.20) for a no hysteresis loss case. Although the airgap was changed by

(a) Pick-up coil set (side view)

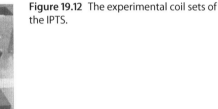

Figure 19.12 The experimental coil sets of the IPTS.

(b) Power supply coil set (front view)

(c) Both power supply and pick-up coil sets (front view)

Figure 19.13 Output voltage characteristics along the airgap.

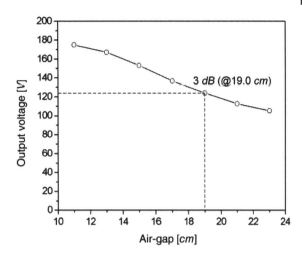

approximately 54%, from 11 cm to 17 cm, the output voltage was reduced by only about 22%, from 175 V to 137 V, due to the increase in the effective area.

The output voltages along the longitudinal and lateral displacement were also measured at the airgap of 15 cm. When the pick-up coil laterally deviated from the center of the power supply coil by 39 cm, the output voltage was reduced by only about 29% from the maximum output voltage 150 V, as shown in Figure 19.14. This large marginal lateral displacement confirms the designed lateral tolerance of (19.3), which was 40 cm.

The output voltage along the longitudinal displacement is shown in Figure 19.15. The output voltage was reduced by 3 dB when the pick-up deviated from the center of the power supply coil by 18 cm. This large marginal longitudinal displacement also confirms the designed longitudinal tolerance of (19.4), which was 20 cm.

The output voltage and power were measured along the output current when $I_S = 100$ A, $N_1 = 3$, and $h = 15$ cm, as shown in Figure 19.16. The source current can be made small for a large number of turns, that is, $I_S = 20$ A, $N_1 = 15$, which may be appropriately designed [30,31].

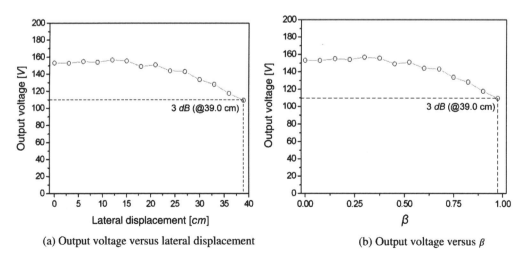

(a) Output voltage versus lateral displacement

(b) Output voltage versus β

Figure 19.14 Output voltage characteristics along the lateral displacement.

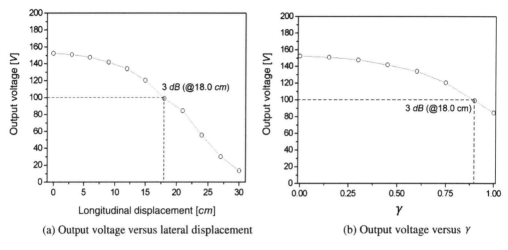

(a) Output voltage versus lateral displacement   (b) Output voltage versus γ

Figure 19.15 Output voltage characteristics along the longitudinal.

Because the output voltage is nearly constant regardless of the output current, which it is for this type of IPTS [27–32], the output power is nearly proportional to the output current, as shown in Figure 19.16(b). The maximum output power was obtained as 15.6 kW for the output current $I_o = 60$ A, where partial core saturation becomes dominant; hence, the power efficiency was as low as 75%, which can be much improved if the output current level is low and a thick cable is used, as can be identified from (19.22) and (19.23).

19.3.2 Equivalent Output Resistance

From the experimental results of the output voltage, as shown in Figure 19.16, a slight voltage drop of linear form is observed. The extracted resistance from the measured I–V curve, as shown in Figure 19.17, is found to be much greater than the internal resistance of the output coil set, which was precisely measured by an LCR meter ($r_2 = 0.178 \ \Omega$).

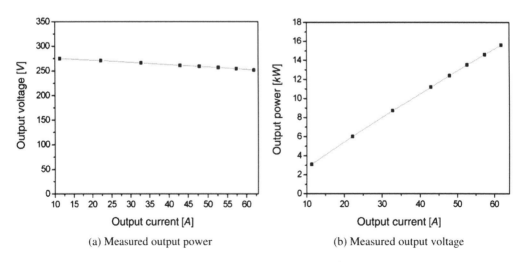

(a) Measured output power   (b) Measured output voltage

Figure 19.16 Output voltage and power characteristics along the output current.

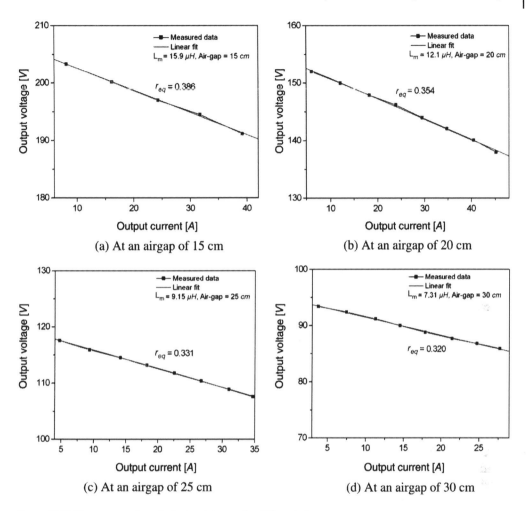

Figure 19.17 The measured equivalent resistances for different airgaps.

From the equivalent circuit of Figure 19.11 and (19.23), the difference between the extracted and measured resistances is attributed to the hysteresis resistance. As shown in Figure 19.17, the equivalent resistance decreases as the airgap is increased, which means that the hysteresis loss is mitigated for higher airgaps.

From (19.21) and (19.23), the hysteresis resistance R_h, as shown in Figure 19.18, can be obtained as follows:

$$R_h = \frac{(n\omega_s L_m)^2}{R_{oh}} = \frac{(n\omega_s L_m)^2}{r_{eq} - r_2} \tag{19.24}$$

In (19.24), L_m was measured for each airgap, as shown in Figure 19.19, and r_{eq} was measured, as shown in Figure 19.17. For a larger airgap, the hysteresis resistance R_h is decreased, as shown in Figure 19.18; however, this does not mean that the increase of hysteresis loss is due to the decreased L_m. From Figures 19.18 and 19.19, it is identified that the assumption $R_h \gg \omega_s L_m$, used in (19.19) to (19.22), is quite reasonable.

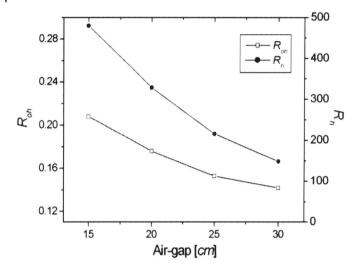

Figure 19.18 The hysteresis loss characteristics along the airgap.

19.3.3 EMF

The EMF generated from the power supply coil set, where the current of the pick-up coil set is zero, is considerably reduced because of the alternating magnetic fluxes of the power supply coils, as discussed. The supply current I_s and the number of a power supply coil set N_1 are 30 A and 3 turns, respectively, which is the normal charging mode. The EMF was measured as low as 6.1 µT at a distance of 1 m from the center of the power supply coil, as shown in Figure 19.20. The ICNIRP Guidelines of 6.25 µT at 20 kHz, therefore, is satisfied. According to the proposed DoFA principle, the EMF generated from the pick-up coil set, which is linearly proportional to the output current, can be separately determined from that of the power supply coil set. The EMF of a pick-up can be effectively mitigated by using an active canceling coil set [29], though this is not introduced in this chapter. An additional EMF cancel coil of the power supply coil may be required to meet the ICNIRP Guidelines for the worst cases of lateral and airgap conditions under the quick charging mode. For detailed designs of active EMF cancel coils for the power supply coil set and pick-up coil set, it is recommended to see references [29] and [42].

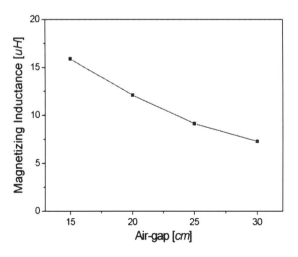

Figure 19.19 The magnetizing inductance of a power supply coil set along the airgap.

Figure 19.20 The EMF characteristics along the lateral direction from the proposed power supply coil.

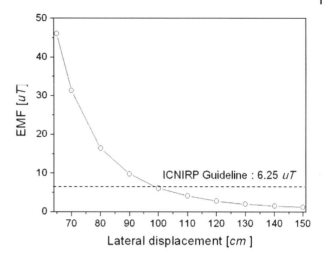

19.4 Conclusion

The prototype of new coil sets for a wireless stationary EV charger has been developed. By applying an extremely asymmetric structure of power supply coil and pick-up coil sets, large lateral and longitudinal tolerances were achieved, which can be enlarged as much as required. The achieved lateral tolerance of 39 cm is significantly larger than previous ones, which is twofold or threefold of any conventional IPTSs [1–24, 33–35] and quite adequate for practical applications. The proposed DoFA method for complex vector domain analysis is applicable to any IPTS as well as CMRS so far as each coil is resonated, which is the most common and requisite case in practice. Furthermore, the proposed hysteresis loss model is quite accurate and successfully reflects the partial core saturation on the IPTS system.

Problems

19.1 As shown in Figure 19.12, the prototype asymmetric coils have unnecessarily large heights h_1 and h_2. Especially h_2 cannot be so large when consideration is taken of the quite tough accommodation space under an EV. Identify the rationales for the heights and suggest the minimum values of them for the proposed example design.

19.2 The tip widths w_{t1} and w_{t2} as well as the tip lengths l_{t1} and l_{t2} are not discussed in detail in the proposed example design. Discuss the side effects and merits of them when they are very large in terms of the coupling factor due to magnetic leakage flux change, self-inductance, and lateral displacements.

References

1 R. Laouamer, M. Brunello, J.P. Ferrieux, O. Normand, and N. Buchheit, "A multi-resonant converter for noncontact charging with electromagnetic coupling," in *Proc. IEEE Industrial Electronics Society (IECON)*, vol. 2, pp. 792–797, 1997.

2 H. Sakamoto, K. Harada, S. Washimiya, and K. Takehara, "Large air-gap coupler for inductive charger," *IEEE Trans. on Magnetics*, vol. 35, no. 5, pp. 3526–3529, September 1999.

3 J. Hirai, K.T. Woong, and A. Kawamura, "Study on intelligent battery charging using inductive transmission of power and information," *IEEE Trans. on Power Electron.*, vol. 15, no. 2, pp. 335–345, 2000.

4 F. Nakao, Y. Matsuo, M. Kitaoka, and H. Salcamoto, "Ferrite core couplers for inductive chargers," in *IEEE Power Conversion Conference*, 2002, pp. 850–854.

5 R. Mecke, and C. Rathge, "High frequency resonant inverter for contactless energy transmission over large air gap," in *Proc. 35th Annual IEEE Power Electron. Spec. Conf.*, June 2004, vol. 3, pp. 1737–1743.

6 Y. Kamiya, Y. Daisho, F. Kuwabara, and S. Takahashi, "Development and performance evaluation of an advanced electric micro bus transportation system," in *JSAE Annual Spring Congress*, 2006, pp. 7–14.

7 M. Budhia, G. Covic, J. Boys, and C.Y. Huang, "Development and evaluation of single sided flux couplers for contactless electric vehicle charging," in *IEEE Energy Conversion Congress and Exposition (ECCE)*, 2011, pp. 614–621.

8 M. Budhia, G. Covic, and J. Boys, "Design and optimization of circular magnetic structures for lumped inductive power transfer systems," *IEEE Trans. on Power Electron.*, vol. 26, no. 11, pp. 3096–3107, 2011.

9 S. Hasanzadeh, S.V. Zadeh, and A.H. Isfahani, "Optimization of a contactless power transfer system for electric vehicles," *IEEE Trans. on Vehicular Tech.*, vol. 51, no. 8, pp. 3566–3573, 2012.

10 Y. Iga, H. Omori, T. Morizane, N. Kimura, Y. Nakamura, and M. Nakaoka, "New IPT-wireless EV charger using single-ended quasi-resonant converter with power factor correction," in *IEEE Renewable Energy Research and Applications (ICRERA)*, 2012, pp. 1–6.

11 H. Wu, A. Gilchrist, K. Sealy, and D. Bronson, "A 90 percent efficient 5 kW inductive charger for EVs," *IEEE Energy Conversion Congress and Exposition (ECCE)*, 2012, pp. 275–282.

12 F. Sato, J. Murakami, H. Matsuki, K. Harakawa, and T. Satoh, "Stable energy transmission to moving loads utilizing new CLPS," *IEEE Trans. on Magnetics*, vol. 32, no. 5, pp. 5034–5036, September 1996.

13 F. Sato, J. Murakami, T. Suzuki, H. Matsuki, S. Kikuchi, K. Harakawa, H. Osada, and K. Seki, "Contactless energy transmission to mobile loads by CLPS-test driving of an EV with starter batteries," *IEEE Trans. on Magnetics*, vol. 33, no. 5, pp. 4203–4205, September 1997.

14 C. Wang, O.H. Stielau, and G.A. Covic, "Design considerations for a contactless electric vehicle battery charger," *IEEE Trans. Ind. Electron.*, vol. 52, no. 5, pp. 1308–1314, October 2005.

15 J. Sallan, J.L. Villa, A. Llombart, and J.F. Sanz, "Optimal design of ICPT systems applied to electric vehicle battery charge," *IEEE Trans. on Ind. Electron.*, vol. 56, no. 6, pp. 2140–2149, June 2009.

16 J.L. Villa, J. Sallan, A. Llombart, and J.F. Sanz, "Design of a high frequency inductively coupled power transfer system for electric vehicle battery charge," *Applic. Energy*, vol. 86, no. 3, pp. 355–363, 2009.

17 T. Maruyama, K. Yamamoto, S. Kitazawa, K. Kondo, and T. Kashiwagi, "A study on the design method of the light weight coils for a high power contactless power transfer system," in *2012 15th International Conference on Electrical Machines and Systems (ICEMS)*, 2012, pp. 1–6.

18 J.G. Bolger and F.A. Kirsten, "Investigation of the feasibility of a dual mode electric transportation system," Lawrence Berkeley Laboratory Report, 1977.

19 J.G. Bolger, F.A. Kirsten, and L.S. Ng, "Inductive power coupling for an electric highway system," in *Proc. IEEE 28th Vehicular Technology Conference*, 1978, pp. 137–144.

20 C.E. Zell and J.G. Bolger, "Development of an engineering prototype of a road powered electric transit vehicle system," in *Proc. 32nd IEEE Vehicular Technology Conference*, 1982, pp. 435–438.

21 M. Eghtesadi, "Inductive power transfer to an electric vehicle – an analytical model," in *Proc. 40th IEEE Vehicular Technology Conference*, 1990, pp. 100–104.

22 A.W. Green and J.T. Boys, "10 kHz inductively coupled power transfer concept and control," in *Proc. 5th Int. Conf. IEEE Power Electron Variable-Speed Drivers,"* October 1994, pp. 694–699.

23 G.A.J. Elliott, J.T. Boys, and A.W. Green, "Magnetically coupled systems for power transfer to electric vehicles," in *Proc. Int. Conf. Power Electron. Drive System*, February 1995, pp. 797–801.

24 G.A. Covic, J.T. Boys, M.L.G. Kissin, and H.G. Lu, "A three-phase inductive power transfer system for road powered vehicles," *IEEE Trans. Ind. Electron.*, vol. 54, no. 6, pp. 3370–3378, December 2007.

25 N.P. Suh, D.H. Cho, and C.T. Rim, "Design of on-line electric vehicle (OLEV)," *Plenary Lecture at the 2010 CIRP Design Conference*, 2010, pp. 3–8.

26 S.W. Lee, J. Huh, C.B. Park, N.S. Choi, G.H. Cho, and C.T. Rim, "On-line electric vehicle using inductive power transfer system," in *IEEE Energy Conversion Congress and Exposition (ECCE)*, 2010, pp. 1598–1601.

27 J. Huh and C.T. Rim, "KAIST wireless electric vehicles – OLEV," in *JSAE Annual Congress*, 2011.

28 J. Huh, S.W. Lee, C.B. Park, G.H. Cho, and C.T. Rim, "High performance inductive power transfer system with narrow rail width for on-line electric vehicles," *IEEE Energy Conversion Congress and Exposition (ECCE)*, 2010, pp. 647–651.

29 S.W. Lee, W.Y. Lee, J. Huh, H.J. Kim, C.B. Park, G.H. Cho, and C.T. Rim, "Active EMF cancellation method for I-type pick-up of on-line electric vehicles," *IEEE Applied Power Electronics Conference and Exposition (APEC)*, 2011, pp. 1980-1983.

30 J. Huh, W.Y. Lee, G.H. Cho, B.H. Lee, and C.T. Rim, "Characterization of novel inductive power transfer systems for on-line electric vehicles," *IEEE Applied Power Electronics Conference and Exposition (APEC)*, 2011, pp. 1975–1979.

31 J. Huh, S.W. Lee, W.Y. Lee, G.H. Cho, and C.T. Rim, "Narrow-width inductive power transfer system for on-line electrical vehicles," *IEEE Trans. on Power Electron.*, vol. 26, no. 12, pp. 3666–3679, December 2011.

32 S.Y. Choi, J. Huh, W.Y. Lee, S.W. Lee and C.T. Rim, "New cross-segmented power supply rails for road powered electric vehicles," *IEEE Trans. on Power Electron.*, vol. 28, no. 12, pp. 5832–5841, December 2013.

33 Y. Nagatsuka, N. Ehara, Y. Kaneko, S. Abe, and T. Yasuda, "Compact contactless power transfer system for electric vehicles," in *International Power Electronics Conference (IPEC)*, 2010, pp. 807–813.

34 M. Chigira, Y. Nagatsuka, Y. Kaneko, S. Abe, T. Yasuda, and A. Suzuki, "Small-size light-weight transformer with new core structure for contactless electric vehicle power transfer system," in *IEEE Energy Conversion Congress and Exposition (ECCE)*, 2011, pp. 260–266.

35 M. Budhia, G.A. Covic, and J.T. Boys, "A new magnetic coupler for inductive power transfer electric vehicle charging systems," in *36th Annual Conference of the IEEE Industrial Electronics Society (IECON)*, 2010, pp. 2481–2486.

36 C.T. Rim, "Unified general phasor transformation for AC converters," *IEEE Trans. on Power Electron.*, vol. 26, pp. 2465–2745, September 2011.

37 S.W. Lee, C.B. Park, and C.T. Rim, "Static and dynamic analyses of three-phase rectifier with LC input filter by Laplace phasor transformation," in *IEEE Energy Conversion Congress and Exposition (ECCE)*, September 2012, pp. 1570–1577.

38 S.W. Lee, B.H. Choi, and C.T. Rim, "Dynamics characterization of the inductive power transfer system for on-line electric vehicles by Laplace phasor transform," *IEEE Trans. on Power Electron.*, vol. 28, no. 12, pp. 5902–5909, December 2013.

39 W.Y. Lee, J. Huh, S.Y. Choi, X.V. Thai, J.H. Kim, E.A. Al-Ammar, M.A. El-Kady, and C.T. Rim, "Finite-width magnetic mirror models of mono and dual coils for wireless electric vehicle," *IEEE Trans. on Power Electron.*, vol. 28, pp. 1413–1428, March 2013.

40 E.S. Lee, J. Huh, X.V. Thai, S.Y. Choi, and C.T. Rim, "Impedance transformers for compact and robust coupled magnetic resonance systems," in *IEEE Energy Conversion Congress and Exposition (ECCE)*, 2013, pp. 2239–2244.

41 ICNIRP Guidelines, "International commission on non-ionizing radiation protection (ICNIRP) guidelines," 1998, www.icnirp.de/documents/emfgdl.pdf.

42 S.Y. Choi, B.W. Gu, S.W. Lee, J. Huh, and C.T. Rim, "Generalized active EMF cancel methods for wireless electric vehicles," *IEEE Trans. on Power Electron.*, vol. 29, no. 11, pp. 5770–5783, November 2014.

20

DQ Coils for Large Tolerance EV Chargers

20.1 Introduction

EVs such as plug-in hybrid EVs (PHEVs) and battery EVs (BEVs) need battery chargers [1–7]. Because conductive chargers [8–10] are unwelcome by users due to inconvenience of connecting power cables as well as safety issues [11, 12], wireless chargers [13–19] are becoming welcome. The wireless stationary EV chargers, which have loop-type coils such as circular- and rectangular-type coils, are widely developed because they have simple and compact structures as well as a low electromagnetic field (EMF) for pedestrians [20]. For these reasons, the loop-type pick-up coils were designated for a low-power application as Standard J2954 of the SAE [21].

However, the magnetic coupling between the pick-up coil set and the power supply coil set rapidly decreases and results in the reduction of its delivered power capacity, that is, narrow tolerances, when the pick-up coil set is misaligned with the power supply coil set. In practice, it is hard for drivers to accurately align their EV for the best wireless power transfer performance. Increasing the size of a pick-up coil set is one of the methods used to increase its tolerance but it is not practical because the available bottom size of a vehicle is generally limited. Even though the tolerance is enhanced to more than 23 cm by adapting a double-sided coil, which has the same structure as both the power supply coil set and pick-up coil set, the EMF problem still remains [17]. Recently asymmetric coil sets that achieved a large lateral displacement of 40 cm and longitudinal displacement of 20 cm at an airgap of 15 cm were proved, which make it possible to enlarge the tolerance of wireless stationary EV chargers [18]. However, this scheme does not follow Standard J2954, which does not permit double-sided coils. Moreover, in practice, not only lateral and longitudinal but also diagonal displacements are so important because longitudinal and lateral displacements occur at the same time, which results in the diagonal displacement.

In this chapter, new DQ power supply coil sets for wireless stationary EV chargers, which have two coil sets having a 90 phase difference with large longitudinal, lateral, as well as diagonal displacements, are explained. Each power supply coil set consists of two same magnetic poles to generate a uniform magnetic flux density over the power supply coil. A prototype IPTS using the proposed DQ power supply coil sets for the wireless stationary EV chargers was fabricated for the experimental verifications with an operating frequency of 85 kHz to satisfy Standard J2954 of the SAE.

Wireless Power Transfer for Electric Vehicles and Mobile Devices, First Edition. Chun T. Rim and Chris Mi.
© 2017 John Wiley & Sons Ltd. Published 2017 by John Wiley & Sons Ltd.

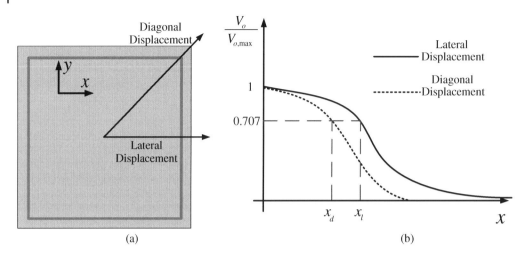

Figure 20.1 (a) Top view of the rectangular type loop coil and (b) its lateral displacement and diagonal displacement.

20.1.1 Problems of the Rectangular-Type Loop Coil

Design of a power supply coil set should be carefully conducted because its power transfer capacity is very sensitive to variations of longitudinal and lateral displacements, as well as an airgap between a pick-up coil and a power supply coil. Here, the conventional rectangular pick-up coil set satisfying Standard J2954 is considered and the size of the pick-up coil set is chosen to be smaller than the size of the power supply coil set to enlarge longitudinal and lateral tolerances.

The conventional rectangular-type power supply loop coil and an induced output voltage of the pick-up coil set V_o along lateral and diagonal displacements are shown in Figure 20.1. From now on, the longitudinal and lateral displacements will be mentioned as a lateral displacement because the conventional square structure shows the same characteristics for longitudinal and lateral displacements.

When the pick-up coil set moves along the diagonal direction, V_o drops more rapidly compared with lateral displacement because the overlap area between the power supply coil set and the pick-up coil set is smaller than that of the lateral displacement. Hence, it is noteworthy to consider a diagonal displacement. In general, the relationship between the diagonal displacement x_d and the lateral displacement x_l can be determined as follows:

$$x_d < x_l \tag{20.1}$$

where x_d and x_l should be designed as large as possible in the wireless stationary EV chargers.

20.1.2 Displacement Extension Methods

There are numerous researches on the lateral displacement enlargement; however, there are few researches on the diagonal misalignment. In this section, a general method to maximize the diagonal displacement is introduced.

First of all, a conventional rectangular-type loop coil is examined by splitting it into two coils where the same ampere-turns are applied to the two coils, as shown in Figure 20.2(a). V_o is the sum of induced voltages of the coils with the same current directions, as shown in Figure 20.2(b).

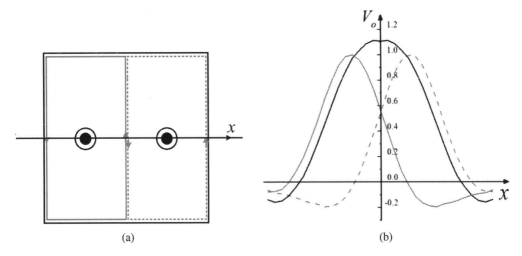

Figure 20.2 (a) Top view of a rectangular type loop coil and (b) its V_o.

If the area of the split rectangular coil is reduced, as shown in Figure 20.3(a), the peak of V_o will drop, as shown in Figure 20.3(b), resulting in tolerance enhancement. The increase in the tolerance can be explained by the increase in the distance between each center of the two coils. Even though V_o reduces by adapting the split-type loop coils, the larger tolerance, which is generally defined as the −3dB point of power, can be achieved. The voltage reduction for the well-aligned case can be compensated by increasing the number of pick-up coil set turns N_2.

In summary, the power supply coil set with the split type loop coil can achieve a larger lateral tolerance.

20.1.3 Configuration of the Proposed Power Supply Coil Sets

As mentioned in the previous section, the larger diagonal displacement can be achieved by adopting the split-type loop coil in the diagonal direction. The area of each coil, controlled

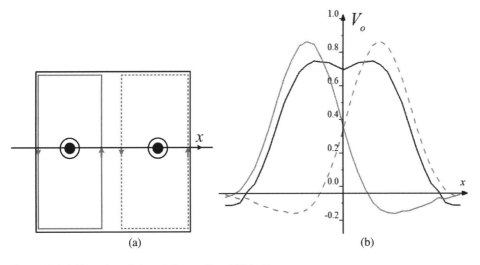

Figure 20.3 (a) Top view of the split loop coil and (b) its V_o.

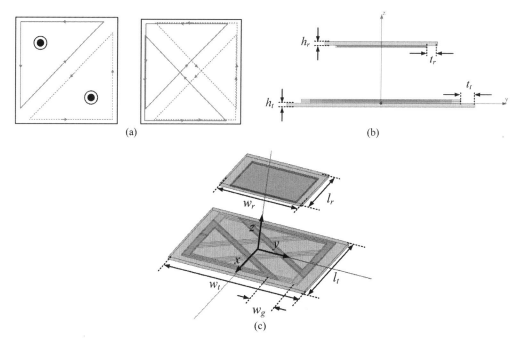

Figure 20.4 Overall configuration of the proposed DQ power supply coil sets with a pick-up coil: (a) top view of the D coil set (left) and DQ coil set (right), (b) its front view, and (c) its bird's eye view.

by the distance w_g between two coils, should be maximized to get more power but should be minimized to get a larger tolerance. Therefore, the distance w_g should be optimally selected, balancing the contradictory performances.

In this proposed design each power supply coil set consists of two magnetic poles, facing diagonally, while each coil set has a 90 degree phase difference, as shown in Figure 20.4. Compared to the square-type coil of Figure 20.3, this triangular-type coil of Figure 20.4 has roughly $\sqrt{2}$ times the distance between two coils for a given coil outer size, which results in more increased lateral tolerance. Moreover, adopting DQ coil sets by overlapping two sets of split coils will result in a symmetric power delivery performance.

For the proposed coil sets, two half-bridge inverters for providing D- and Q-phase currents are used, as shown in Figure 20.5.

Since the proposed DQ power supply coil sets are magnetically coupled with each other, the half-bridge inverter should operate as an independent constant current source regardless of induced voltage due to the magnetic coupling. This can be achieved by controlling the duty of each leg to eliminate current unbalancing, which becomes severe for a large misalignment between the power supply coils and pick-up coil.

The induced voltages V_d and V_q due to the linked magnetic flux of each power supply coil set can be calculated as follows:

$$V_d = jN_2\omega_s\Phi_d \tag{20.2a}$$
$$V_q = jN_2\omega_s\Phi_q \tag{20.2b}$$

where N_2 and ω_s are the number of turns of the pick-up coil and the switching angular frequency, respectively. Analytic calculation of the intersecting magnetic flux Φ_d and Φ_q, generated from the D- and Q-power supply coils, respectively, is quite challenging; Φ_d and Φ_q vary

Figure 20.5 Proposed DQ coil circuits, including two half-bridge inverters, power supply coil sets, and compensation circuits.

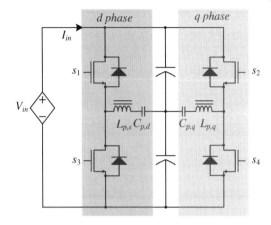

for airgap change, lateral misalignments, tilting, and rotation. If only the intersecting magnetic fluxes are determined, V_o is simply obtained by summing the V_d and V_q of (20.2) as follows:

$$V_o = |V_d + V_q| = N_2 \omega_s |\Phi_d + \Phi_q| \tag{20.3}$$

20.2 Example Design and Simulation Verifications

Parameters of the proposed DQ power supply coil sets are summarized in Table 20.1. A simulation was performed using ANSOFT Maxwell ver. 15.1 to verify the larger lateral tolerances of the proposed power supply coil sets for the airgap of 150 mm.

The simulation results for various positions of the pick-up coil along the x-axis and the diagonal direction are plotted, as shown in Figure 20.6. As mentioned in the previous section, the larger diagonal displacement was achieved in the case of a 90 degree phase difference.

It is important to determine the proper distance w_g between the coils in order to maximize the tolerance. When the pick-up coil set is well aligned with the power supply coil set, V_o should show a maximum value. The tolerance is improved compared with the conventional rectangular

Table 20.1 Parameters of the proposed DQ coil sets

Parameter	Value	Parameter	Value
I_s	5 A$_{rms}$	w_t	500 mm
f_s	85 kHz	l_t	500 mm
N_1	5 turns	w_r	300 mm
N_2	5 turns	l_r	300 mm
L_{p1}	43.7 μH	h_t	10 mm
L_{p2}	52.4 μH	h_r	10 mm
L_s	26.5 μH	w_g	60 mm
C_{p1}	81.2 nF	t_t	50 mm
C_{p2}	69.2 Nf	t_r	30 mm

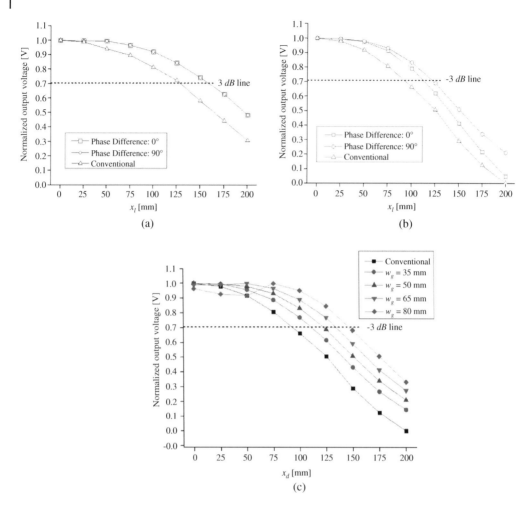

Figure 20.6 FEA simulation results: (a) the lateral displacement along the *x*-axis at *y* = 0, (b) the diagonal displacement, and (c) the diagonal displacement with respect to w_g.

loop-type coil by changing only the coil structure of the power supply coil sets without changing the structure of the pick-up coil.

20.3 Experimental Verification for the Example Design

In order to verify the proposed DQ power supply coil sets, which allows not only large lateral and longitudinal displacements but also a large diagonal displacement, an experimental kit was fabricated, as shown in Figure 20.7, where the resonant frequency of the power supply coil sets f_r is a little smaller than f_s to guarantee zero voltage switching (ZVS) [19].

The input current of the power supply coil set was set to 5 A_{rms} and the series resonant compensation is adapted for the power supply coil windings to reduce reactive power and voltage stresses of power cables [13]. The pick-up coil structure is similar to that of the power supply coil of Figure 20.7(a) except in size. The airgap, defined as the top of the core of the power supply

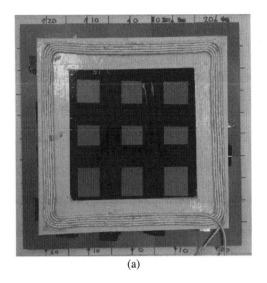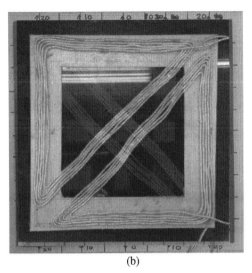

(a)   (b)

Figure 20.7 An experimental power supply coil set: (a) a conventional rectangular coil and (b) the proposed DQ power supply coil sets.

coil set and the bottom of the core of the pick-up coil, is set at 150 mm, while other parameters are listed in Table 20.1.

The measured V_o is shown in Figure 20.8. The lateral and diagonal displacements of the conventional one are just about 130 mm and 90 mm, respectively, whereas those of the proposed one are about 160 mm and 130 mm, respectively. The lateral and diagonal displacements increase by about 35% and 19%, respectively, compared to the conventional rectangular loop-type coil for the airgap of 150 mm.

The prototype IPTS using the proposed DQ power supply coil sets showed fairly good agreement with the simulations.

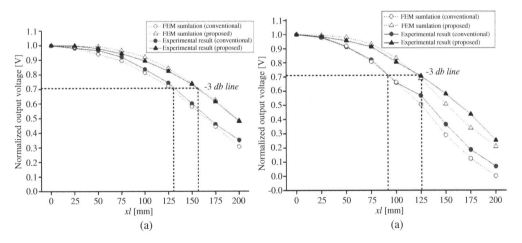

Figure 20.8 Measured normalized V_o: (a) the lateral displacement for $w_g = 50$ mm and (b) the diagonal displacement for $w_g = 50$ mm.

20.4 Conclusion

The proposed DQ power supply coil sets with dual triangular coils show the large diagonal displacement as well as the lateral displacement at the same time. From the simulation and experimental results, it is found that the lateral and diagonal displacements increase by about 35% and 19%, respectively, compared to the conventional rectangular loop-type coil for an airgap of 150 mm.

Problems

20.1 In Figure 20.4, what is the optimum space between the center of each winding and the outer core?

20.2 In Figure 20.4, calculate the mutual inductance that simplifies magnetic resistances for a very small airgap condition.

20.3 In Figure 20.6(c), what is your selection between $w_g = 65$ mm and $w_g = 80$ mm? What is the rationale for your choice?

20.4 In Figure 20.6, what is the effect of the phase angle on the IPT performances?

References

1 J.T. Salihi, P.D. Agarwal, and G.J. Spix, "Induction motor control scheme for battery-powered electric car (GM-Electrovair I)," *IEEE Trans. on Ind. General Applic.*, vol. IGA-r3, no. 5, pp. 463–469, September 1967.

2 J.R. Bish and G.P. Tietmeyer, "Electric vehicle field test experience," *IEEE Trans. on Veh. Technol.*, vol. 32, no. 1, pp. 81–89, February 1983.

3 J. Dixon, I. Nakashima, E.F. Arcos, and M. Ortuzar, "Electric vehicle using a combination of ultra capacitors and ZEBRA battery," *IEEE Trans. on Ind. Electron.*, vol. 57, no. 3, pp. 943–949, March 2010.

4 C.H. Kim, M.Y. Kim, and G.W. Moon, "A modularized charge equalizer using a battery monitoring IC for series-connected Li-ion battery string in electric vehicles," *IEEE Trans. on Power Electron.*, vol. 28, no. 8, pp. 3779–3787, August 2013.

5 F.L. Mapelli, D. Tarsitano, and M. Mauri, "Plug-in hybrid electric vehicle: modeling, prototype, realization, and inverter losses reduction analysis," *IEEE Trans. on Ind. Electron.*, vol. 57, no. 2, pp. 598–607, February 2010.

6 E. Tara, S. Shahidinejad, and E. Bibeau, "Battery storage sizing in a retrofitted plug in hybrid electric vehicle," *IEEE Trans. on Veh. Technol.*, vol. 59, no. 6, pp. 2786–2794, July 2010.

7 S.G. Li, S.M. Sharkh, F.C. Walsh, and C.N. Zhang, "Energy and battery management of a plug-in series hybrid electric vehicle using fuzzy logic," *IEEE Trans. on Veh. Technol.*, vol. 60, no. 8, pp. 3571–3585, October 2011.

8 N.H. Kutkut, D.M. Divan, D.W. Novotny, and R.H. Marion, "Design considerations and topology selection for a 120-kW IGBT converter for EV fast charging," *IEEE Trans. on Power Electron.*, vol. 13, no. 1, pp. 169–178, January 1998.

9 C. Praisuwanna and S. Khomfoi, "A quick charger station for EVs using a pulse frequency technique," in *Proc. IEEE Energy Conversion Congress and Exposition ECCE)*, September 2013, pp. 3595–3599.

10 J.D. Marus and V.L. Newhouse, "Method for charging a plug-in electric vehicle," U.S. Patent 20 130 234 664, September 12, 2013.

11 C.Y. Huang, J.T. Boys, G.A. Covic, and M. Budhia, "Practical considerations for designing IPT system for EV battery charging," in *IEEE Vehicle Power and Propulsion Conference (VPPC)*, 2009, pp. 402–407.

12 H.H. Wu, A. Gilchrist, K. Sealy, P. Israelsen, and J. Muhs, "A review on inductive charging for electric vehicles," in *IEEE International Electric Machines and Drives Conference (IEMDC)*, 2011, pp. 143–147.

13 J. Huh, S.W. Lee, W.Y. Lee, G.H. Cho, and Chun T. Rim, "Narrow-width inductive power transfer system for on-line electrical vehicles (OLEV)," *IEEE Trans. on Power Electron.*, vol. 26, no. 12, pp. 3666–3679, December 2011.

14 S.Y. Choi, J. Huh, W.Y. Lee, S.W. Lee, and Chun T. Rim, "New cross-segmented power supply rails for road powered electric vehicles," *IEEE Trans. on Power Electron.*, vol. 28, no. 12, pp. 5832–5841, December 2013.

15 W.Y. Lee, B. Choi, and Chun T. Rim, "Dynamics characterization of the inductive power transfer system for online electric vehicles by Laplace phasor transform," *IEEE Trans. on Power Electron.*, vol. 28, no. 12, pp. 5902–5909, December 2013.

16 S.Y. Choi, B.W. Gu, S.W. Lee, W.Y. Lee, J. Huh, and Chun T. Rim, "Generalized active EMF cancel methods for wireless electric vehicles," *IEEE Trans. on Power Electron.*, vol. 29, no. 11, pp. 5770–5783, November 2014.

17 Chun T. Rim, "The development and deployment of on-line electric vehicles (OLEV)," in *Proc. IEEE Energy Conversion Congress and Exposition (ECCE)*, September 2013.

18 S.Y. Choi, J. Huh, W.Y. Lee, J.G. Cho, and Chun T. Rim, "Asymmetric coil sets for wireless stationary EV chargers with large lateral tolerance by dominant field analysis," *IEEE Trans. on Power Electron.*, to be published, doi: 10.1109/TPEL.2014.2305172.

19 S.Y. Choi, B.W. Gu, S.Y. Jeong, and Chun T. Rim, "Advances in wireless power transfer systems for road powered electric vehicles," *IEEE Journal of Emerging and Selected Topics in Power Electronics*, vol. 3, no. 1, pp. 18–36, March 2015.

20 M. Budhia, G.A. Covic, and J.T. Boys, "Design and optimization of circular magnetic structures for lumped inductive power transfer systems," *IEEE Trans. on Power Electron.*, vol. 26, no. 11, pp. 3096–3108, November 2011.

21 "SAE J2954 overview and path forward," http://www.sae.org/smartgrid/sae-j2954-status_1-2012.pdf.

21

Capacitive Power Transfer for EV Chargers Coupler

21.1 Introduction

IPT has been widely applied in the charging of portable devices [1, 2] and electric vehicles [3]. The efficiency of an IPT system from the DC source to the DC load has reached 96% with 7 kW output power [4], which is already comparable to that of the traditional plug-in charger. However, the drawback of IPT technology lies in its sensitivity to conductive objects, such as metal debris in the airgap. The magnetic fields generate eddy current losses in the metals near the system, causing a significant temperature increase, which is dangerous in practice [5].

CPT technology is an alternative solution to replace the IPT system. It utilizes electric fields, instead of magnetic fields, to transfer power [6]. Unlike magnetic fields, electric fields can pass through metal barriers without generating significant power losses. Therefore, CPT technology is suitable for electric vehicle charging applications [7].

As shown in Figure 21.1, CPT has better lateral tolerance than IPT due to the non-canceling characteristic of the E-field compared to the B-field. Another advantage of the CPT system is its low cost. In CPT systems, metal plates are used to form capacitors to transfer power [8, 9], while in IPT systems, the coils are made of expensive Litz wire [10]. An aluminum plate is a cost-efficient option as it has good conductivity, low weight, and low cost.

Most of the recent CPT systems focus on low-power or short-distance applications, such as the LED driver [11], soccer robot charging [12], and synchronous motor excitation [13]. In particular, the transfer distance is usually around 1 mm, which is far less than the ground clearance of electric vehicles, limiting the application of CPT technology.

This limitation of current CPT systems comes from the circuit topologies working with the coupling capacitors, which are classified into two categories: non-resonant and resonant topologies. The non-resonant topology is a PWM converter, such as the SEPIC converter. The coupling capacitors work as power storage components to smooth the power in the circuit [14]. Therefore, the system requires large capacitances, usually in the tens of the nF range, and the transfer distance is less than 1 mm. The resonant topologies include the series resonance converter [15] and the class-E converter [16], in which the coupling capacitors resonate with the inductors in the compensation circuit. The benefit of these topologies is that the coupling capacitance can be reduced as long as the resonant inductance or switching frequency is high enough. However, the inductance is limited by its self-resonant frequency [17] and the switching frequency is limited by the efficiency and power capability of the converter [18]. Another problem is that the resonant topology is sensitive to the parameter variations caused by misalignment, which is unacceptable in some critical applications. In general, all these topologies require either too large a capacitance or too high a switching frequency, making them difficult to realize. Therefore, better circuit topologies need to be proposed for the CPT system.

Wireless Power Transfer for Electric Vehicles and Mobile Devices, First Edition. Chun T. Rim and Chris Mi.
© 2017 John Wiley & Sons Ltd. Published 2017 by John Wiley & Sons Ltd.

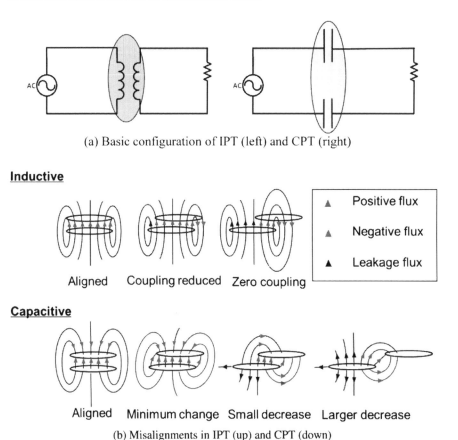

(a) Basic configuration of IPT (left) and CPT (right)

Inductive

Aligned   Coupling reduced   Zero coupling

▲ Positive flux

▲ Negative flux

▲ Leakage flux

Capacitive

Aligned   Minimum change   Small decrease   Larger decrease

(b) Misalignments in IPT (up) and CPT (down)

Figure 21.1 Fundamental principles of IPT and CPT.

A double-sided LCLC-compensated circuit has been proposed in [7] for high power and large airgap applications. The transfer distance is 150 mm and the output power reaches 2.4 kW with an efficiency of 91%. Although the coupling capacitor is around tens of pF, there is a 100 pF capacitor connected in parallel with the coupling plates, which reduces the resonant inductor to hundreds of μH and the switching frequency to 1 MHz. Therefore, the resonances are not affected by parameter variations and misalignments. However, there are eight external components in the compensation network that increase the complexity of the system and are difficult to construct. Also, the two pairs of plates are horizontally separated by 500 mm to eliminate the coupling between the adjacent plates, meaning the plates take up more space than necessary.

In this chapter, a more compact four-plate structure is proposed for high-power CPT applications. In this structure, all the plates are vertically arranged to save space, as shown in Figure 21.2(a). At each side, two plates are placed close to each other to maintain a large coupling capacitance, which is used to replace two external compensation capacitors in the LCLC topology. Therefore, the LCLC compensation topology can be simplified, as in the LCL topology. Transferring high power through the coupler requires generating high voltage between the plates to build up electric fields. For each two plates on the same side, the coupling capacitance can be adjusted through regulating the distance. Then the switching frequency and system power can be controlled. Since the distance can be reduced to maintain a large capacitance, the system frequency can be reduced to a reasonable range. In addition, the LCL topology can

Figure 21.2 Proposed four-plate CPT.

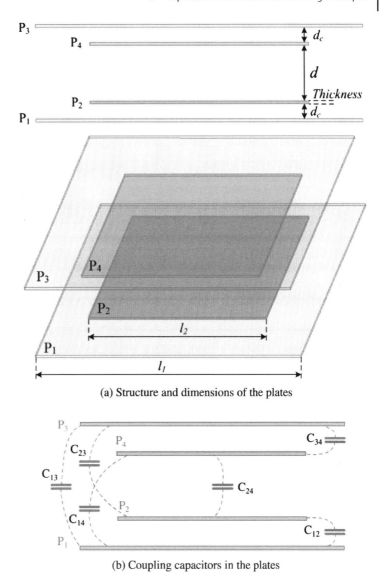

(a) Structure and dimensions of the plates

(b) Coupling capacitors in the plates

resonate with the plates to provide high voltage and acts as a constant current source for both the input and output, which is suitable for the battery load.

Another benefit of the vertical structure is its misalignment ability. As shown in Figure 21.2(a), the two plates on the same side are of different sizes and the outer plate is larger to maintain the coupling with the plates on the other side. For the horizontal structure, the rotary misalignment, which is the mismatch between the primary and secondary plates, can reduce the output power. For the vertical structure, the rotation in the horizontal plane does not cause too much mismatch and the coupling capacitors remain nearly unchanged. Therefore, it is meaningful to replace the horizontal structure with the proposed vertical one.

Although the vertical structure has been applied in low-power commercial products, there is no overlap between the two plates on the same side in these applications. Therefore, the large capacitance considered in this chapter is not built into the coupler in reference [19], which means the circuit model of [19] is the same as that of the asymmetric horizontal plates. The function of the asymmetric structure is to reduce the voltage stress on the large plate, compared to the small one. Reference [20] attempts to model a vertical structure with plate overlap, but the Π diagram structure is not sufficient to model the plates. The method of moment (MoM) is used in [20], but it does not provide details, and nor does it consider the capacitances between each of the two plates. Reference [21] tries to introduce the concept of mutual capacitance in the plate model, but the model does not have duality with the classic transformer model and the capacitances between each of the two plates are not studied. Therefore, more accurate circuit models of the coupling plates are needed for the CPT system design. Throughout this chapter, all the circuits are sinusoidal in the steady state, operating at the switching frequency.

21.2 Four-Plate Structure and Its Circuit Model

21.2.1 Plate Structure

Figure 21.2(a) illustrates the structure and dimensions of the plates. Both the three-dimensional view and front view are provided. The plates are designed to be symmetric from the primary to the secondary side. P_1 and P_2 are embedded in the ground as the power transmitter. P_3 and P_4 are installed on the vehicle as the power receiver. In Figure 21.2(a), P_1 and P_3 are larger than P_2 and P_4. Therefore, the coupling between P_1 and P_3 cannot be eliminated by P_2 and P_4. The plate shape does not affect coupling, so all the plates are designed to be square to simplify the analysis. In practical applications, the plates can be designed to be any shape to fit the installation on the vehicle. The only principle is to maintain the area of the plates to transfer sufficient power. The length of P_1 and P_3 is l_1, the length of P_2 and P_3 is l_2, the distance of P_1–P_2 and P_3–P_4 is d_c, and the distance of P_2–P_4 is d, which is the airgap between the primary and secondary sides. The thickness of all the plates is the same.

21.2.2 Circuit Model of the Plates

There is a coupling capacitance between each of two plates, as shown in Figure 21.2(b). The airgap, d, is much larger than the plate distance, d_c, in electric vehicle charging applications, so C_{13} and C_{24} are much smaller than C_{12} and C_{34}. The cross-couplings of C_{14} and C_{23} are generated by the edge effects of P_1–P_4 and P_2–P_3, so they are usually smaller than C_{12} and C_{34}. However, they cannot be neglected in an accurate circuit model. The resulting circuit model of the four-plate vertical structure is shown in Figure 21.3. The equivalent input capacitances of

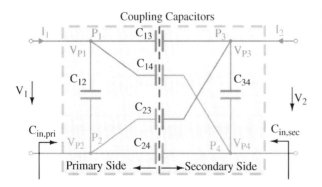

Figure 21.3 Circuit model of the coupling capacitors.

the plates from the primary and secondary sides are defined as $C_{in,pri}$ and $C_{in,sec}$, respectively, which are mainly determined by C_{12} and C_{34}. Since C_{12} and C_{34} are primarily determined by the distance d_c and do not relate to the misalignment between the primary and secondary sides, the resonance of this coupler is not sensitive to the misalignment.

In Figure 21.3, two independent sinusoidal voltage sources, V_1 and V_2, are applied to the plates to derive the relationship between the input and output. All the capacitors correspond to the couplings in Figure 21.2(b). The voltage on each plate is defined as V_{p1}, V_{p2}, V_{p3}, and V_{p4}, respectively. Plate P_2 is selected as the reference, so $V_{p2} = 0$, $V_1 = V_{p1}$, and $V_2 = V_{p3} - V_{p4}$. Therefore, Kirchhoff's current equations are expressed in the steady state as

$$\begin{cases} (C_{12} + C_{13} + C_{14}) \cdot V_{P1} - C_{13} \cdot V_{P3} - C_{14} \cdot V_{P4} = I_1/(j\omega) \\ \qquad -C_{12} \cdot V_{P1} - C_{23} \cdot V_{P3} - C_{24} \cdot V_{P4} = -I_1/(j\omega) \\ -C_{13} \cdot V_{P1} + (C_{13} + C_{23} + C_{34}) \cdot V_{P3} - C_{34} \cdot V_{P4} = I_2/(j\omega) \\ -C_{14} \cdot V_{P1} - C_{34} \cdot V_{P3} + (C_{14} + C_{24} + C_{34}) \cdot V_{P4} = -I_2/(j\omega) \end{cases} \quad (21.1)$$

where I_1 and I_2 are the currents injected into the plates from the primary and secondary sides, respectively, $\omega = 2\pi f_{sw}$ and f_{sw} is the frequency of the input and output AC sources.

The plates are modeled as a two-port network with V_1 and V_2 as the input and I_1 and I_2 as the output variables. There are four equations in (21.1) and any three of them are independent. The relationship between voltage and current can be derived from the equations in (21.1). Considering the first two equations in (21.1), V_{P3} and V_{P4} can be eliminated, as shown by

$$\begin{cases} [C_{24}(C_{12} + C_{13} + C_{14}) + C_{12}C_{14}] \cdot V_{P1} - (C_{13}C_{24} - C_{14}C_{23}) \cdot V_{P3} = (C_{24} + C_{14})I_1/(j\omega) \\ [C_{23}(C_{12} + C_{13} + C_{14}) + C_{12}C_{13}] \cdot V_{P1} + (C_{13}C_{24} - C_{14}C_{23}) \cdot V_{P4} = (C_{23} + C_{13})I_1/(j\omega) \end{cases} \quad (21.2)$$

Since $V_1 = V_{P1}$ and $V_2 = V_{P3}\text{-}V_{p4}$, the relationship between V_1, I_1, and V_2 can be expressed as

$$\begin{aligned} V_1 = I_1 \cdot &\frac{1}{j\omega \left[C_{12} + \dfrac{(C_{13} + C_{14}) \cdot (C_{23} + C_{24})}{C_{13} + C_{14} + C_{23} + C_{24}} \right]} \\ &+ V_2 \cdot \frac{C_{24}C_{13} - C_{14}C_{23}}{C_{12} \cdot (C_{13} + C_{14} + C_{23} + C_{24}) + (C_{13} + C_{14}) \cdot (C_{23} + C_{24})} \end{aligned} \quad (21.3)$$

Similarly, using the other two equations in (21.1), the relationship between V_2, I_2, and V_1 can be expressed as

$$\begin{aligned} V_2 = I_2 \cdot &\frac{1}{j\omega \left[C_{34} + \dfrac{(C_{13} + C_{23}) \cdot (C_{14} + C_{24})}{C_{13} + C_{14} + C_{23} + C_{24}} \right]} \\ &+ V_1 \cdot \frac{C_{24}C_{13} - C_{14}C_{23}}{C_{34} \cdot (C_{13} + C_{14} + C_{23} + C_{24}) + (C_{13} + C_{23}) \cdot (C_{14} + C_{24})} \end{aligned} \quad (21.4)$$

From (21.3) and (21.4), capacitances C_1, C_2, and C_M can be defined as

$$\begin{cases} C_1 = C_{12} + \dfrac{(C_{13} + C_{14}) \cdot (C_{23} + C_{24})}{C_{13} + C_{14} + C_{23} + C_{24}} \\[2mm] C_2 = C_{34} + \dfrac{(C_{13} + C_{23}) \cdot (C_{14} + C_{24})}{C_{13} + C_{14} + C_{23} + C_{24}} \\[2mm] C_M = \dfrac{C_{24}C_{13} - C_{14}C_{23}}{C_{13} + C_{14} + C_{23} + C_{24}} \end{cases} \tag{21.5}$$

Therefore, (21.3) and (21.4) can be rewritten as

$$\begin{cases} V_1 = I_1 \cdot \dfrac{1}{j\omega C_1} + V_2 \cdot \dfrac{C_M}{C_1} \\[2mm] V_2 = I_2 \cdot \dfrac{1}{j\omega C_2} + V_1 \cdot \dfrac{C_M}{C_2} \end{cases} \tag{21.6}$$

Move I_1 and I_2 to the left-hand side, and the relationship between the current and voltage is shown as

$$\begin{cases} I_1 = j\omega C_1 \cdot V_1 - j\omega C_M \cdot V_2 \\ I_2 = j\omega C_2 \cdot V_2 - j\omega C_M \cdot V_1 \end{cases} \tag{21.7}$$

According to (21.7), the simplified equivalent model of coupling capacitors with behavior sources can be shown as in Figure 21.4(a). Both of the current sources depend on the voltage at the other side and are separated by a dashed line.

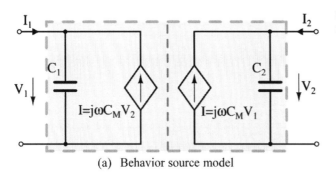

(a) Behavior source model

Figure 21.4 Simplified equivalent model of coupling capacitors.

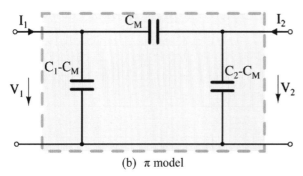

(b) π model

Equation (21.7) is further rewritten as

$$\begin{cases} I_1 = j\omega(C_1 - C_M) \cdot V_1 + j\omega C_M \cdot (V_1 - V_2) \\ I_2 = j\omega(C_2 - C_M) \cdot V_2 + j\omega C_M \cdot (V_2 - V_1) \end{cases} \qquad (21.8)$$

Then the equivalent model of the capacitors is simplified to a π shape, as shown in Figure 21.4(b). This model is suitable to simplify the parameter calculation in the circuit. It needs to be emphasized that the primary and secondary sides are not separated in the π shape model.

Similar to that of the coils, the capacitive coupling coefficient k_c of the plates is defined with the parameters from (21.5), as shown by

$$k_C = \frac{C_M}{\sqrt{C_1 \cdot C_2}} = \frac{\sqrt{C_{24}C_{13} - C_{14}C_{23}}}{\sqrt{C_{12}(C_{13} + C_{14} + C_{23} + C_{24}) + (C_{13} + C_{14})(C_{23} + C_{24})}}$$

$$\cdot \frac{\sqrt{C_{24}C_{13} - C_{14}C_{23}}}{\sqrt{C_{34}(C_{13} + C_{14} + C_{23} + C_{24}) + (C_{13} + C_{23})(C_{14} + C_{24})}} \qquad (21.9)$$

From the simplified model in Figure 21.4, the self-capacitance of the primary side is C_1, the self-capacitance of the secondary side is C_2, the mutual capacitance between the primary and secondary sides is C_M, and the capacitive coupling coefficient is k_c.

For the horizontal structure in reference [7], the plates on the same side are placed 500 mm away, so coupling capacitances C_{12} and C_{34} are both close to zero. Cross-coupling capacitances C_{14} and C_{23} are also close to zero. As a result, the capacitive coupling coefficient is $k_c \approx 1$.

For the vertical structure shown in Figure 21.2(b), the plate distance d_c is much smaller than the airgap distance d, so C_{12} and C_{34} are much larger than C_{13} and C_{24}. As a result, the capacitive coupling coefficient $k_c \ll 1$, which means it is a loosely coupled CPT system.

Since the plates work as a single capacitor to resonate with the inductor in the circuit, it is important to calculate the equivalent input capacitances of the plates. Figure 21.3 shows capacitances $C_{in,pri}$ and $C_{in,sec}$ from the primary and secondary sides, respectively. It is convenient to use the simplified capacitor model in Figure 21.4 to perform the calculations. Therefore, the equivalent input capacitances are

$$\begin{cases} C_{in,pri} = \left.\frac{I_1}{j\omega \cdot V_1}\right|_{I_2=0} = C_1 - C_M + \frac{C_M(C_2 - C_M)}{C_2} = \left(1 - k_c^2\right) C_1 \\ C_{in,sec} = \left.\frac{I_2}{j\omega \cdot V_2}\right|_{I_1=0} = C_2 - C_M + \frac{C_M(C_1 - C_M)}{C_1} = \left(1 - k_c^2\right) C_2 \end{cases} \qquad (21.10)$$

The transfer function between the input and output voltage is also an important parameter used to determine the amount of transferred power. The voltage transfer function from the primary to the secondary side is defined as $H_{1,2}$ and the voltage transfer function from the secondary to the primary is defined as $H_{2,1}$. They are expressed as

$$\begin{cases} H_{1,2} = \left.\frac{V_2}{V_1}\right|_{I_2=0} = \frac{C_M}{C_2} = k_c\sqrt{\frac{C_1}{C_2}} \\ H_{1,2} = \left.\frac{V_1}{V_2}\right|_{I_1=0} = \frac{C_M}{C_1} = k_c\sqrt{\frac{C_2}{C_1}} \end{cases} \qquad (21.11)$$

The plate model in Figure 21.4 includes the voltage stress between plates P_1 and P_2 and the stress between P_3 and P_4. However, it does not consider the voltage between P1 and P3, nor the voltage between P_2 and P_4, which are also important in the system design. Since P_2 is set to be the reference, the second equation in (21.2) is used to calculate the voltage between P_2 and P_4:

$$V_{P4-P2} = V_{P4} = \frac{(C_{23} + C_{13}) \cdot I_1}{j\omega(C_{13}C_{24} - C_{23}C_{14})} - \frac{C_{12}(C_{13} + C_{23}) + C_{23}(C_{13} + C_{14})}{(C_{13}C_{24} - C_{23}C_{14})} \cdot V_1 \quad (21.12)$$

Using (1), the voltage between P_1 and P_3 is expressed as

$$V_{P1-P3} = V_{P1} - V_{P3} = \frac{-(C_{23} + C_{24}) \cdot I_2}{j\omega(C_{13}C_{24} - C_{23}C_{14})} + \frac{C_{34}(C_{23} + C_{24}) + C_{23}(C_{14} + C_{24})}{(C_{13}C_{24} - C_{23}C_{14})} V_2 \quad (21.13)$$

21.2.3 Plate Dimensions

Using the circuit model of the plates, the dimensions of the plates can be determined for electric vehicle charging applications. All the variables are shown in Figure 21.2(a). The purpose of dimension design is to calculate all the capacitances in (21.5) and analyze the behavior of the plates. Since the plate structure is designed to be symmetric from the primary to the secondary side, C is defined as $C = C_1 = C_2$.

Considering the space limitation, the length l_1 of P_1 and P_3 is 914 mm. The airgap is set to be 150 mm, which is the ground clearance of the electric vehicle. Therefore, there are only two parameters, d_c and l_2, that need to be designed. The plate ratio r_p is defined as $r_p = l_2/l_1$.

The plate structure in Figure 21.2(b) is much more complex than the parallel plates in reference [22]. Since the cross-coupling is usually small, it can be neglected in the beginning stage of the plate design. The empirical formula of the parallel plates in [22] can be used to estimate the capacitance based on the system dimensions in order to accelerate the design process. This estimation can provide a reasonable range of the dimensions. Then, finite element analysis (FEA) by Maxwell can be used to accurately determine the final dimensions and the corresponding circuit model of the plates.

The FEA simulation provides a capacitance matrix with all six mutual capacitors shown in Figure 21.2(b). Using the capacitance matrix, the equivalent capacitances C and C_M are further calculated according to (21.5) and the capacitive coupling coefficient is obtained from (21.9). When the plate ratio r_p and distance d_c vary, all the plate dimensions are analyzed in Maxwell. The equivalent parameters (C_M, C, and k_C) in the plate model are shown in Figure 21.5 as a function of r_p and d_c.

Figure 21.5(a) shows that the mutual capacitance C_M is only sensitive to the plate ratio r_p. Figure 21.5(b) shows that the increase of r_p and the decrease of d_c both cause the increase of the self-capacitance C. For the capacitive coupling coefficient k_C, Figure 21.5(c) shows that it is smaller than 10%, which indicates that it is a loosely coupled CPT system.

According to [7], there is an external inductor resonating with the self-capacitance. The self-capacitance should be large enough to reduce the inductor's value and volume. At the same time, the coupling coefficient should be large enough to maintain the system power as well. Therefore, considering Figure 21.5(b) and (c), the plate distance d_c is set to be 10 mm and the plate ratio r_p is 0.667. All the capacitances for this scenario are shown in Table 21.1.

The misalignment ability of the plates is also analyzed in Maxwell. When there is a rotation of the secondary plates in the horizontal plane, as shown in Figure 21.2(a), the variation of self-capacitance $C_{1,2}$ is within 1% of the well-aligned value and the variation of mutual capacitance C_M is within 10%. When there is displacement misalignment, the variation of self-capacitance is negligible and the variation of mutual inductance is as shown in Figure 21.6. This figure shows that C_M can maintain higher than 50% of the nominal value when the misalignment increases to 250 mm.

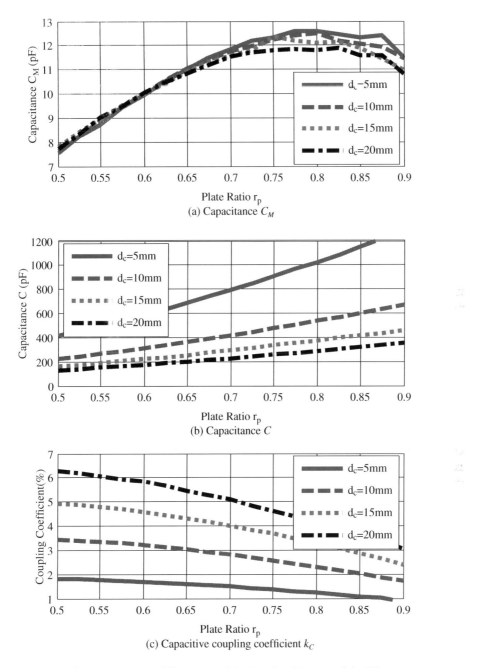

(a) Capacitance C_M

(b) Capacitance C

(c) Capacitive coupling coefficient k_C

Figure 21.5 Plate parameters at different r_p and d_c, when $l_1 = 914$ mm and $d = 150$ mm.

21.3 Double-Sided LCL Compensation Topology

A double-sided LCL compensation circuit is proposed to work with the plates, as shown in Figure 21.7. The plates are arranged in the vertical structure shown in Figure 21.2(b). There are multiple resonances in the circuit. At the primary side, there is a full-bridge inverter,

Table 21.1 Capacitances of plates, $l_1 = 914$ mm, $l_2 = 610$ mm, $d_c = 10$ mm, $d = 150$ mm

Parameter	Value	Parameter	Value
C_{12}	366 pF	C_{34}	366 pF
C_{13}	42.4 pF	C_{24}	19.5 pF
C_{14}	4.72 pF	C_{23}	4.72 pF
C_1	381 pF	C_2	381 pF
C_M	11.3 pF	k_C	2.90 %

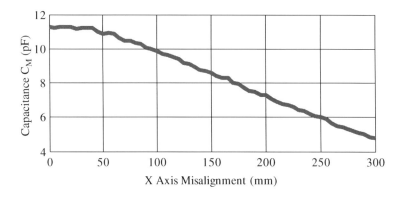

Figure 21.6 Mutual capacitance C_M at displacement misalignment conditions.

generating excitation V_{in} to the resonant tank. At the secondary side, a full-bridge rectifier is used to provide DC current to the output battery. In Figure 21.7, all the components are assumed to have a high-quality factor and the parasitic resistances are neglected in the analysis process.

L_{f1} and C_{f1} work as a low-pass filter at the front end. Similarly, L_{f2} and C_{f2} work as a low-pass filter at the back end. Therefore, there is no high-order harmonics current injected into

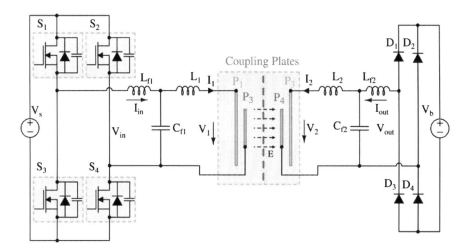

Figure 21.7 Double-sided LCL compensated circuit.

Figure 21.8 FHA analysis of the CPT system.

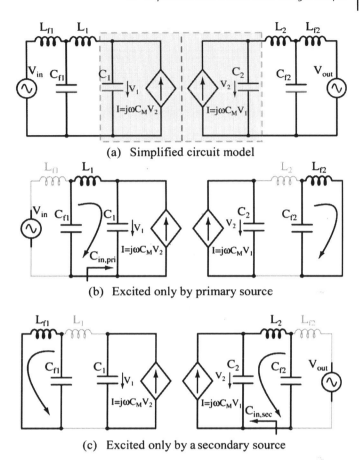

(a) Simplified circuit model

(b) Excited only by primary source

(c) Excited only by a secondary source

the plates. The fundamental harmonics approximation (FHA) method is used to analyze the working principle of the system. Figure 21.8(a) shows the simplified circuit topology of the CPT system, with the equivalent circuit model of the plates shown in Figure 21.4(a). The input and output square-wave sources are represented by two sinusoidal AC sources. Since the circuit in Figure 21.8(a) is linear, the superposition theorem can be used to analyze the two AC sources separately, as shown in Figure 21.8(b) and (c).

Figure 21.8(b) shows that the resonant circuit is excited only by the primary source. L_{f2} and C_{f2} form a parallel resonance and their impedance is infinite. L_2 is treated as an open circuit. L_1, C_{f1}, and $C_{in,pri}$ form another parallel resonance, so there is no current flowing through L_{f1}, which means the input current does not rely on the input voltage V_{in}. Therefore, the relationship between the circuit parameters is

$$\begin{cases} \omega = 2\pi f_{sw} = 1/\sqrt{L_{f2}C_{f2}} \\ L_1 = 1/(\omega^2 C_{f1}) + 1/(\omega^2 C_{in,pri}) \end{cases} \tag{21.14}$$

The output current depends on the input voltage. Since L_{f1} and L_2 are treated as open circuit, then $V_{Cf1} = V_{in}$ and $V_{Cf2} = V_2$. The transfer function between the primary and secondary voltages in (21.11) is used to calculate the voltage and current. Therefore, the output current is calculated as

$$
\begin{cases}
V_1 = V_{Cf1} \cdot \dfrac{C_{f1}}{C_{in,pri}} = V_{in} \cdot \dfrac{C_{f1}}{\left(1 - k_C^2\right) C_1} \\[3mm]
V_2 = H_{1,2} \cdot V_1 = \dfrac{C_M \cdot C_{f1} \cdot V_{in}}{\left(1 - k_C^2\right) \cdot C_1 C_2} \\[3mm]
I_{Lf2} = V_2 \cdot \dfrac{1}{j\omega L_{f2}} = V_2 \cdot \dfrac{\omega \cdot C_{f2}}{j} = \dfrac{\omega \cdot C_M \cdot C_{f1} C_{f2} \cdot V_{in}}{j \left(1 - k_C^2\right) \cdot C_1 C_2}
\end{cases}
\tag{21.15}
$$

There is a full-brige rectifier at the secondary side, so the output voltage and current are in phase. Figure 21.8(c) indicates that the output voltage does not affect the output current. Therefore, the output power can be expressed as

$$
P_{out} = |V_{out}| \cdot |I_{Lf2}| = \frac{\omega \cdot C_M \cdot C_{f1} C_{f2}}{\left(1 - k_C^2\right) \cdot C_1 C_2} \cdot |V_{in}| \cdot |V_{out}|
\tag{21.16}
$$

Figure 21.8(c) shows that the resonant circuit is excited only by the secondary source. Similar to the analysis of Figure 21.8(b), there are two parallel resonances: L_{f1} and C_{f1} form one resonance, and L_2, C_{f2}, and $C_{in,sec}$ form the other resonance. Because of the infinite impedance of the parallel resonances, L_1 and L_{f2} are treated as open circuit. In addition, the input current only depends on the output voltage. Therefore, the relationship between the circuit parameters is

$$
\begin{cases}
\omega = 2\pi f_{sw} = 1/\sqrt{L_{f1} C_{f1}} \\[3mm]
L_2 = 1/(\omega^2 C_{f2}) + 1/(\omega^2 C_{in,sec})
\end{cases}
\tag{21.17}
$$

Since L_1 and L_{f2} are open circuit, then $V_{Cf1} = V_1$ and $V_{Cf2} = V_{out}$. Considering the equivalent capacitance $C_{in,sec}$ in (21.10) and the voltage transfer function $H_{2,1}$ in (21.11), the input current can be calculated as

$$
\begin{cases}
V_2 = V_{Cf2} \cdot \dfrac{C_{f2}}{C_{in,sec}} = V_{out} \cdot \dfrac{C_{f2}}{\left(1 - k_C^2\right) C_2} \\[3mm]
V_1 = H_{2,1} \cdot V_2 = \dfrac{C_M \cdot C_{f2} \cdot V_{out}}{\left(1 - k_C^2\right) \cdot C_1 C_2} \\[3mm]
I_{Lf1} = V_1 \cdot \dfrac{1}{j\omega L_{f1}} = V_1 \cdot \dfrac{\omega \cdot C_{f1}}{j} = \dfrac{\omega \cdot C_M \cdot C_{f1} C_{f2} \cdot V_{out}}{j \left(1 - k_C^2\right) \cdot C_1 C_2}
\end{cases}
\tag{21.18}
$$

Equation (21.18) shows that I_{Lf1} is 90° lagging V_{out} and (21.15) shows that I_{Lf2} is 90° lagging V_{in}. Since V_{out} and I_{Lf2} are in phase, I_{Lf1} is 180° lagging V_{in}. The input current direction is opposite to that of I_{Lf1}, so it is in phase with V_{in}. Therefore, the input power is expressed as

$$
P_{in} = |V_{in}| \cdot |-I_{Lf1}| = \frac{\omega \cdot C_M \cdot C_{f1} C_{f2}}{\left(1 - k_C^2\right) \cdot C_1 C_2} \cdot |V_{in}| \cdot |V_{out}|
\tag{21.19}
$$

A comparison of (21.16) and (21.19) shows that when the parasitic resistances are neglected, the input and output power are the same, which also supports the previous assumption.

Table 21.2 Voltage stress on circuit components

Component	Voltage stress
L_{f1}, L_{f2}	$V_{Lf1} = \dfrac{C_M \cdot C_{f2} \cdot V_{out}}{\left(1 - k_C^2\right) \cdot C_1 C_2}, \; V_{Lf2} = \dfrac{C_M \cdot C_{f1} \cdot V_{in}}{\left(1 - k_C^2\right) \cdot C_1 C_2}$
C_{f1}, C_{f2}	$V_{Cf1} = V_{in} + V_{Lf1}, \; V_{Cf2} = V_{out} + V_{Lf2}$
L_1, L_2	$V_{L1} = \omega^2 L_1 C_{f1} \cdot V_{in}, \; V_{L2} = \omega^2 L_2 C_{f2} \cdot V_{out}$
$P_1 - P_2$	$V_{P1-P2} = \dfrac{C_{f1} \cdot V_{in}}{\left(1 - k_C^2\right) \cdot C_1} + \dfrac{C_M C_{f2} \cdot V_{out}}{\left(1 - k_C^2\right) \cdot C_1 C_2}$
$P_3 - P_4$	$V_{P3-P4} = \dfrac{C_M C_{f1} \cdot V_{in}}{\left(1 - k_C^2\right) \cdot C_1 C_2} + \dfrac{C_{f2} \cdot V_{out}}{\left(1 - k_C^2\right) \cdot C_2}$
$P_1 - P_3$	$\dfrac{C_{34}(C_{23} + C_{24}) + C_{23}(C_{14} + C_{24})}{(C_{13}C_{24} - C_{23}C_{14})} V_{P3-P4} - \dfrac{(C_{23} + C_{24})C_{f2}V_{out}}{(C_{13}C_{24} - C_{23}C_{14})}$
$P_2 - P_4$	$\dfrac{(C_{23} + C_{13})C_{f1}V_{in}}{(C_{13}C_{24} - C_{23}C_{14})} - \dfrac{C_{12}(C_{13} + C_{23}) + C_{23}(C_{13} + C_{14})}{(C_{13}C_{24} - C_{23}C_{14})} V_{P1-P2}$

Equation (21.19) shows that the system power is propotional to the mutual capacitance C_M, the filter capacitances $C_{f1,2}$, the voltages V_{in} and V_{out}, and the switching frequency f_{sw}. According to the plate design in Section 21.2, the capacitive coupling coefficient k_C is usually much smaller than 10%, so $(1 - k_C^2) \approx 1$. Therefore, the system power can be simplified as

$$P_{in} = P_{out} \approx \frac{\omega \cdot C_M \cdot C_{f1} C_{f2}}{C_1 C_2} \cdot |V_{in}| \cdot |V_{out}| \tag{21.20}$$

Considering the input DC voltage V_s and the output battery voltage V_b in Figure 21.7, (21.20) can be rewritten as

$$P_{in} = P_{out} \approx \frac{\omega \cdot C_M \cdot C_{f1} C_{f2}}{C_1 C_2} \cdot \frac{2\sqrt{2}}{\pi} V_s \cdot \frac{2\sqrt{2}}{\pi} V_b. \tag{21.21}$$

In a high-power CPT system, the voltage stress on the circuit component, especially the metal plates, is an important concern. The voltages on inductors $L_{f1,2}$ and $L_{1,2}$ and capacitors $C_{f1,2}$ can be calculated using the current flowing through them. The voltage between each of the two plates can be calculated according to (21.12), (21.13), (21.15), and (21.18), which are shown in Table 21.2.

21.4 Prototype Design

After the plate structure and LCL compensation circuit topology have been proposed, a prototype of the CPT system is designed. According to (21.14), (21.17), and (21.19), all the circuit parameters are designed as shown in Table 21.3.

Because of the symmetry of the plate structure, the other circuit parameters are also designed to be symmetric. Considering the limitation of the semiconductor devices, the switching frequency is 1 MHz. This circuit topology is similar to the LCLC compensation topology in [7] and L_2 is larger than L_1 to provide a soft-switching condition to the input inverter.

Table 21.3 System specifications and circuit parameter values

Parameter	Design value	Parameter	Design value
V_{in}	270 V	V_{out}	270 V
l_1	914 mm	l_2	610 mm
d_c	10 mm	r_p	0.667
f_{sw}	1 MHz	C_M	11.3 pF
L_{f1}	2.90 μH	L_{f2}	2.90 μH
C_{f1}	8.73 nF	C_{f2}	8.73 nF
L_1	69.4 μH	L_2	70.0 μH
C_1	381 pF	C_2	381 pF

The CPT system is simulated in LTspice with the parameter values shown in Table 21.3. The simulated waveforms of the CPT system are shown in Figure 21.9. The input voltage, V_{in}, is almost in phase with the input current, I_{in}. The cut-off current at the switching transient is about 6 A, which is enough to provide a soft-switching condition to the switches. The input voltage is 90° lagging the output voltage V_{out}, which coincides with the FHA analysis in Section 21.3.

The rms value of the voltage stress on each component is calculated using Table 21.2, which also agrees with the LTspice simulation. The simulated results are shown in Table 21.4, which shows that the voltage stress on the filter component is relatively low, but that the voltage stress on the inductor $L_{1,2}$ is higher than 5 kV, so the insulation between each turn of the inductor should be considered in the manufacturing process. The voltage between P_1 and P_2 is 5.12 kV and the plate distance $d_c = 10$ mm. The breakdown voltage of air is about 3 kV/mm. Therefore, there is no concern with arcing.

The radiation of leakage electric flux is an important safety concern in a high-power CPT system. The electric fields around the plates can be analyzed in Maxwell, using the voltage stresses in Table 21.4. The simulation result is shown in Figure 21.10. According to the IEEE Standard, the leakage electric field should be lower than 614 V/m at 1 MHz due to the concern

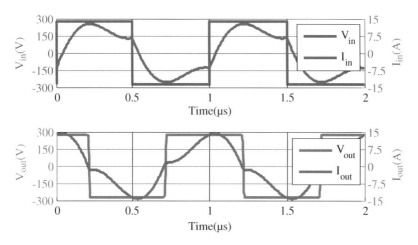

Figure 21.9 Simulated input and output voltage and current waveforms.

Table 21.4 The rms value of the voltage stress on each component

Parameter	Voltage	Parameter	Voltage
V_{Lf1}	211 V	V_{Lf2}	211 V
V_{Cf1}	278 V	V_{Cf2}	331 V
V_{L1}	5.34 kV	V_{L2}	5.36 kV
V_{P1-P2}	5.12 kV	V_{P3-P4}	5.08 kV
V_{P1-P3}	2.44 kV	V_{P2-P4}	5.29 kV

Figure 21.10 Electric field distribution of the plates.

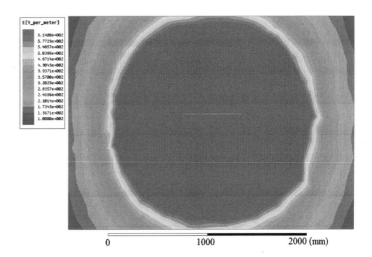

for human safety [23]. The simulation result indicates that the required safety distance for this system is about 1 m. Future research will optimize the plate structure to reduce the leakage electric field.

21.5 Experimental Verifications

21.5.1 Experiment Setup

Using the parameters in Table 21.3, a CPT system prototype is constructed as shown in Figure 21.11. Four aluminum plates are vertically arranged to form the capacitive coupler. The ceramic spacers are used to separate the inner and outer plates on the same side. The white PVC tubes are used to hold the plates as shown in Figure 21.11. The skin depth of copper is 65 μm at 1 MHz, so the 3000-strand AWG 46 Litz wire with a diameter of 40 μm is used to make the inductors, thereby reducing the skin effect losses. Since the inductors are air-cored and wound on PVC tubes, the magnetic losses are also eliminated. High-power, high-frequency polypropylene thin-film capacitors from KEMET are used to resonate with the inductor and the dissipation factor is 0.18% at 1 MHz. The connections between the inductors, capacitors, and plates are also shown in Figure 21.11, in which the components are connected to the edge of the plates.

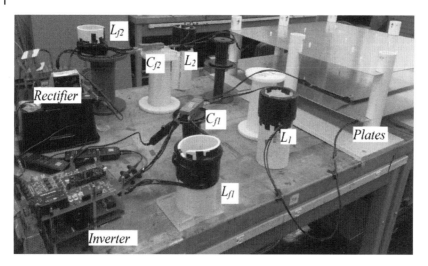

Figure 21.11 The prototype of a CPT system with a vertical plate structure.

Silicon carbide (SiC) MOSFETs C2M0080120D from CREE are used in the input inverter. The datasheet shows that the output parasitic capacitance between the drain and source is 110 pF at 270 V. As mentioned in Section 21.4, the MOSFETs can achieve zero-voltage switching and only the conduction losses are considered. The output rectifier utilizes SiC Diode IDW30G65C from Infineon and the forward voltage of the diode is used to estimate the power losses.

21.5.2 Experiment Results

The experimental waveforms, which are similar to those of the simulation results in Figure 21.9, are shown in Figure 21.12. The input voltage and current are almost in phase with each other. V_{out} is 180° inverted, so it is lagging V_{in} in Figure 21.9. The switch current at the switching transient is about 6 A and the zero-voltage switching conditon is achieved. Although there is noise on the driver signal at the switching transient, the magnitude of the noise is within 3 V, which is lower than the threshold voltage of the SiC MOSFET, so it is still acceptable for the safe operation of the MOSFETs.

When the input and ouput voltages increase, the relationship between the output power and efficiency is as shown in Figure 21.13. This shows that the system efficiency keeps increasing with the increase in output power. For the no-misalignment case, the system can maintain an efficiency higher than 85% when the power is higher than 600 W. When the input and output voltages are both 270 V, the system input power reaches a maximum of 2.17 kW. The output power is 1.88 kW with an efficiency of 85.9%.

The misalignment ability of the prototype is also tested, as shown in Figure 21.13. When there is a 15 cm misalignment between the primary and secondary plates, the maximum output power of the system drops to 1.60 kW with an efficiency of 85.4%. When the misalignment increases to 30 cm, the maximum output power drops to 1.06 kW with an efficeincy of 84.7%. At the maximum misalignment, the system power drops to about 56.4% of the well-aligned case. A rotatory misalignment test is also conducted, in which the primary plates are fixed and the secondary plates are rotated in the horizontal plane to different angles. In this experiment,

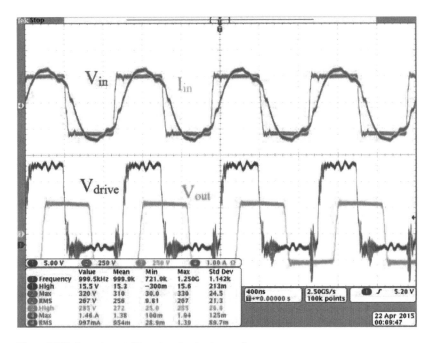

Figure 21.12 Experimental input and output waveforms.

the system output power is maintained and the power ripple is within ±5.0% of the nominal power, which shows that this system has good misalignment ability in relation to the horizontal plane rotation.

The distribution of the power losses in the circuit components is also estimated for the prototype. The circuit component models discussed in references [24] and [25] are used to calculte the power losses. For all the inductors, the AC resistance at 1 MHz is 3.4 times the DC resistance. For all the capacitors, the parasitic resistance is calculated according to the dissipation factor, which is 0.18% at 1 MHz. For the MOSFETs in the inverter, since they work under soft-switching conditions, only the conduction losses are considered. For the diodes in the rectifier, the forward voltage is used to calculate the losses. With all these models, the remaining losses are estimated to be from the coupling plates. Therefore, the power loss distribution is as shown in Figure 21.14.

Figure 21.13 System output power and efficiency at different misalignment conditions.

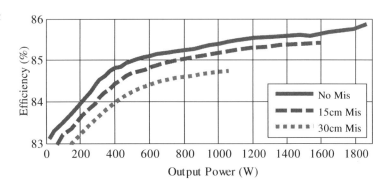

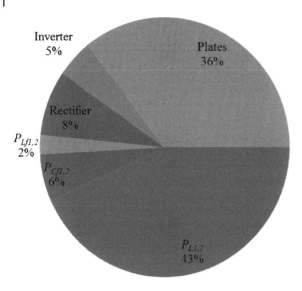

Figure 21.14 Power loss distribution of the components.

This shows that 43% of the power consumption is from the inductors L_1 and L_2. The aluminum plates consume 36% of the total power losses. Considering the power distribution, the structures of the inductors and plates need to be optimized in future research to reduce the total power loss.

21.5.3 Discussion and Comparison with the LCLC-Compensated System

The experiment results of this LCL-compensated CPT system with vertically arranged plates is compared with the previously published LCLC-compensated CPT system with horizontally arranged plates in reference [7]. The LCL system has a better misalignment ability in response to rotation, whereas the LCLC system has a better ability in response to displacement misalignment. When there is a 90° rotation between the primary and secondary plates, the LCL system can maintain the original power, while the power in the LCLC system drops to zero. When there is a 30 cm displacement between the plates, the LCLC system can maintain 87.5% of the origianl power, but the power in the LCL system drops to 56.4%. Therefore, the proposed vertically arranged plate structure is more sutiable for the situation where rotatory misalignment is unavoidable and significant.

The experiments also show that the efficiency of the LCLC system in reference [7] is about 5% higher than that of the LCL system explained in this chapter, which is because there is more power consumed by the two inductors L_1 and L_2. For a given system, the power loss distribution relates to the parameter values. Compared to the LCLC system, although inductances L_1 and L_2 are reduced, the currents flowing through them are increased. Therefore, all the power losses in the components need to be considered together. In future research, the system efficiency will be optimized through appropriately designing the parameter values.

The power transfer density is also an important specification for evaluating the CPT system. For the LCL-compensated system in this chapter, the power transfer density is calculated as

$$P_{D,LCL} = \frac{P_{out}}{l_1^2(d + 2d_c)} = \frac{1.88 \text{ kW}}{0.914^2 \times 0.17 \text{ m}^3} = 13.24 \text{ kW/m}^3 \tag{21.22}$$

For the LCLC-compensated system in reference [7], the space between the two pairs of plates should be considered, so its power transfer density is calculated as

$$P_{D,LCLC} = \frac{P_{out}}{l(2l + d_1)d} = \frac{2.4 \text{ kW}}{0.61(2 \times 0.61 + 0.5)0.15 \text{ m}^3} = 15.25 \text{ kW/m}^3 \qquad (21.23)$$

The power transfer density of the inductive power tranfer system is also compared. For the LCC-compensated IPT system in reference [24], the power transfer density is calculated as

$$P_{D,IPT} = \frac{P_{out}}{l_1 l_2 d} = \frac{5.7 \text{ kW}}{0.60 \times 0.80 \times 0.18 \text{ m}^3} = 65.97 \text{ kW/m}^3. \qquad (21.24)$$

This shows that the power transfer density of the two CPT systems is comparable, while the IPT system has a much higher power transfer density than either of the CPT systems. In future research, the power density of the CPT system can be improved by increasing the voltage on the plates. Meanwhile, the safety and radiation issues should be considered in the system densign.

21.6 Conclusion

This chapter proposed a vertical structure of plates and the corresponding LCL compensation circuit topology for high-power capacitive power transfer. The equivalent circuit model of the plates is derived using the coupling capacitance between each of two plates. The circuit model is described by a voltage-controlled current source, which is the duality of the classic transformer model. The voltage and current of each component are calculated using the fundamental harmonic approximation (FHA) method, with which the system power is derived. A prototype of the CPT system is designed and built to validate the proposed plate structure and compensation circuit topology. The system efficiency reaches 85.9% with 1.87 kW output power and a 150 mm airgap. Future research will focus on the study of electric field radiation to make the proposed system safer for use in electric vehicle charging applciations.

Problems

21.1 Explain how the proposed CPT can successfully deliver such a large power of a few kW using key equations and values.

21.2 Determine the voltage maximum ratings of capacitances and inductors for the proposed CPT design.

21.3 Discuss the merits and demerits of CPT and IPT, considering compensation circuit components such as capacitors and inductors.

References

1 K. Chang-Gyun, S. Dong-Hyun, Y. Jung-Sik, P. Jong-Hu, and B.H. Cho, "Design of a contactless battery charger for cellular phone," *IEEE Trans. on Ind. Electron.*, vol. 48, no. 6, pp. 1238–1247, 2001.

2 S. Raabe and G.A. Covic, "Practical design considerations for contactless power transfer quadrature pick-ups," *IEEE Trans. on Ind. Electron.*, vol. 60, no. 1, pp. 400–409, 2013.

3 S. Mohagheghi, B. Parkhideh, and S. Bhattacharya, "Inductive power transfer for electric vehicle: potential benefits for the distribution grid," in *IEEE Int. Electric Vehicle Conference (IEVC)*, 2012, pp. 1–8.

4 J. Deng, F. Lu, S. Li, T. Nguyen, and C. Mi, "Development of a high efficiency primary side controlled 7 kW wireless power charger," in *Proc. IEEE Electric Vehicle Conference (IEVC)*, 2014, pp. 1–6.

5 D. Chen, L. Wang, C. Liao, and Y. Guo, "The power loss analysis for resonant wireless power transfer," in *Proc. IEEE Transport and Electrics Asia-Pacific Conference*, 2014, pp. 1–4.

6 J. Dai and D. Ludois, "A survey of wireless power transfer and a critical comparison of inductive and capacitive coupling for small gap applications", *IEEE Trans. on Power Electron.*, vol. 30, pp. 6017–6029, 2015.

7 F. Lu, H. Zhang, H. Hofmann, and C. Mi, "A double-sided *LCLC*-compensated capacitive power transfer system for electric vehicle charging," *IEEE Trans. on Power Electron.*, vol. 30, pp. 6011–6014, 2015.

8 C. Liu, A.P. Hu, G.A. Covic and N.K.C. Nair, "Comparative study of CCPT system with two different inductor tuning positions," *IEEE Trans. on Power Electron.*, vol. 27, pp. 294–306, 2012.

9 M. Kline, I. Izyumin, B. Boser, and S. Sanders, "Capacitive power transfer for contactless charging," in *Proc. IEEE Applied Power Electrics Conference (APEC)*, 2011, pp. 1398–1404.

10 S. Li, W. Li, J. Deng, T.D. Nguyen, and C.C. Mi, "A double-sided LCC compensation network and its tuning method for wireless power transfer," *IEEE Trans. on Veh. Technol.*, pp. 1–12, 2014.

11 D. Shmilovitz, A. Abramovitz, and I. Reichman, "Quasi resonant LED driver with capacitive isolation and high PF," *IEEE Journal of Emerging and Selective Topics in Power Electrics*, vol. 3, pp. 633–641, 2015.

12 A.P. Hu, C. Liu, and H. Li, "A novel contactless battery charging system for soccer playing robot," *IEEE Int. Mechanical and Machine Vision in Practice Conference (M2VIP)*, 2008, pp. 646–650.

13 D.C. Ludois, M.J. Erickson, and J.K. Reed, "Aerodynamic fluid bearings for translational and rotating capacitors in noncontact capacitive power transfer systems," *IEEE Trans. on Ind. Applic.*, pp. 1025–1033, 2014.

14 J. Dai and D.C. Ludios, "Single active switch power electronics for kilowatt scale capacitive power transfer," *IEEE Journal of Emerging and Selective Topics in Power Electrics*, vol. 3, pp. 315–323, 2015.

15 L. Huang, A.P. Hu, A. Swwain, and X. Dai, "Comparison of two high frequency converters for capacitive power transfer," in *Proc. IEEE Energy Conversion Congress and Exposition*, pp. 5437–5443, 2014.

16 B.H. Choi, D.T. Dguyen, S.J. Yoo, J.H. Kim, and C.T. Rim, "A novel source-side monitored capacitive power transfer system for contactless mobile charger using class-E converter," in *Proc. IEEE Vehicle Technology Conference*, 2014, pp. 1–5.

17 S. Pasko, M. Kazimierczuk, and B. Grzesik, "Self-capacitance of coupled toroidal inductors for EMI filters," *IEEE Trans. on Electromagnetic Compatability*, vol. 57, pp. 216–223, 2015.

18 P. Srimuang, N. Puangngernmak, and S. Chalermwisutkul, "13.56 MHz Class E power amplifier with 94.6% efficiency and 31 watts output power for RF heating applications," in *IEEE Electrical Engineering Composite Telecommunication and Information Technical Conference (ECTI-CON)*, 2014, pp. 1–5.

19 S. Goma, "Capacitive coupling powers transmission module," http://www.mutata.com/~/media/webrenewal/about/newsroom/tech/power/wptm/ta1291.ashx.

20 T. Komaru and H. Akita, "Positional characteristics of capacitive power transfer as a resonance coupling system," in *Proc. IEEE Wireless Power Transfer Conference*, 2013, pp. 218–221.

21 C. Liu, A.P. Hu, and M. Budhia, "A generalized coupling model for capacitive power transfer system," in *IEEE Industrial Electronics Conference (IECON)*, 2010, pp. 274–279.

22 H. Nishiyama and M. Nakamura, "Form and capacitance of parallel plate capacitor", *IEEE Transaction on Components, Packaging, and Manufacturing Technology – Part A*, vol. 17, no. 3, pp. 477–484, 1994.

23 IEEE Standard for Safety Levels with Respect to Human Exposure to Radio Frequency Electromagnetic Fields, 3 kHz to 300 GHz, C95.1, 2005.

24 F. Lu, H. Hofmann, J. Deng, and C. Mi, "Output power and efficiency sensitivity to circuit parameter variations in double-sided LCC-compensated wireless power transfer system," in *Proc. IEEE Applied Power Electrics Conference (APEC)*, 2015, pp. 597–601

25 F. Lu, H. Zhang, H. Hofmann, and C. Mi, "A high efficiency 3.3 kW loosely-coupled wireless power transfer system without magnetic material," in *Proc. IEEE Energy Conversion Congress and Exposition (ECCE)*, 2015, pp. 1–5.

22

Foreign Object Detection

22.1 Introduction

Dynamic or static wireless charging is becoming more important than ever due to its convenience, automatic operation, and safety [1–8]. The IPT system is comprised of Tx and Rx coils and a strong magnetic field, which is formed between the coils. This may cause a fire when the magnetic induction current comes in contact with a metal piece, called a foreign object Foreign object detection (FOD) is an essential function to prevent a potential fire, owing to its eddy current loss during EV charging. FOD has been widely used in other fields such as metal debris detection in airports [25] and factories [26]. The underlying principles of detecting a foreign object should be different for different applications. For example, a 78 GHz radio frequency (RF) is used for wide-area inspection at airports [25].

The principles of FOD methods in wireless EV chargers include inductive, capacitive, RF, ultrasonic, optical vision, infrared (IR) sensors, and even mechanical sensors. The FOD for EVs can be further classified, in this chapter, into metal object detection (MOD) and live object detection (LOD). Quite often, FOD means MOD and is sometimes called LOD as opposed to FOD (not MOD). To avoid this sort of confusion, "FOD = MOD + LOD" will be used throughout the chapter. In this chapter, however, LOD is not dealt with in detail but MOD is highly focused, that is, FOD ≈ MOD throughout the chapter.

Let me first explain a few potential FOD (= MOD) methods applicable to IPT for EVs. One of them is to compare power losses with and without conductive objects, which is said to be relatively simple and cheap [9]. However, this type is not applicable to high-power applications such as wireless stationary EV chargers, where the portion of power loss due to conductive debris is too small compared to total power transfer. Another FOD method based on the variation of the quality factor (Q) for an Rx coil is introduced [10]; however, this is valid only for firmly fixed wireless power transfer applications because the Q value changes when an Rx coil moves on a Tx coil. Witricity has developed the overlapped detection coils on the Tx coil to sense the imbalance of induced voltages due to metal debris [24], as shown in Figure 22.1.

As identified from Figure 22.1(a), the detection voltage becomes zero when there is no metal object due to cancelation of generated induced voltages of each loop coil. One of the problems of single detection coil patterns of Figure 22.1(a) is blind zones at every intersecting point of coils, which can be solved by adopting interleaved detection coil patterns, as shown in Figure 22.1(b). Surveying all potential FOD methods, as far as I know, this induced voltage detection method of Witricity is a potential candidate for the FOD (= MOD) solution of EVs.

In addition to the FOD, the position detection (POD) of EVs to align an Rx coil with a Tx coil for its high efficient power transfer is also one of the important functions for wireless stationary EV chargers. For POD, several candidates have been proposed so far, which include a video

Wireless Power Transfer for Electric Vehicles and Mobile Devices, First Edition. Chun T. Rim and Chris Mi.
© 2017 John Wiley & Sons Ltd. Published 2017 by John Wiley & Sons Ltd.

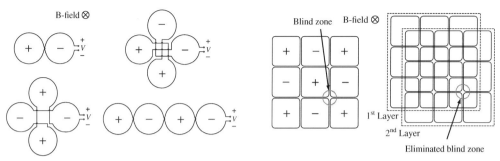

(a) Overlapped detection coils for
canceling induced voltages

(b) Two patterns for eliminating blind zones

Figure 22.1 Witricity patent for overlapped detection coil patterns [25].

camera, an RFID, and optical sensors in the SAE J2954 [11–23], but those are complex systems requiring a high cost to implement.

In this chapter, a novel dual-purpose non-overlapped detection coil (also briefly referred as a "detection coil") set for both FOD and POD, which is located on a Tx coil, is explained. The proposed detection coil set can be easily fabricated on the printed circuit board, which is not possible for the Witricity FOD method. As shown in Figure 22.2, the proposed detection coil set has no self-overlap between coils, which makes it possible to print the detection coil patterns with great ease and low price. Both the detection of conductive debris on a Tx coil and the position of them can be simultaneously found by the proposed coil sets using an induced voltage difference of non-overlapped coil sets for FOD. Moreover, by measuring an induced voltage of non-overlapped coil sets, displacements between a Tx coil and an Rx coil can also be expected when an Rx coil moves on a Tx coil. The proposed FOD and POD methods do not bring any meaningful power loss in the IPT system of EV chargers. The proposed non-overlapped coil sets have been demonstrated by simulations and experiments through a prototype non-overlapped coil set.

22.2 Non-overlapped Coil Sets for FOD and POD

22.2.1 Overall Configuration of the Proposed Coil Sets

In general, the IPTS for wireless stationary EV chargers consist of two subsystems. One is the transmitter (Tx) subsystem to provide power and consists of a utility frequency rectifier, a high-frequency inverter, a primary capacitor bank, and a Tx coil including cores and power cables. The other one is the on-board (Rx) subsystem to receive the required power from the transmitter subsystem and includes an Rx coil with cores and power cables, a secondary capacitor bank, a high-frequency rectifier, and a DC–DC regulator.

The proposed non-overlapped coil sets for FOD and POD are installed on a Tx coil. As shown in Figure 22.2(a), the coil sets are categorized into two coil sets based on their different missions: one is lateral coil sets to obtain the lateral position information of conductive objects on a Tx coil and the other is for the longitudinal position information. In practice, lateral and longitudinal coil sets overlap each other and are located on a power supply rail simultaneously to obtain both lateral and longitudinal information. In Figure 22.2(a), those two different coil sets (left and center) are spatially separated for the better understanding of readers.

Figure 22.2 Proposed non-overlapped detection coil set.

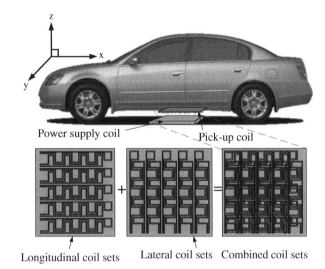

(a) Overall configuration (power supply coil = Tx, pick-up coil = Rx)

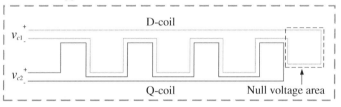

(b) Proposed non-overlap detection coil set

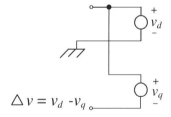

(c) Connection of D-coil and Q-coil to get a detection voltage

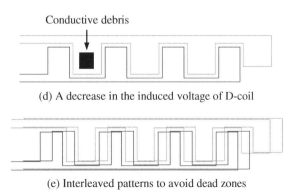

(d) A decrease in the induced voltage of D-coil

(e) Interleaved patterns to avoid dead zones

22.2.2 Proposed Non-overlapped Coil Sets for FOD and POD

As shown in Figure 22.2(b), the proposed non-overlapped detection coil set consists of two non-overlapped symmetric coils, D- and Q-coils. Their magnetic poles are symmetrically arranged to obtain the same induced voltages when a Tx coil generates magnetic flux during or before its EV charging. Then a reference voltage, which is ideally zero from (22.1) below, can be obtained when there are no conductive objects on the Tx coil, as shown in Figure 22.2(c). In practice, the magnetic flux generated by the Tx coil is not uniform through the D- and Q-coils. Therefore, a null-voltage area of each detection coil, which can be manually controlled to increase or decrease, is needed so that the reference voltage can be set to zero or nearly zero by hand, as depicted in Figure 22.2(b):

$$\Delta v = v_d - v_q \tag{22.1}$$

The induced voltages of the proposed detection coil set can be calculated from Faraday's law as follows:

$$v_d = \frac{d\phi_d}{dt} \tag{22.2a}$$

$$v_q = \frac{d\phi_q}{dt} \tag{22.2b}$$

where the time-varying induced voltages of the D- and Q-coils are v_d and v_q, respectively, and ϕ_d and ϕ_q denote the magnetic fluxes generated by a Tx coil to pass through the D- and Q-coils, respectively.

When a metal debris on the Tx coil exists as well as on the detection coil set, the voltage difference $\Delta v = v_d - v_q$ would have a non-zero value because the debris disturbs magnetic flux passing through one of the D- and Q-coils. The voltage difference $\Delta v = v_d - v_q$ could be zero even though conductive materials exist on a Tx coil when conductive materials cover the same area of both the D- and Q-coils, which is different from Figure 22.2(d). Similar to Figure 22.1(b), an additional detection coil set can be interleaved as a solution for this problem, as shown in Figure 22.2(e).

As shown in Figure 22.3, vertical and horizontal detection coil patterns can be constructed by assembling the detection coil set of Figure 22.2. In this way, conductive object detection as well as position detection can be performed from the vertical and horizontal detection voltages

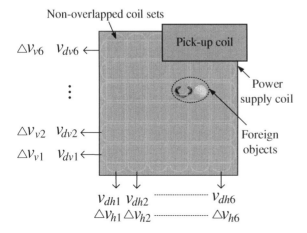

Figure 22.3 Proposed vertical and horizontal detection coil patterns for both FOD and POD.

Figure 22.4 Principle of determining the threshold voltage of the proposed detection coil set for POD.

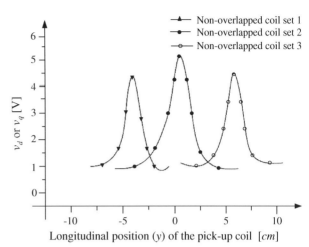

such as Δv_{v4}, Δv_{h4}, and Δv_{v5} for the example of foreign objects located at the (4, 4) and (5, 4) matrix, as shown in Figure 22.3.

In addition, the position of an Rx coil for POD can be determined by measuring D- or Q-coil voltages such as v_{dv5}, v_{dv6}, v_{dh3}, v_{dh4}, v_{dh5}, and v_{dh6}, as shown in Figure 22.3. The magnetic flux passing through each coil set changes and leads to the variation of the induced voltages of each detection coil when the Rx coil moves on the Tx coil, as shown in Figure 22.3. The vehicle position is thus identified by measuring one of D- or Q-coils. The threshold voltages of D- or Q-coils, which determine whether the Rx coil covers the Tx coil or not, can be obtained by comparing the induced voltage variation as the Rx coil moves over the Tx coil, as shown in Figure 22.4. In general, the induced voltage of the detection coil has its minimum value when the Rx coil is totally out of the Tx coil, while the induced voltage increases as the Rx coil moves close to the Tx coil.

22.2.3 Simulations for Foreign Object Detection

In order to figure out the feasibility of the proposed detection coil sets for FOD, an FEA simulation model is used, as shown in Figure 22.5. In the simulation, only a detection coil set is examined to see its feasibility and to find appropriate dimensions, as listed in Table 22.1.

As shown in Figure 22.6, the voltage difference between the D- and Q-coils increases as the number of coins increases. The reference voltage is initially adjusted to 7.5 mV and the voltage difference increases up to 22.4 mV for eight coins on a Tx coil.

22.2.4 FOD Operating Algorithm

The overall operating sequence of the EV charger using the proposed detection coil set for FOD can be explained by using the flowchart shown in Figure 22.7. You may suggest another operating sequence if necessary.

First of all, check a vehicle in the parking area. If there is no vehicle on the Tx coil, then turn off the Tx inverter.

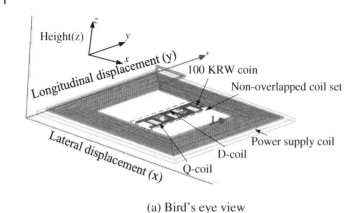

(a) Bird's eye view

Figure 22.5 FEA simulation model of the proposed detection coil set for FOD, together with a Tx coil.

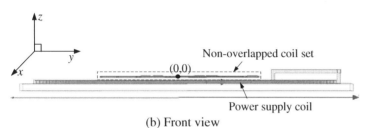

(b) Front view

If there is an EV, it starts to flow a small current into the Tx coil to check conductive objects. Then the difference voltages of the coil sets are measured to check non-zero values among the coil sets.

If one of the difference voltages is higher than the threshold value, then stop the current supply into the Tx coil and send a notification message to a driver to remove the foreign objects on a Tx coil.

On the other hand, if all the difference voltages show zero, which means that all the detection voltages are lower than a reference voltage, check the charging state of a battery. If not fully charged yet, increase the charging current of the Tx coil.

Finally, when the battery is fully charged, then stop the EV charging process.

22.2.5 Simulation for Position Detection

In order to figure out the feasibility of the detection coil set for POD, another simulation model is designed, as shown in Figure 22.8. In the simulation, ten non-overlapped coil sets, that, CS1,

Table 22.1 Simulation parameters of the proposed detection coil set

Parameter	Value
Size of Tx coil	$46 \times 46 \times 0.5$ (cm^3)
Size of detection coil set	$24 \times 4.3\ 0.2$ (cm^3)
Tx coil current	20 A
Size of coin (area)	4.5 (cm^2)

Figure 22.6 Simulation results of the voltage difference of the detection coil set w.r.t. the number of coins.

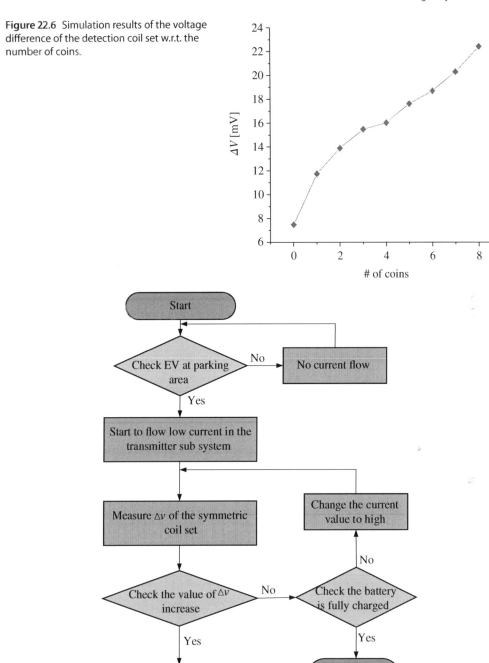

Figure 22.7 Operating diagram of non-overlapped coil sets for FOD.

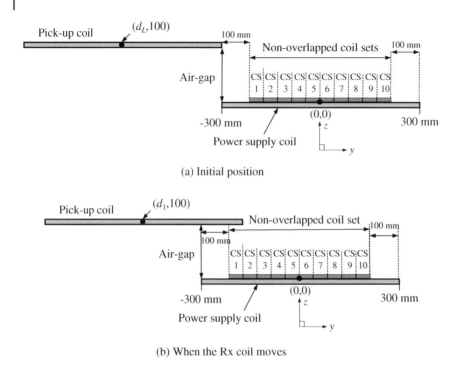

Figure 22.8 Position of the Rx coil with ten detection coil sets for POD (units are mm).

CS2, …, CS10 are used, and the parameters of the Tx coil and the detection coil set are listed in Table 22.2.

At the initial position, the center point of the Rx coil along the y-axis is $d_L = -600$ mm, as shown in Figure 22.8(a). As explained in the previous section, only the induced voltage v_q for the Q-coil is used for POD. When the Rx coil moves close to the Tx coil, v_q steadily increases. For example, if the center point of the Rx coil is at the position $(d_1, 100)$, as shown in Figure 22.8(b), the Rx coil can fully cover the CS1. Therefore, the CS1 gets higher voltage than the other detection coils. By checking the peak voltage of the detection coils, the Rx coil position can be identified.

As shown in Figure 22.9, the simulation for POD shows that v_q of each Q-coil increases when the Rx coil approaches each detection coil and reaches its peak value at the position where the Rx coil fully covers the Q-coil. Using this voltage profile data over different positions, it is possible to determine the position. Note that the detected voltage change reduces drastically

Table 22.2 Simulation parameters of non-overlapped coil EET

Parameter	Value
Size of Tx coil	$60 \times 60 \times 1$ (cm^3)
Size of detection coil set	$40 \times 4 \times 0.1$ (cm^3)
Ampere-turn of Tx coil	1 kA-turn
Size of Rx coil	$60 \times 60 \times 1$ (cm^3)

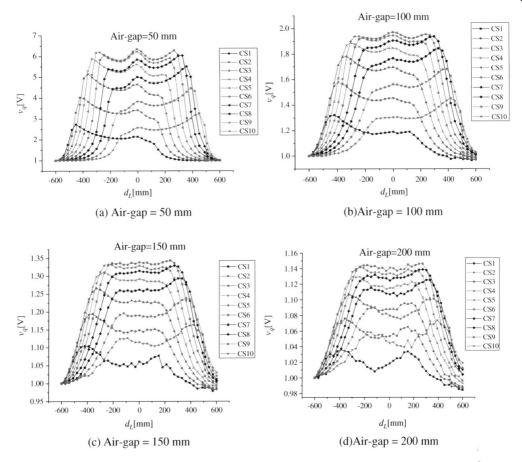

Figure 22.9 Simulation results of the POD, obtaining voltages of Q-coils for different airgaps.

for a higher airgap, which means that the proposed detection coil is not sensitive for high airgap applications.

Different from FOD, there are possibly a few practical limitations on POD for the proposed detection coil set in terms of implementation. One of them is the insensitivity for a large airgap and another is too many datum on the detected information.

22.3 Example Design and Experimental Verifications

In order to verify the feasibility of the proposed detection coil set, an experimental setup was fabricated on a Tx coil, as shown in Figure 22.10.

To mimic the film coil structure, which can be used for the commercialization, a very thin copper wire with a diameter of 0.25 mm has been selected to fabricate the detection coil set. The dimension of the Tx coil was designed as 50 cm × 50 cm 1 cm, respectively, while that of the acryl sheet was selected as 60 cm × 10 cm × 1 cm to separate the coil set from the Tx coil. Here the operating frequency of the inverter is selected as 80 kHz. Other circuit parameters are appropriately selected as listed in Table 22.3.

Power supply coil

Figure 22.10 Experimental setup of the proposed detection coil set with a Tx coil.

Non-overlapped coil set

22.3.1 Low-Pass Filter

During experiments, it was found that there is a large amount of noise in the induced voltages of the detection coil sets, produced by the inverter switching voltage through the capacitive coupling between the detection coil set and the Tx coil. Therefore, it was difficult to set the reference difference voltage at a zero value. In order to reduce the noise, a simple low-pass RC filter with a cut-off frequency of 81 kHz was chosen, as shown in Figure 22.11, which is slightly higher than the operating frequency, to effectively reduce high frequency harmonics. The values R and C are 3.9 kΩ and 0.49 nF, respectively, to get the cut-off frequency.

As shown in Figure 22.12, the high-frequency components of switching noise were significantly reduced by 10 mV due to the RC low-pass filter.

22.3.2 Foreign Object Detection

In order to verify the proposed detection coil set for FOD, the reference voltage is set to less than 10 mV by changing the null voltage area of the D-coil. Two different coins of 10 and 100 Korean won (KRW), which are composed of irons and coppers, respectively, were used as foreign objects.

As shown in Figure 22.13, the experimental results show a good linearity with respect to the number of coins on a Tx coil. However, there is not much voltage difference between two coins even though their areas are different from each other. This is because there is only a small amount of area difference between them, 4.5 cm^2 for 100 KRW and 4.1 cm^2 for 10 KRW.

Table 22.3 Parameters of the proposed detection coil set

Parameter	Value
l_c	40 cm
w_c	5 cm
h_c	8 cm

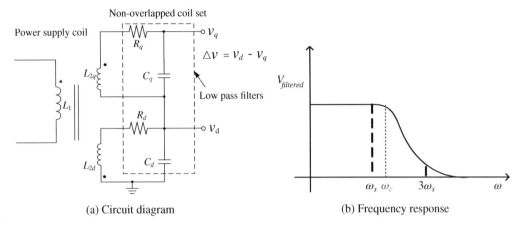

(a) Circuit diagram   (b) Frequency response

Figure 22.11 Proposed RC low-pass filter.

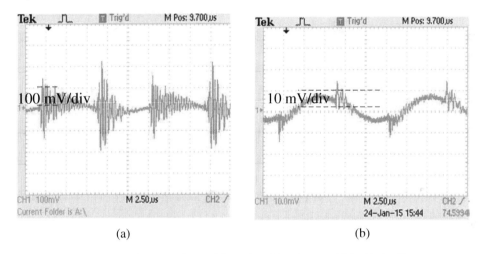

(a)   (b)

Figure 22.12 Measured difference voltage of the coil set (a) without and (b) with the filter.

Figure 22.13 Measured difference voltage of the detection coil set w.r.t. the number of coins.

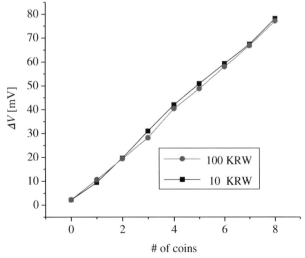

22.4 Conclusion

The proposed non-overlapped detection coil set has been verified for both FOD and POD. FOD is relatively well proved but POD needs a lot of studies to be implemented. Big data handling for different airgaps and positions will be a challenging work in POD of this type.

Problems

22.1 Estimate the stray capacitance between the proposed detection coil set and Tx coil of the example design in this chapter and explain the switching ringing noise with an appropriate equivalent circuit.

22.2 Explain why the proposed induced voltage method is not sensitive to detection of live objects. Actually, this is one of the reasons why capacitive detection is more preferred for LOD.

References

1 S.Y. Choi, S.Y. Jeong, E.S. Lee, B.W. Gu, S.W. Lee, and C.T. Rim, "Generalized models on self-decoupled dual pick-up coils for large lateral tolerance," *IEEE Trans. on Power Electronics*, accepted for publication.

2 S.Y. Choi, B.W. Gu, and S.Y. Jeong, and C.T. Rim, "Trends of wireless power transfer systems for road powered electric vehicles," in *IEEE Vehicular Technology Conference (VTC Spring)*, 2014, pp. 1–5.

3 S.Y. Choi, B.W. Gu, S.Y. Jeong, and C.T. Rim, "Ultra-slim S-type inductive power transfer system for road powered electric vehicles," in *EVTeC and APE Japan*, 2014.

4 A. Shafiei and S.S. Williamson, "Plug-in hybrid electric vehicle charging: current issues and future challenges," in *IEEE Vehicle Power and Propulsion Conference (VPPC)*, 2010, pp. 1–8.

5 C.Y. Huang, J.T. Boys, G.A. Covic, and M. Budhia, "Practical considerations for designing IPT system for EV battery charging," in *IEEE Vehicle Power and Propulsion Conference (VPPC)*, 2009, pp. 402–407.

6 H.H. Wu, A. Gilchrist, K. Sealy, P. Israelsen, and J. Muhs, "A review on inductive charging for electric vehicles," *IEEE International Electric Machines and Drives Conference (IEMDC)*, 2011, pp. 143–147.

7 K. Aditya and S.S. Williamson "Design considerations for loosely coupled inductive power transfer (IPT) system for electric vehicle battery charging – a comprehensive review," *IEEE Transportation Electrification Conference and Expo (ITEC)*, 2014, pp. 1–6.

8 S. Li and C.C. Mi, "Wireless power transfer for electric vehicle applications," *IEEE Journal of Emerging and Selected Topics in Power Electronics*, vol. 3, no. 1, pp. 4–17, March 2015.

9 N. Kuyvenhoven, C. Dean, J. Melton, J. Schwannecke, and A.E. Umenei, "Development of a foreign object detection and analysis method for wireless power systems," in *IEEE Symposium on Product Compliance Engineering (IPSES)*, 2011, pp. 1–6.

10 S. Fukuda, H. Nakano, Y. Murayama, T. Murakami, O. Kozakai, and K. Fujimaki, "A novel metal detector using the quality factor of the secondary coil for wireless power transfer systems," in *IEEE International Microwave Workshop Series on Innovative Wireless Power Transmission: Technologies, Systems, and Applications (IMWS)*, 2012, pp. 241–244.

11 "SAE J2954 overview and path forward," http://www.sae.org/smartgrid/sae-j2954-status_1-2012.pdf.

12 X. Qunyu, N. Huansheng, and C. Weishi, "Video-based foreign object debris detection," in *IEEE International Workshop on Imaging Systems and Techniques*, 2009, pp. 119–122.

13 S. Futatsumori, K. Morioka, A. Kohmura, and N. Yonemoto, "Design and measurement of W-band offset stepped parabolic reflector antennas for airport surface foreign object debris detection radar systems," in *IEEE International Workshop on Antenna Technology: (iWAT)*, 2014, pp. 51–52.

14 T. Kato, Y. Ninomiya, and I. Masaki, "An obstacle detection method by fusion of radar and motion stereo," *IEEE Transactions on Intelligent Transportation Systems*, vol. 3, no. 3, pp. 182–188, September 2002.

15 A. Kohmura, S. Futatsumori, N. Yonemoto, and K. Okada, "Fiber connected millimeter-wave radar for FOD detection on runway," in *IEEE European Radar Conference (EuRAD)*, 2013, pp. 41–44.

16 Z.N. Low, J.J. Casanova, P.H. Maier, J.A. Taylor, R.A. Chinga, and J. Lin, "Method of load/fault detection for loosely coupled planar wireless power transfer system with power delivery tracking," *IEEE Trans. on Industrial Electronics*, vol. 57, no. 10, pp. 1478–1486, April 2010.

17 G. Ombach, "Design considerations for wireless charging system for electric and plug-in hybrid vehicles," in *IEEE Hybrid and Electric Vehicles Conference* (*HEVC 2013*), 2013, pp.1–4.

18 F. Dan, Z. Qi, and Y. Xuelian, "Electromagnetic characteristics simulation of airport runway FOD," in *IEEE International Workshop on Microwave and Millimeter Wave Circuits and System Technology*, 2013, pp. 13–16.

19 H. Kikuchi, "Metal-loop effects in wireless power transfer systems analyzed by simulation and theory," in *IEEE Electrical Design of Advanced Packaging and Systems Symposium (EDAPS)*, 2013, pp. 201–204.

20 J. Svatos, J. Vedral, and P. Novacek, "Metal object detection and discrimination using Sinc signal," in *IEEE 13th Biennial Baltic Electronics Conference (BEC2012)*, 2012, pp. 307–310.

21 L.S. Riggs and J.E. Mooney, "Identification of metallic mine-like objects using low frequency magnetic fields," *IEEE Transactions on Geoscience and Remote Sensing*, vol. 39, no. 1, January 2001.

22 D.C. Chin, R. Srinivasan, and R.E. Ball, "Discrimination of buried plastic and metal objects in subsurface soil," in *IEEE Geoscience and Remote Sensing Symposium Proceedings, IGARSS '98*, 1998, vol. 1, pp. 505–508.

23 H. Kudo, K. Ogawa, N. Oodachi, and N. Deguchi, "Detection of a metal obstacle in wireless power transfer via magnetic resonance," in *IEEE 33rd International Telecommunications Energy Conference (INTELEC)*, 2011, pp. 1–6.

24 S. Verghese, M.P. Kesler, K.L. Hall, and H.T. Lou, "Foreign object detection in wireless energy transfer systems," Patent US 20130069441 A1 (Witricity Corporation), filed on September 9, 2011.

25 P. Feil1, W. Menzel, T.P Nguyen, Ch. Pichot, and C. Migliaccio, "Foreign objects debris detection (FOD) on airport runways using a broadband 78 GHz sensor," in *Proceedings of the 38th European Microwave Conference*, 27–31 October 2008, pp. 1608–1611.

26 R.W. Engelbart, R. Hannebaum, S. Schrader, S.T. Holmes, and C. Walters, "Systems and method for identifying foreign objects and debris (FOD) and defects during fabrication of a composite structure," Patent US 7236625 B2 (The Boeing Company), filed on July 28, 2003.

Part V

Mobile Applications for Phones and Robots

In this part of the book, relatively small power applications of a few W to a few tens of W are explained.

First, reviews on coupled magnetic resonance systems are given. Then mid-range inductive power transfer (IPT) by dipole coils as well as long-range IPT by dipole coils are suggested. Free-space omnidirectional mobile chargers are beginning to get public attention, so this issue is addressed. Lastly, two-dimensional omnidirectional IPT for robots is explained.

Wireless Power Transfer for Electric Vehicles and Mobile Devices, First Edition. Chun T. Rim and Chris Mi.
© 2017 John Wiley & Sons Ltd. Published 2017 by John Wiley & Sons Ltd.

23

Review of Coupled Magnetic Resonance System (CMRS)

23.1 Introduction

CMRS has attracted strong public attention due to its remarkably long wireless power transfer of 60 W up to 2.1 m [1]. A conventional CMRS transfers its wireless power through three major magnetic couplings: a source coil to the transmitter (Tx) coil, the Tx coil to the receiver (Rx) coil, and the Rx coil to the load coil [1–21], as shown in Figure 23.1. Since the CMRS has been proposed [1], numerous studies have followed [2–21]: analysis and improvement of the efficiency by tuning matching circuits [2–13] and equivalent circuit modeling for the analysis of CMRS [2–7, 14–21].

CMRS coils in most previous works, however, are inevitably bulky because four coils with large diameters should be used for a long distance wireless power transfer. Furthermore, open-ended resonant coils, which utilize the stray capacitance of the coil, are often adopted for Tx and Rx coils because of the lack in a high-frequency capacitor with tens of kV ratings [1, 2, 8]. This resonant coil has a much larger size than that of a lumped coil and is inferior under ambient changes such as temperature, humidity, and proximity. The stray inductance and capacitance of a resonant coil may change due to variations in permittivity and adjacent objects, which may eventually result in being out of resonance.

For a long-distance wireless power delivery, an extremely high quality factor of resonant tank Q_i of 2000 [1] was used for previous CMRSs in order to drastically increase the magnetic flux between Tx and Rx coils. The magnetic coupling factor of the Tx–Rx coils is very low when the distance between the coils is large, and a low resistance of the coil is requisite for building a large current in the resonant coil, which results in an extremely large Q_i. One of the problems of adopting high Q_i in a resonant circuit is a narrow operating frequency bandwidth [1,21], which is determined as follows:

$$\Delta f_s = \frac{f_s}{Q_i} \tag{23.1}$$

where f_s is the source frequency, which is tuned to the resonant frequency of the Tx or Rx resonant coil. For example, the frequency tolerance $\Delta f_s/f_s$ becomes merely 0.05% for a $Q_i = 2000$, that is, $\Delta f_s = 6.78$ kHz when $f_s = 13.56$ MHz. Together with the inherently vulnerable structure of conventional CMRS coils, this high Q_i makes the coils very sensitive to environmental changes. Furthermore, the voltage ratings of Tx and Rx resonant coils $V_{L,Q}$ become tremendously large compared to the non-resonant case $V_{L,1}$ by a factor of Q_i as follows:

$$V_{L,Q} = Q_i V_{L,1} \tag{23.2}$$

As identified from (23.2), the power rating of the Tx or Rx coil as well as resonant capacitors for a 60 W power delivery becomes as high as 300 kVA if $Q_i = 2000$ and the power efficiency is 40%.

Wireless Power Transfer for Electric Vehicles and Mobile Devices, First Edition. Chun T. Rim and Chris Mi.
© 2017 John Wiley & Sons Ltd. Published 2017 by John Wiley & Sons Ltd.

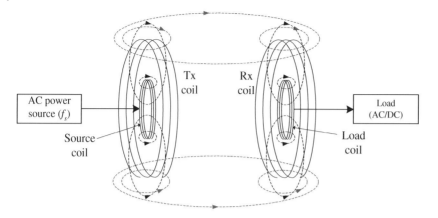

Figure 23.1 Diagram of the conventional CMRS composed of three major magnetic couplings.

Because of the limitations discussed above, CMRS has not been widely accepted, except for some lowpower and high-frequency applications. To keep tuning several resonant coils is extremely painful work and the design of the coils requires elaborative skills.

The understanding of CMRS is complicated in practice due to multiple magnetic couplings that are often omitted for simplification purposes. The source coil is usually embodied in the Tx coil and is coupled with not only the Tx coil but also the Rx coil, and this is also true for the load coil; hence, the exact analysis of CMRS is much more difficult if all coupling effects are considered. Most equivalent models consider the couplings between two adjacent coils by neglecting others.

It is noteworthy that the source and load coils do not directly contribute to the long-distance power delivery and their roles are to transfer power to the Tx coil or fetch power from the Rx coil, respectively. Hence, the critical part of the long-distance wireless power delivery in CMRS is the Tx–Rx coils, and the operation of CMRS is inherently the same as that of the conventional IPTS [22–34]. In other words, the characteristic of the long-distance wireless power delivery of conventional CMRSs does not stem from "coupled magnetic resonance" but from the extremely high Q_i (or low resistance) and large diameter of Tx and Rx coils. Therefore, the design of the Tx and Rx coils of CMRS can be done in the same way as IPTS, and the source and load coils are not necessarily coreless bulky coils; moreover, the source and load coils could even be completely eliminated from CMRS because they have inherently nothing to do with wireless power delivery except for current or voltage level change.

Contrary to popular belief, there is no fundamental difference between CMRS and IPTS, and the conventional CMRS is just a kind of IPTS with a very large Q_i and multiple resonances. Furthermore, CMRS is no longer an exclusive candidate for long- distance wireless power delivery. For example, the IPTS with two dipole coils could transfer 209 W for a long distance of 5 m [34, 35] or 10 W for 7 m [36]. The loop coil that has been used for canonical CMRS has been proven to be no longer effective for long-distance power delivery compared to the dipole coil [36].

Based on this new viewpoint on CMRS, the bulky coils for non-critical magnetic coupling regarding source and load coils have been replaced in this chapter by their corresponding compact lumped transformers. As discussed, these transformers are no longer mandatory for CMRS, but they are adopted here because they can change the high circulating current of Tx and Rx coils to a low current, which is appropriately manageable from the source and load sides. The role of the proposed transformers matches the impedance between the source and Tx coil as well as the Rx coil and load. Lumped capacitors are also adopted in this chapter to nullify the reactance of the corresponding coil and transformer. By using these lumped components, the proposed CMRS is no longer highly sensitive to ambient conditions, which was the major drawback of conventional CMRS for practical applications. A high efficiency class-E inverter [37–41] is used in this chapter due to its simplicity and zero voltage switching (ZVS) capability. The explicit transformer model [19] is used to analyze the proposed CMRS; hence, high power and efficiency conditions are explicitly derived in closed form. In the following sections, the operation principle and design procedure are described in detail, and experimental verifications for both 1 W and 10 W prototypes at 500 kHz switching frequency are provided, where very low Q_i of less than 100 is intentionally used to show that Q_i is no longer crucial for CMRS.

23.2 Static Analysis and Design of the Proposed CMRS with Impedance Transformers

23.2.1 Overall Structure

As shown in Figure 23.2, the proposed CMRS is composed of a class-E inverter, a Tx coil, an Rx coil, lumped resonating capacitors C_T, C_R, C_L, a load resistor R_L, and the proposed source and load transformers. For simplicity, it is assumed that the Tx and Rx coils are symmetric, and that all switching harmonics are sufficiently eliminated by resonant tanks. All parasitic capacitances and resistances, except for internal resistances of Tx and Rx coils together with the equivalent series resistances (ESRs) of resonant capacitors, are considered in this chapter. The impedance transformer is made with a high permeability core. Besides the conventional CMRS, there is little magnetic flux leakage from the proposed impedance transformers. Therefore, the coupling factor of this transformer is nearly unity and there is no cross-coupling with other coils, which was a cumbersome problem in the previous CMRS.

Figure 23.3 shows the explicit circuit schematic of the proposed CMRS, where each of the Tx and Rx coils and source and load transformers is modeled as a leakage inductance L_l and a magnetizing inductance L_m with a corresponding turn ratio, neglecting core loss. The resistance r_T represents the sum of the internal resistance of the Tx coil and the ESR of the resonant capacitors C_T, whereas r_R is similarly defined for the Rx coil and C_R. The leakage inductance L_{lT} is the sum of the secondary leakage inductance of the source impedance transformer and the leakage inductance of the Tx coil, whereas L_{lR} is similarly defined for the load impedance

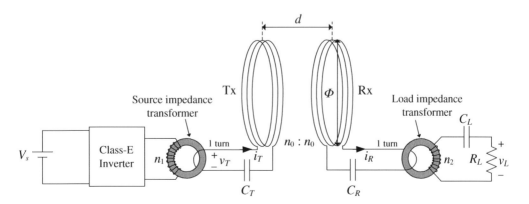

Figure 23.2 Overall configuration of the proposed CMRS including source and load impedance transformers.

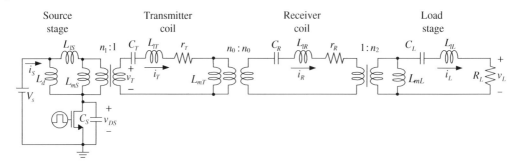

Figure 23.3 Overall circuit schematic of the proposed CMRS, valid for static and dynamic operations.

transformer and Rx coil. The internal resistance of impedance transformers is not included in the model because their power loss is negligible.

As shown in Figure 23.3, a class-E inverter is used to drive the proposed CMRS, where the source impedance transformer is connected in parallel with an external inductor L_d. The detail design of the class-E inverter is available in numerous studies in the literature [37–41] and is omitted here. Instead, the fundamental frequency component of the output of the class-E inverter v_T, which is the input voltage of the proposed CMRS, is exploited, and is represented as a phasor voltage source V_T in the steady state, as shown in Figures 23.4 and 23.5. Under the following resonant conditions, which are fairly unchanged with regard to distance d, each simplified equivalent static circuit from the load side viewpoint and source side viewpoint can be obtained, as shown in Figures 23.4 and 23.5, respectively:

$$j\omega_s(L_{mT} + L_{lT}) + \frac{1}{j\omega_s C_T} = 0 \tag{23.3}$$

$$j\omega_s\left(L_{mT} + L_{lR} + \frac{L_{mL}}{n_2^2}\right) + \frac{1}{j\omega_s C_R} = 0 \tag{23.4}$$

$$j\omega_s(L_{mL} + L_{lL}) + \frac{1}{j\omega_s C_L} = 0 \tag{23.5}$$

In (23.3) to (23.5), note that all the resonant angular frequencies are the same as the source angular frequency ω_s. All the variables in Figures 23.4 and 23.5 are represented in static phasor form. By the successive application of the Thevenin theorem to the resonant tanks, an equivalent voltage source and pure resistance is obtained, as shown in Figures 23.4 and 23.5. Note that neither matrix manipulation nor complicated analysis is required to analyze the eighth-order complex resonant filter circuits with three transformers, which is one of the merits of the proposed circuit oriented analysis. Under the resonant conditions of (23.3) to (23.5), the power factor of the CMRS from the viewpoint of the class-E inverter is unity, as identified from Figure 23.5(d). The voltage gain G_V of the proposed CMRS can be determined from Figure 4(d) as follows:

$$\begin{aligned}
G_V &\equiv \frac{V_L}{V_T} = \frac{V_{Lth}}{V_T} \cdot \frac{V_L}{V_{Lth}} \\
&= \frac{-\omega_s^2 L_{mT} L_{mL}}{n_2\left(r_T r_R + \omega_s^2 L_{mT}^2\right)} \cdot \frac{R_L}{R_{Lth} + R_L} \\
&= \frac{-\omega_s^2 L_{mT} L_{mL} R_L n_2}{r_T \omega_s^2 L_{mL}^2 + R_L n_2^2\left(r_T r_R + \omega_s^2 L_{mT}^2\right)}
\end{aligned} \tag{23.6}$$

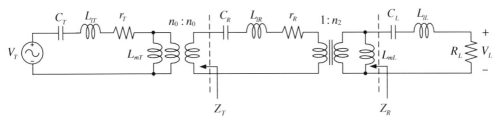

(a) An equivalent static circuit of the CMRS, simplifying the source stage as a sinusoidal voltage source of the fundamental switching frequency

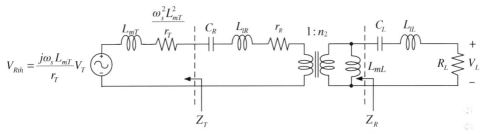

(b) Simplified circuit including the Thevenin equivalent circuit of V_{Rth} and Z_T under resonant condition (23.3)

$$R_{Lth} = \frac{r_T \omega_s^2 L_{mL}^2}{n_2^2 (r_T r_R + \omega_s^2 L_{mT}^2)}$$

$$V_{Lth} = \frac{-\omega_s^2 L_{mT} L_{mL}}{n_2 (r_T r_R + \omega_s^2 L_{mT}^2)} V_T$$

(c) Simplified circuit including the Thevenin equivalent circuit of V_{Lth} and Z_R under resonant condition (23.4)

$$R_{Lth} = \frac{r_T \omega_s^2 L_{mL}^2}{n_2^2 (r_T r_R + \omega_s^2 L_{mT}^2)}$$

$$V_{Lth} = \frac{-\omega_s^2 L_{mT} L_{mL}}{n_2 (r_T r_R + \omega_s^2 L_{mT}^2)} V_T$$

(d) Final equivalent circuit of the CMRS, composed of pure resistance and a voltage source, under resonant condition (23.5)

Figure 23.4 Simplification of static circuits from the load side viewpoint.

Note that the voltage gain has a negative sign, which means that the load phase is the opposite of the input voltage phase. This phase inversion is the result of two consecutive LC resonances, each of which imposes a 90 degree phase difference to the next stage. G_V can be further simplified in case there is no loss in CMRS, which becomes the maximum available

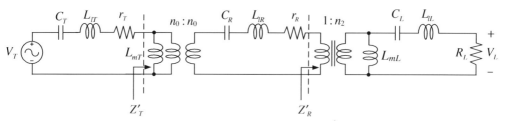

(a) An equivalent static circuit of the CMRS, simplifying the source stage as a sinusoidal voltage source of the fundamental switching frequency

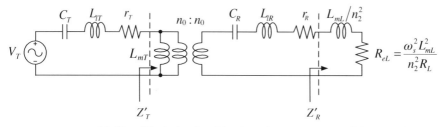

(b) Simplified circuit with Z'_R under resonant condition (23.5)

(c) Simplified circuit with Z'_R under resonant condition (23.4)

(d) Final equivalent circuit of the CMRS under resonant condition (23.3)

Figure 23.5 Simplification of static circuits from the source side viewpoint.

value of (23.6) as follows:

$$\rightarrow G_{V,\max} = \frac{-L_{mL}}{n_2 L_{mT}} = \frac{-n_2^2 L_{mLo}}{n_2 n_0^2 L_{mTo}} = \frac{-n_2 L_{mLo}}{n_0^2 L_{mTo}}$$
$$\text{if } r_T = r_R = 0 \tag{23.7a}$$

$$\because L_{mL} = n_2^2 L_{mLo} \text{ and } L_{mT} = n_0^2 L_{mTo} \tag{23.7b}$$

In (23.7), L_{mL} and L_{mT} are defined using the unit magnetizing inductances of the load transformer and Tx coil (or Rx coil), L_{mLo} and L_{mTo}, respectively. Note that $G_{V,\max}$ is independent of

ω_s and R_L, which is contrary to the general belief, but $G_{V,\max}$ is proportional to n_2 and inversely proportional to n_0^2. Therefore, it is possible to lower the operating frequency if needed and the load voltage can be arbitrarily increased by appropriate selection of n_2. Of course, too low n operating frequency is not recommended when the parasitic resistance effect becomes dominant, as identified from (23.6).

23.2.2 Static Analyses: Power and Efficiency

The load power and efficiency of the system can also be calculated from Figures 23.4 and 23.5, as follows:

$$P_L \equiv I_L^2 R_L = \left(\frac{V_{Lth}}{R_{Lth} + R_L} \right)^2 R_L = \frac{n_2^2 \omega_s^4 L_{mT}^2 L_{mL}^2 V_T^2 R_L}{\left\{ n_2^2 R_L \left(r_T r_R + \omega_s^2 L_{mT}^2 \right) + r_T \omega_s^2 L_{mL}^2 \right\}^2} \tag{23.8}$$

$$\eta \equiv \frac{P_L}{P_i} = \left(\frac{V_{Lth}}{R_{Lth} + R_L} \right)^2 R_L \Big/ \left(\frac{V_T^2}{r_T + R_{eT}} \right)$$

$$= \frac{n_2^2 \omega_s^4 L_{mT}^2 L_{mL}^2 R_L}{\left(n_2^2 r_R R_L + \omega_s^2 L_{mL}^2 \right) \left\{ n_2^2 R_L \left(r_T r_R + \omega_s^2 L_{mT}^2 \right) + r_T \omega_s^2 L_{mL}^2 \right\}} \tag{23.9}$$

In order to determine the optimum number of turns, parameters in (23.8) and (23.9) are rewritten using (23.7b) and $r_T = r_R \cong n_0 r_o$, as follows:

$$P_L = \frac{n_0^2 n_2^2 \omega_s^4 L_{mTo}^2 L_{mLo}^2 V_T^2 R_L}{\left\{ n_0 R_L \left(r_o^2 + n_0^2 \omega_s^2 L_{mTo}^2 \right) + n_2^2 r_o \omega_s^2 L_{mLo}^2 \right\}^2} \tag{23.10}$$

$$\eta = \frac{n_2^2 n_0^3 \omega_s^4 L_{mTo}^2 L_{mLo}^2 R_L}{\left(n_0 r_o R_L + n_2^2 \omega_s^2 L_{mLo}^2 \right) \left\{ n_0 R_L \left(r_o^2 + n_0^2 \omega_s^2 L_{mTo}^2 \right) + n_2^2 r_o \omega_s^2 L_{mLo}^2 \right\}} \tag{23.11}$$

where r_o is the unit internal resistance of either the T_X or R_X coil, and the ESRs of the resonant capacitors and the internal resistance of impedance transformers are omitted in r_T and r_R at the moment due to their negligible power loss compared to that of Tx and Rx coils. As identified from (23.11), efficiency drops as n_2 becomes either too small or too large; hence, a maximum point of efficiency can be found at n_{2m} by taking the derivative of (23.11) with respect to n_2:

$$\left. \frac{\partial \eta}{\partial n_2} \right|_{n_2 = n_{2m}} = 0 \quad \Rightarrow \quad n_{2m} = \frac{\sqrt[4]{n_0^2 R_L^2 \left(r_o^2 + n_0^2 \omega_s^2 L_{mTo}^2 \right)}}{\omega_s L_{mLo}} \tag{23.12}$$

Thus, the maximum load power $P_{L,m}$ and maximum efficiency $\eta_{\max}$ can be derived, respectively, as follows:

$$P_{L,m} \equiv P_L|_{n_2 = n_{2m}} = \frac{V_T^2}{n_0 r_o} \cdot \frac{Q_c^2}{\sqrt{1 + Q_c^2} \left(1 + \sqrt{1 + Q_c^2} \right)^2} = \frac{V_T^2}{r_T} \cdot \frac{\eta_{\max}}{\sqrt{1 + Q_c^2}} \tag{23.13}$$

$$\eta_{\max} \equiv \eta|_{n_2 = n_{2m}} = \frac{n_0^2 \omega_s^2 L_{mTo}^2}{\left(r_o + \sqrt{r_o^2 + n_0^2 \omega_s^2 L_{mTo}^2} \right)^2} = \frac{Q_c^2}{\left(1 + \sqrt{1 + Q_c^2} \right)^2} \tag{23.14}$$

$$\because Q_c \equiv \frac{n_0 \omega_s L_{mTo}}{r_o} = \frac{\omega_s L_{mT}}{r_T} \tag{23.15}$$

In (23.15), Q_c is named the "coupled quality factor" of the Tx or Rx coil in this chapter and is defined as the ratio of magnetizing inductance reactance and internal resistance. Note that Q_c is not the same as the conventional quality factor Q_i determined by the self-inductance and internal resistance, but is related to a coupling factor κ between Tx and Rx as follows:

$$Q_c = \frac{\omega_s(L_{mT} + L_{lT})}{r_T} \cdot \frac{L_{mT}}{L_{mT} + L_{lT}} = \kappa \, Q_i$$

$$\text{for} \quad Q_i \equiv \frac{\omega_s(L_{mT} + L_{lT})}{r_T} \quad \text{and} \quad \kappa \equiv \frac{L_{mT}}{L_{mT} + L_{lT}} \tag{23.16}$$

As pointed out in (23.14), η_{max} is solely determined by Q_c and increases as Q_c increases; hence, a high value of Q_c is crucial for a high η_{max}, as anticipated. For example, Q_c must be 2.8 for $\eta_{max} = 50\%$ and 19.0 for $\eta_{max} = 90\%$, respectively.

For convenience, the normalized maximum load power $P_{L,n}$ is defined from (23.13) and (23.15), for given V_T and r_o, as follows:

$$P_{L,n} \equiv \frac{P_{L,m}}{V_T^2/r_o} = \frac{1}{n_0} \cdot \frac{Q_c^2}{\sqrt{1 + Q_c^2}\left(1 + \sqrt{1 + Q_c^2}\right)^2}$$

$$= \frac{n_0 Q_0^2}{\sqrt{1 + n_0^2 Q_0^2}\left(1 + \sqrt{1 + n_0^2 Q_0^2}\right)^2} \tag{23.17a}$$

$$\because Q_c \equiv n_0 Q_0, \quad Q_0 = \frac{\omega_s L_{mTo}}{r_o} \tag{23.17b}$$

As observed in (23.17), $P_{L,n}$ is a function of not only n_0 but also Q_0, that is, $\omega_s L_{mTo}$, and reaches its maximum at an appropriate value of either n_0 or Q_0. Therefore, an optimum $P_{L,n}$ can be found by taking the derivative of (23.17) with respect to n_0 as follows:

$$\left.\frac{\partial P_{L,n}}{\partial n_0}\right|_{n_0=n_{0m}} = 0 \quad \Rightarrow \quad n_{0m} = \sqrt[4]{\frac{3}{4}} \cdot \frac{1}{Q_0} \cong \frac{0.931}{Q_0} \tag{23.18}$$

Applying (23.18) to (23,17), noticing that $Q_c = n_{0m}Q_0 = 0.931$, results in the normalized maximum power as follows:

$$P_{L,n,max} = 0.122 Q_0 \tag{23.19}$$

Applying (23.18) to (23.14) for the same $Q_c = n_{0m}Q_0 = 0.931$ results in the maximum efficiency as follows:

$$\eta_{max}|_{Q_c=0.931} = 15.5\% \tag{23.20}$$

As identified from (23.20), the theoretical maximum efficiency under the optimum power condition of (23.18) is quite low and is not likely to be preferred by engineers.

In a similar fashion with (23.18), another optimum $P_{L,n}$ can be found by taking the derivative of (23.17) with respect to Q_0 as follows:

$$\left.\frac{\partial P_{L,n}}{\partial Q_0}\right|_{Q_0=Q_{0m}} = 0 \quad \Rightarrow \quad Q_{0m} = \frac{\sqrt{2(1 + \sqrt{2})}}{n_0} \cong \frac{2.20}{n_0} \tag{23.21}$$

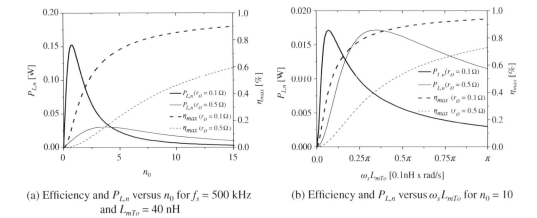

(a) Efficiency and P_{Ln} versus n_0 for $f_s = 500$ kHz and $L_{mTo} = 40$ nH

(b) Efficiency and P_{Ln} versus $\omega_s L_{mTo}$ for $n_0 = 10$

Figure 23.6 Theoretical results for the normalized load power P_{Ln} and maximum efficiency versus n_0 and $\omega_s L_{mTo}$ for $r_o = 0.1\,\Omega$ and $0.5\,\Omega$, respectively.

Applying (23.21) to (23.17), noticing that $Q_c = n_0 Q_{0m} = 2.20$, results in the normalized maximum power as follows:

$$P_{L,n,\text{max}} = \frac{0.172}{n_0} \tag{23.22}$$

Applying (23.21) to (23.14), for the same $Q_c = n_0 Q_{0m} = 2.20$, results in the maximum efficiency as follows:

$$\eta_{\text{max}}|_{Q_c=2.20} = 41.5\% \tag{23.23}$$

Comparing (23.23) with (23.20), the theoretical maximum efficiency under the optimum load power condition of (23.21) is found to be much better and is likely to be preferred by some engineers who want to maximize the transfer power with reasonable efficiency.

Figure 23.6(a) shows the calculation results of (23.14) and (23.17) w.r.t. n_0, where a maximum load power exists at an optimum n_0. For a higher efficiency than the theoretical limit of (23.23), the maximum load power condition (23.18) must be discarded and a large value of n_0 should be chosen. Figure 23.6(b) also presents the calculation results of (23.14) and (23.17) by changing $\omega_s L_{mTo}$, that is, Q_0, where the curve patterns are similar to that in Figure 23.6(a) except for the increased efficiency. The efficiency at each peak load power is always 41.5%, regardless of r_o, as predicted from (23.23). The maximum efficiency and load power are in a trade-off relation; hence, high efficiency and large power cannot be simultaneously achieved by just increasing n_0 and $\omega_s L_{mTo}$ when all other parameters but n_{2m} are fixed. By selecting the parameters n_{2m} and n_0, which are easily manageable, the desired power and optimum efficiency can be appropriately designed.

23.2.3 Design of the Proposed CMRS

The design of a CMRS involves an enormous number of circuit parameters even though the Tx and Rx of the system are symmetrical, as shown in Figure 23.3. They can be classified into the hard parameters that are difficult to change and the soft parameters that are easy to manage. The former includes the distance d between Tx and Rx coils, the diameter Φ, and the source angular frequency ω_s, which are usually determined by other system requirements. The latter includes the numbers of turns of the source, load, and Tx and Rx coils, that is n_1, n_2, and n_0, respectively. Therefore, the design is focused on determining the numbers of turns for maximizing the efficiency or load power under the given source voltage V_s and load resistor R_L.

The assumptions for the design of the proposed CMRS are as follows: source voltage V_s and load resistor R_L are given; the distance d between Tx and Rx coils, the diameter Φ, the internal resistance for a unit turn of the coils r_o, and the switching angular frequency ω_s are also given; the load power and efficiency are specified for the application; and the number of turns n_1 is greater than 1 for reducing the current in the class-E inverter.

Under these assumptions, the design procedure of the CMRS is established as six steps, as shown below.

1. Calculate Q_c from (23.14) for the given efficiency as follows:

$$Q_c \equiv \frac{2\sqrt{\eta_{max}}}{1 - \eta_{max}} \quad (0 \leq \eta_{max} < 1) \tag{23.24}$$

2. Determine the unit magnetizing inductance L_{mTo} and unit internal resistance r_o by simulation, design theory, or measurement for the given dimensional parameters d and Φ
3. Calculate n_0 from (23.15) by using the Q_c, L_{mTo}, and r_o of steps 1 and 2, as follows:

$$n_0 = \frac{r_o Q_c}{\omega_s L_{mTo}} \tag{23.25}$$

4. Calculate n_2 from (23.12), where the unit magnetizing inductance L_{mLo} is determined for a selected impedance transformer.
5. Calculate $|V_T|$ from (23.13) by using the n_o, r_o, and Q_c of steps 1, 2, and 3, as follows:

$$|V_T| = \sqrt{\frac{n_o r_o P_{L,m} \sqrt{1 + Q_c^2}}{\eta_{max}}} \tag{23.26}$$

6. Determine n_2 from the class-E inverter design [37–41], using the $|V_T|$ of step 5.

The design procedure is summarized as shown in Figure 23.7.

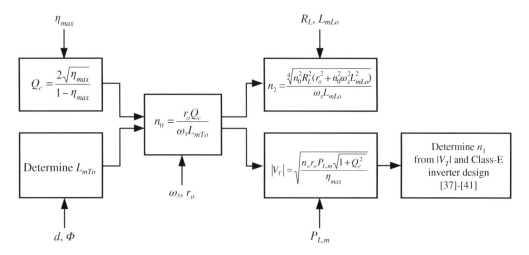

Figure 23.7 Design procedure of the proposed CMRS.

A design example, which is the baseline of the 1 W experimental kit, is shown here. First, the power, efficiency, source voltage V_s, and load resistor R_L are selected as 1 W, 50%, 20 V, and 50 Ω, respectively. The diameter Φ and the distance between Tx and Rx coils d are given as 20 cm and 13 cm, respectively; hence, the unit magnetizing inductance is determined as 45 nH by measurement. Considering commercially available components of capacitors, cores, and power switches, f_s is selected as 500 kHz. Unlike conventional CMRS, a thin litz wire of 0.5 mm diameter, which brings $r_o = 0.5$ Ω at 500 kHz, is intentionally used to see the practical impact of high internal resistance on the system performance. From (23.24) and (23.25), $Q_c = 2.8$ and $n_0 = 10$ are obtained, which are quite acceptable for practical implementation. Obviously, the efficiency may reach to 80% for $Q_c = 8.9$, that is, $n_0 = 10$ and $r_o = 0.18$ Ω, which are still reasonable to implement. Assuming that an Mn–Zn-type toroid ferrite core PL-F2, applicable for 1 MHz, is selected to implement impedance transformers, the unit magnetizing inductance L_{mLo} is determined to be 2.50 μH by measurement. Using these parameters, the optimum number of turns n_2 is calculated from (23.12) to be 3.48; hence, an integer number n_2 needs to be determined between 3 and 4 by an experiment.

23.3   Example Design and Experimental Verifications

The analysis and design results of the proposed CMRS in the previous sections are verified at 500 kHz switching frequency with two experimental kits of 1 W and 10 W load powers, where Q_c is 2.8 and 8.9, respectively. The intrinsic quality factor Q_i of the Tx and Rx coils for the experimental kits are 26.3 and 76.8, respectively. Note that this Q_i is much smaller than 2000 of the conventional CMRS. A class-E inverter was fabricated based on the design guidelines [37–41] at 500 kHz, where the soft switching characteristic is guaranteed over a wide range of experimental conditions.

23.3.1   Experimental CMRS: Fixed Distance and Variable Load Conditions

A prototype CMRS, as shown in Figure 23.8, was implemented for the design verification of the previous section. Compact impedance transformers were used and found to be quite robust to ambient operating conditions such as proximity to the human body. The self-inductances of Tx and Rx were measured as 58.1 μH and 60.3 μH, respectively. A little discrepancy between

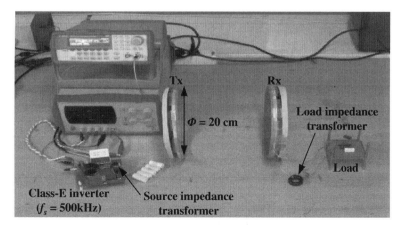

Figure 23.8 A prototype fabrication of the proposed CMRS for 1 W load power.

Table 23.1 Circuit parameters for the 1 W experimental kit.

Parameter	Value	Parameter	Value	Parameter	Value	Parameter	Value
L_{mLo}	2.50 μH	L_{mTo}	0.045 μH	C_T	1.74 nF	C_R	1.68 nF
L_{mS}	40.0 μH	L_T	58.1 μH	r_T	5.37 Ω	C_L	2.23 nF
L_{mL}	40.0 μH	L_R	60.3 μH	r_R	6.92 Ω	C_S	15.0 nF

them may be due to mismatch in winding. The resonant capacitors C_T and C_R for Tx and Rx coils were accordingly determined from (23.3) and (23.4) as 1.74 nF and 1.68 nF, respectively, and the load side capacitance C_L was determined from (23.5) as 2.53 nF. The turn ratio of the load impedance transformer n_2 was experimentally determined as "4," because it gives a higher power delivery than "3." The turn ratio of the source impedance transformer n_1 was determined, considering the prototype class-E inverter voltage ratings, as "4" as well. The unit magnetizing inductance of the impedance transformers was measured experimentally to be 2.5 μH, where their leakage inductance is negligible. The parameters of the proposed experimental CMRS are summarized in Table 23.1.

Figure 23.9 shows the experimental voltage and current waveforms for the condition of $V_s = 20$ V$_{dc}$, $f_s = 500$ kHz, $R_L = 50$ Ω, and $n_1 = n_2 = 4$. The gate and drain-source voltages of the MOSFET of the class-E inverter ensure the zero voltage switching operation, as shown in Figure 23.9(a). The waveforms for v_T and each coil current are shown in Figure 23.9(b), where the phase difference between successive coil currents is $\pi/2$, which matches well with the theory [19].

The load power and efficiency were measured for a wide range of load resistances from 10 Ω to 400 Ω, as shown in Figure 23.10. It is found that the theoretical results of (23.10) and (23.11) matched quite well with the experimental measurements. A slight discrepancy in the efficiency was observed, which is partially due to parasitic resistances. The remaining discrepancy is mainly due to the resistance and core loss of impedance transformers. In practice, it is quite difficult to have exact tuning for the resonance condition (23.3) to (23.5) in a high frequency because of limited values of the capacitor and parasitics. The maximum power of 1.4 W was achieved for $R_L = 30$ Ω, whereas the maximum efficiency of 40% was obtained for $R_L = 75$ Ω.

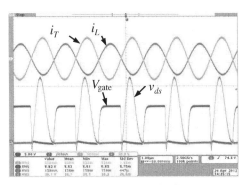

(a) The inverter voltages and coil currents of Tx and the load

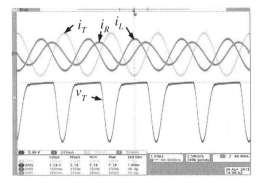

(b) v_T of the CMRS and each coil current in the resonant mode

Figure 23.9 The voltages of the class-E inverter and each coil current in the resonant mode: $V_s = 20$ V, $f_s = 500$ kHz, $R_L = 50$ Ω, $n_1 = n_2 = 4$, $I_T = 525$ mA, $I_R = 280$ mA, $I_L = 157$ mA, and $V_T = 7.19$ V.

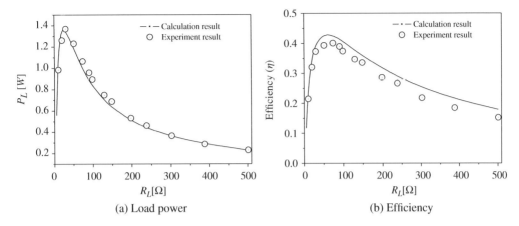

(a) Load power (b) Efficiency

Figure 23.10 Measured load power and efficiency of 1 W experimental kit, compared to the theoretical analyses results.

The measured maximum efficiency is quite close to the theoretical limit of 41.5% because the Q_c of the experimental kit is 2.8, which is close to the optimum condition $Q_c = 2.2$ from (23.21).

23.3.2 Experimental CMRS: Variable Distance and Fixed Load Condition

Another experimental kit for 10 W level power delivery was fabricated to see the effect of distance between Tx and Rx coils for higher efficiency, as shown in Figure 23.11. The internal resistance r_R, which includes not only the Rx coil resistance but also the capacitor ESR and ferrite core loss of the load impedance transformer in practice, was measured as 1.8 Ω with a thicker litz wire of 1.5 mm diameter. This experimental kit was designed to have a maximum efficiency of 80% at $d = 13$ cm, where the load power P_L, the source voltage V_s, and the load resistance R_L were selected as 10 W, 44 V, and 50 Ω, respectively. Other design parameters such as n_0, n_1, n_2, and Φ are the same as that of the 1 W case except the unit magnetizing inductance

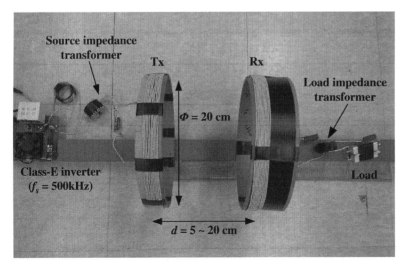

Figure 23.11 A prototype fabrication of the proposed CMRS.

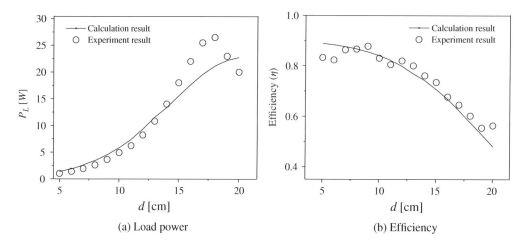

(a) Load power

(b) Efficiency

Figure 23.12 Measured load power and efficiency of the 10 W experimental kit, compared to the theoretical analyses results.

L_{mTo}, which is subtle, decreased to 44 nH at $d = 13$ cm due to the change in thickness of the Tx and Rx coils.

The load power P_L and efficiency η were measured w.r.t. the d from 5 cm to 20 cm and compared with the calculation results, as shown in Figure 23.12. As identified from (23.8) or (23.10), P_L decreases from its maximum as the magnetizing inductance of Tx or Rx coils L_{mT} deviates from its optimum value determined from (23.15) and (23.21). The optimum distance d to give the optimum L_{mT} was found to be 23 cm, but the measured value of d was 18 cm, as shown in Figure 23.12(a). The discrepancy is mostly due to stray capacitances of Tx and Rx coils as well as impedance transformers. As identified from (23.9) or (23.11), efficiency η increases and becomes saturated as L_{mT} increases. Therefore, η gradually decreases as d increases, which results in a decrease of L_{mT}. As shown in Figure 23.12(b), the efficiency η at $d = 13$ cm, excluding the class-E power consumption, was measured as 80.2%, while the total system efficiency including the class-E inverter was 34.5%. Comparing Figure 23.12(a) with (b), a trade-off relationship between maximum efficiency and load power is found, where all parameters except L_{mTo} are fixed.

23.4 Discussion on Phase and EMF Cancel

It is noteworthy that the phases between adjacent coils are found to be sequentially different by $\pi/2$ since the resonant current built up the induced voltage at the magnetizing inductance of the adjacent coil by $\pi/2$ (see Figure 23.13). These phase relationships should be considered for lowering the EMF levels. As the phase of the current in each coil differs by $\pi/2$, an EMF cancelation scheme should be employed for each coil. In other words, there is no simple way of canceling the EMF generated from four coils by an EMF cancel coil, for example.

23.5 Conclusion

A compact and robust CMRS with two lumped impedance transformers was presented in this chapter. Throughout the detailed analysis, a complete but simple design guide was explained.

Figure 23.13 Phasor vector plots of all coils.

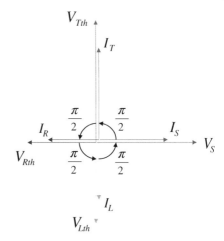

The proposed CMRS was verified by two experimental kits, 1 W and 10 W, with good agreement with theoretical load power and efficiency. Contrary to conventional CMRS, which is hypersensitive to surroundings due to its excessively high quality factor, a very low quality factor of less than 100 was implemented for achieving a 10 W power transfer with a reasonably high efficiency of 80.2% at a distance of 13 cm. The proposed impedance transformer not only drastically simplified the design and implementation of a CMRS but also made the CMRS degenerate into an ordinary IPTS. Therefore, it can be concluded that the conventional CMRS, generally speaking, is just a special form of IPTS whose magnetic coupling is very weak and quality factor is extremely high for a long-distance wireless power delivery.

Problems

23.1 It was first proved by the author that CMRS is a sort of IPTS, as explained in this chapter. In order to demonstrate that the so-called long-distance WPT characteristic of CMRS does not come from a WPT type but come from a coil design only, I have developed the dipole coils for a 5 m long-distance IPT, as will be introduced in subsequent chapters. Nevertheless, a few problems still remain.

    (a) Is there no real benefit for the four-coil system of CMRS compared to an existing IPTS with the same coil design? If yes, how can it be proved in general?

    (b) How about the use of resonant relay coils in CMRS compared to the corresponding IPTS? By successive use of the resonant relay coils, you can extend the WPT distance up to even 10 m.

23.2 The merits of CMRS come from the air coil structure, which has a very high operating frequency due to its coreless structure. More focus on CMRS researches should be made in the air coil design and magnetic field shield design, rather than compensation circuit design. Suggest any design idea on the more fruitful areas mentioned above.

References

1 A. Kurs, A. Karalis, R. Moffatt, J.D. Joannopoulos, P. Fisher, and M. Soljacic, "Wireless power transfer via strongly coupled magnetic resonance," *Science*, vol. 317, no. 5834, pp. 83–86, June 2007.

2 A.P. Sample, D.A. Meyer, and J.R. Smith, "Analysis, experimental results, and range adaption of magnetically coupled resonators for wireless power transfer," *IEEE Trans. on Ind. Electron.*, vol. 58, no. 2, pp. 544–554, February 2011.

3 T. Imura and Y. Hori, "Maximizing air gap and efficiency of magnetic resonant coupling for wireless power transfer using equivalent circuit and Neumann formula," *IEEE Trans. on Ind. Electron.*, vol. 58, no. 10, pp. 4746–4752, October 2011.

4 S.G. Lee, H. Hoang, Y.H. Choi, and F. Bien, "Efficiency improvement for magnetic resonance based wireless power transfer with axial-misalignment," *Electron. Letters*, vol. 48, no. 6, pp. 339–340, March 2012.

5 T.C. Beh, T. Imura, and Y. Hori, "Basic study of improving efficiency of wireless power transfer via magnetic resonance coupling based on impedance matching," in *2010 ISIE Conference*, pp. 2100–2016.

6 M. Zargham and P.G. Gulak, "Maximum achievable efficiency in near-field coupled power transfer systems," *IEEE Trans. on Biomed. Circuits Syst.*, vol. 6, no. 3, pp. 228–245, June 2012.

7 A.K. Ramrakhyani, S. Mirabbasi, and M. Chiao, "Design and optimization of resonance-based efficient wireless power delivery systems for biomedical implant," *IEEE Trans. on Biomed. Circuits Syst.*, vol. 5, no. 1, pp. 48–63, February 2011.

8 J. Park, Y. Tak, Y. Kim, Y. Kim, and S. Nam, "Investigation of adaptive matching methods for near-field wireless power transfer," *IEEE Trans. on Antennas Propagation*, vol. 59, no. 5, pp. 1769–1773, May 2011.

9 C.K. Lee, W.X. Zhong, and S.Y.R. Hui, "Effects of magnetic coupling of nonadjacent resonators on wireless power transfer domino–resonator systems," *IEEE Trans. on Power Electron.*, vol. 27, no. 4, pp. 1905–1916, April 2012.

10 D.K. An and S.C. Hong, "A study on magnetic field repeaters in wireless power transfer," *IEEE Trans. on Ind. Electron.*, vol. 60, no. 1, pp. 360–371, January 2013.

11 T. Mizuno, S. Yachi, A. Kamiya, and D. Yamamoto, "Improvement in efficiency of wireless power transfer of magnetic resonant coupling using magnetoplated wire," *IEEE Trans. on Magnetics*, vol. 47, no. 10, pp. 4445–4448, October 2011.

12 E.A. Setiawan, A. Qolbi, F. Kawolu, and I. Jotaro, "Analysis of the effect of nickel electroplating layer addition on receiver coil of wireless power transfer system," in *2011 TENCON Confence*, pp. 964–967.

13 J. Kim, H. Son, K. Kim, and Y. Park, "Efficiency analysis of magnetic resonance wireless power transfer with intermediate resonant coil," *Antennas and Wireless Propagation Letters*, vol. 10, pp. 389–392, May 2011.

14 B.L. Cannon, J.F. Hoburg, D.D. Stancil, and S.C. Goldstein, "Magnetic resonant coupling as a potential means for wireless power transfer to multiple small receivers," *IEEE Trans. on Power Electron.*, vol. 24, no. 7, pp. 1819–1825, July 2009.

15 E.M. Thomas, J.D. Heebl, C. Pfeiffer, and A. Grbic, "A power link study of wireless non-radiative power transfer systems using resonant shielded loops," *IEEE Trans. on Circuits Syst.*, vol. 59, no. 9, pp. 2125–2136, September 2012.

16 M. Kiani, and M. Ghovanloo, "The circuit theory behind coupled-mode magnetic resonance-based wireless power transmission," *IEEE Trans. on Circuits Syst.*, vol. 59, no. 9, pp. 2065–2074, September 2012.

17 S.H. Cheon, Y.H. Kim, S.Y. Kang, M.L. Lee, J M. Lee, and T.Y. Zyung, "Circuit-model-based analysis of a wireless energy-transfer system via coupled magnetic resonances," *IEEE Trans. on Ind. Electron.*, vol. 58, no. 7, pp. 2906–2914, July 2011.

18 J.A. Faria, "Pointing vector flow analysis for contactless energy transfer in magnetic systems," *IEEE Trans. on Power Electron.*, vol. 27, no. 10, pp. 4292–4300, October 2012.

19 J. Huh, W.Y. Lee, S.Y. Choi, G.H. Cho, and C.T. Rim, "Explicit static circuit model of coupled magnetic resonance system," in *2011 ECCE-Asia Conference*, pp. 2233–2240.

20 E. Lee, J. Huh, X.V. Thai, S. Choi, and C. Rim, "Impedance transformers for compact and robust coupled magnetic resonance systems," in *2013 ECCE Conference*, pp. 2239–2244.

21 B. Luo, S. Wu, and N. Zhou, "Flexible design method for multi-repeater wireless power transfer system based on coupled resonator bandpass filter model," *IEEE Trans. on Circuits Syst.*, accepted for publication.

22 B. Lee, H. Kim, S. Lee, C. Park, and C. Rim, "Resonant power shoes for humanoid robots," in *2011 ECCE Conference*, pp. 1791–1794.

23 W.X. Zhong, X. Liu, and S.Y. Hui, "A novel single-layer winding array and receiver coil structure for contactless battery charging systems with free-positioning and localized charging features," *IEEE Trans. on Ind. Electron.*, vol. 58, no. 9, pp. 4136–4144, September 2011.

24 W.P. Choi, W.C. Ho, X. Liu, and S.Y.R. Hui, "Bidirectional communication technique for wireless battery charging systems and portable consumer electronics," in *2010 APEC Conference*, pp. 2251–2259.

25 P. Arunkumar, S. Nandhakumar, and A. Pandian, "Experimental investigation on mobile robot drive system through resonant induction technique," in *2010 ICCCT Conference*, pp. 699–705.

26 G. Elliott, S. Raabe, G. Covic, and J. Boys, "Multiphase pickups for large lateral tolerance contactless power-transfer systems," *IEEE Trans. on Ind. Electron.*, vol. 57, no. 5, pp. 1590–1598, May 2010.

27 N. Keeling, G. Covic, and J. Boys, "A unity-power-factor IPT pickup for high-power applications," *IEEE Trans. on Ind. Electron.*, vol. 57, no. 2, pp. 744–751, February 2010.

28 H. Wu, J. Boys, and G. Covic, "An AC processing pickup for IPT systems," *IEEE Trans. on Power Electron.*, vol. 25, no. 5, pp. 1275–1284, May 2010.

29 H. Wu, G. Covic, J. Boys, and D. Robertson, "A series-tuned inductive-power-transfer pickup with a controllable AC-voltage output," *IEEE Trans. on Power Electron.*, vol. 26, no. 1, pp. 98–109, January 2011.

30 J. Huh, S. Lee, W. Lee, G. Cho, and C. Rim, "Narrow-width inductive power transfer system for on-line electrical vehicles," *IEEE Trans. on Power Electron.*, vol. 26, no. 12, pp. 3666–3679, December 2011.

31 M. Budhia, G. Covic, and J. Boys, "Design and optimization of circular magnetic structures for lumped inductive power transfer systems," *IEEE Trans. on Power Electron.*, vol. 26, no. 11, pp. 3096–3108, November 2011.

32 H. Li, A. Hu, and G. Covic, "A direct AC–AC converter for inductive power transfer systems," *IEEE Trans. on Power Electron.*, vol. 27, no. 2, pp. 661–668, February 2012.

33 H. Matsumoto, Y. Neba, K. Ishizaka, and R. Itoh, "Model for a three-phase contactless power transfer system," *IEEE Trans. on Power Electron.*, vol. 26, no. 9, pp. 2676–2678, September 2011.

34 C. Park, S. Lee, and C. Rim, "5 m-off-long-distance inductive power transfer system using optimum shaped dipole coils," in *2012 IPEMC Conference*, pp. 1137–1142.

35 C. Park, S. Lee, G. Cho, and C. Rim, "Innovative 5 m-off-distance inductive power transfer systems with optimally shaped dipole coils," *IEEE Trans. on Power Electron.*, 2014, accepted for publication.

36 B. Choi, E. Lee, J. Kim, and C. Rim, "7 m-off-long-distance extremely loosely coupled inductive power transfer systems using dipole coils," in *2014 ECCE Conference*, accepted for publication.

37 N.O. Sokal and A.D. Sokal, "Class E-A new class of high-efficiency tuned single-ended switching power amplifiers," *IEEE Journal on Solid-State Circuits*, vol. 10, no. 3, pp. 168–176, June 1975.

38 J. Garnica, J. Casanova, and J. Lin, "High efficiency midrange wireless power transfer system," in *2011 IMWS Conference*, pp. 73–76.

39 Z.N. Low, R.A. Chinga, R. Tseng, and J. Lin, "Design and test of a high-power high efficiency loosely coupled planar wireless power transfer system," *IEEE Trans. on Ind. Electron.*, vol. 56, no. 5, pp. 1801–1812, May 2011.

40 F.H. Raab, "Idealized operation of the class E tuned power amplifier," *IEEE Trans. on Circuits Syst.*, vol. 24, no. 12, pp. 725–735, December 1977.

41 B. Choi, D.N. Tan, J. Kim, and C. Rim, "A novel source-side capacitive power transfer system for contactless mobile charger using class-E converter," in *2014 VTC Conference (Workshop on Emerging Technologies: Wireless Power)*, pp. 6–10.

24

Mid-Range IPT by Dipole Coils

24.1 Introduction

Extending the distance of wireless power has a long history beginning with Nikola Tesla trying to make an electric power grid without wires [1]. In 2007, a wireless power transfer scheme using strongly coupled magnetic resonance systems (CMRS) was introduced whose power transfer level and coil-to-coil efficiency are 60 W and 45%, respectively, at a distance of 2 m [2]. CMRS adopted large self-resonant coils at each primary and secondary side to induce a large magnetic field to obtain an extended transfer range. For the high current in these self-resonant coils, the internal resistances of the coils must be very small. This means that the coils must have very high Q factors, which consequently result in very thick wires. The high Q factors also result in substantial voltage stresses on the coils because the coils should sustain Q times larger reactive current or voltage than the corresponding current or voltage of the real power. For example, a 1 MVA rating of coil is required to deliver 400 W for $Q = 2500$. To sustain high-voltage stress among wires in the coil, the coils cannot help being bulky with a large airgap between adjacent wires. Furthermore, the resonant frequencies of the coils are not set by lumped capacitors and inductors but by their inherent stray capacitances and inductances. The stray capacitances and inductances are too sensitive to surroundings such as temperature, humidity, and human proximity [3, 4]. With high Q factors that result in an extremely narrow resonant frequency bandwidth and moving resonant frequency due to environmental sensitivity, a complicated automated matching system is needed to track and tune-up the resonant condition of the high Q coils using switched inductors and capacitors [5, 6]. Even though a tuning-up scheme is applied, matching the multiple resonant coils with high environmental sensitivity is extremely difficult in practice. Due to the distributed coil structure of parasitic capacitance and inductance, the operating frequency of the CMRS tends to be of the order of 10 MHz, which results in the use of RF power amplifiers rather than efficient switching converters [6]. The CMRS for 60 W power transferred over the distance of 50 cm has an apparently high coil-to-coil efficiency of 80% [7]; however, its system efficiency including power source and AC–DC conversion would be quite low. These characteristics are why the well-known CMRS is seldom used in high-power applications. Therefore, inductive power transfer systems (IPTS) have been widely used [8–35] for applications that consume more than tens of watts.

In this chapter, an IPTS driven by an inverter of 20 kHz switching frequency for 5 m off distance is explained. Magnetic dipole coils of narrow and long structure, having ferrite cores inside, are adopted for the primary and secondary coils, minimizing parasitic effects [35]. Optimum-stepped core structures that minimize core loss for a given amount of ferrite material are explained. It is verified by simulations, analyses, and experiments for 20 kHz and 105 kHz

Wireless Power Transfer for Electric Vehicles and Mobile Devices, First Edition. Chun T. Rim and Chris Mi.
© 2017 John Wiley & Sons Ltd. Published 2017 by John Wiley & Sons Ltd.

that the IPTS, which has been, so far, believed to be adequate for proximity wireless power transfer only, is quite suitable for long-distance power delivery as well.

24.2 Primary and Secondary Coil Design

24.2.1 Overall Coil Configuration

The overall configuration of the IPTS, which is composed of an inverter, capacitor banks, a rectifier, and load as well as proposed primary and secondary coils, is shown in Figure 24.1. The primary and secondary windings are wound around the center of primary and secondary cores, and its winding shape is analogous to a helical coil. Each winding is composed of a litz wire to reduce the AC series resistance of the coils. The current of the primary winding generates a magnetic field and then the linkage magnetic flux induces the voltage at the secondary winding.

If an air coil were to be used, the inner part of the coil would have a large magnetic reluctance whereas the outer part of it would have a relatively very small magnetic reluctance because of the large effective field crossing area of the outer part of the coil. Strong magnetic field generation, which is crucial for longer distance power delivery, is limited by the large magnetic reluctance of the inner part of the air coil. To reduce this magnetic reluctance, a long rod-type ferrite core is inserted into the air coil, as shown in Figure 24.1. As a rule of thumb, this ferrite coil generates about a 50 times stronger magnetic field intensity than the air coil by means of the magnetic reluctance reduction.

Magnetic flux lines between the primary and secondary coils are shown in Figure 24.2, showing that parts of the magnetic flux lines are effectively interlinked to the secondary coil. To make

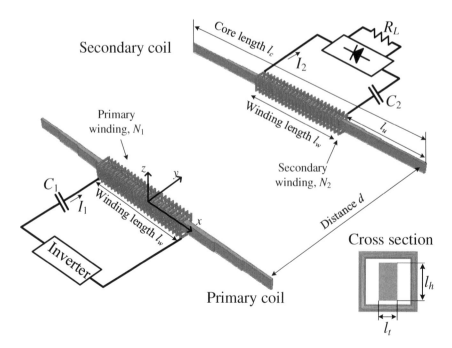

Figure 24.1 Overall configuration of the proposed IPTS including primary and secondary coils.

Figure 24.2 Simulation result of the magnetic flux lines of the proposed coil configuration, where $d = 3$ m and $I_1 = 10$ A.

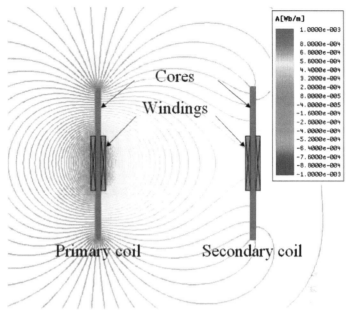

a larger linkage flux, a longer ferrite rod should be inserted. The simulations, throughout this chapter, were performed by Ansoft Maxwell ver. 14.0.

The rms value of the induced voltage of the secondary coil V_2 is proportional to the rms value of the magnetic flux crossing its winding $B_2(x)$ as follows:

$$V_2 = \omega \overline{B_2(x)} A_2 N_2 \tag{24.1}$$

where the winding is evenly distributed, ω is the angular switching frequency, A_2 is the cross-section area of the secondary coil near the center $l_{h2}-l_{t2}$, N_2 is the number of turns of the secondary coil, and the averaged magnetic flux density over the winding length l_w is determined as follows:

$$\overline{B_2(x)} \equiv \frac{1}{l_w} \int_{-l_w/2}^{l_w/2} B_2(x)dx \tag{24.2}$$

The simulated magnetic flux density at the center of the secondary coil $B_2(0)$ versus the primary and secondary core length l_c for various distances d is shown in Figure 24.3, where the core lengths of the primary and secondary cores are assumed to be same. The longer the core length is, the larger is the magnetic flux density in the secondary core. For a longer distance of power transfer, the core should be lengthened. As shown in Figure 24.3, the core length of $1-2$ m is too short for a 5 m off-power transfer because of the very low magnetic flux density; however, it becomes considerably increased if the core length is 3 m.

Concerning the winding length l_w, the magnetic linkage flux passing through the secondary winding increases as l_w decreases, as shown in Figure 24.2; hence, $l_w = 0$ is the optimum condition for maximizing the induced voltage, as was identified from (24.2) as follows:

$$Max\{\overline{B_2(x)}\} = \lim_{l_w \to 0} \frac{1}{l_w} \int_{-l_w/2}^{l_w/2} B_2(x)dx = B_2(0) = Max\{B_2(x)\} \tag{24.3}$$

In practice, (24.3) cannot be realized because the narrowed winding length causes deterioration in the frequency response due to parasitic capacitances between each coil winding and

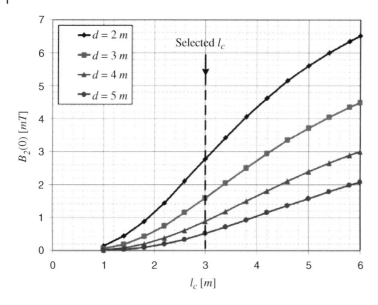

Figure 24.3 Simulation results of the magnetic flux density at the center of the secondary coil for the primary and secondary core length l_c (1–6 m) and various distances d (2–5 m); $l_c = 3$ m was selected as the baseline design in this chapter.

between the core and winding, respectively, as shown later in Figure 24.9. Furthermore, local core saturation may occur for a large secondary current due to concentrated magnetic flux if l_w is too small.

24.2.2 Optimized Design of the Proposed Stepped Core

Conventional ferrite material has a saturation flux density of about 300 mT at room temperature, but, practically, the maximum should be less than 200 mT considering the core loss and temperature increase of the material due to the loss. The higher temperature of the ferrite material decreases the saturation flux density. If the thickness of the ferrite core along the x-axis is even, the magnetic flux density profile in the ferrite core is not uniform along the longitudinal line (x-axis), as shown in Figure 24.4. This uneven profile can be easily anticipated from Figure 24.2, where the magnetic linkage of each side is concentrated at the center of the core; hence,

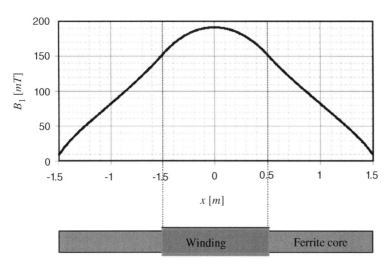

Figure 24.4 Simulated magnetic flux density for the primary coil with an even thickness of ferrite core $l_{f1} = 7$ cm, when $l_c = 3$ m and $l_w = 1$ m. The number of turns is 22 and the primary current I_1 is 40 A_{rms} to reach the saturation level of 190 mT.

the magnetic flux density becomes highest at the center of the core and gradually decreases for the outer section. Therefore, the lower magnetic field section of other parts does not have to be thick. Even though applicable to the secondary coil, this optimum core design, in this chapter, is focused on the primary coil, which undergoes severe core saturation for generating high magnetic flux.

As shown in Figure 24.4, the ampere-turn of the primary coil is found to be 880 (amp-turn) by simulation with the configuration of $l_{h1} = 20$ cm and $l_{t1} = 7$ cm, that is, $A_1 = 140$ cm^2, to reach the magnetic saturation level of 190 mT. Now the current level of the primary coil I_1 was chosen as 40 A_{rms}, considering the current rating of an available inverter; however, it could be as long as the ampere-turn value is met. Then, the number of turns of the primary coil N_1 was determined as 22.

The ferrite core thickness should be optimized when considering the magnetic field profile. If the total amount of the ferrite material is given, the outer section needs to be thinner so that the magnetic flux density can be uniform. A simple optimum design rule for the uniform magnetic flux density is to make the core cross-section area as follows:

$$A_{1opt}(x) = \frac{A_1(0)}{B_1(0)}B_1(x) \tag{24.4}$$

where $A_1(0)$ and $B_1(x)$ are the cross-section area and magnetic flux density, respectively, of the even thickness core, as shown in Figure 24.5. Under the condition of (24.4), the magnetic flux density of the proposed optimized core becomes

$$B_{1opt}(x) = \frac{B_1(x)A_1(0)}{A_{1opt}(x)} = \frac{B_1(0)}{A_1(0)} \equiv \frac{B_0}{A_1(0)} = \text{constant} \tag{24.5}$$

In other words, if the cross-section area is of the same form of magnetic flux density profile as the even thickness core, then a uniform magnetic flux density can be obtained.

For fabrication purposes, however, the core length should be finitely segmented, as shown in Figure 24.6, where half of the core is presented due to symmetry of the coil. This stepped shape configuration can be easily implemented by small-size ferrite blocks, where the thickness of a block is now 2 cm. It has been assumed that the magnetic flux density profile $B_1(x)$ is unchanged even though $A_{1opt}(x)$ is changed, which will be valid for the case in this chapter where the core thickness is much less than the core length.

The optimization can be done by finding the longitudinal (x-axis) points x_1, x_2, x_3, and x_4, where the magnetic flux density at each stepped junction reaches the maximum value B_0. In this chapter, five segmentations were assumed.

Figure 24.5 Magnetic flux distribution at the stepped junction of the ferrite core. Magnetic flux density change is plotted at the bottom.

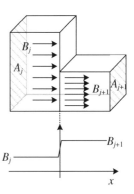

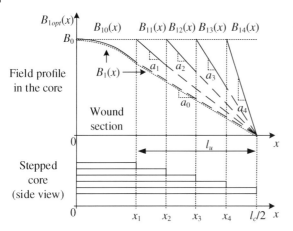

Figure 24.6 Magnetic flux density profile of the proposed stepped core, where x_1, x_2, x_3, and x_4 denote the junctions of each stepped core.

Different from conventional air coils, the exact calculation of the magnetic flux density in the proposed ferrite-core coil is hardly possible and measuring the magnetic flux density in the ferrite core is impossible. In this chapter, the simulated magnetic flux density profile, as shown in Figure 24.4, is numerically modeled, considering the curved and straight line portions, as follows:

$$B_1(x) = \begin{cases} B_w(x) = B_0\{1 - c_0|x|^n\} & \text{for} \quad |x| < l_w/2 \text{ wound section} & (24.6a) \\ B_u(x) = a_0(|x| - x_1) + b_0 & \text{for} \quad l_w/2 < |x| < l_c/2 \text{ unwound section} & (24.6b) \end{cases}$$

where $x_1 \approx l_w/2$ and the coefficients are determined from Figure 24.6 as follows:

$$a_0 = -\frac{B_w(x_1)}{l_c/2 - x_1}, \quad b_0 = B_w(x_1) \tag{24.7a}$$

$$c_0 = 0.9, \quad n = 2.0 \tag{24.7b}$$

In (24.7b), the c_0 and n are obtained by curve fitting from Figure 24.6.

Similar to (24.5), an optimized magnetic flux density profile for the stepped core can also be found by approximating the profile to the even magnetic flux density $B_1(0)$. The optimized magnetic flux density profile function is then as follows:

$$B_{1opt}(x) = B_{10}(x) + B_{11}(x) + B_{12}(x) + B_{13}(x) + B_{14}(x) \tag{24.8}$$

where

$$B_{10}(x) = B_w(x) \quad \text{for } 0 \leq |x| < x_1 \tag{24.9a}$$
$$B_{11}(x) = a_1(|x| - x_1) + b_1 \quad \text{for } x_1 \leq |x| < x_2 \tag{24.9b}$$
$$B_{12}(x) = a_2(|x| - x_2) + b_2 \quad \text{for } x_2 \leq |x| < x_3 \tag{24.9c}$$
$$B_{13}(x) = a_3(|x| - x_3) + b_3 \quad \text{for } x_3 \leq |x| < x_4 \tag{24.9d}$$
$$B_{14}(x) = a_4(|x| - x_4) + b_4 \quad \text{for } x_4 \leq |x| < l_c/2 \tag{24.9e}$$

In (24.9), what should be determined are 12 constants, that is, a_i's, b_i's, and x_i's.

First, b_i's are easily found, considering the fact that the initial value of each $B_{1i}(x_i)$ should be the same as B_0, as shown in Figure 24.6. From (24.9b) to (24.9e), $B_{1i}(x_i)$ is found as follows:

$$B_{1i}(x_i) = a_i(|x_i| - x_i) + b_i = b_i = B_0 \quad \text{for } i = 1, 2, 3, 4 \tag{24.10}$$

The magnetic flux should be continuous at the stepped junction of the core, as shown in Figure 24.5. The adjacent magnetic flux density B_{j+1} can be found as follows:

$$B_{j+1} = \frac{A_j}{A_{j+1}} B_j, \quad \because \phi_{j+1} = A_{j+1} B_{j+1} = \phi_j = A_j B_j \tag{24.11}$$

At $x = x_1$, the number of core stacks is changed from five to four; therefore, the magnetic flux density is increased by this ratio of cross-section area $A_{10}/A_{11} = 5/4$, as identified from (24.11). Moreover, the magnetic field density of $B_{11}(x_1)$ should be the same as the maximum allowable magnetic field density B_0, as follows:

$$B_{11}(x_1) = B_0 = \frac{A_{10}}{A_{11}} B_{10}(x_1) = \frac{5}{4} B_{10}(x_1) \tag{24.12}$$

From (24.12), x_1 can be calculated by using (24.6a) and (24.9a) as follows:

$$x_1 = \frac{1}{(5c_0)^{1/n}} \tag{24.13}$$

From (24.13), x_1 is calculated from (24.7b) as 0.49 m.

It is noteworthy that the magnetic flux density at the ends of the ferrite core, that is, $|x| = l_c/2$, is nearly zero, as shown in Figures 24.4 and 24.6. This means that each $B_{1i}(x)$ has the same zero value at $|x| = l_c/2$, as shown in Figure 24.6, because each segmented stacked core assumes an even thickness to the ends, that is,

$$B_{1i}(l_c/2) = a_i(|l_c/2| - x_i) + B_0 = 0 \quad \text{for } i = 1, 2, 3, 4 \tag{24.14}$$

From (24.14), a_i's can be determined as follows:

$$a_i = \frac{-B_0}{l_c/2 - x_i} \quad \text{for } i = 1, 2, 3, 4 \tag{24.15}$$

Applying (24.7a) and (24.12) to (24.15), a_1 can be determined as follows:

$$a_1 = \frac{-B_0}{l_c/2 - x_1} = -\frac{5}{4} \frac{B_{10}(x_1)}{l_c/2 - x_1} = \frac{5}{4} a_0 \tag{24.16}$$

Now the coefficients of (24.9b) are completely determined and at $x = x_2$ the number of core stacks is changed from four to three; therefore, the magnetic flux density is increased by $A_{11}/A_{12} = 4/3$. Similar to (24.12), the magnetic field density of $B_{12}(x_2)$ should be the same as the maximum allowable magnetic field density B_0, as identified from (24.11), as follows:

$$B_{12}(x_2) = B_0 = \frac{A_{11}}{A_{12}} B_{11}(x_2) = \frac{4}{3} B_{11}(x_2) = \frac{4}{3} \{a_1(x_2 - x_1) + B_0\} \tag{24.17}$$

From (24.17), x_2 can be determined from the following:

$$x_2 = x_1 + \frac{-B_0}{4a_1} = x_1 + \frac{1}{4}(l_c/2 - x_1) \equiv x_1 + \frac{l_u}{4} \tag{24.18}$$

In (24.18), it was found that the second segmented position corresponds to a fourth of the unwound core length l_u.

By recursive applications of this procedure of finding a_i's and x_i's, as shown in (24.12) to (24.18), a complete determination of the coefficients becomes

$$a_2 = \frac{5}{3}a_0 \tag{24.19a}$$

$$a_3 = \frac{5}{2}a_0 \tag{24.19b}$$

$$a_4 = \frac{5}{1}a_0 \tag{24.19c}$$

$$x_3 = x_2 + \frac{l_u}{4} \tag{24.20a}$$

$$x_4 = x_3 + \frac{l_u}{4} \tag{24.20b}$$

In general, the coefficients for an arbitrary m-segmentation are found to be, as long as the magnetic flux density of the unwound core area is a straight line as in Figure 24.4, as follows:

$$a_i = \frac{m}{m-i}a_0 \qquad \text{for } i = 1, 2, \ldots, m-1 \tag{24.21a}$$

$$b_i = B_0 \qquad \text{for } i = 1, 2, \ldots, m-1 \tag{24.21b}$$

$$x_{i+1} - x_i = \frac{l_u}{m-1} \qquad \text{for } i = 1, 2, \ldots, m-1 \tag{24.21c}$$

Using the designed parameters, the calculated magnetic flux density of (24.8) and the simulated one were compared with each other, as shown in Figure 24.7. As anticipated, the magnetic flux density has the peak value B_0 at each stepped junction point (x_1, x_2, x_3, x_4).

Figure 24.8 shows the simulation results of the normalized magnetic flux densities for the optimized stepped core and even core, respectively. The amount of used ferrite core for each case is assumed to be the same, for fair comparison. It is found that the peak magnetic field of the stepped core is reduced to 65% of that of the even core, which means that 35% of the core can be saved by the proposed stepped core design to achieve the same magnetic flux density profile.

The magnetic flux in the core is proportional to the inductance of a coil in general for a given current. Hence, the inductance of the stepped core and even core, L_{1opt} and L_1, respectively, are calculated by simulations for comparison, which are 942 μH and 991 μH, respectively. Because

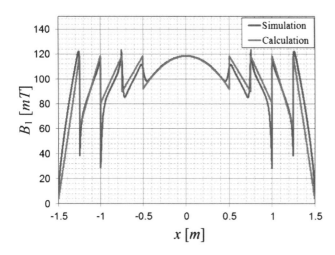

Figure 24.7 Comparison of the simulated magnetic flux density profile with the calculated one for the proposed stepped core, when $l_c = 3$ m and $l_w = 1$ m. The number of turns is 22 and the primary current I_1 is 40 A_{rms}.

Figure 24.8 Simulation results of the normalized magnetic flux density for the stepped core and even core, where $l_c = 3$ m and $l_w = 1$ m. The number of turns is 22 and the primary current I_1 is 40 A$_{rms}$.

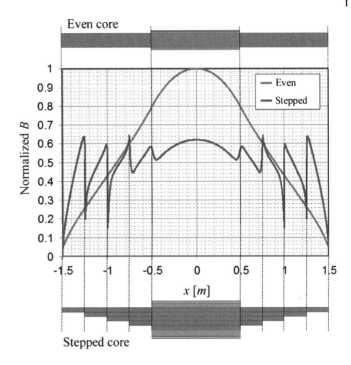

the inductance of the proposed stepped core is only 4.9% less than the even core, the induced voltage drop is not as significant. Therefore, it can be concluded that the proposed scheme can deliver 2.13 ($= 1.46^2$) times the wireless power than the even core type for same amount of core, considering that the delivered wireless power is proportional to the square of the induced voltage of (24.1) or the magnetic flux density, as follows:

$$\frac{P_{1,opt}}{P_1} = \left(\frac{V_{1,opt}}{V_2}\right)^2 = \left(\frac{L_{1,opt}}{L_1} \cdot \frac{B_{0,opt}}{B_0}\right)^2 \tag{24.22}$$

24.2.3 Core Loss Calculation

The magnetic flux is significantly intensified and the coil size becomes quite compact by using the core; however, the penalty of using the core is core loss. As identified from the side view of the proposed stepped core, as shown in Figure 24.8, both the total volume and maximum magnetic flux density of a core should be minimized to mitigate the core loss. Hysteresis loss, which is the major source of core loss, can be modeled in W per unit volume by using the following Steinmetz equation [30]:

$$P_{cv} = C_m C_T f^p B^q_{peak} [\text{W/m}^3], \quad \because C_T = C_{T0} - C_{T1}T + C_{T2}T^2 \tag{24.23}$$

where C_m and C_T are core loss coefficient and temperature correction parameter, respectively.

Quantitative core loss can be assessed for the even core and optimized stepped core when the amount of core is given. Applying (24.8) to (24.23), the ferrite core loss can be analytically

Table 24.1 Fit parameters to calculate the hysteresis core loss density

Parameter	Value
C_m	7.13×10^3
p	1.42
q	3.02
C_{T2}	3.65×10^{-4}
C_{T1}	6.65×10^{-2}
C_{T0}	4.00
T	100 °C

calculated as follows:

$$P_{1h} = \int\int\int P_{cv}\,\mathrm{d}x\mathrm{d}y\mathrm{d}z = \int C_m C_T f^p B_{peak}^q A_1(x)\mathrm{d}x$$
$$= \begin{cases} C_m C_T f^p \int B_1^q(x)A_1(x)\mathrm{d}x & \text{for even core} \\ C_m C_T f^p \int B_{1opt}^q(x)A_{1opt}(x)\mathrm{d}x & \text{for stepped core} \end{cases} \tag{24.24}$$

where $A_1(x)$ and $A_{1opt}(x)$ are the cross-section area of the even core and stepped core, respectively. Parameters for the power loss density of the ferrite core from the maker are summarized in Table 24.1. From (24.24), the losses for the even core and stepped core were calculated with the temperature of 100 °C and I_1 of 40 A_{rms} by the mathematical software Maple ver. 16.0, and the losses were calculated to be 1340 W and 550 W, respectively. Adopting the proposed optimizing technique, the core loss became just 41% of the unoptimized even core.

24.2.4 Winding Methods and Parasitic Capacitances

Compared with air coils, the proposed core wound coils have relatively large inductances of the order of mH; therefore, small parasitic capacitances of the coils may affect the self-resonant frequencies of the coils. Two major parasitic capacitances of the core wound coil are shown in Figure 24.9. Different from conventional air coils that have just parasitic capacitances C_w between adjacent wires, the proposed core wound coils have the parasitic capacitances C_f between the wire and core. The C_f of the proposed coil is not negligible because the wire length of the coil is several tens of meters. These parasitic capacitances, C_w and C_f, constitute a parallel capacitance C_p in an equivalent circuit model, as shown in Figure 24.10.

Figure 24.10 shows the simplified equivalent circuit model of a resonant tank, which is composed of the ferrite wound coils including its parallel parasitic capacitor C_p and a series resonant capacitor C_s. Considering that L_s represents either L_1 or L_2, the impedance of the circuit Z is as follows:

$$Z = \frac{1}{j\omega C_s} + \frac{1}{j\omega C_p}||j\omega L_s$$
$$= \frac{1 - L_s(C_s + C_p)\cdot\omega^2}{j\omega C_s(1 - L_s C_p\omega^2)} = \frac{1 - (\omega/\omega_s)^2}{j\omega C_s\{1 - (\omega/\omega_p)^2\}}, \tag{24.25}$$

Figure 24.9 Parallel parasitic capacitances of ferrite wound coils (upper) and the fabricated coil for minimizing parasitics (lower) for hundred kHz operation.

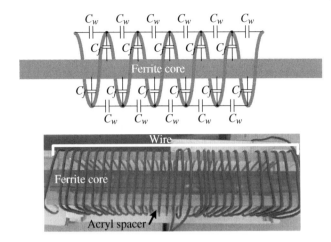

where the parallel angular resonant frequency ω_p and series angular resonant frequency ω_s are as follows:

$$\omega_p = \frac{1}{\sqrt{L_s C_p}} \tag{24.26a}$$

$$\omega_s = \frac{1}{\sqrt{L_s(C_p + C_s)}} \tag{24.26b}$$

To make the secondary induced voltage large, the number of turns of the secondary winding should be large and the operating frequency should be as high as several hundreds of kHz, as identified in (24.1). To meet these conditions, C_s should be as small as a few nanofarad because of the large L_s of mH. As identified from (24.25), C_p should be at least a few times smaller than C_s in order to separate ω_s far enough from ω_p. To reduce C_f and C_w, an acryl spacer is inserted between the wire and ferrite core, and the wire is wound with a sufficient interwire gap based on the given winding space and the number of turns, as shown in Figure 24.9. Thus, the coil can operate over a hundred kHz without deteriorating resonance characteristics.

Figure 24.10 A resonant tank model comprises a ferrite wound coil with a parasitic capacitance and a series resonant capacitor.

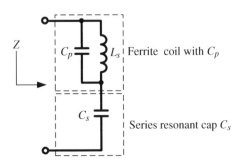

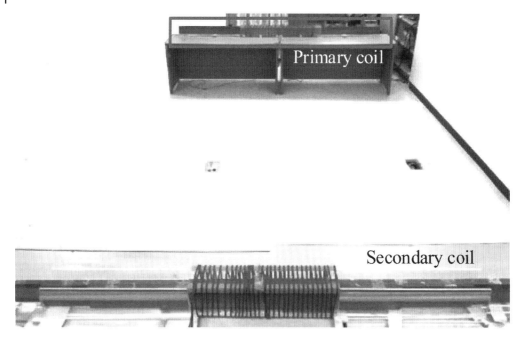

Figure 24.11 A picture of the proposed primary and secondary coils with stepped cores and acryl spacers for experiments.

24.3 Example Design and Experimental Verifications of the IPTS

24.3.1 Overall Configuration

Experimental verifications of the proposed coils were made in a laboratory, as shown in Figure 24.11, where the primary and secondary coils were placed on the tables.

The circuit diagram of the proposed IPTS is shown in Figure 24.12, where the primary coil and its series resonant capacitor C_1 are driven by a full-bridge inverter. To guarantee the zero voltage switching (ZVS) operation of the inverter, the switching frequency of the inverter was

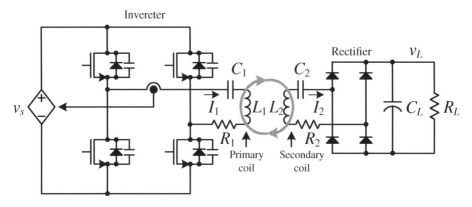

Figure 24.12 Circuit diagram of the proposed IPTS for experiments, which includes an inverter for the primary coil driving and a rectifier for the secondary coil.

Table 24.2 Parameters of the proposed IPTS including coils

Parameters	Values	Parameters	Values
f	20 kHz	l_c	3 m
d	3 m, 4 m, 5 m	l_w	1 m
L_m	18.5 μH, 10.6 μH, 6.94 μH	x_1	0.5 m
κ	0.68%, 0.39%, 0.26%	x_2	0.75 m
L_1	832 μH	x_3	1 m
L_2	8.78 mH	x_4	1.25 m
C_1	80 nF	l_{t1}	0.1 m
C_2	7.1 nF	l_{h1}	0.2 m
N_1	22	l_{t2}	0.05 m
N_2	86	l_{h2}	0.1 m
R_1	0.63 Ω	C_L	220 μF
R_2	4.17 Ω	Q_2	30.2
R_L	40 Ω		

selected to be slightly higher than the primary side resonant frequency determined by C_1 and L_1 [11]. On the other hand, the secondary side resonant frequency determined by C_2 and L_2 was tuned exactly to the switching frequency. These are summarized as follows:

$$\omega \cong 1.05\,\omega_1 = \frac{1.05}{\sqrt{L_1 C_1}} \tag{24.27a}$$

$$\omega = \omega_2 = \frac{1}{\sqrt{L_2 C_2}} \tag{24.27b}$$

From (24.27), C_1 and C_2 were determined by using the measured values of L_1 and L_2, which are listed in Table 24.2.

A full-bridge rectifier converts the induced AC voltage of the secondary coil to DC voltage. The R_1 and R_2 represent primary and secondary effective series AC resistances, respectively, which include the equivalent series resistances (ESR) of the resonant capacitors and equivalent AC resistances of the coils, comprise the conduction loss and eddy current loss. R_L was fixed to 40 Ω throughout the experiments to give 1 kW at 200 V load voltage.

24.3.2 Fabrication of Coils

The proposed coils designed in the previous sections were fabricated, as shown in Figures 24.9 and 24.11, which are the stepped core type and even core type. The lengths of primary and secondary cores are all $l_c = 3$ m, and other parameters for fabrication are listed in Table 24.2. For fabrication, we selected a low-priced Mn–Zn-type soft ferrite material named PL-7 from Samwha Electronics, South Korea, because the material's loss characteristic is similar to an Mn–Zn-type material called 3C30 produced by Ferrox Cube.

The primary coil inductance L_1 was measured as 832 μH, which is 12% less than the simulated value of 942 μH; this is mainly due to thin isolation films between each ferrite block, which were inserted to mitigate eddy current loss inside the cores.

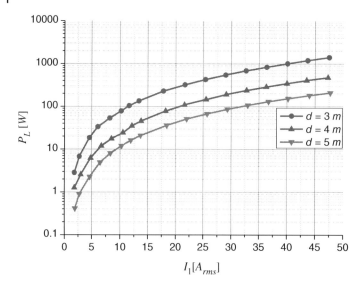

Figure 24.13 Measured output power versus the primary rms current I_1 for various distances (20 kHz).

The acryl spacers were also inserted between the wire and ferrite core to reduce C_f in order to obtain about a 3 cm gap. Thus, the measured parasitic capacitances of the primary and secondary coils, C_{p1} and C_{p2}, were merely 95 pF and 44 pF, respectively. Comparing these values of C_{p1} and C_{p2} with $C_1 = 80$ nF and $C_2 = 7.1$ nF, they are 0.12% and 0.62%, which correspond to 28.9 and 12.7 times the frequency separation of (24.26), respectively. In other words, the parallel resonance frequencies of the primary and secondary coils are 566 kHz and 256 kHz, respectively; hence, the fabricated coils can be used up to about a hundred kHz in practice.

24.3.3 Efficiency Measurements

To measure the power efficiency of the proposed IPTS, the AC input power of the primary coil was measured by a precision digital power analyzer, Yokogawa WT1800, whereas the DC output power of the load resistor was measured by multimeters. The measured output power versus the primary current I_1 for various distances at 20 kHz is shown in Figure 24.13. The maximum output powers for distances of 3 m, 4 m, and 5 m were 1403 W, 471 W, and 209 W at the maximum primary current $I_1 = 47$ A$_{rms}$, respectively.

Figure 24.14(a) shows the power efficiency measured from the primary coil to the load resistor versus the output power P_L for various distances d. The maximum output powers and efficiencies for 3 m, 4 m, and 5 m at 20 kHz were 1403 W, 471 W, and 209 W and 29%, 16%, and 8%, respectively. The power efficiency decreases as P_L or, correspondingly, I_1 increases due to the substantially increased hysteresis core losses in the coils. The efficiencies of the unoptimized even core and optimized stepped core are compared in Figure 24.14(b). For a low-load power, the core loss difference between the two cases is not substantial because of the undistinguishable core loss difference. As the load power increases, the core loss difference between the optimized case and unoptimized case gets larger because the efficiency of the optimized case is improved due to a smaller core loss than the unoptimized case.

A much higher operating frequency of 105 kHz was also tried, as shown in Figures 24.15 and 24.16, where a high-priced Mn–Zn-type soft ferrite material for higher frequency application up to 1 MHz, named PL-F1, from Samwha Electronics was used. Now the input power was measured at the DC input side of the inverter because the power measurements for high frequency with digital power meters incurred numerous errors. As shown in Figure 24.15, the maximum

Figure 24.14 (a) Measured efficiency from the primary coil to the load resistor versus output power P_L for various distances (20 kHz). (b) Measured efficiency comparison between an unoptimized even core and optimized stepped core, $d = 3$ m.

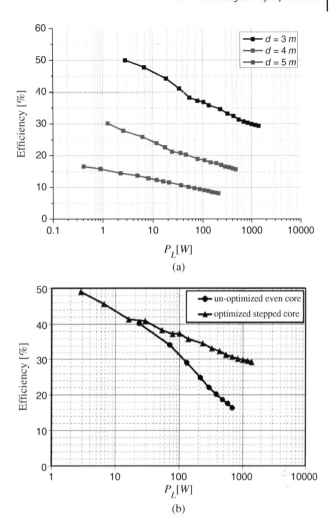

(a)

(b)

Figure 24.15 Measured output power versus primary current I_1 for various distances (105 kHz).

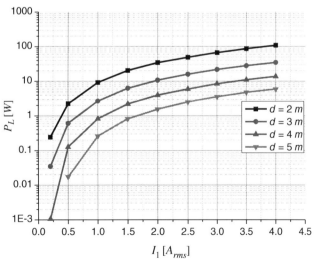

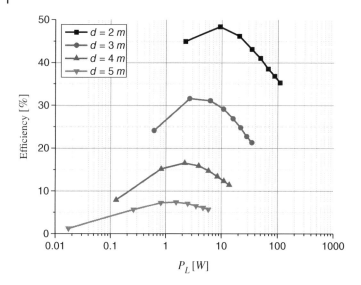

Figure 24.16 Measured power efficiency from inverter input to load resistor versus output power P_L for various distances (105 kHz).

output powers for each distance of 2 m, 3 m, 4 m, and 5 m were 109 W, 34.8 W, 13.8 W, and 5.93 W, respectively, where the primary current I_1 was less than 4 A_{rms} due to the power limit of the prototype high-frequency inverter specially made for a hundred kHz.

Figure 24.16 shows the power efficiency from an inverter input to a load resistor versus the output power for various distances. For the output power of 5 W, the power efficiencies were 46%, 31%, 15%, and 6% for the distances of 2 m, 3 m, 4 m, and 5 m, respectively. Comparing these results of Figure 24.16 with those of Figure 24.14, the efficiencies for 105 kHz are apparently lower than that for 20 kHz; however, it is not very obvious because the power measurement method, inverter fabrication, and current level were drastically changed. A higher operating frequency may make the overall system compact; however, it is not straightforward to say that a higher operating frequency is optimal in terms of the overall power efficiency, output power level, and total cost due to the limited capacitor selection, increased core loss, and parasitic effects.

Detailed studies on the operating frequency selection together with the much longer wireless power delivery by shorter l_c of coils and higher efficiency are left for further work.

24.3.4 Loss Measurements

To verify the calculated hysteresis loss in the ferrite core of (24.24), the loss difference of the primary coils between the optimized stepped core and unoptimized even core was measured, as follows:

$$\Delta P_{1h} = C_m C_{Tf} f^p \int \left\{ B_1^q(x) A_1(x) - B_{1opt}^q(x) A_{1opt}(x) \right\} dx \tag{24.28}$$

Except for the hysteresis losses in the optimized core and unoptimized core, other losses are almost the same. Therefore, other loss components are canceled out and only the difference of hysteresis core losses is measured, which is found to be quite similar to (24.28), as shown in Figure 24.17.

The measured surface temperatures of the even core and optimized stepped core for the same amount of ferrite material by using a thermographic camera is shown in Figure 24.18. The temperature of the optimized stepped core was much lower than the even core and the temperature distribution of the optimized stepped core was more uniform than the even core. From this

Figure 24.17 Measured (red curve) and calculated (blue curve) loss differences of the primary coils between optimized stepped cores and unoptimized even cores at 20 kHz.

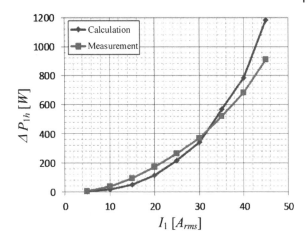

temperature measurement, power loss minimization using the proposed optimization method has been further verified.

The measured surface temperature of a segment of the stepped core is shown in Figure 24.19. The surface temperature distribution of the core reflects the magnetic flux density profile, which cannot be directly measured in a core. The simulated and calculated magnetic flux density profiles, as shown in Figure 24.7, correspond well to this measured temperature profile.

Figure 24.18 Measured surface temperature using a thermographic camera after a 20 min operation of the IPTS at 20 kHz and $I_1 = 40\ A_{rms}$.

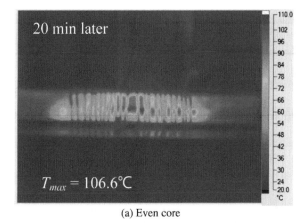

(a) Even core

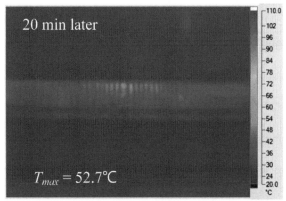

(b) Optimized stepped core

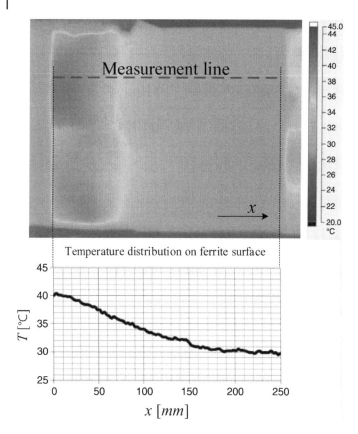

Figure 24.19 Measured surface temperature of a segment of the stepped core (side view) showing internal magnetic flux density.

Though it is not possible to measure all the detail losses, the loss analysis of the proposed IPTS with stepped cores for an example of the power transfer distance of 5 m and primary current I_1 of 40 A_{rms} at 20 kHz is shown in Figure 24.20. Except for the term P_{misc}, which is the remaining unexplained discrepancy, the powers were either measured or calculated; P_{1h} and P_{2h} were calculated from (24.24) and all the other powers were measured. The dominant losses are found to be the hysteresis and eddy current losses of the primary coil. Therefore, it should be possible to develop a better ferrite core for a long-distance wireless power transfer with a higher system efficiency.

The parameters of the proposed IPTS are summarized in Table 24.2. The effective AC resistances R_1 and R_2 were measured by tuning the resonant circuits of the primary and secondary coils at the operating frequency of 20 kHz, respectively. The mutual inductance L_m was calculated from the measured induced voltage V_2 for a given primary current I_1. The mutually coupling factor κ was calculated from L_m, L_1, and L_2.

Note that the quality factor of the secondary circuit Q_2 for 20 kHz, considering the effective AC side resistance of the DC side resistor [31], is merely 30.2 as follows:

$$Q_2 \equiv \frac{\omega L_2}{R_e} = \frac{\omega L_2}{R_{L,eff} + R_2} = \frac{\omega L_2}{R_L \frac{8}{\pi^2} + R_2}, \quad \because R_{L,eff} = \left(\frac{2\sqrt{2}}{\pi}\right)^2 R_L \qquad (24.29)$$

Comparing the very large Q of the CMRS, the proposed Q_2 is about 100 times smaller; hence, the proposed IPTS is well within the practical Q_2 of less than 100, which has been verified by several applications [11, 32, 33].

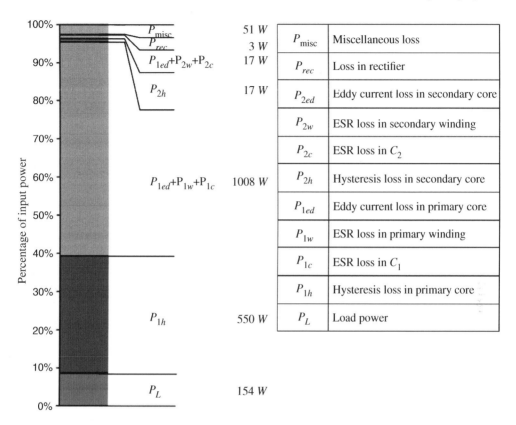

The following table appears within the figure:

P_{misc}	Miscellaneous loss	51 W
P_{rec}	Loss in rectifier	3 W
P_{2ed}	Eddy current loss in secondary core	17 W
P_{2w}	ESR loss in secondary winding	
P_{2c}	ESR loss in C_2	
P_{2h}	Hysteresis loss in secondary core	17 W
P_{1ed}	Eddy current loss in primary core	
P_{1w}	ESR loss in primary winding	
P_{1c}	ESR loss in C_1	1008 W
P_{1h}	Hysteresis loss in primary core	550 W
P_L	Load power	154 W

Figure 24.20 An example of the power loss analysis of the proposed IPTS with stepped cores at 20 kHz when $d = 5$ m and $I_1 = 40$ A$_{rms}$.

24.4 Conclusion

The 5 m off-long-distance IPTS has been demonstrated, introducing the possibility of a new remote power delivery mechanism that has so far never been implemented, even by the CMRS. The dipole structure coil with a ferrite core instead of conventional loop-type coils used in the CMRS is found to be quite effective for a longer power delivery. Its shape is not bulky but narrow and long so that it can be installed in the corner or ceiling of a room. The proposed optimized stepped core has been experimentally verified to have only 41% core loss compared with the unoptimized even core but delivers 2.1 times more wireless power for a given amount of core. Experimentally obtained maximum output powers and primary-coil-to-load-power efficiencies for 3 m, 4 m, and 5 m at 20 kHz were 1403 W, 471 W, and 209 W, and 29%, 16%, and 8%, respectively. The proposed IPTS is under development as a potential back-up power of essential sensors during severe accidents in a nuclear power plant.

Problems

24.1 A fundamental principle of the proposed dipole coil IPTS compared to conventional loop coil IPTS is "dimension-less." In other words, a dipole coil has one dimension like a line but a loop coil has two dimensions like a plane. Therefore, it is easy for the dipole coil to be installed in a room, where corners can be used for fitting the dipole coil. However, it is said that there are a few demerits of the dipole coils, such as a large core loss and large inductance for longer distance WPTs.

(a) For a fair comparison, can you roughly show that the inductance of a dipole is similar to that of a loop coil so long as the WPT distance is same? If it is true, then the large inductance of a dipole is not a demerit but an inevitable parameter for a long-distance WPT.

(b) Suggest any idea for improving the performances of dipole coils in terms of high-frequency operation and magnetic field shielding.

24.2 It is sometimes said that CMRS has a very weak magnetic field though transferring wireless power over a long distance. If this is true, then the same argument will be true for the IPTS, as discussed in the previous chapter.

(a) Can you prove that the minimum magnetic field between two identical loop coils (for a Tx coil and an Rx coil) must be large enough for transferring wireless power?

(b) Can you generally prove that the minimum magnetic field between a Tx coil and an Rx coil of any type must be large enough for transferring wireless power?

References

1 N. Tesla, "Apparatus for transmitting electrical energy," US Patent 1 119 732, December 1, 1914.

2 A. Kurs, A. Karalis, R. Moffatt, J.D. Joannopoulos, P. Fisher, and M. Soljacic, "Wireless power transfer via strongly coupled magnetic resonances," *Science*, vol. 317, no. 5834, pp. 83–86, July 2007.

3 L.H. Ford, "The effect of humidity on the calibration of precision air capacitors," *Journal of the Institution of Power Engineering*, vol. 95, no. 48, pp. 709–712, December 1948.

4 V.J. Brusamarello, Y.B. Blauth, R. Azambuja, I. Muller, and F.R. Sousa, "Power transfer with an inductive link and wireless tuning," *IEEE Trans. on Instrumentation and Measurement*, vol. 62, no. 5, pp. 924–931, May 2013.

5 T.C. Beh, M. Kato, T. Imura, S. Oh, and Y. Hori, "Automated impedance matching system for robust wireless power transfer via magnetic resonance coupling," *IEEE Trans. on Ind. Electron.*, vol. 60, no. 9, pp. 3689–3698, September 2013.

6 A.P. Sample, D.T. Meyer, and J.R. Smith, "Analysis, experimental results, and range adaptation of magnetically coupled resonators for wireless power transfer," *IEEE Trans. on Ind. Electron.*, vol. 58, no. 2, pp. 544–554, February 2010.

7 Sony Corp. (2009, Oct. 2). "Sony develops highly efficient wireless power transfer system based on magnetic resonance," October 2, 2009 [online], available at: http://www.sony.net/SonyInfo/News/Press/200910/09-119E/index.html.

8 J. Hirai, T.-W. Kim, and A. Kawamura, "Study on intelligent battery charging using inductive transmission of power and information," *IEEE Trans. on Power Electron.*, vol. 15, no. 2, pp. 335–345, March 2000.

9 B.L. Cannon, J.F. Hoburg, D.D. Stancil, and S.C. Goldstein, "Magnetic resonant coupling as a potential means for wireless power transfer to multiple small receivers," *IEEE Trans. on Power Electron.*, vol. 24, no. 7, pp. 1819–1825, July 2009.

10 M. Budhia, G.A. Covic, and J.T. Boys, "Design and optimization of circular magnetic structures for lumped inductive power transfer systems," *IEEE Trans. on Power Electron.*, vol. 26, no. 11, pp. 3096–3108, November 2011.

11 J. Huh, S.W. Lee, W.Y. Lee, G.H. Cho, and C.T. Rim, "Narrow-width inductive power transfer system for online electrical vehicles," *IEEE Trans. on Power Electron.*, vol. 26, no. 12, pp. 3666–3679, December 2011.

12 S.H. Lee and R.D. Lorenz, "Development and validation of model for 95%-efficiency 220-W wireless power transfer over a 30-cm air gap," *IEEE Trans. on Ind. Applic.*, vol. 47, no. 6, pp. 2495–2504, November–December 2011.

13 H. Matsumoto, Y. Neba, K. Ishizaka, and R. Itoh, "Comparison of characteristics on planar contactless power transfer systems," *IEEE Trans. on Power Electron.*, vol. 27, no. 6, pp. 2980–2993, June 2012.

14 Z. Pantic and S.M. Lukic, "Framework and topology for active tuning of parallel compensated receivers in power transfer systems," *IEEE Trans. on Power Electron.*, vol. 27, no. 11, pp. 4503–4513, November 2012.

15 J.P.C. Smeets, T.T. Overboom, J.W. Jansen, and E.A. Lomonova, "Comparison of position-independent contactless energy transfer systems," *IEEE Trans. on Power Electron.*, vol. 28, no. 4, pp. 2059–2067, April 2013.

16 M. Pinuela, D.C. Yates, S. Lucyszyn, and P.D. Mitcheson, "Maximizing DC-to-load efficiency for inductive power transfer," *IEEE Trans. on Power Electron.*, vol. 28, no. 5, pp. 2437–2447, May 2013"

17 S. Lee, B. Choi, and C.T. Rim, "Dynamics characterization of the inductive power transfer system for online electric vehicles by laplace phasor transform," *IEEE Trans. on Power Electron.*, vol. 28, no. 12, pp. 5902–5909, December 2013.

18 Y. Zhang, Z. Zhao, and K. Chen, "Frequency decrease analysis of resonant wireless power transfer," *IEEE Trans. on Power Electron.*, vol. 29, no. 3, pp. 1058–1063, March 2014.

19 H. Hao, G.A. Covic, and J.T. Boys, "A parallel topology for inductive power transfer power supplies," *IEEE Trans. on Power Electron.*, vol. 29, no. 3, pp. 1140–1151, March 2014.

20 C.-S. Wang, G.A. Covic, and O.H. Stielau, "Power transfer capability and bifurcation phenomena of loosely coupled inductive power transfer systems," *IEEE Trans. on Ind. Electron.*, vol. 51, no. 1, pp. 148–157, February 2004.

21 Z.N. Low, R.A. Chinga, R. Tseng, and J. Lin, "Design and test of a high-power high-efficiency loosely coupled planar wireless power transfer system," *IEEE Trans. on Ind. Electron.*, vol. 56, no. 5, pp. 1801–1812, May 2009.

22 J. Sallan, J.L. Villa, A. Llombart, and J.F. Sanz, "Optimal design of ICPT systems applied to electric vehicle battery charge," *IEEE Trans. Ind. Electron.*, vol. 56, no. 6, pp. 2140–2149, June 2009.

23 J.U.W. Hsu, A.P. Hu, and A. Swain, "A wireless power pickup based on directional tuning control of magnetic amplifier," *IEEE Trans. on Ind. Electron.*, vol. 56, no. 7, pp. 2771–2781, July 2009.

24 G. Elliott, S. Raabe, G.A. Covic, and J.T. Boys, "Multiphase pickups for large lateral tolerance contactless power-transfer systems," *IEEE Trans. on Ind. Electron.*, vol. 57, no. 5, pp. 1590–1598, May 2010.

25 M.L.G. Kissin, G.A. Covic, and J.T. Boys, "Steady-state flat-pickup loading effects in polyphase inductive power transfer systems," *IEEE Trans. on Ind. Electron.*, vol. 58, no. 6, pp. 2274–2282, June 2011.

26 F.F.A. van derPijl, M. Castilla, and P. Bauer, "Adaptive sliding-mode control for a multiple-user inductive power transfer system without need for communication," *IEEE Trans. on Ind. Electron.*, vol. 60, no. 1, pp. 271–279, January 2013.

27 S. Chopra and P. Bauer, "Driving range extension of EV with on-road contactless power transfer – a case study," *IEEE Trans. on Ind. Electron.*, vol. 60, no. 1, pp. 329–338, January 2013.

28 D. Kurschner, C. Rathge, and U. Jumar, "Design methodology for high efficient inductive power transfer systems with high coil positioning flexibility," *IEEE Trans. on Ind. Electron.*, vol. 60, no. 1, pp. 372–381, January 2013.

29 J. Shin, S. Shin, Y. Kim, S. Ahn, S. Lee, G. Jung, S.-J. Jeon, and D.-H. Cho, "Design and implementation of shaped magnetic-resonance-based wireless power transfer system for road powered moving electric vehicles," *IEEE Trans. on Ind. Electron.*, vol. 61, no. 3, pp. 1179–1192, March 2014.

30 Ferrox Cube, "Design of planar power transformers," [online], available at: http://www .ferroxcube.com/appl/info/plandesi.pdf.

31 C.T. Rim and G.H. Cho, "Phasor transformation and its application to the DC/AC analyses of frequency/phase controlled series resonant converters (SRC)," *IEEE Trans. on Power Electron.*, vol. 5, no. 2, pp. 201–211, April 1990.

32 W.Y. Lee, J. Huh, S.Y. Choi, X.V. Thai, J.H. Kim, E.A. Al-Ammar, M.A. El-Kady, and C.T. Rim, "Finite-width magnetic mirror models of mono and dual coils for wireless electric vehicles," *IEEE Trans. on Power Electron.*, vol. 28, no. 3, pp. 1413–1428, March 2013.

33 S. Choi, J. Huh, S. Lee, and C.T. Rim, "New cross-segmented power supply rails for road powered electric vehicles," *IEEE Trans. on Power Electron.*, vol. 28, no. 12, pp. 5832–5841, December 2013.

34 G.A. Covic and J.T. Boys, "Modern trends in inductive power transfer for transportation applications," *IEEE Journal of Emerging and Selected Topics in Power Electron.*, vol. 1, no. 1, pp. 28–41, March 2013.

35 C.B. Park, S.W. Lee, and C.T. Rim, "5 m-off-long-distance inductive power transfer system using optimum shaped dipole coils," *ECCE-ASIA*, June 2012, pp. 1137–1142.

25

Long-Range IPT by Dipole Coils

25.1 Introduction

Wireless power transfer (WPT) has been successfully applied to many technologies such as electric vehicles, consumer electronics, medical devices, wireless sensors, Internet of things (IoT) devices, and defense systems [1–36]. Wireless power delivery of a few W up to several tens of meters could be a promising solution for battery-free wireless sensors in ubiquitous networks and in plant monitoring instruments, where the wireless sensors consist of the three major parts of a sensing device, a communication device, and an energy source [3,31,32].

As possible energy source candidates for use in various wireless sensors and IoT applications, several WPT methods have been suggested, in which hypersensitive and vulnerable powering characteristics are strongly prohibited, as listed in Table 25.1. Photovoltaic power generation devices such as solar cells are easily installable and represent suitably mature technology; however, power generation with these devices is strongly affected by the weather conditions. Acoustic and/or vibration power generation using piezoelectric effects involves lightweight materials and is generally associated with extremely low power density levels and durability problems. Thermoelectric generation using the Seebeck effect can convert thermal energy into electric energy without any EMI/EMF issues, similar to photovoltaic and piezoelectric types, but this process is applicable to only a few special areas having high-temperature differences [33–36]. Radio frequency (RF) and optical power transfer methods can deliver energy to selected loads with high efficiency; however, neither method is suitable for powering various loads simultaneously [32].

Recently, it was found that the coupled magnetic resonance system (CMRS) is simply a special form of an IPTS [26–30] where a source coil and a transmitting (Tx) coil constitute a strongly coupled transformer, a load coil and a receiving (Rx) coil constitute another strongly coupled transformer, and the transmitting coil and receiving coil constitute a loosely coupled transformer. Therefore, it is no longer necessary for us to consider the CMRS as an independent candidate, as it generally requires a quality factor that is too high (~2000) and complicated multiresonance tunings with bulky volumetric loop coils.

Instead, an IPTS based on two dipole coils has been proven to be quite suitable for long-distance power delivery and therefore needs to be considered [28–30]. Two 3 m long slim dipole coils with plate cores are adopted for a 209 W power transfer over a 5 m distance in [28] and [29] and a feasibility of small power transfer for a longer distance of 7 m is experimentally studied with 2 m long dipole coils in [30]. In spite of the electromagnetic interference issue due to the omnidirectional field distribution from the primary and secondary coils, the IPTS can be a suitable solution for long-distance applications considering its robust power and simple coil structure. For practical applications such as backup power for monitoring instruments during

Wireless Power Transfer for Electric Vehicles and Mobile Devices, First Edition. Chun T. Rim and Chris Mi.
© 2017 John Wiley & Sons Ltd. Published 2017 by John Wiley & Sons Ltd.

Table 25.1 Comparative classifications of possible power sources for wireless sensors

Methods	Advantages	Disadvantages	Remarks
Photovoltaic	• No EMI/EMF • Easy installation	• High dependence on weather or shading conditions • Fragile structure	Unacceptable
Acoustic (vibration)	• No EMI/EMF • Light weight	• Wear-out problem • Low power density (W/cm^2)	Unacceptable
Thermoelectric	• No EMI/EMF • No vibration and/or audible noise • Relatively long life span	• Low conversion efficiency • Needs a high temperature difference	Unacceptable
RF	• Relatively good in environment containing steam, vapor, and/or debris (as compared to the laser type)	• Impossible to penetrate water layers and/or metal objects • High EMF/EMI	Unacceptable
Laser	• No EMI/EMF due to the highest straightness	• Impossible to penetrate obstacles • Difficult to overcome environment containing steam, vapor, and/or debris	Unacceptable
CMRS	• Relatively good for long-distance wireless power transfers (due to the high quality factor)	• Large diameter of coils • Complex structure • Hypersensitive performance	Unacceptable
IPTS	• Simple structure • High output power • Long-distance power transfer with a low quality factor	• High EMF/EMI • An EMI/EMF shield can be achieved by using a metal plate	Acceptable

severe accidents in nuclear power plants (NPPs), the physical size of the coils is restricted due to the limited space in NPPs. Moreover, the ambiguities increase severely in wireless power delivery due to unpredictable conducting objects that may be randomly distributed between the primary and secondary coils of the IPTS. As a part of an effort to achieve higher power transfer capability, it is proven that properly located metal plates or conductive coils can increase flux density of between IPTS coils in [17].

It is somewhat challenging to deliver wireless power over long distances with high efficiency, especially when the primary and secondary coils are extremely loosely coupled with a coupling coefficient of less than 0.01. Conventionally, the efficiency of wireless power transfer methods has been defined from a single primary coil to a single secondary coil, and it decreases sharply with an increase in the distance, especially for low-power applications such as wireless sensors. Hence, conventional designs aiming at high efficiency between the two coils are no longer valid when the single primary coil must work with a number of unpredictably distributed secondary coils some distance away. Instead, the maximum load power condition for each secondary coil should be separately evaluated, where the total efficiency increases incrementally as the number of secondary coils increases.

In this chapter, a new IPTS applicable to low-power applications in volatile environments is proposed, as shown in Figure 25.1. It is proven, in general, that the dipole coil structure is superior to the loop coil type for long-distance power delivery with the given configuration

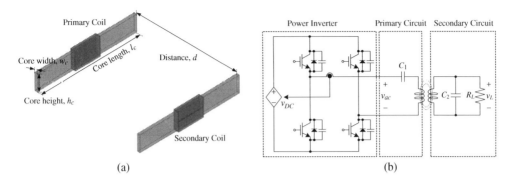

Figure 25.1 Schematics of the proposed extremely loosely coupled inductive power transfer system (IPTS): (a) overall primary and secondary coil configuration and (b) circuit diagram showing series–parallel resonance circuits.

of a square core by comparing the magnetizing inductance between the primary and the secondary coils. A comparative analysis of the two coil structures is conducted using a finite element method (FEM) simulation. Series–parallel resonances instead of series–series resonances are used to achieve a higher load voltage. The quality factor Q is set toa relatively low value throughout the proposed IPTSs design to guarantee frequency tolerance, which is essential for IoT and wireless sensor applications. The coupling coefficient between the primary and secondary coils was measured over distances ranging from 2 m to 12 m for different secondary positions and inside and outside environments surrounded by metal were compared. A comparative evaluation of different core material is also conducted and experimentally verified to provide design guidelines for a long-distance extremely loosely coupled IPTS. A design example shown in this chapter is developed as a backup power unit for several wireless sensors aiming to take emergency measurements under an NPP severe accident, where a prime requirement of such an application is the robust powering over 7 m with less consideration of efficiency.

25.2 Analysis and Design of the Proposed Extremely Loosely Coupled Dipole Coils

The proposed IPTS consists of a power inverter, a primary circuit, and a secondary circuit, as shown in Figure 25.1. The power inverter provides a regulated constant current for the primary circuit by the controlled DC voltage source v_{DC}, and the switching angular frequency of the inverter ω_s is slightly higher than the resonant frequency of the primary circuit ω_{r1} such that the inverter switches can operate under the zero voltage switching condition [22–25]. The secondary circuit resonates at ω_{r2}, which is identical to ω_s, to maximize the load voltage V_L at the operating frequency of the inverter. The relationship between ω_s ω_{r1}, and ω_{r2} is defined as (25.1a), where the resonant frequencies of the primary and secondary circuits are given as (25.1b). The load resistance R_L represents a diode rectifier circuit in practice; however, a resistor is used for the experiments in this chapter to focus on the behavior of the coils.

$$\omega_{r1} < \omega_s, \quad \omega_{r2} = \omega_s \tag{25.1a}$$

$$\omega_{r1} = \frac{1}{\sqrt{L_1 C_1}}, \quad \omega_{r2} = \frac{1}{\sqrt{L_2 C_2}} \tag{25.1b}$$

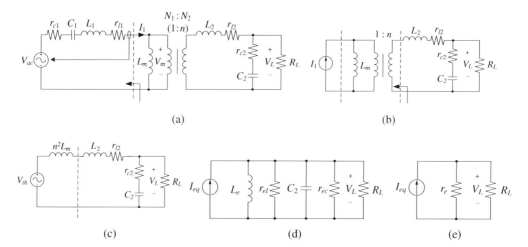

Figure 25.2 Simplified circuit of the proposed IPTS reflected to the load in the resonant condition: (a) equivalent circuit of the proposed IPTS, (b) controlled source current circuit with I_1, (c) equivalent Thevenin circuit with V_{th}, (d) approximated circuit reflected to the secondary coil, and (e) equivalent circuit of (d) at the resonant frequency ω_s.

25.2.1 Explicit Circuit Model and Static Analysis of the Proposed IPTS

The explicit coupled transformer circuit models used in CMRSs or IPTSs [19–30] are adopted in the proposed extremely loosely coupled case, as shown in Figure 25.2, where a circuit simplification process is shown for a series–parallel tuned IPTS. A magnetizing inductance L_m is used instead of a coupling inductance considering the constant L_m regardless of the number of turns of a secondary coil, which brings simplicity in the analysis. Parallel tuning of the secondary circuit is used here to increase the load voltage V_L because the induced voltage is quite small for this extremely loosely coupled case and therefore must be substantially amplified by resonance, as shown in Figure 25.2(c). Every circuit component is reflected to the secondary side of the ideal transformer of the turn ratio n, where the parasitic resistances are approximated in cases.

As shown in Figure 25.2(a) and (b), the controlled current source I_1 is regulated to be constant in a steady state [19–24] as follows:

$$I_1 = \frac{V_{ac} - V_m}{r_{l1} + r_{c1} + j(\omega_s L_1 - 1/\omega_s C_1)} \tag{25.2}$$

The Thevenin's equivalent voltage V_{th} in Figure 25.2(c) and the equivalent current source I_{eq} in Figure 25.2(d) are then derived, respectively, as follows:

$$V_{th} = j\omega_s L_m I_1 \frac{N_2}{N_1} \tag{25.3a}$$

$$I_{eq} = \frac{V_{th}}{r_{l2} + j\omega_s L_e} = \frac{j\omega_s L_m I_1 \dfrac{N_2}{N_1}}{j\omega_s L_e \left(1 + \dfrac{r_{l2}}{j\omega_s L_e}\right)} \cong \frac{L_m I_1 N_2}{N_1 L_2}\left(1 - \frac{r_{l2}}{j\omega_s L_2}\right) \cong \frac{L_m I_1 N_2}{N_1 L_2} \tag{25.3b}$$

$$\because r_{l2} \ll \omega_s L_e, \quad L_e \equiv n^2 L_m + L_2 \cong L_2 \tag{25.3c}$$

In (25.3b), the imaginary current component is neglected because it is very small compared to the real current component, which results in only a slight contribution to the absolute value of I_{eq}. In (25.3c), the contribution of the coupled magnetizing inductance L_m to the equivalent inductance L_e is negligible owing to the extremely low magnetic coupling. As shown in Figure 25.2(d), the parasitic inductor resistance r_{l2} and the parasitic capacitor resistance r_{c2} of the secondary circuit, which are in series, can be equivalently converted into r_{el} and r_{ec} in parallel, respectively, as follows:

$$\frac{1}{r_{l2} + j\omega_s L_2} = \frac{r_{l2} - j\omega_s L_2}{r_{l2}^2 + \omega_s^2 L_2^2} \cong \frac{r_{l2}}{\omega_s^2 L_2^2} + \frac{1}{j\omega_s L_2} \equiv \frac{1}{r_{el}} + \frac{1}{j\omega_s L_2} \quad \rightarrow \quad r_{el} \cong \frac{(\omega_s L_2)^2}{r_{l2}} \tag{25.4a}$$

$$\frac{1}{r_{c2} - j/\omega_s C_2} = \frac{r_{c2} + j/\omega_s C_2}{r_{c2}^2 + 1/\omega_s^2 C_2^2} \cong r_{c2}\omega_s^2 C_2^2 + j\omega_s C_2 \equiv \frac{1}{r_{ec}} + j\omega_s C_2 \quad \rightarrow \quad r_{ec} \cong \frac{1}{r_{c2}(\omega_s C_2)^2} \tag{25.4b}$$

where the values of the parasitic resistances are much smaller than those of the impedances of L_2 and C_2, respectively. The total equivalent parasitic resistance r_e in Figure 25.2(e) is therefore written as follows using (25.4a) and (25.4b):

$$r_e \equiv r_{el} // r_{ec} = \frac{(\omega_s L_2)^2}{r_{l2} + r_{c2}(\omega_s^2 L_2 C_2)^2} \tag{25.5}$$

25.2.2 Analysis and Design of the Load Voltage and Load Power

The quality factor Q of the secondary circuit, which characterizes the bandwidth relative to the resonant frequency ω_{r2}, becomes

$$Q = \frac{\omega_{r2}}{\Delta\omega} = \frac{R_L // r_e}{\omega_s L_2} \tag{25.6}$$

To achieve a robust load power characteristic regardless of the surroundings without an auto-tuning system, Q should not be too high. For instance, $Q = 100$ is selected in this chapter, which gives a frequency tolerance level of 1%. Under this constraint, the load voltage V_L and load power P_L are derived from Figure 25.2 and (25.3) and (25.6) as follows:

$$V_L = I_{eq}(R_L // r_e) = \omega_s L_m I_1 Q \frac{N_2}{N_1} \tag{25.7a}$$

$$P_L = \frac{V_L^2}{R_L} = \frac{V_{th}^2 Q(r_e - \omega_s L_2 Q)}{\omega_s L_2 r_e} \tag{25.7b}$$

As identified from (25.7b), the load power drops as the Q of the secondary circuit becomes either too small or too large; hence, a maximum point of efficiency can be found at Q_m by taking the derivative of (25.7b) with respect to Q:

$$\frac{\partial P_L}{\partial Q}\bigg|_{n=Q_m} = 0 \quad \Rightarrow \quad Q_m = \frac{r_e}{2\omega_s L_2} \tag{25.8}$$

Thus, the maximum load power $P_{L,m}$ and the maximum load resistance $R_{L,m}$ can be derived from (25.6), (25.7), and (25.8), as follows:

$$P_{L,m} \equiv P_L\big|_{Q=Q_m} = \frac{V_{th}^2 r_e}{4\omega_s^2 L_2^2} \tag{25.9a}$$

$$R_{L,m} \equiv R_L\big|_{Q=Q_m} = r_e \tag{25.9b}$$

According to (25.9), the maximum load power is limited by the equivalent internal resistances of the secondary circuit. For example, the proposed IPTS has a maximum load power of 12.6 W when $V_{th} = 3.69$ V for $I_1 = 40$ A, $L_2 = 460$ µH for $N_2 = 20$, and $r_e = 250$ mΩ for Mn–Zn-type ferrite cores with resonant capacitors in the secondary circuit, which constitutes one of the experimental conditions in this chapter. The equivalent series resistance of L_2, which consists of the copper loss and the core losses, is increased by N_2, while the equivalent series resistance of C_2 varies nonlinearly with its operating frequency and temperature. Therefore, N_2 should be carefully selected considering the effects on r_e and the minimum load voltage in order to mitigate the diode conduction loss.

Assuming that the core losses of the secondary coil are dominant, where the equivalent series resistance of the resonant capacitors and the copper losses of the wires used are suitably mitigated by selecting appropriate capacitors and wires that can withstand high-frequency usage, (25.5) and (25.9a) can be rewritten as follows:

$$r_e \equiv r_{el}//r_{ec} \cong r_{el} = \frac{(\omega_s L_2)^2}{r_{l2}} \cong \frac{(\omega_s k_1 N_2)^2}{k_2} \tag{25.10a}$$

$$P_{L,m} = \frac{V_{th}^2 r_e}{4\omega_s^2 L_2^2} \cong \frac{V_{th}^2}{4} \cdot \frac{1}{r_{l2}} = \frac{V_{th}^2}{4} \cdot \frac{1}{k_2 N_2^2} = L_1 \cdot \frac{k_1}{k_2} \cdot \left(\frac{\omega_s I_1 \kappa}{2}\right)^2 \quad \text{for} \quad \kappa \equiv \frac{N_2 L_m}{N_1 \sqrt{L_1 L_2}} \tag{25.10b}$$

where k_1 and k_2 are the coefficients of L_2 and r_{l2}, respectively, as functions of the number of turns of the secondary coil N_2, and κ is a coupling factor between the primary and secondary coils:

$$L_2 = g_1(N_2) \cong k_1 N_2^2 \tag{25.11a}$$
$$r_{l2} = g_2(N_2) \cong k_2 N_2^2 \tag{25.11b}$$

Note that the coefficients k_1 and k_2 are determined by the real part μ' and the imaginary part μ'', respectively, of the relative complex permeability of the secondary core material used in this study. Thus, $P_{L,m}$ can be simplified further with μ' and μ'', where the relationships among the relative complex permeability and L_2 and r_{l2} are identified in [37] and [38], as follows:

$$P_{L,m} \cong L_1 \cdot \frac{k_1}{k_2} \cdot \left(\frac{\omega_s I_1 \kappa}{2}\right)^2 = \omega_s L_1 \cdot \frac{\mu'}{\mu''} \cdot \left(\frac{I_1 \kappa}{2}\right)^2 \tag{25.12}$$

As indicated in (25.12), the maximum load power is limited by the ratio of the real to the imaginary parts of the relative complex permeability and the coupling factor, which is determined by the complex relationship between the permeability and the structures of the given IPTSs, where the switching frequency and certain characteristics of the primary coil design, such as ampere-turns and core shapes, are predetermined. Because the relative complex permeability of core materials varies in terms of the frequency, not only the well-known parasitic capacitance issue [29, 30] but also variations of the relative permeability should be carefully considered when adopting a higher switching frequency to achieve both the maximum load power and the most compact size for the proposed IPTS.

25.2.3 Comparison of the Proposed Dipole Coils with Conventional Loop Coils

According to (25.3a), a higher magnetizing inductance L_m results in a higher induced voltage across the secondary coil. However, L_m is not simply calculated for the case when the coils include ferromagnetic material. A comparative study of two coil structures of a dipole and a loop

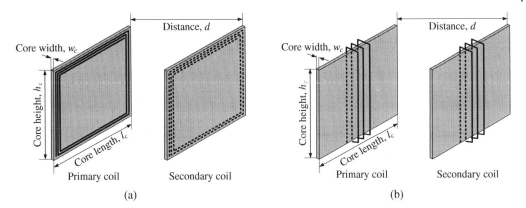

Figure 25.3 Configurations of the coil structures for comparison: (a) loop type and (b) dipole type.

Figure 25.4 The magnetizing inductances versus the distance *d* for various coil structures (log scale).

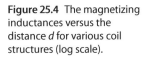

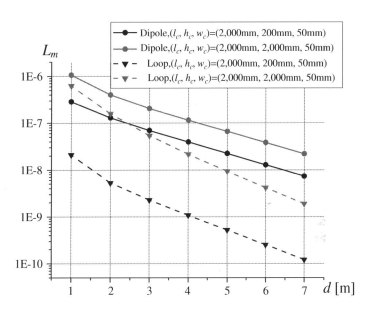

was therefore performed with an FEM simulation tool, as shown in Figures 25.3 and 25.4, where the length, height, and width of the cores are l_c, h_c, and w_c, respectively. Unlike the conventional structure, a loop coil structure with ferrite cores was evaluated since it can be attached to walls that include steel frames and/or pipes. As identified from Figure 25.4, the dipole structure with a narrow core has a higher magnetizing inductance by 60 times compared to that of a loop structure of the same size.

25.3 Example Design and Experimental Verifications of the Proposed IPTS

The analysis and design considerations of the proposed wide-range IPTS are verified with an experimental kit of 10.3 W maximum load power at 7 m distance and with a comparative evaluation of the maximum load power among different core materials. The primary core size is

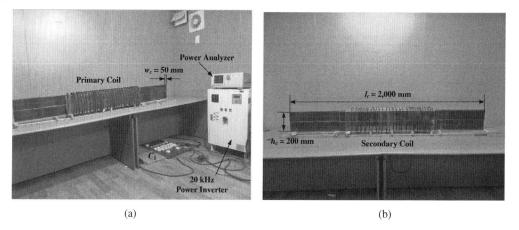

(a)   (b)

Figure 25.5 Fabricated prototype of the proposed extremely loosely coupled IPTS: (a) primary coil and (b) secondary coil.

$l_c = 2000$ mm, $h_c = 200$ mm, and $w_c = 50$ mm, as shown in Figure 25.5(a), where N_1 is set to 30 when considering saturation of the core.

25.3.1 10 W Load Power with a Fixed Quality Factor

A prototype 7 m off-distance IPTS, as shown in Figure 25.5, was fabricated with the secondary core having a structure symmetrical to the primary core. Large numbers of planner-shaped Mn–Zn-type ferrite cores, where the unit size of the ferrite core is 100 mm × 100 mm × 10 mm, commonly applicable to applications operating in the mid-to-high frequencies, were assembled for both the primary and secondary coils. The measured and calculated load voltage and load power versus the primary current I_1 for various values of N_2 are shown in Figure 25.6. The load resistance was selected for each N_2 and r_e to maintain the $Q = 100$ condition, with other circuit parameters for each value of N_2 tabulated in Table 25.2. The discrepancies between the calculations and the experimental results in Figure 25.6 are due to the increased parasitic

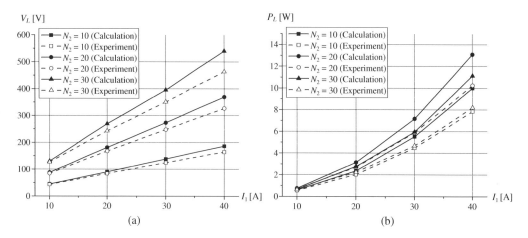

(a)   (b)

Figure 25.6 Comparisons of the analyses results with the experiment results for 20 kHz operation when $Q = 100$: (a) load voltage V_L and (b) load power P_L.

Table 25.2 Circuit parameters of the secondary circuit for the proposed IPTS when $Q = 100$

N_2 (turns)	L_2 (μH)	r_{l2} (mΩ)	k_1 ($\times 10^{-6}$)	k_2 ($\times 10^{-3}$)	C_2 (nH)	r_{c2} (Ω)	R_L (kΩ)
10	120	80	1.20	0.800	530	0.005	3.46
20	460	250	1.15	0.625	138	0.02	10.8
30	930	560	1.03	0.622	67.8	0.09	26.3

capacitance and the subtle changes in the inductances as both the number of turns and the flux density increase. The obtained experimental load power was 10.3 W, which is sufficient for two or more wireless sensor sets in Table 25.3 [31], while the low coil-to-coil power efficiency of 0.35% does not present a problem.

For practical applications, a high-step down-switching regulator can be adopted to achieve equivalently high load resistance for satisfying the maximum powering condition, where the load estimation in Table 25.3 includes the typical switching regulator efficiency of 70%. The higher the operating frequency, the more compact the system; however, the parasitic capacitances between the windings of each coil and the core loss would increase with the increased frequency.

25.3.2   Comparative Evaluation of the Metal-Surrounded Environment

In most practical applications, the proposed IPTS is highly likely to be surrounded by non-deterministic metal objects during its operation. Considering that most wireless sensors are attached on-to walls and/or ceilings that have steel frames inside, the effects of such environments, surrounded by metal, on the proposed IPTS are experimentally evaluated in this section.

To mimic this type of environment, a steel container made of SS400, which has conductivity of 7.51×10^6 S/m and relative permeability of 1000, was used [39]. The rectangular cuboid container size is 8 m $\times$ 4 m $\times$ 2.5 m and the thickness of each face is 1.4 mm, as shown in Figure 25.7(a). Both inside and outside the container, each coil was installed 1 m apart from the bottom and the ground, respectively. To conduct experiments with various operating frequencies, a current-controlled half-bridge inverter was fabricated, as shown in Figure 25.7(b), where a high-frequency applicable metal capacitor in Figure 25.7(c) was used together with the secondary coil for achieving negligibly small parasitic ESR at the high frequencies.

To analyze the effects of the metal container, V_L and P_L were measured over a frequency range of 20 kHz to 150 kHz when I_1 was fixed at 10 A, as shown in Figure 25.8(a) and (b). In

Table 25.3 Load power estimation for driving a single wireless sensor et

Component	Power consumption	Remarks
Microprocessor	1.20 W	Dspic30F6012A
Wi-Fi module	0.75 W	RN-171XVWI/RM
Sensor and transducer	0.90 W	Rosemount 1154 pressure meter
Switching regulator	1.22 W	Efficiency = 70%
Total	4.07 W	Assuming 20% margin, 5 W in total

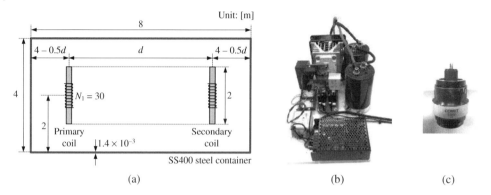

(a)　　　　　　　　(b)　　　　　　　　(c)

Figure 25.7 Experimental conditions for the comparative evaluations of the metal-surrounded environments: (a) overall configuration, (b) fabricated half-bridge inverter, and (c) used resonant capacitor.

spite of the higher r_{l2} at the secondary coil in Figure 25.8(a), a higher P_L can be achieved in the container due to the higher κ as compared to the outside in Figure 25.8(b). Inside the container, the fixed Q of 100 was suitably matched with Q_m at each frequency; hence, the proposed IPTS could operate under the maximum powering conditions, as shown in Figure 25.8(c); r_{l2},

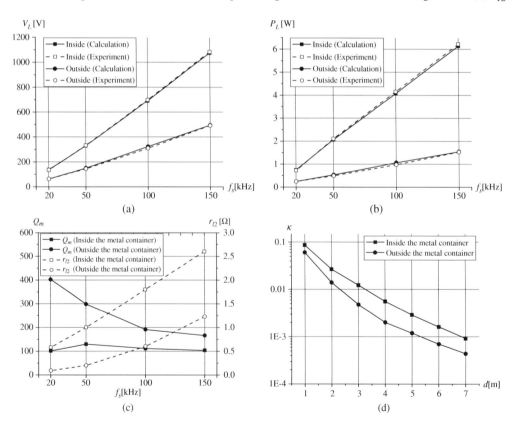

Figure 25.8 Comparisons of the experimental results inside the metal container with the outside case for different operation frequencies when $N_1 = N_2 = 30$: (a) load voltage V_L when $Q = 100$ and $I_1 = 10$ A, (b) load power P_L when $Q = 100$ and $I_1 = 10$ A, (c) Q_m for the maximum efficiency and parasitic inductor resistance r_{l2}, and (d) coupling coefficient κ.

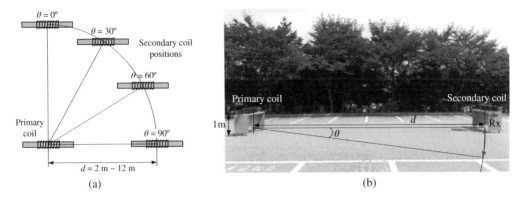

Figure 25.9 Measurement conditions of the variation of coupling coefficient according to the distance and alignment: (a) definition of the distance and the alignment angles and (b) fabricated primary and secondary coils when $d = 10$ m and $\theta = 0°$.

which represents the summation of the hysteresis losses and the eddy-current losses in both the secondary core and the metal container, increased as the operation frequency increased. Due to the losses in the SS400 material, r_{l2} nearly doubled in the container compared to the outside. Figure 25.8(d) shows the measured κ according to the distance between the primary and secondary coils inside and outside the container at 20 kHz. It is noteworthy that both κ and r_{l2} doubled in the container, which explains the linear increase in P_L in Figure 25.8(b) and the constant Q_m in Figure 25.8(c) regarding the increase in the frequency, based on the results of the analysis by (25.7) and (25.8), respectively.

25.3.3 Variation of the Coupling Coefficient According to the Distance and Alignment

To verify the powering conditions for various secondary coil positions over a wide area, the coupling coefficient κ was measured in free space for distances ranging from 2 m to 12 m and alignment angles of 0° to 90° between the proposed primary and secondary coils. As shown in Figure 25.9, the alignment angle θ ranged from 0° to 90° according to the secondary coil positions, where the distance d was defined as the distance from the primary coil center to the secondary coil center. As shown from Figure 25.10, κ decreased logarithmically as d increased, where the lowest κ was obtained when two dipole coils were aligned with $\theta = 30°$. Note that κ is much less than 0.01 regardless of θ when d is more than double the dipole coil length; thus, both the load variation and the secondary resonant condition scarcely affect the driving of the primary coil. Therefore, it is possible to use a fixed resonant condition on the primary side for this long-distance application, resulting in robust power transfer characteristics.

25.3.4 Comparative Evaluation of the Core Materials for Maximizing the Load Power

A comparative evaluation of the three different core materials, that is, two Mn–Zn-type ferrite cores from SAMHWA electronics and an amorphous core from AMOGREENTECH, was conducted to determine the effect of the core losses for a higher maximum load power. A thinner secondary core size, that is, $l_c = 2000$ mm, $h_c = 100$ mm, and $w_c = 20$ mm, was adopted with N_2 of 20 for each material in common, where the primary coil ampere-turn was fixed as 200 A-turns. The effects of core losses were highlighted by selecting a litz wire with a diameter of 15 mm and high-frequency film capacitors, where the internal resistances of the wire and

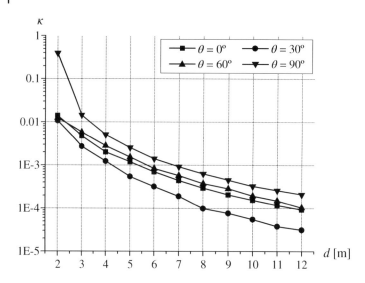

Figure 25.10 Measured coupling coefficient κ w.r.t. the distance and the alignment angle.

resonant capacitors were negligible. The load resistance was selected for the required Q_m condition that brings the maximum load power. The other parameters for each core material are tabulated in Table 25.4. The measured maximum load power and quality factor are quite close to the theoretical limit of each material, as shown in Table 25.4. The discrepancies between the calculations and the experimental results in Table 25.4, which are mitigated for cases with low quality factors, are mostly due to the slight mismatch between Q_m and Q. In practice, it is quite difficult to ensure precise tuning of the resonance condition at a high quality factor due to the limited values of the capacitor and parasitics. The coupling factor and the relative complex permeability, which mainly determine the maximum load power for the given experimental conditions, are measured, as shown in Table 25.4.

The largest $P_{L,m}$ of 0.398 W was experimentally obtained with a PL-13 ferrite core having the highest coupling factor and μ'/μ'', which in this case has a theoretical value of 0.581 W. However, the required Q_m was too high at 312, which would result in an overly sensitive power transfer characteristic in the event of a severe accident at an NPP. The PL-F2 ferrite core, which is known to have better frequency characteristics than PL-13, showed a slightly lower load power performance with a moderated Q_m due to the lower relative permeability. The AMLB-8320 amorphous core, which is formed by layers of laminated 40 μm thin amorphous-metal film, showed the lowest maximum load power due to the high core loss when the required Q_m was only 20.1.

Table 25.4 Circuit parameters of the secondary circuit for different core materials

Material	L_2 (μH)	r_{l2} (mΩ)	$\kappa(\times 10^{-4})$	μ'	μ''	μ'/μ''	$P_{L,m}$ (W)*	P_L (W)	Q_m*	Q
Ferrite 1 (PL-13)	230	46	8	3200	5	640	0	0	312	258
Ferrite 2 (PL-F2)	141	39	5	2200	5	440	0	0	224	170
Amorphous	174	545	6	695	17	40	0	0	20	20

*Calculation results (the other parameters were obtained by measurements).

25.4 Conclusion

In this chapter, a broadly applicable IPTS was demonstrated for extremely loosely coupled dipole coils. The superiority of a dipole structure coil with a ferrite core compared to the conventional loop structure was verified in terms of the magnetizing inductance. Through a detailed analysis, a design guideline was proposed for the extremely loosely coupled case. A 10.3 W power delivery at a distance of 7 m was achieved with a fabricated prototype. To prevent hypersensitive power transfers to the surrounding materials, a relatively low Q factor ($Q =$ 100) was tested. The enhanced powering performance of the proposed IPTS in an environment surrounded by metal was experimentally verified using a steel container. The dependence of the coupling factor on the secondary coil positions was measured for an overly wide range of 2 m to 12 m for different alignment cases. The available maximum load powers depending on the secondary core materials were evaluated in a comparative experiment involving two ferrite cores and an amorphous core. The proposed IPTS is under development for NPPs as an emergency backup power unit of the essential instruments at these facilities. Detail comparisons on the powering characteristics, such as efficiency, size, EMI/EMF issues among the proposed IPTS, the RF power transfer, and the laser power transfer, are left for further works.

Problem

25.1 Suggest a simplified dipole model for a long-distance IPT. In other words, can you analytically determine self-inductance of a dipole coil and mutual inductance of two dipole coils? No such theory has yet been found.

References

1 A. Kurs *et al.*, "Wireless power transfer via strongly coupled magnetic resonances," *Science*, vol. 317, pp. 83–86, July 2007.

2 K. Wan, Q. Xue, X. Liu, and S. Hui, "Passive radio-frequency repeater for enhancing signal reception and transmission in a wireless charging platform," *IEEE Trans. on Ind. Electron.*, vol. 61, no. 4, pp. 1750–1757, April 2014.

3 R. Hui, W. Zhong, and C. Lee, "A critical review of recent progress in mid-range wireless power transfer," *IEEE Trans. on Power Electron.*, vol. 29, no. 9, pp. 4500–4511, September 2014.

4 X. Ju, L. Dong, X. Huang, and X. Liao, "Switching technique for inductive power transfer at high-Q regimes," *IEEE Trans. on Ind. Electron.*, vol. 62, no. 4, pp. 2164–2173, April 2015.

5 Z.N. Low, R.A. Chinga, R. Tseng, and J. Lin, "Design and test of a high-power high-efficiency loosely coupled planar wireless power transfer system," *IEEE Trans. on Ind. Electron.*, vol. 56, no. 5, pp. 1801–1812, May 2009.

6 W. Hsu, A. Hu, and A. Swain, "A wireless power pickup based on directional tuning control of magnetic amplifier," *IEEE Trans. on Ind. Electron.*, vol. 56, no. 7, pp. 2771–2781, July 2009.

7 A.P. Sample, D.A. Meyer, and J.R. Smith, "Analysis, experimental results, and range adaption of magnetically coupled resonators for wireless power transfer," *IEEE Trans. on Ind. Electron.*, vol. 58, no. 2, pp. 544–554, February 2011.

8 T. Imura and Y. Hori, "Maximizing air gap and efficiency of magnetic resonant coupling for wireless power transfer using equivalent circuit and Neumann formula," *IEEE Trans. on Ind. Electron.*, vol. 58, no. 10, pp. 4746–4752, October 2011.

9 W. Zhong, C. Lee, and S. Hui, "General analysis on the use of Tesla's resonators in domino forms for wireless power transfer," *IEEE Trans. on Ind. Electron.*, vol. 60, no. 1, pp. 261–270, January 2013.

10 Z. Low *et al.*, "Method of load/fault detection for loosely coupled planar wireless power transfer system with power delivery tracking," *IEEE Trans. on Ind. Electron.*, vol. 57, no. 4, pp. 1478–1486, April 2010.

11 C.K. Lee, W.X. Zhong, and S.Y.R. Hui, "Effects of magnetic coupling of nonadjacent resonators on wireless power transfer domino-resonator systems," *IEEE Trans. on Power Electron.*, vol. 27, no. 4, pp. 1905–1916, April 2012.

12 J. Shin *et al.*, "Design and implementation of shaped magnetic-resonance-based wireless power transfer system for road powered moving electric vehicles," *IEEE Trans. on Ind. Electron.*, vol. 61, no. 3, pp. 1179–1192, March 2014.

13 S.H. Cheon, Y.H. Kim, S.Y. Kang, M.L. Lee, J.M. Lee, and T.Y. Zyung, "Circuit-model-based analysis of a wireless energy-transfer system via coupled magnetic resonances," *IEEE Trans. on Ind. Electron.*, vol. 58, no. 7, pp. 2906–2914, July 2011.

14 W.X. Zhong, X. Liu, and S.Y. Hui, "A novel single-layer winding array and receiver coil structure for contactless battery charging systems with free-positioning and localized charging features," *IEEE Trans. on Ind. Electron.*, vol. 58, no. 9, pp. 4136–4144, September 2011.

15 G. Elliott, S. Raabe, G. Covic, and J. Boys, "Multiphase pickups for large lateral tolerance contactless power-transfer systems," *IEEE Trans. on Ind. Electron.*, vol. 57, no. 5, pp. 1590–1598, May 2010.

16 N. Keeling, G. Covic, and J. Boys, "A unity-power-factor IPT pickup for high-power applications," *IEEE Trans. Ind. Electron.*, vol. 57, no. 2, pp. 744–751, February 2010.

17 Y. Sohn, B. Choi, E. Lee, and Chun T. Rim, "Comparisons of magnetic field shaping methods for ubiquitous wireless power transfer," in *2015 IEEE WoW*, pp. 1–6.

18 B. Lee, H. Kim, S. Lee, C. Park, and Chun T. Rim, "Resonant power shoes for humanoid robots," in *2011 ECCE Conference*, pp. 1791–1794.

19 J. Huh, S. Lee, W. Lee, G. Cho, and Chun T. Rim, "Narrow-width inductive power transfer system for on-line electrical vehicles," *IEEE Trans. on Power Electron.*, vol. 26, no. 12, pp. 3666–3679, December 2011.

20 S. Lee, B. Choi, and Chun T. Rim, "Dynamics characterization of the inductive power transfer system for online electric vehicles by Laplace phasor transform," *IEEE Trans. on Power Electron.*, vol. 28, no. 12, pp. 5902–5909, December 2013.

21 S. Choi, and J. Hun, W. Lee, and Chun T. Rim, "Asymmetrical coil sets for wireless stationary EV chargers with large lateral tolerance by dominant field analysis," *IEEE Trans. on Power Electron.*, vol. 29, no. 12, pp. 6406–6420, December 2014.

22 S. Lee *et al.*, "On-line electric vehicle using inductive power transfer system," in *2010 ECCE Conference*, pp. 1598–1601.

23 J. Huh *et al.*, "Characterization of novel inductive power transfer systems for on-line electric vehicles," in *2011 APEC Conference*, pp. 1975–1979.

24 J. Huh *et al.*, "High performance inductive power transfer system with narrow rail width for on-line electric vehicles," in *2010 ECCE Conference*, pp. 647–651.

25 J. Huh, W.Y. Lee, S.Y. Choi, G.H. Cho, and Chun T. Rim, "Explicit static circuit model of coupled magnetic resonance system," in *2011 ECCE-Asia Conference*, pp. 2233–2240.

26 E. Lee, J. Huh, X.V. Thai, S. Choi, and Chun T. Rim, "Impedance transformers for compact and robust coupled magnetic resonance systems," in *2013 ECCE Conference*, pp. 2239–2244.

27 B. Choi, E. Lee, J. Huh, and Chun T. Rim, "Lumped impedance transformers for compact and robust coupled magnetic resonance systems," *IEEE Trans. on Power Electron.*, vol. 30, no. 11, pp. 6046–6056, November 2015.

28 C. Park, S. Lee, and Chun T. Rim, "5 m-off-long-distance inductive power transfer system using optimum shaped dipole coils," in *2012 IPEMC Conference*, pp. 137–1142.

29 C. Park, S. Lee, G. Cho, and Chun T. Rim, "Innovative 5 m-off-distance inductive power transfer systems with optimally shaped dipole coils," *IEEE Trans. on Power Electron.*, vol. 30, no. 2, pp. 817–827, February 2015.

30 B. Choi, E. Lee, J. Kim, and Chun T. Rim, "7 m-off-long-distance extremely loosely coupled inductive power transfer systems using dipole coils," in *2014 ECCE Conference*, pp. 858–863.

31 S. Yoo, B. Choi, S. Jung, and Chun T. Rim, "Highly reliable power and communication system for essential instruments under a severe accident of NPP," *Trans. Korean Nucl. Soc.*, vol. 2, pp. 1005–1006, October 2013.

32 B. Choi, E. Lee, Y. Sohn, G. Jang, and Chun T. Rim, "Six degrees of freedom mobile inductive power transfer by crossed dipole Tx and Rx coils," *IEEE Trans. on Power Electron.*, vol. PP, no. 99, pp. 1, June 2015 (rapid post article).

33 S. Kim *et al.*, "Ambient RF energy-harvesting technologies for self-sustainable standalone wireless sensor platforms," *Proc. IEEE*, vol. 102, no. 11, pp. 1649–1666, November 2014.

34 M. Danesh and J.R. Long, "Photovoltaic antennas for autonomous wireless systems," *IEEE Trans. on Circuits Syst. II, Exp. Briefs*, vol. 58, no. 12, pp. 807–811, November 2011.

35 G. Mahan, B. Sales, and J. Sharp, "Thermoelectric materials: new approaches to an old problem," *Physics Today*, vol. 50, no. 3, pp. 42–47, March 1997.

36 S. Roundy, P.K. Wright, and J. Rabaey, "A study of low level vibrations as a power source for wireless sensor nodes," *Comput. Commun.*, vol. 26, no. 11, pp. 1131–1144, July 2003.

37 K. Shin, Y. Kim, and S. Kim, "AC permeability of Fe–Co–Ge/WC/phenol magnetostrictive composites," *IEEE Trans. on Magnetics*, vol. 41, no. 10, pp. 2784–2786, October 2005.

38 J. Fuzerova *et al.*, "Analysis of the complex permeability versus frequency of soft magnetic composites consisting of iron and Fe73Cu1Nb3Si16B7," *IEEE Trans. on Magnetics*, vol. 48, no. 4, pp. 1545–1548, April 2012.

39 Y. Gotoh, A. Kiyal, and N. Takahashi, "Electromagnetic inspection of outer side defect on steel tube with steel support using 3-D nonlinear FEM considering non-uniform permeability and conductivity," *IEEE Trans. on Magnetics*, vol. 46, no. 8, pp. 3145–3148, August 2010.

26

Free-Space Omnidirectional Mobile Chargers

26.1 Introduction

As introduced in Chapter 1, power transfers (PTs) can be classified as stationary and mobile ones depending on the movement of loads, as shown in Figure 26.1. The wireless PT (WPT) is nowadays the most preferable way for a mobile PT that embraces IPT, capacitive PT [1, 2], RF PT, and optical PT [3, 4], whereas the wired PT can also provide power over a flexibly long distance if properly designed [5, 6]. In the era of ubiquitous mobile chargers, the IPT is the most widely used among the WPTs [7–58]. The WPT will play a significantly important role in the realization of the Internet of things (IoT), which includes compact communication devices, sensors, and power sources.

As a power source of IoT, the degrees of freedom (DoF) of an Rx load in position and rotation are of paramount importance, as shown in Figure 26.2. The six DoF in three-dimensional (3-D) space are composed of the position vector $\vec{P}(x, y, z)$ and rotation vector $\vec{R}(\theta_x, \theta_y, \theta_z)$ in this chapter, where the rotation of an Rx normal vector $\vec{n}$ is defined as pitch (θ_x), roll (θ_y), and yaw (θ_z). To ensure the full mobility of an Rx load, two characteristics of "free-positioning" and "omnidirectional powering" should be guaranteed, which corresponds to $\vec{P}(x, y, z)$ and $\vec{R}(\theta_x, \theta_y, \theta_z)$, respectively. Confining the discussions in this chapter to IPT only, the free-positioning of an Rx coil is obtained by an evenly distributed magnetic field intensity of a transmission (Tx) coil, while the omnidirectional powering of an Rx coil is guaranteed by a non-zero induced load voltage over any rotation angles. To cope with the free-positioning and omnidirectional powering problems, multiple planar coil structures have been proposed in wireless charging pad applications [8–11]; however, only lateral free positioning with one rotation over a Tx coil surface can be obtained at a near distance. To achieve the omnidirectional powering, multiple coils are combined in perpendicular to allow two or three rotations in a 3-D space [12–18]; however, volumetric coil structures are hardly implementable to mobile devices in practice. On the other hand, long- and narrow-shaped dipole coils with ferrite cores were used to achieve a dimension reduction in coil structures, where the overall superiority of dipole coils for a long-distance power delivery over loop coils has been verified in [19] and [20].

As a candidate of IPTs solving both free-positioning and omnidirectional powering, new crossed dipole Tx and Rx coils having six DoF with a DQ rotating magnetic field are explained in this chapter, as shown in Figure 26.3. Two crossed dipole coils with orthogonal phase differences generate a rotating magnetic field to provide the Rx coil with the highest DoF. The plane geometry of the proposed coils can resolve the chronic installation problems associated with the volumetric coil structures in conventional omnidirectional IPTs. The operating frequency of the proposed Tx and Rx coils was selected as 280 kHz, meeting the international guideline of Power Matters Alliance (PMA). A simulation-based design of the proposed crossed dipole coils

Wireless Power Transfer for Electric Vehicles and Mobile Devices, First Edition. Chun T. Rim and Chris Mi.
© 2017 John Wiley & Sons Ltd. Published 2017 by John Wiley & Sons Ltd.

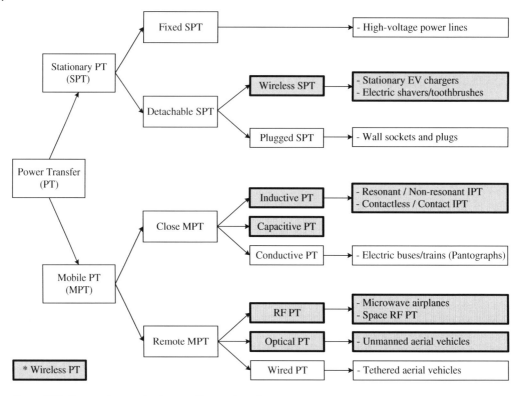

Figure 26.1 Proposed general category of power transfers.

for a uniform magnetic field distribution is provided, and the 3-D omnidirectional IPT is experimentally verified by prototype Rx coils for a wireless power- zone of 1 m^3 with a prototype Tx coil of 1 m^2.

The rest of this chapter is organized as follows. Section 26.2 presents analyses on omnidirectional powering IPT coil structures such as loop coils and dipole coils. Section 26.3 shows the characteristics of the proposed crossed dipole coils having six DoF with a DQ rotating magnetic

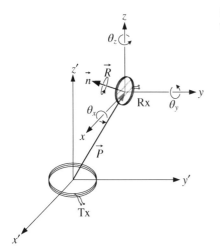

Figure 26.2 Definition of six degrees of freedom of an Rx load w.r.t. Tx in 3-D space.

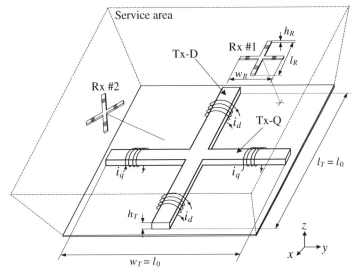

(a) Overall Tx and Rx coil configuration

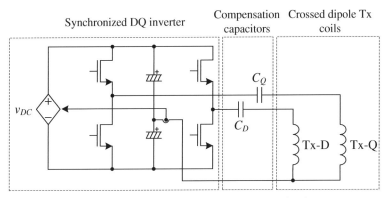

(b) Circuit diagram showing series resonance circuits

Figure 26.3 Proposed crossed dipole coils for six DoF.

field based on analyses and simulations. Section 26.4 presents experimental verifications of the omnidirectional powering using fabricated Tx and Rx coils. The conclusions are provided in Section 26.5.

26.2 Evaluation of DoF on Various Power Transfers

In this section, various PTs in Figure 26.1 are comparatively evaluated according to the DoF of the Rx load in position and rotation. As listed in Table 26.1, the PTs are categorized into seven configurations, as shown in Figure 26.4, according to the proposed classification of Figure 26.1. Neither fixed nor detachable SPTs allow any mobility of te Rx load in general. For example, a high-voltage power line is fixed to metal towers and a wall socket does not allow any DoF of the connected wall plug, as shown in Figure 26.4(a) and (b). On the other hand, at least one DoF in position or rotation is guaranteed for MPTs in general to ensure the mobility of the Rx load.

Table 26.1 The DoF of power transfer according to the proposed seven classifications

Power transfer methods		Maximum allowable number of DoF		Remarks
		Position (x, y, z)	Rotation $(\theta_x, \theta_y, \theta_z)$	
SPT	Fixed SPT	0	0	Wired
	Detachable SPT	0	0	Wired/wireless
MPT	Inductive PT	3	3	Wireless
	Capacitive PT	2	1	Wireless
	Conductive PT	1	0	Wired
	RF/Optical PT	3	3	Wireless
	Wired PT	3	3	Wired

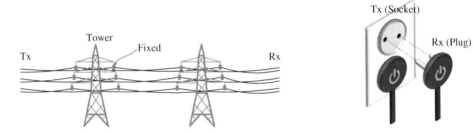

(a) Fixed SPT: High-voltage power line and tower (0P-0R) (b) Detachable SPT: A set of wall socket and plug (0P-0R)

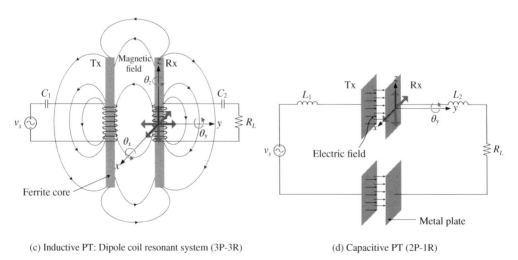

(c) Inductive PT: Dipole coil resonant system (3P-3R) (d) Capacitive PT (2P-1R)

Figure 26.4 Examples of PTs with respect to a six DoF viewpoint: "m_1P- m_2R" means the degrees of freedom of position and rotation are m_1 and m_2, respectively.

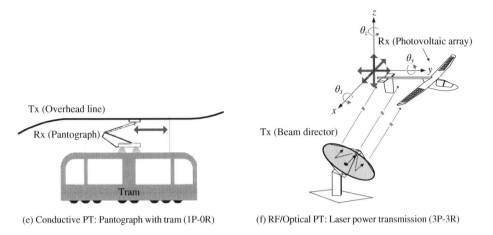

(e) Conductive PT: Pantograph with tram (1P-0R)   (f) RF/Optical PT: Laser power transmission (3P-3R)

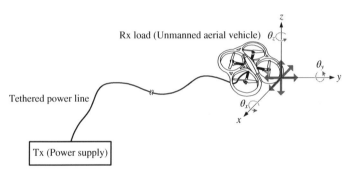

(g) Wired PT: Tethered aerial vehicle (3P-3R)

Figure 26.4 (*Continued*)

As an example of IPTs, a dipole coil resonance system (DCRS) is illustrated in Figure 26.4(c). The maximum allowable DoF for both the position and rotation is three when an omnidirectional Tx coil or Rx coil is used, which will be discussed in detail in this chapter. The capacitive PTs using two capacitive coupled metal plates, however, can transfer power only when the two metal plates are located in parallel, as shown in Figure 26.4(d); hence, the maximum DoF of the capacitive PTs is two for position and one for rotation, respectively. An example of a conductive PT is the pantograph of a tram, which collects power through a contact with an overhead wire; thus, the maximum DoF in position and rotation are one and zero, respectively, as shown in Figure 26.4(e). Note that the heading of a tram has three DoF in position and rotation, but a rail is fixed and not arbitrarily changeable. A laser power transmission, as depicted in Figure 26.4(f), can support all DoF of an Rx load in position and rotation over an arbitrary 3-D space. A tethered aerial vehicle connected with a flexible power line also has six DoF within the range of an available power line, as shown in Figure 26.4(g).

26.3 Omnidirectional Wireless Power Transfers by Loop Coils

26.3.1 Analysis on Rx Coils under Evenly Distributed Magnetic Flux Density

As the IPT has been identified as a candidate for omnidirectional powering, the DoF for three rotations θ_x, θ_y, and θ_z in IPTs are evaluated for loop coil structures first, which have been

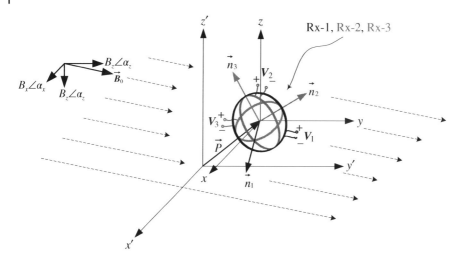

Figure 26.5 Three Rx loop coils under the evenly distributed magnetic flux density.

widely used for ubiquitous powering [12–18]. While the omnidirectional IPT was experimentally verified in [12] to [18], the omnidirectional powering conditions in the IPT are theoretically analyzed in detail in this chapter. The minimum physical dimensions of Tx and Rx coils allowing arbitrary rotations in the 3-D position are also investigated. It is assumed that the Rx coil is quite small compared to the Tx coil and that the magnetic field intensity from the Tx coil is evenly distributed for a specified 3-D space; thus, the free-positioning of the Rx coil is inherently guaranteed. The number of Tx coils, which are orthogonal to each other, is defined as m_T and that of Rx coils is defined as m_R, respectively. For example, a single loop coil or a dipole coil corresponds to $m_T = 1$ and two orthogonal loop coils correspond to $m_T = 2$ when they are used for a Tx purpose.

Considering typical ubiquitous powering coil structures, it is assumed that the maximum m_R is 3 and all symmetric Rx coils, namely Rx-1, Rx-2, and Rx-3, are orthogonal to each other, as shown in Figure 26.5. The induced voltage of Rx coils in the steady state, which has area A_{Rx} and number of turns N_{Rx}, are calculated for an arbitrary rotation when the evenly distributed sinusoidal magnetic flux density vector $\vec{B}_0$ of phasor form, having angular frequency of ω_s, is externally applied to the 3-D space. Throughout this chapter, bold letters are used to represent phasors and an upper arrow is used to represent a vector. The magnetic flux ϕ_1 of phasor form, penetrating Rx-1, can be determined as follows:

$$\phi_1 \equiv \int_S \vec{B}_0 \cdot d\vec{S} \cong A_{Rx}\vec{B}_0 \cdot \vec{n}_1 = A_{Rx}(n_{1x}B_x\angle\alpha_x + n_{1y}B_y\angle\alpha_y + n_{1z}B_z\angle\alpha_z) \tag{26.1}$$

where $\vec{B}_0$ and $\vec{n}_1$ are represented as a phasor vector and a normalized scalar vector, respectively, as shown below:

$$\vec{B}_0 \equiv (B_x\angle\alpha_x, B_y\angle\alpha_y, B_z\angle\alpha_z) \tag{26.2a}$$

$$\vec{n}_1 \equiv (n_{1x}, n_{1y}, n_{1z}) \tag{26.2b}$$

Then the induced Rx coil voltage V_1 for Rx-1 of phasor form can be determined by Faraday's law from (26.1) as follows:

$$V_1 = j\omega_s N_{Rx}\phi_1 = jV_m(n_{1x}B_x\angle\alpha_x + n_{1y}B_y\angle\alpha_y + n_{1z}B_z\angle\alpha_z) \quad \because V_m \equiv \omega_s N_{Rx}A_{Rx} \tag{26.3}$$

In (26.3), V_1 can be determined for an arbitrary direction defined by $\vec{n}_1 \equiv (n_{1x}, n_{1y}, n_{1z})$, which is determined by a transformation matrix R_{ijk} [59] and an x-directional normal vector $\vec{n}_{10}$ as follows:

$$\begin{bmatrix} n_{1x} \\ n_{1y} \\ n_{1z} \end{bmatrix} = R_i R_j R_k \begin{bmatrix} 1 \\ 0 \\ 0 \end{bmatrix} = R_{ijk} \begin{bmatrix} 1 \\ 0 \\ 0 \end{bmatrix} \quad \text{for} \quad R_{ijk} \equiv R_i R_j R_k, \tag{26.4}$$

where $\vec{n}_1 \equiv (1, 0, 0)$ is a reference vector of $\vec{n}_1$ in this chapter before any rotating of Rx-1. Note that R_{ijk} is defined as a matrix multiplication of R_i, R_j, and R_k when "ijk" is a permutation of a set $\{x, y, z\}$, namely $xyz, xzy, yxz, yzx, zxy,$ and zyx. For simplicity, R_{zyx} is selected among R_{ijk} in this chapter, which denotes a sequential rotation of Rx-1 w.r.t. pitch (θ_x), roll (θ_y), and yaw (θ_z), where R_x, R_y, and R_z are defined as follows [59]:

$$R_x = \begin{bmatrix} 1 & 0 & 0 \\ 0 & \cos\theta_x & -\sin\theta_x \\ 0 & \sin\theta_x & \cos\theta_x \end{bmatrix}, R_y = \begin{bmatrix} \cos\theta_y & 0 & \sin\theta_y \\ 0 & 1 & 0 \\ -\sin\theta_y & 0 & \cos\theta_y \end{bmatrix}, \text{ and } R_z = \begin{bmatrix} \cos\theta_z & -\sin\theta_z & 0 \\ \sin\theta_z & \cos\theta_z & 0 \\ 0 & 0 & 1 \end{bmatrix}$$
$$\tag{26.5}$$

Applying (26.4) to (26.5), V_1 in (26.3) becomes

$$V_1|_{\vec{n}=\vec{n}_1} = jV_m(\cos\theta_y \cos\theta_z B_x \angle\alpha_x + \cos\theta_y \sin\theta_z B_y \angle\alpha_y - \sin\theta_y B_z \angle\alpha_z) \tag{26.6}$$

Expanding the discussion in (26.1) to (26.3) to the multiple orthogonal Rx coils, the V_2 and V_3, which are the induced Rx coil voltages of Rx-2 and Rx-3, respectively, can be derived as follows:

$$V_2 = jV_m(n_{2x}B_x\angle\alpha_x + n_{2y}B_y\angle\alpha_y + n_{2z}B_z\angle\alpha_z) \tag{26.7a}$$
$$V_3 = jV_m(n_{3x}B_x\angle\alpha_x + n_{3y}B_y\angle\alpha_y + n_{3z}B_z\angle\alpha_z) \tag{26.7b}$$

where the normal vectors of each coil are defined as $\vec{n}_2 \equiv (n_{2x}, n_{2y}, n_{2z})$ and $\vec{n}_3 \equiv (n_{3x}, n_{3y}, n_{3z})$. In a similar fashion to (26.4) to (26.6), Equation (26.7) becomes as follows for the y- and z-directional reference vectors of $\vec{n}_{20} \equiv (0, 1, 0)$ and $\vec{n}_{30} \equiv (0, 0, 1)$, respectively, in this chapter:

$$V_2|_{\vec{n}=\vec{n}_2} = jV_m\{(\sin\theta_x \sin\theta_y \cos\theta_z - \cos\theta_x \sin\theta_z)B_x\angle\alpha_x$$
$$+ (\sin\theta_x \sin\theta_y \sin\theta_z + \cos\theta_x \cos\theta_z) B_y\angle\alpha_y + \sin\theta_x \cos\theta_y B_z\angle\alpha_z\} \tag{26.8a}$$

$$V_3|_{\vec{n}=\vec{n}_3} = jV_m\{(\cos\theta_x \sin\theta_y \cos\theta_z + \sin\theta_x \sin\theta_z)B_x\angle\alpha_x$$
$$+ (\cos\theta_x \sin\theta_y \sin\theta_z - \sin\theta_x \cos\theta_z) B_y\angle\alpha_y + \cos\theta_x \cos\theta_y B_z\angle\alpha_z\} \tag{26.8b}$$

In the following sections, three possible configurations of a few Tx and Rx coils guaranteeing the omnidirectional powering are evaluated in detail based on the above analysis. The evaluation results for the first two coil configurations of "1Tx–3Rx" having $(m_T, m_R) = (1, 3)$ and "3Tx–1Rx" having $(m_T, m_R) = (3, 1)$ match well with the results of [12] to [18], which experimentally verified the omnidirectional powering characteristics. The "2Tx–2Rx" configuration having $(m_T, m_R) = (2, 2)$ is first explained in this chapter and also guarantees omnidirectional wireless powering, that, non-zero induced load voltage over any rotation angles.

26.3.2 Single Tx Coil and Triple Rx Coils (1Tx–3Rx)

It is well known that a set of triple orthogonal Rx coils enables us to achieve omnidirectional powering with a single Tx coil, that is, $(m_T, m_R) = (1, 3)$ [14–18]. Two possible coil configurations achieving $(m_T, m_R) = (1, 3)$ are illustrated in Figure 26.6(a) and (b). As shown in Figure 26.6(b), a pair of loop coils facing each other, namely "Tx-1a and Tx-1b," is used instead of

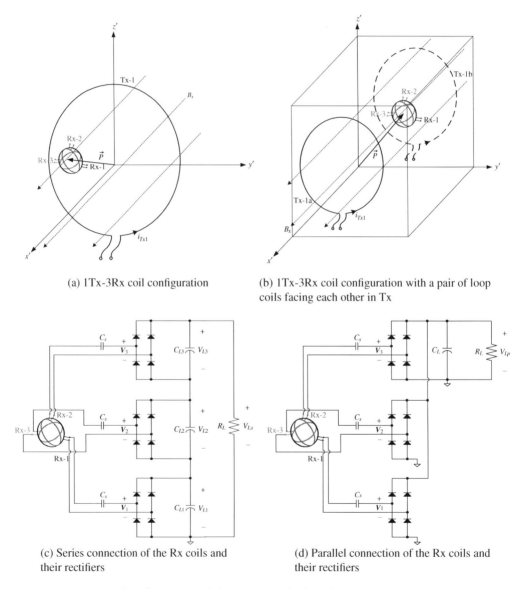

(a) 1Tx-3Rx coil configuration

(b) 1Tx-3Rx coil configuration with a pair of loop coils facing each other in Tx

(c) Series connection of the Rx coils and their rectifiers

(d) Parallel connection of the Rx coils and their rectifiers

Figure 26.6 1Tx–3Rx coil configurations and their normalized values of induced Rx coil voltages when (m_T, m_R) is (1, 3).

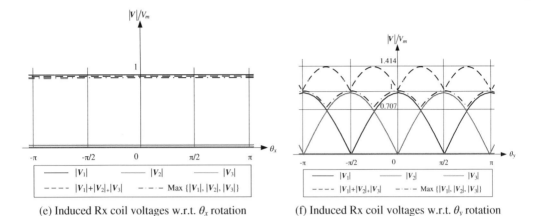

(e) Induced Rx coil voltages w.r.t. θ_x rotation   (f) Induced Rx coil voltages w.r.t. θ_y rotation

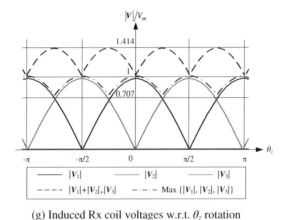

(g) Induced Rx coil voltages w.r.t. θ_z rotation

Figure 26.6 *(Continued)*

the single loop coil in Figure 26.6(a) for higher uniformity of magnetic flux. For simplicity, it is assumed that the magnetic flux density from all the Tx coils in Figure 26.6 has only x-directional components, that is, $B_x \neq 0$, $B_y = B_z = 0$. Then the Rx coil voltages in (26.6) and (26.8) can be rewritten as follows:

$$V_1|_{\vec{n}=\vec{n}_1} = jV_m \cos\theta_y \cos\theta_z B_x \angle\alpha_x \tag{26.9a}$$

$$V_2|_{\vec{n}=\vec{n}_2} = jV_m(\sin\theta_x \sin\theta_y \cos\theta_z - \cos\theta_x \sin\theta_z) \quad B_x \angle\alpha_x \tag{26.9b}$$

$$V_3|_{\vec{n}=\vec{n}_3} = jV_m(\cos\theta_x \sin\theta_y \cos\theta_z + \sin\theta_x \sin\theta_z) \quad B_x \angle\alpha_x \tag{26.9c}$$

Practically, three induced Rx coil voltages should be connected in series or in parallel to drive a single load R_L, as shown in Figure 26.6(c) and (d). For convenience of analysis, it is assumed that the diode rectifiers are operating in a continuous conduction mode with no loss. In addition, the reactance of each Rx coil is completely compensated by a series resonant capacitor C_s so that the induced coil voltages can be entirely applied to the diode rectifiers. A parallel resonant capacitor for each Rx coil may also be used for the compensation, though this is not shown here. For the series connection of Figure 26.6(c), each Rx coil voltage is rectified and stacked at the DC stage for providing load voltage. On the other hand, the parallel connection of Figure 26.6(d) automatically delivers the highest Rx coil voltage to the load. Then, the relationship between the Rx coil voltages and DC load voltage can be obtained for both the series and parallel connections as follows:

$$V_{Ls} \equiv V_{L1} + V_{L2} + V_{L3} = \frac{\pi}{2\sqrt{2}}(|V_1| + |V_2| + |V_3|) \quad \text{for the series connection} \quad (26.10a)$$

$$V_{Lp} = \frac{\pi}{2\sqrt{2}} \cdot \text{Max}\{|V_1|, |V_2|, |V_3|\} \quad \text{for the parallel connection} \quad (26.10b)$$

For an intuitive understanding of the influence of rotations, the magnitude portions of phasor voltages in (26.9) and (26.10) are depicted in Figure 26.6(e) to (g). It is noteworthy that the normalized values of both $|V_1| + |V_2| + |V_3|$ and Max $\{|V_1|, |V_2|, |V_3|\}$ in Figure 26.6(e) to (g) are always higher than $1/\sqrt{2}$ ($\sim$70.7%) over any rotation angles.

26.3.3 Triple Tx Coils and Single Rx Coils (3Tx–1Rx)

Another coil configuration offering omnidirectional powering is the triple orthogonal Tx coils with a single Rx coil, where $(m_T, m_R) = (3, 1)$ [13]. Two possible coil configurations achieving $(m_T, m_R) = (3, 1)$ are illustrated in Figure 26.7(a) and (b). For omnidirectional powering, the three Tx coil currents should be appropriately different from each other in order not to generate a constant magnetic field for a fixed direction. There are three possible modulation methods in general: (1) phase domain modulation (PDM), (2) time domain modulation (TDM), and (3) frequency domain modulation (FDM). However, the TDM requires complex controls dealing with dynamic responses of the Tx and Rx resonant circuits. The FDM is also not a viable solution because it requires three resonant circuits of different frequencies for an Rx. Hence, the PDM is mostly preferred, considering the simplicities of control and structure. Assuming that each Tx current has same magnitude with a $2\pi/3$ phase difference from each other, the Rx coil voltage of (26.6) can be rewritten when $B_x = B_y = B_z = B_0$, $\alpha_x = 0$, $\alpha_y = 2\pi/3$, and $\alpha_z = 4\pi/3$, as follows:

$$V_1|_{\vec{n}=\vec{n}_1} = jV_m(\cos\theta_y \cos\theta_z \angle 0 + \cos\theta_y \sin\theta_z \angle 2\pi/3 - \sin\theta_y \angle 4\pi/3) \quad (26.11)$$

As shown in Figure 26.7(c) to (e), the magnitude of the phasor voltage in (26.11) can be rewritten as follows:

$$V_{1x} \equiv |V_1||_{\vec{n}=\vec{n}_1, \theta_y=\theta_z=0} = V_m \quad (26.12a)$$

$$V_{1y} \equiv |V_1||_{\vec{n}=\vec{n}_1, \theta_x=\theta_z=0} = V_m|\cos\theta_y \angle 0 - \sin\theta_y \angle 4\pi/3| = V_m\sqrt{1 + 0.5\sin 2\theta_y} \quad (26.12b)$$

$$V_{1z} \equiv |V_1||_{\vec{n}=\vec{n}_1, \theta_x=\theta_y=0} = V_m|\cos\theta_z \angle 0 + \sin\theta_z \angle 2\pi/3| = V_m\sqrt{1 - 0.5\sin 2\theta_z} \quad (26.12c)$$

Though there are voltage fluctuations w.r.t. θ_y and θ_z rotations, the lowest value of $|V_1|$ is $1/\sqrt{2}$ ($\sim$70.7%) times V_m over any rotation angles.

26.3.4 Double Tx Coils and Double Rx Coils (2Tx–2Rx)

Another candidate coil configuration of omnidirectional powering is the double Tx coils with double Rx coils, where both coils are composed of two orthogonal coils when both m_T and m_R are 2. As shown in Figure 26.8(a) and (b), the Tx coils can consist of either a single set of two

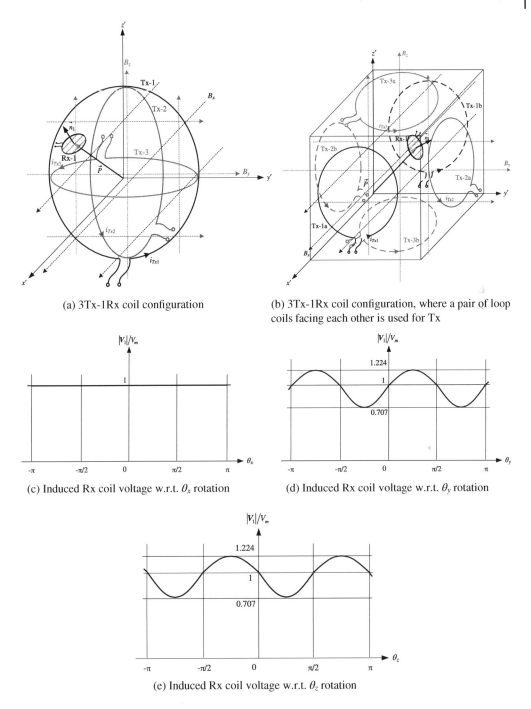

(a) 3Tx–1Rx coil configuration

(b) 3Tx–1Rx coil configuration, where a pair of loop coils facing each other is used for Tx

(c) Induced Rx coil voltage w.r.t. θ_x rotation

(d) Induced Rx coil voltage w.r.t. θ_y rotation

(e) Induced Rx coil voltage w.r.t. θ_z rotation

Figure 26.7 3Tx–1Rx coil configurations and their normalized values of induced Rx coil voltages when (m_T, m_R) is (3, 1).

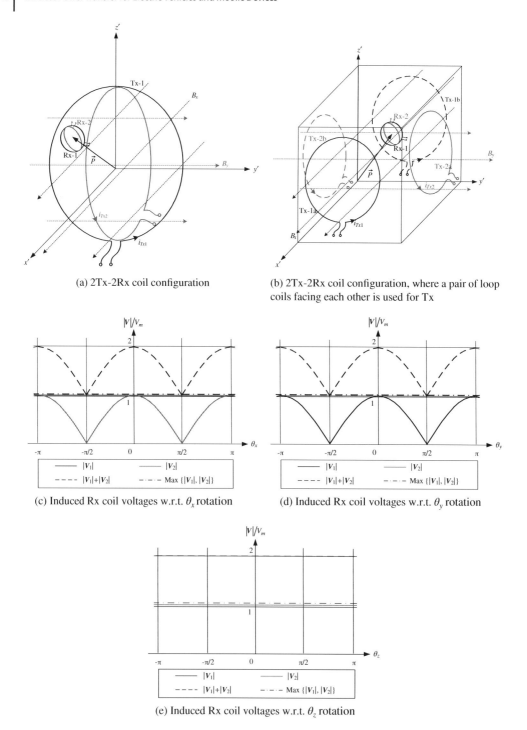

(a) 2Tx-2Rx coil configuration

(b) 2Tx-2Rx coil configuration, where a pair of loop coils facing each other is used for Tx

(c) Induced Rx coil voltages w.r.t. θ_x rotation

(d) Induced Rx coil voltages w.r.t. θ_y rotation

(e) Induced Rx coil voltages w.r.t. θ_z rotation

Figure 26.8 2Tx–2Rx coil configurations and their normalized values of induced Rx coil voltages when (m_T, m_R) is $(2, 2)$.

orthogonal coils or a dual set of two orthogonal coils, respectively. In a similar fashion to the 3Tx–1Rx configuration, the PDM is preferred to drive each Tx current.

Assuming that each Tx current has the same magnitude with a $\pi/2$ phase difference from each other, the Rx coil voltage can be rewritten from (26.6) and (26.8) when $B_x = B_y = B_0$, $B_z = 0$, $\alpha_x = 0$, and $\alpha_y = \pi/2$, as follows:

$$V_1|_{\vec{n}=\vec{n}_1} = V_m \cos\theta_y(-\sin\theta_z + j\cos\theta_z) \tag{26.13a}$$

$$V_2|_{\vec{n}=\vec{n}_2} = V_m(-\sin\theta_x \sin\theta_y \sin\theta_z - \cos\theta_x \cos\theta_z + j\sin\theta_x \sin\theta_y \cos\theta_z - j\cos\theta_x \sin\theta_z) \tag{26.13b}$$

As shown in Figure 26.8(c) to (e), the magnitude of phasor voltage in (26.13) can be rewritten as follows:

$$V_{1x} \equiv |V_1||_{\vec{n}=\vec{n}_1,\theta_y=\theta_z=0} = V_m \tag{26.14a}$$

$$V_{1y} \equiv |V_1||_{\vec{n}=\vec{n}_1,\theta_x=\theta_z=0} = \omega_s N_{Rx} A_{Rx} B_0 |\cos\theta_y| = V_m|\cos\theta_y| \tag{26.14b}$$

$$V_{1z} \equiv |V_1||_{\vec{n}=\vec{n}_1,\theta_x=\theta_y=0} = \omega_s N_{Rx} A_{Rx} B_0 |\cos\theta_z + j\sin\theta_z| = V_m \tag{26.14c}$$

$$V_{2x} \equiv |V_2||_{\vec{n}=\vec{n}_2,\theta_y=\theta_z=0} = \omega_s N_{Rx} A_{Rx} B_0 |j\cos\theta_x| = V_m|\cos\theta_x| \tag{26.14d}$$

$$V_{2y} \equiv |V_2||_{\vec{n}=\vec{n}_2,\theta_x=\theta_z=0} = V_m \tag{26.14e}$$

$$V_{2z} \equiv |V_2||_{\vec{n}=\vec{n}_2,\theta_x=\theta_y=0} = \omega_s N_{Rx} A_{Rx} B_0 |-\sin\theta_z + j\cos\theta_z| = V_m \tag{26.14f}$$

Under the same assumptions used in the 3Tx–1Rx configuration analysis, the Rx coil voltages and DC load voltage can be obtained for both the series and parallel connections as follows:

$$V_{Ls} \equiv V_{L1} + V_{L2} = \frac{\pi}{2\sqrt{2}}(|V_1| + |V_2|) \quad \text{for the series connection} \tag{26.15a}$$

$$V_{Lp} = \frac{\pi}{2\sqrt{2}} \cdot \text{Max}\{|V_1|, |V_2|\} \quad\quad \text{for the parallel connection} \tag{26.15b}$$

As identified from (26.15) and Figure 26.8, the DC load voltages are always higher than 0.5 times their maximum values over any rotation angle. Note that the series connection provides a fluctuated but higher DC load voltage than that of the parallel connection, whereas the parallel connection provides a relatively constant but lower DC load voltage.

Note that the voltage profiles shown in Figures 26.6 to 26.8 are specific cases of only one rotation among θ_x, θ_y, and θ_z rotations. In practice, the rotation of an Rx coil could be arbitrary and θ_x, θ_y, and θ_z are not correlated with each other in general so they may all be non-zeros.

26.3.5 Omnidirectional Powering Conditions in IPT

Based on the analyses results in the previous sections, the necessary condition of the omnidirectional powering in the IPT can be derived as follows:

$$m_T + m_R \geq 4 \tag{26.16}$$

where each Tx current is not correlated when m_T is higher than 1. Note that the abovementioned cases of 1Tx–3Rx, 3Tx–1Rx, and 2Tx–2Rx are the only three cases that satisfy the condition $m_T + m_R = 4$, which means that they are the minimum number of cases of Tx and Rx coils.

Table 26.2 Possible loop coil combinations for omnidirectional powering

Numbers of orthogonal coils			Loop coil dimension	
m_T	m_R	$m_T + m_R$	n_{TL}	n_{RL}
1	3	4	2	3
2	2	4	3	3
2	3	5	3	3
3	1	4	3	2
3	2	5	3	3
3	3	6	3	3

Based on the above discussions so far, six possible loop–coil combinations for omnidirectional powering are listed in Table 26.2, where the physical dimensions of Tx and Rx coils are defined as n_{TL} and n_{RL}, respectively. Note that, as is found from Figures 26.6 to 26.8, the volumetric coil structure is inevitable for at least one of Tx and Rx coils with the loop coils, which is one of the motivations to propose the crossed dipole coil structure in the subsequent section.

26.4 Analysis and Design of the Proposed Crossed Dipole Coils

26.4.1 Omnidirectional Wireless Power Transfers by Dipole Coils

In order to reduce the physical dimension of omnidirectional powering coils, the dipole coil structure [19, 20] is adopted instead of the loop coils, as shown in Figure 26.9. Using a ferromagnetic core such as ferrite, amorphous, and silicon steel in the dipole coils, the physical dimensions of Tx and Rx are reduced from volume to plane and from plane to line compared to the loop coils. Thus, the coil structures for omnidirectional powering are not necessarily volumetric when the plane-type crossed dipole coils in Figure 26.9(c) are used for both Tx and Rx coils to accomplish (m_T, m_R) of (2, 2).

Unlike the loop coil configurations, the theoretical derivation of the induced Rx coil voltage for a given Tx coil structure and current is not straightforward because of the complicated magnetic field distorted by the ferromagnetic core. Therefore, a simulation-based coil design is inevitably adopted in this chapter.

In a similar fashion to the previous section, several possible combinations for omnidirectional powering are listed in Table 26.3, where the physical dimensions of Tx and Rx coils are defined as n_{TD} and n_{RD}, respectively. Though not shown here, the combinations of Tx loop coils and Rx dipole coils, and vice versa, are also possible to implement the six DoF IPT.

26.4.2 Simulation-Based Design of Proposed Crossed Dipole Tx and Rx Coils

For most practical applications, neither Tx nor Rx coils should be in a volumetric configuration. Excluding the cases of either $n_{TD} = 3$ or $n_{RD} = 3$ from Table 26.3, the 2Tx–2Rx configuration is the only viable case where both Tx and Rx coils are of the plane type. The synchronized DQ

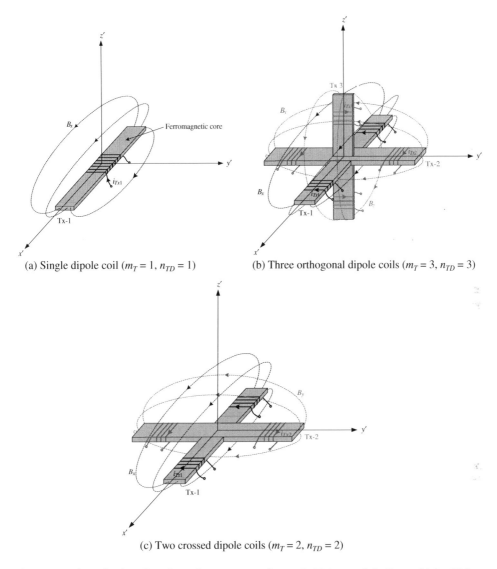

(a) Single dipole coil ($m_T = 1$, $n_{TD} = 1$)

(b) Three orthogonal dipole coils ($m_T = 3$, $n_{TD} = 3$)

(c) Two crossed dipole coils ($m_T = 2$, $n_{TD} = 2$)

Figure 26.9 Three dipole coil configurations corresponding to the Tx loop coils in Figures 26.6 to 26.8.

inverter is introduced in this chapter to provide two orthogonal phased Tx currents, as shown in Figure 26.3. The switching angular frequency of the DQ inverter ω_s is a little higher than the resonant frequency determined by the Tx coils and compensating capacitors C_D and C_Q so that the inverter switches may operate at a zero voltage switching condition [41–44]. By using the Rx coil of identical shape but smaller size than the Tx coil, both free-positioning and omnidirectional powering are achieved.

As shown in Figure 26.10, the rotating magnetic field in the steady state is formed by two orthogonal dipole coils of Tx-D and Tx-Q having an identical magnitude of currents I_d and I_q, respectively, as follows:

$$I_d \equiv I_d \angle 0 \tag{26.17a}$$

$$I_q \equiv I_d \angle \pi/2 = jI_d \tag{26.17b}$$

Table 26.3 Possible dipole coil combinations for omnidirectional powering

Numbers of orthogonal coils			Dipole coil dimension	
m_T	m_R	$m_T + m_R$	n_{TD}	n_{RD}
1	3	4	1	3
2	2	4	2	2
2	3	5	2	3
3	1	4	3	1
3	2	5	3	2
3	3	6	3	3

Both magnetic flux density vectors $\vec{B}_d$ and $\vec{B}_q$, which are generated by the two orthogonal dipole Tx currents in (26.17), can be described in phasor form as follows:

$$\vec{B}_d \equiv (\boldsymbol{B}_{dx}, \boldsymbol{B}_{dy}, \boldsymbol{B}_{dz}) = (B_{dx}\angle 0, B_{dy}\angle 0, B_{dz}\angle 0) = (B_{dx}, B_{dy}, B_{dz}) \tag{26.18a}$$

$$\vec{B}_q \equiv (\boldsymbol{B}_{qx}, \boldsymbol{B}_{qy}, \boldsymbol{B}_{qz}) = (B_{qx}\angle \pi/2, B_{qy}\angle \pi/2, B_{qz}\angle \pi/2) = (jB_{qx}, jB_{qy}, jB_{qz}) \tag{26.18b}$$

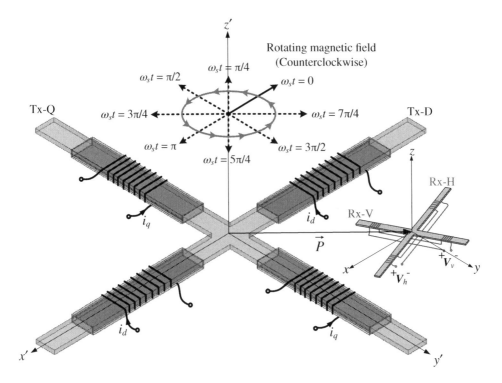

Figure 26.10 DQ Rotating magnetic field generation of the proposed crossed dipole Tx coil.

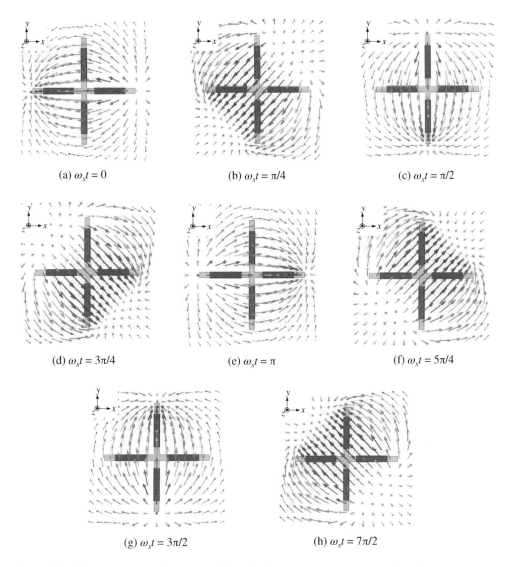

Figure 26.11 FEM simulations of the proposed Tx coil with two orthogonal currents i_d and i_q.

Note that the phasor of $\vec{B}_q$ is $\pi/2$ ahead of $\vec{B}_d$; however, the magnitude of $\vec{B}_q$ is irrelevant to $\vec{B}_d$ although the magnitudes of two currents I_d and I_q are identical.

Then the polarities of the summed magnetic field $\vec{B}_t$ on the xy-plane rotate around the coils, as shown in Figure 26.11, where finite element method (FEM) simulations of the magnetic field line by ANSYS Maxwell ver. 15 are shown for a switching period. Omnidirectional powering can be obtained by the proposed crossed dipole Tx and Rx coils of plane type as a result of this rotating magnetic field.

From (26.18), $\vec{B}_t$ can be derived as follows:

$$\vec{B}_t \equiv \vec{B}_d + \vec{B}_q = (B_{dx} + B_{qx}, B_{dy} + B_{qy}, B_{dz} + B_{qz}) = (B_{dx} + jB_{qx}, B_{dy} + jB_{qy}, B_{dz} + jB_{qz}) \quad (26.19)$$

Then the magnitude of (26.19) can be determined as follows:

$$B_t \equiv |\vec{B}_t| = |\vec{B}_d + \vec{B}_q| = \sqrt{|\boldsymbol{B}_{dx} + \boldsymbol{B}_{qx}|^2 + |\boldsymbol{B}_{dy} + \boldsymbol{B}_{qy}|^2 + |\boldsymbol{B}_{dz} + \boldsymbol{B}_{qz}|^2}$$

$$= \sqrt{|B_{dx} + jB_{qx}|^2 + |B_{dy} + jB_{qy}|^2 + |B_{dz} + jB_{qz}|^2}$$

$$= \sqrt{B_{dx}^2 + B_{qx}^2 + B_{dy}^2 + B_{qy}^2 + B_{dz}^2 + B_{qz}^2} = \sqrt{B_{dx}^2 + B_{dy}^2 + B_{dz}^2 + B_{qx}^2 + B_{qy}^2 + B_{qz}^2}$$

$$= \sqrt{|\vec{B}_d|^2 + |\vec{B}_q|^2} = \sqrt{B_d^2 + B_q^2} \quad \because B_d \equiv |\vec{B}_d|, B_q \equiv |\vec{B}_q| \tag{26.20}$$

As identified from (26.20), the magnitude of the total magnetic field density B_t is determined only by B_d and B_q, and is irrelevant to the vector direction. As shown in Figure 26.12, B_q can be found from B_d, considering the symmetric structures of the Tx coils and identical magnitude of DQ currents as follows:

$$B_q(x_1, y_1, z_1) = |\vec{B}_q(x_1, y_1, z_1)| = |\vec{B}_d(x_2, y_2, z_2)| = B_d(y_1, -x_1, z_1)$$
$$\therefore B_q(x, y, z) = B_d(y, -x, z) \quad \text{for arbitrary} \quad x, y, z \tag{26.21}$$

In (26.21), the fact that the magnitude of the magnetic flux density from a Tx-Q at a position $\vec{P}_1$ should be the same as that from a Tx-D at the same relative position $\vec{P}_2$ is used. Applying (26.21) to (26.20) results in the following formula:

$$B_t(x, y, z) = \sqrt{B_d^2(x, y, z) + B_q^2(x, y, z)} = \sqrt{B_d^2(x, y, z) + B_d^2(y, -x, z)} \tag{26.22}$$

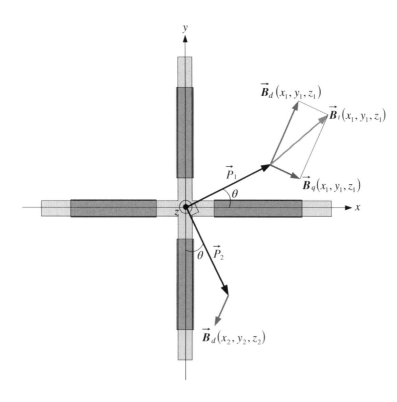

Figure 26.12 The magnetic flux densities generated from Tx-D (*x*-axis) and Tx-Q (*y*-axis) coils.

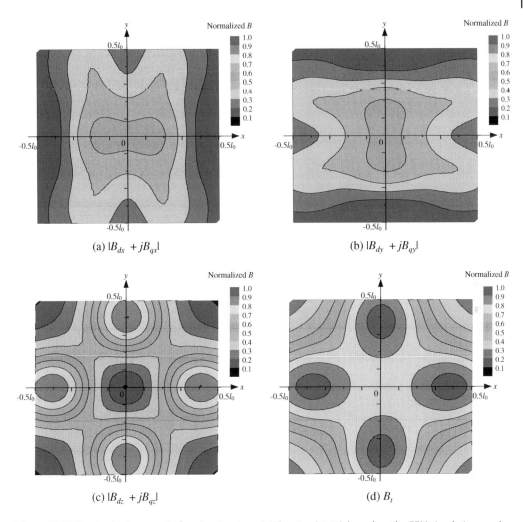

Figure 26.13 Synthesized magnetic flux density at $z = 0.25l_0$ using (26.22), based on the FEM simulation results.

As identified from (26.22), B_t can be completely determined by B_d, which can be found from simulation or experiment. Therefore, the total rms value of the magnetic flux density can be simply found from a single axis DC analysis instead of a DQ axis AC time-domain analysis, which is time consuming and does not give us the useful rms value, as shown in Figure 26.11.

Using (26.22) and the FEM simulation result for the magnetic field density from a Tx-D coil, the effective powering area for Rx is evaluated for $z = 0.25l_0$, $z = 0.5l_0$, and $z = 0.75l_0$ over the xy-plane, as shown in Figures 26.13 to 26.15, where l_0 is the length of a Tx coil. An aluminum plate is displaced at the bottom of the Tx to block the magnetic field. The magnitudes of the magnetic field density in Figures 26.13 to 26.15 are normalized with respect to their maximum values at each height. The x- and y-directional magnetic field densities reach their maximum at the center of the Tx coils and decrease away from the center on the xy-plane, whereas the z-directional magnetic field density reaches its maximum at both ends of each Tx coil. Therefore, a relatively even powering area can be achieved from the summed magnetic field, which is the result of averaging over the three-directional magnetic flux density.

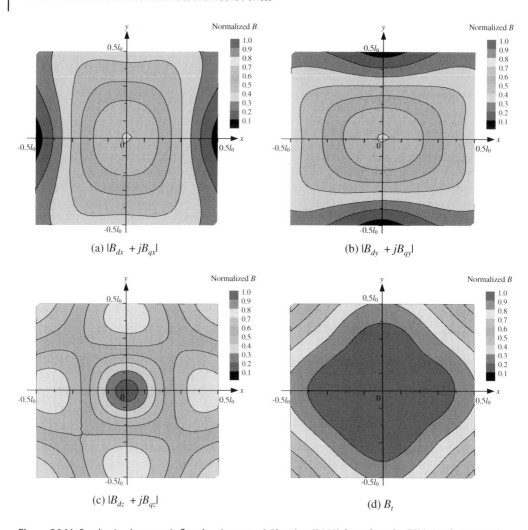

(a) $|B_{dx} + jB_{qx}|$

(b) $|B_{dy} + jB_{qy}|$

(c) $|B_{dz} + jB_{qz}|$

(d) B_t

Figure 26.14 Synthesized magnetic flux density at $z = 0.5l_0$ using (26.22), based on the FEM simulation results.

26.5 Example Design and Experimental Verifications

26.5.1 Prototype Fabrication of 1 m²-Sized Crossed Dipole Tx Coil

The design principle of the previous section has been applied to the prototype Tx and Rx coils operating at 280 kHz, where the Tx coil size is $l_T = w_T = 1000$ mm and $h_T = 30$ mm and the Rx coil size is $l_R = w_R = 100$ mm and $h_R = 1$ mm, as shown in Figure 26.16(a) and (b). The selection of design parameters in this chapter is just for demonstration of the proposed concept and is not intended for optimization of cost, mass, volume, and efficiency. For practical applications, the thicknesses of Tx and Rx coils h_T and h_R were selected as small as possible. The plane-shaped Mn–Zn ferrite cores having a high permeability of 2000 were assembled for the Tx coil, where the unit size of the ferrite core is 100 mm × 50 mm × 10 mm. An experimental DQ inverter having the operating frequency of 280 kHz, which meets the guideline of Power Matters Alliance (PMA), was also fabricated to drive Tx currents with an efficiency of 97%, as

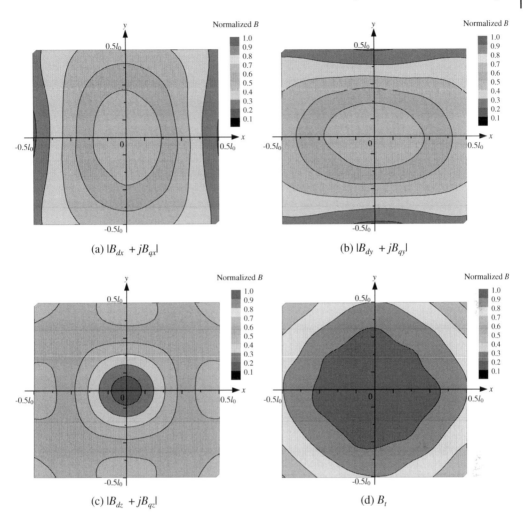

Figure 26.15 Synthesized magnetic flux density at $z = 0.75l_0$ using (26.22), based on the FEM simulation results.

shown in Figure 26.16(c). Litz wires having a total diameter of 1.8 mm were selected for the Tx coil, where the ampere-turn of each Tx coil was 120 A-turn and the number of turns was 60. An aluminum plate of 1.44 m^2 area and 2 mm thickness with a wooden insulator of 3 mm was displaced under the Tx coil. The total power loss including the DQ inverter was 185 W, where the core loss was relatively high compared to the litz wire and high-frequency-applicable film capacitors, which have negligible conduction losses.

The experimental results are illustrated in Figure 26.17, where B_t is determined as in (26.20). The measured effective powering area agreed fairly well with the simulation results in Figure 26.14, where the maximum B_t was measured as 10.4 µT at $z = l_0/2 = 500$ mm.

26.5.2 Verification of Omnidirectional Powering

Figure 26.18(a) shows seven measuring positions, which were arbitrarily selected to verify the free-positioning and omnidirectional powering characteristics. Both horizontal and vertical Rx

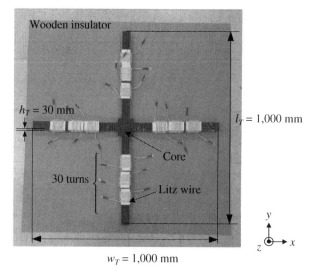

(a) Tx coil with an aluminum plate

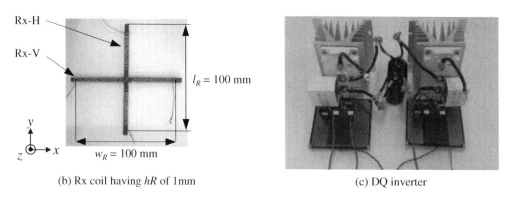

(b) Rx coil having hR of 1mm

(c) DQ inverter

Figure 26.16 Fabricated prototypes.

coil voltages of $|V_h|$ and $|V_v|$ were measured along the z-axis from $z = 100$ mm to 1000 mm at point P_A, as shown in Figure 26.18(b). The slightly lower coil voltage at $z = 100$ mm compared to that at $z = 200$ mm was due to the low x- and y-directional magnetic flux densities close to the center of the dipole coils. The Rx coil voltages with respect to five possible rotations, which are pitch, roll, yaw, 90-degree-pitched roll, and 90-degree-rolled pitch, were measured at three different positions of P_A, P_B, and P_C at $z = 500$ mm, as shown in Figures 26.19 to 26.23. As shown in Figure 26.18(a), before any rotating of the Rx coils, the Rx-H and Rx-V are in parallel with the x-axis and y-axis, respectively. The 90-degree-pitched roll denotes the roll rotation of the Rx coils after the Rx coils are 90-degree-pitch rotated. Likewise, 90-degree-rolled pitch denotes the pitch rotation of the Rx coils after the Rx coils are 90-degree-roll rotated. The series connected voltage $|V_h| + |V_v|$ ranged from 1.0 V to 3.8 V over wide rotations for the selected points. Therefore, both free-positioning and omnidirectional powering were verified with the fabricated Tx and Rx coils.

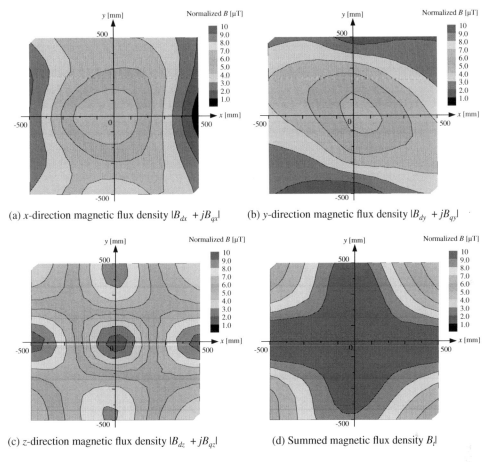

(a) *x*-direction magnetic flux density $|B_{dx} + jB_{qx}|$

(b) *y*-direction magnetic flux density $|B_{dy} + jB_{qy}|$

(c) *z*-direction magnetic flux density $|B_{dz} + jB_{qz}|$

(d) Summed magnetic flux density B_t

Figure 26.17 Measured magnetic flux density at $z = l_0/2 = 500$ mm.

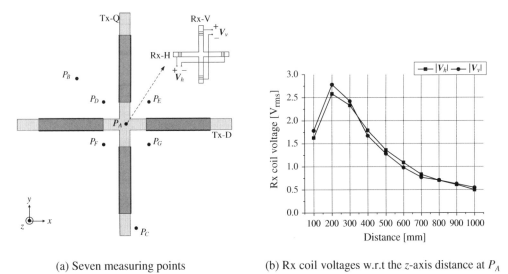

(a) Seven measuring points

(b) Rx coil voltages w.r.t the *z*-axis distance at P_A

Figure 26.18 Experimental conditions and Rx coil voltages along the *z*-axis.

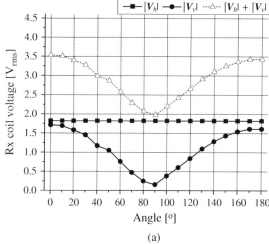

(a)

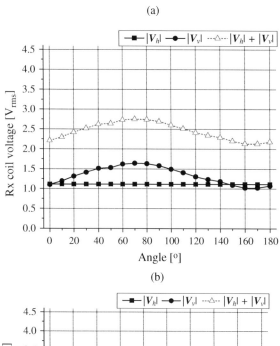

(b)

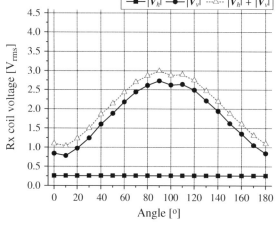

(c)

Figure 26.19 Measured Rx coil voltages for pitch rotation: (a) P_A, (b) P_B, and (c) P_C.

Figure 26.20 Measured Rx coil voltages for roll rotation: (a) P_A, (b) P_B, and (c) P_C.

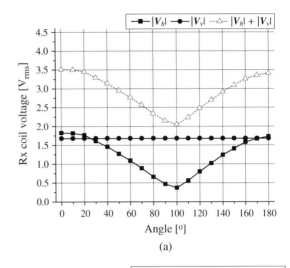

(a)

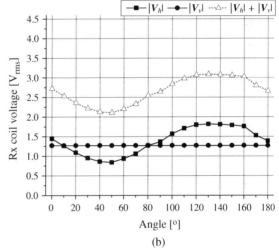

(b)

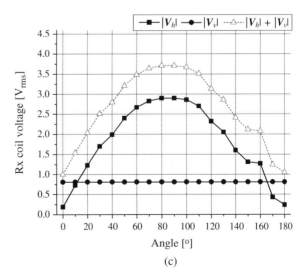

(c)

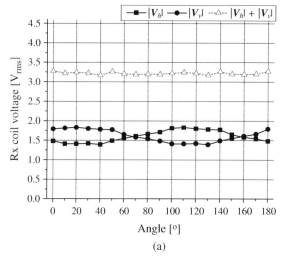

(a)

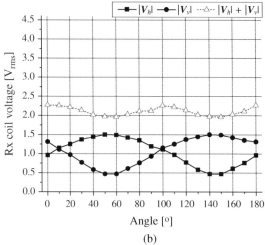

(b)

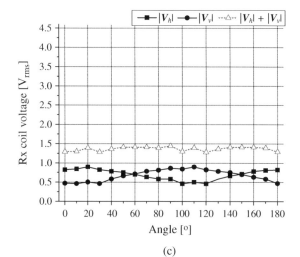

(c)

Figure 26.21 Measured Rx coil voltages for yaw rotation: (a) P_A, (b) P_B, and (c) P_C.

Figure 26.22 Measured Rx coil voltages for 90-degree-pitched roll rotation: (a) P_A, (b) P_B, and (c) P_C.

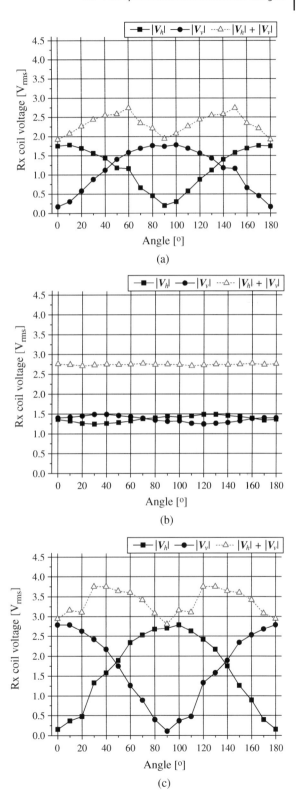

(a)

(b)

(c)

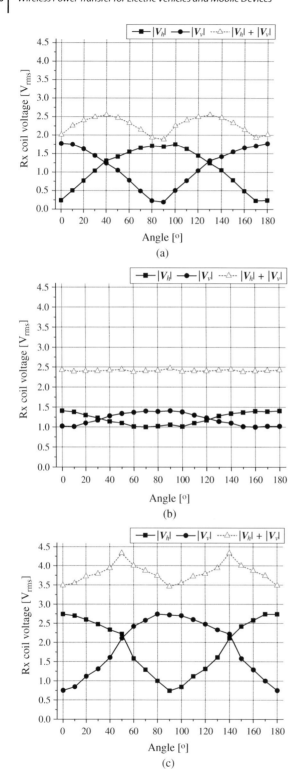

(a)

(b)

(c)

Figure 26.23 Measured Rx coil voltages for 90-degree-rolled pitch rotation: (a) P_A, (b) P_B, and (c) P_C.

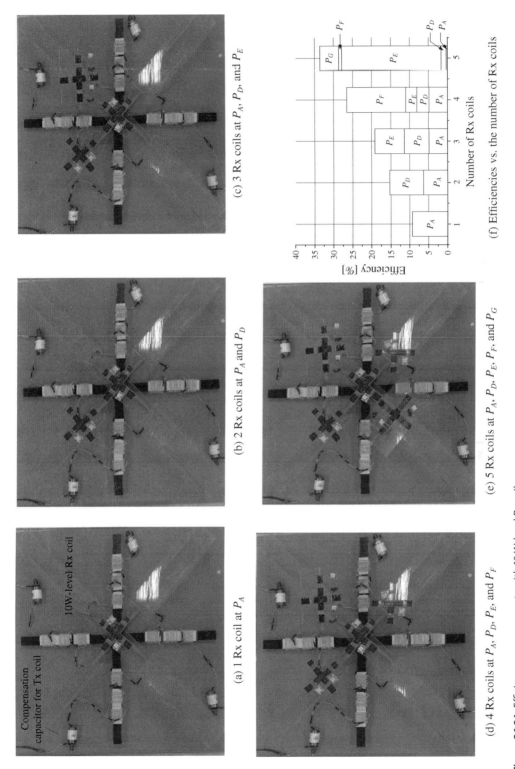

(a) 1 Rx coil at P_A

(b) 2 Rx coils at P_A and P_D

(c) 3 Rx coils at P_A, P_D, and P_E

(d) 4 Rx coils at P_A, P_D, P_E, and P_F

(e) 5 Rx coils at P_A, P_D, P_E, P_F, and P_G

(f) Efficiencies vs. the number of Rx coils

Figure 26.24 Efficiency measurements with 10 W-level Rx coils.

26.5.3 Efficiency Measurements with 10 W-level Rx Coils

The efficiency of the prototype Tx and Rx coils including the DQ inverter was measured for a few Rx coils, as shown in Figure 26.24. For the efficiency measurements, 10 W-level Rx coils having $l_R = 200$ mm, $w_R = 200$ mm, and $h_R = 5$ mm were located at P_A, P_D, P_E, P_F, and P_G, as shown in Figure 26.24(a) to (e). The measured efficiency increased as the number of Rx coils increased, where the maximum efficiency was 33.6% when the input power of the DQ inverter was fixed to 100 W for the case of five 10 W-level Rx coils at $z = 200$ mm. Due to the cross-coupling between adjacent Rx coils having fixed compensation circuits, the output powers of each Rx coil were changed as the number of Rx coils increased.

26.6 Conclusion

The crossed dipole Tx and Rx coils having six DoF in mobile IPTs with a DQ rotating magnetic field have been verified both by simulations and experiments throughout this chapter. A general classification of various IPTs in terms of DoF was first presented and a thorough analysis of loop coil configurations for omnidirectional powering was also provided so that the minimum physical dimensions for ubiquitous powering can be found for both Tx and Rx coils. Throughout this general survey on omnidirectional powering, the proposed crossed dipole of Tx and Rx coils was found to be the only viable configuration that enables us both plane-type Tx and Rx coils. By virtue of a rotating magnetic field, both free-positioning and omnidirectional powering were guaranteed over a wide area of the Tx coil. The maximum efficiency of the prototype including the DQ inverter was measured as 33.6% when the input power was fixed to 100 W. Due to its plane geometry and the wide powering area, the proposed crossed dipole coil can be applied to mobile devices as well as IoT, where less dependency on a battery will be required. An optimized design of the crossed dipole coil is left for future work.

Problems

26.1 Can you suggest a multiple dipole structure to have better evenly distributed magnetic-field characteristics?

26.2 Can you suggest a modular Tx coil that can be arbitrarily extendable to any area guaranteeing evenly distributed magnetic-field characteristics?

References

1 M. Kline, I. Izyumin, B. Boser, and S. Sanders, "Capacitive power transfer for contactless charging," in *2011 ECCE Conference*, pp. 1398–1404.

2 B. Choi, D. Nguyen, S. Yoo, J. Kim, and Chun T. Rim, "A novel source-side monitored capacitive power transfer system for contactless mobile charger using class-E converter," in *2014 VTC Conference*, pp. 1–5.

3 E.Y. Chow, "Wireless powering and the stud of RF propagation through ocular tissue for development of implantable sensors," *IEEE Trans. Antennas Propagation*, vol. 59, no. 6, pp. 2379–2387, June 2011.

4 N. Wang *et al.*, "One-to-multipoint laser remote power supply system for wireless sensor networks," *IEEE Sensors Journal*, vol. 12, no. 2, pp. 389–396, February 2012.

5 I. Shnaps and E. Rimon, "Online coverage by a tethered autonomous mobile robot in planar unknown environments," *IEEE Trans. on Robotics*, vol. 30, no. 4, pp. 966–974, August 2014.

6 S. Choi *et al.*, "Tethered aerial robots using contactless power systems for extended mission time and range," in *2014 ECCE Conference*, pp. 912–916.

7 O.C. Onar, J. Kobayashi, and A. Khaligh, "A fully directional universal power electronic interface for EV, HEV, and PHEV applications," *IEEE Trans. on Power Electron.*, vol. 28, no. 12, pp. 5489–5498, December 2013.

8 E. Waffenschmidt, "Free positioning for inductive wireless power system," in *2011 ECCE Conference*, pp. 3481–3487.

9 W. Zhong, X. Liu, and S. Hui, "A novel single-layer winding array and receiver coil structure for contactless battery charging systems with free-positioning and localized charging features," *IEEE Trans. on Ind. Electron.*, vol. 58, no. 9, pp. 4136–4143, September 2011.

10 C. Park, S. Lee, G. Cho, S. Choi, and Chun T. Rim, "Omni-directional inductive power transfer system for mobile robots using evenly displaced multiple pick-ups," in *2012 ECCE Conference*, pp. 2492–2497.

11 C. Park, S. Lee, G. Cho, S. Choi, and Chun T. Rim, "Two-dimensional inductive power transfer system for mobile robots using evenly displaced multiple pickups," *IEEE Trans. on Ind. Applic.*, vol. 50, no. 1, pp. 538–565, June 2013.

12 B. Che *et al.*, "Omnidirectional non-radiative wireless power transfer with rotating magnetic field and efficiency improvement by metamaterial," *Appl. Phys. A*, vol. 116, no. 4, pp. 1579–1586, April 2014.

13 W. Ng, C. Zhang, D. Lin, and S. Hui, "Two- and three-dimensional omnidirectional wireless power transfer," *IEEE Trans. on Power Electron.*, vol. 29, no. 9, pp. 4470–4474, January 2014.

14 H. Li, G. Li, X. Xie, Y. Huang, and Z. Wang, "Omnidirectional wireless power combination harvest for wireless endoscopy," in *2014 BioCAS Conference*, pp. 420–423.

15 X. Li *et al.*, "A new omnidirectional wireless power transmission solution for the wireless endoscopic micro-ball," in *2011 ISCAS Conference*, pp. 2609–2612.

16 R. Carta *et al.*, "Wireless powering for a self-propelled and steerable endoscopic capsule for stomach inspection," *Biosens. Bioelectron.*, vol. 25, no. 4, pp. 845–851, December 2009.

17 T. Sun *et al.*, "Integrated omnidirectional wireless power receiving circuit for wireless endoscopy," *Electron. Lett.*, vol. 48, no. 15, pp. 907–908, July 2012.

18 B. Lenaerts and R. Puers, "An inductive power link for a wireless endoscope," *Biosens. Bioelectron.*, vol. 22, no. 7, pp. 1390–1395, February 2007.

19 B. Choi, E. Lee, J. Kim, and Chun T. Rim, "7 m-off-long-distance extremely loosely coupled inductive power transfer system using dipole coils," in *2014 ECCE Conference*, pp. 858–863.

20 C. Park, S. Lee, G. Cho, and Chun T. Rim, "Innovative 5-m-off-distance inductive power transfer systems with optimally shaped dipole coils," *IEEE Trans. on Power Electron.*, vol. 30, no. 2, pp. 817–827, November 2014.

21 Chun T. Rim and G. Cho, "New approach to analysis of quantum rectifier-inverter," *Electron. Lett.*, vol. 25, no. 25, pp. 1744–1745, December 1989.

22 Chun T. Rim, "Unified general phasor transformation for AC converters," *IEEE Trans. on Power Electron.*, vol. 26, no. 9, pp. 2465–2475, September 2011.

23 J. Huh, W. Lee, S. Choi, G. Cho, and Chun T. Rim, "Frequency-domain circuit model and analysis of coupled magnetic resonance systems," *J. Power Electron.*, vol. 13, no. 2, pp. 275–286, March 2013.

24 A. Kurs, A. Karalis, R. Moffatt, J.D. Joannopoulos, P. Fisher, and M. Soljacic, "Wireless power transfer via strongly coupled magnetic resonance," *Science*, vol. 317, no. 5834, pp. 83–86, June 2007.

25 A.P. Sample, D.A. Meyer, and J.R. Smith, "Analysis, experimental results, and range adaption of magnetically coupled resonators for wireless power transfer," *IEEE Trans. on Ind. Electron.*, vol. 58, no. 2, pp. 544–554, February 2011.

26 T. Imura and Y. Hori, "Maximizing air gap and efficiency of magnetic resonant coupling for wireless power transfer using equivalent circuit and Neumann formula," *IEEE Trans. on Ind. Electron.*, vol. 58, no. 10, pp. 4746–4752, October 2011.

27 T.C. Beh, T. Imura, and Y. Hori, "Basic study of improving efficiency of wireless power transfer via magnetic resonance coupling based on impedance matching," in *2010 ISIE Conference*, pp. 2011–2016.

28 J. Park, Y. Tak, Y. Kim, Y. Kim, and S. Nam, "Investigation of adaptive matching methods for near-field wireless power transfer," *IEEE Trans. on Antennas Propagation*, vol. 59, no. 5, pp. 1769–1773, May 2011.

29 J. Huh, W.Y. Lee, S.Y. Choi, G.H. Cho, and Chun T. Rim, "Explicit static circuit model of coupled magnetic resonance system," in *2011 ECCE-Asia Conference*, pp. 2233–2240.

30 E. Lee, J. Huh, X.V. Thai, S, Choi, and Chun T. Rim, "Impedance transformers for compact and robust coupled magnetic resonance systems," in *2013 ECCE Conference*, pp. 2239–2244.

31 R. Hui, W. Zhong, and C. Lee, "A critical review of recent progress in mid-range wireless power transfer," *IEEE Trans. on Power Electron.*, vol. 29, no. 9, pp. 4500–4511, September 2014.

32 G. Covic, M. Kissin, D. Kacprzak, N. Clausen, and H. Hao, "A bipolar primary pad topology for EV stationary charging and highway power by inductive coupling," in *2011 ECCE Conference*, pp. 1832–1838.

33 S. Li and C. Mi, "Wireless power transfer for electric vehicle applications," *IEEE Trans. on Emerg. Sel. Topics Power Electron.*, vol. 3, no. 1, pp. 4–17, March 2015.

34 S. Choi, J. Huh, W. Lee, and Chun T. Rim, "Asymmetric coil sets for wireless stationary EV chargers with large lateral tolerance by dominant field analysis," *IEEE Trans. on Power Electron.*, vol. 29, no. 12, pp. 6406–6420, December 2014.

35 M. Budhia, G. Covic, and J. Boys, "Design and optimization of circular magnetic structures for lumped inductive power transfer systems," *IEEE Trans. on Power Electron.*, vol. 26, no. 11, pp. 3096–3108, November 2011.

36 M. Budhia, J. Boys, G. Covic, and C. Huang, "Development of a single-sided flux magnetic coupler for electric vehicle IPT charging systems," *IEEE Trans. on Ind. Electron.*, vol. 60, no. 1, pp. 318–328, January 2013.

37 T. Nguyen, S. Li, W. Li, and C. Mi, "Feasibility study on bipolar pads for efficient wireless power chargers," in *2014 APEC Conference*, pp. 1676–1682.

38 P. Meyer, P. Germano, M. Markovic, and Y. Perriard, "Design of a contactless energy-transfer system for desktop peripherals," *IEEE Trans. on Ind. Applic.*, vol. 47, no. 4, pp. 1643–1651, July 2011.

39 J. Shin *et al.*, "Design and implementation of shaped magnetic-resonance-based wireless power transfer system for road powered moving electric vehicles," *IEEE Trans. on Power Electron.*, vol. 61, no. 3, pp. 1179–1192, March 2014.

40 G. Elliott, J. Boys, and G. Covic, "A design methodology for flat pick-up ICPT systems," in *2006 ICIEA Conference*, pp. 1–7.

41 S. Lee *et al.*, "On-line electric vehicle using inductive power transfer system," in *2010 ECCE Conference*, pp. 1598–1601.

42 J. Huh, S. Lee, C. Park, G. Cho, and Chun T. Rim, "High performance inductive power transfer system with narrow rail width for on-line electric vehicles," in *2010 ECCE Conference*, pp. 647–651.

43 J. Huh, W. Lee, B. Lee, G. Cho, and Chun T. Rim, "Characterization of novel inductive power transfer systems for on-line electric vehicles," in *2011 APEC Conference*, pp. 1975–1979.

44 J. Huh, S. Lee, W. Lee, G. Cho, and Chun T. Rim, "Narrow-width inductive power transfer system for on-line electrical vehicles," *IEEE Trans. on Power Electron.*, vol. 26, no. 12, pp. 3666–3679, December 2011.

45 S. Lee *et al.*, "Active EMF cancellation method for I-type pickup of on-line electric vehicles," in *2011 APEC Conference,* pp. 1980–1983.

46 W. Lee *et al.*, "Finite-width magnetic mirror models of mono and dual coils for wireless electric vehicles," *IEEE Trans. on Power Electron.*, vol. 28, no. 3, pp. 1413–1428, March 2013.

47 S. Choi, J. Huh, W. Lee, S. Lee, and Chun T. Rim, "New cross-segmented power supply rails for road powered electric vehicles," *IEEE Trans. on Power Electron.*, vol. 28, no. 12, pp. 5832–5841, December 2013.

48 S. Lee, B. Choi, and Chun T. Rim, "Dynamic characterization of the inductive power transfer system for online electric vehicles by Laplace phasor transform," *IEEE Trans. on Power Electron.*, vol. 28, no. 12, pp. 5902–5909, December 2013.

49 S. Choi, B. Gu, S. Jeong, and Chun T. Rim, "Ultra-slim S-type inductive power transfer system for road powered electric vehicles," in *2014 EVTeC Conference*, pp. 1–7.

50 S. Choi *et al.*, "Generalized active EMF cancel methods for wireless electric vehicles," *IEEE Trans. on Power Electron.*, vol. 29, no. 11, pp. 5770–5783, November 2014.

51 C. Wang, O. Stielau, and G. Covic, "Design considerations for a contactless electric vehicle battery charger," *IEEE Trans. on Ind. Electron.*, vol. 52, no. 5, pp. 1308–1314, October 2005.

52 C. Wang, G. Covic, and O. Stielau, "Power transfer capability and bifurcation phenomena of loosely coupled inductive power transfer systems," *IEEE Trans. on Ind. Electron.*, vol. 51, no. 1, pp. 148–157, February 2004.

53 G. Covic and J. Boys, "Modern trends in inductive power transfer for transportation applications," *IEEE Trans. on Emerg. Sel. Topics Power Electron.*, vol. 1, no. 1, pp. 28–41, March 2013.

54 O. Onar *et al.*, "A novel wireless power transfer for in-motion EV/PHEV charging," in *2013 APEC Conference*, pp. 2073–3080.

55 S. Choi, B. Gu, S. Jeong, and Chun T. Rim, "Advances in wireless power transfer systems for road powered electric vehicles," *IEEE Trans. on Emerg. Sel. Topics Power Electron.*, vol. 3, no. 1, pp. 18–35, March 2015.

56 B. Lee, H. Kim, S. Lee, C. Park, and Chun T. Rim, "Resonant power shoes for humanoid robots," in *2011 ECCE Conference*, pp. 1791–1794.

57 B. Choi, E. Lee, J. Huh, and Chun T. Rim, "Lumped impedance transformers for compact and robust coupled magnetic resonance systems," *IEEE Trans. on Power Electron.*, vol. PP, no. 99, pp. 1, January 2015 (early access article).

58 J. Kim *et al.*, "Coil design and shielding methods for a magnetic resonant wireless power transfer system," *Proc. IEEE*, vol. 101, no. 6, pp. 1332–1342, June 2013.

59 J. Craig, "Spatial descriptions and transformations," in *Introduction to Robotics*, 3rd edn, Prentice Hall, New Jersey, 2004, Ch. 2, Sec. 8, pp. 41–51.

27

Two-Dimensional Omnidirectional IPT for Robots

27.1 Introduction

Mobile robots are becoming widely used in many areas such as home automation, factory automation, and hazardous and dirty missions in military or nuclear power plants. Mobile robots may bring any appropriate mission payloads including environmental monitoring sensors and vision systems. Most robots rely on the on-board battery as their power source; hence, long and heavy missions of the robots could be a burden on the battery. Due to the limited energy capacity of the battery, the robots cannot carry on their missions continuously. To recharge the battery, the docking systems with conductive contacts [1–3] were introduced. These systems require complicated docking process, control strategy, and long charging time, which results in high cost and frequent mission cease of the robots. To overcome the problems of the conductive contact-type charging systems, such as a spark among contacts and a narrow docking margin, contactless inductive power transfer systems (IPTSs) were proposed [4–9]. Because this kind of power transfer mechanism does not have any mechanical contact, there is no spark at the contact surface and it can be operated in dirty and explosive environments such as a coal mine. The coaxial winding transformer with power of 120 kW for fast charging of electric vehicles was proposed [5]. The contactless charger for electric shavers that operate in wet circumstances was also proposed [8]. The wireless power transfer system with an efficiency of 95% at the transferred power of 220 W over 30 cm was recently proposed [9], where the remarkably high efficiency is due to the very thick cable and large coil size.

The plug-in type of power systems such as docking systems and contactless inductive power transfer systems [1–9] are a kind of zero-dimensional powering method with no degree of freedom. To increase the freedom of powering the robots, wireless power transfer technologies as well as wide-area primary power feeding systems must be used. The contactless power delivery system for mining [10] is an example of a one-dimensional powering method, where the movable pick-up winding wraps around the fixed straight source wire. Because of this fixed wire, this system does not allow any lateral displacement of the pick-up. The IPTS for electric vehicles [11, 12] is another example optimized for a vehicle one-dimensional method that runs along a lane, allowing small lateral displacement. For two-dimensional powering, capacitive or inductive wireless power floors were proposed [13–17]. The capacitive coupling type [13, 14] has a relatively shorter airgap than the inductive coupling type [15–17] due to the inherently small relative permittivity ε_r compared to the relative permeability μ_r.

Hence, the IPTS type is adequate for the two-dimensional wireless robot power systems; however, the power floor involves an enormous number of turns of subwindings [15, 17] or multilayer structure [10, 17], which causes a high manufacturing cost and complicated design.

Wireless Power Transfer for Electric Vehicles and Mobile Devices, First Edition. Chun T. Rim and Chris Mi.
© 2017 John Wiley & Sons Ltd. Published 2017 by John Wiley & Sons Ltd.

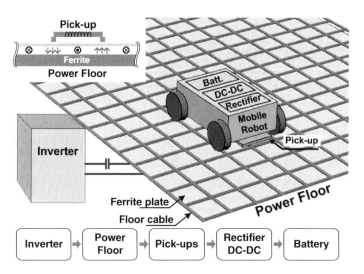

Figure 27.1 Overall system configuration of the proposed IPTS for mobile robots.

Furthermore, the tolerance of powering positions and directions of pick-ups are small, which results in large fluctuations of delivering power.

In this chapter, the IPTS for mobile robots that can be used in the continuous monitoring of nuclear power plants for 24 hours a day is explained. The power floor has the simplest structure of a single-layer winding. Evenly displaced multiple pick-ups are used to reduce the power fluctuation. The system performances are verified by simulations and experiments.

27.2 Overall System Configuration

The overall configuration of the proposed IPTS for mobile robots is shown in Figure 27.1. The IPTS consists of an inverter, a power floor, and the components in the robots, which are pick-ups, rectifiers, a DC–DC converter, and a battery pack. The IPTS provides power for the four-wheel-drive two-dimensional mobile robots via the power floor and on-board pick-ups. The targeted mission of the robots is the observation of surroundings using a wireless camera. The robot specification is summarized in Table 27.1.

Table 27.1 Specifications of the mobile robots

Model	PFindBot, Roboblock systems
Robot type	Four-wheel-drive two-dimensional mobile robots
Average power consumption	10 W
Battery	Li-ion, 11.1 V
On-board devices	A 2.4 GHz wireless camera
	An autonomous driving controller
Weight	4 kg
Dimension ($W \times H \times D$)	400 mm × 200 mm × 400 mm

27.3 Design and Fabrication of the IPTS for Mobile Robots

27.3.1 Power Floor Winding without Crossing Over Points

For the wide area adoption of the proposed power floor, a simple power floor configuration is crucial for reducing the cost and mass production. Extremely many magnetic poles with alternating polarity can be crossly deployed [15] to provide two-dimensional wireless power for the robots, as shown in Figure 27.2(a). It was found that there are groups that have same magnetic polarity along the diagonal direction; therefore, a bundle of floor cable instead of an enormous number of subwindings [16] can be wound like that shown in Figure 27.2(b). Furthermore, the

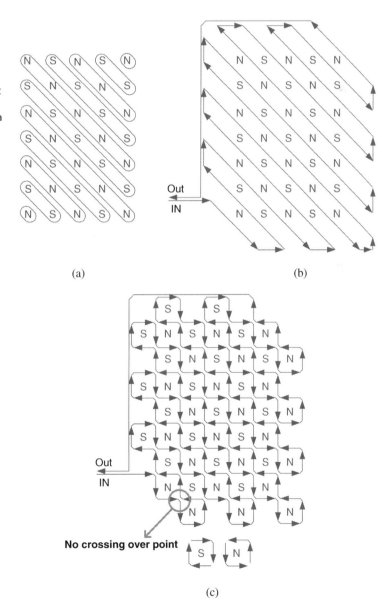

Figure 27.2 Power floor winding methods: (a) grouping the same magnetic polarity of the power floor composed of many alternating magnetic poles [10], (b) meandrous winding of each group by a bundle of cable, and (c) proposed single-layer winding without any crossing-over points.

(a)

(b)

(c)

cable can be bent to a rectangular form to give better magnetic flux distribution, as shown in Figure 27.2(c).

Since the winding layer of the power floor is just single, the power floor is slim and the fabrication of it is drastically simplified. In the proposed power floor, the adjacent windings have opposite magnetic polarity such that the magnetic flux can form interlink paths. Therefore the EMF for pedestrians at a distance of several times that of the subwinding size is very small. Thus the power floor winding has a small inductance of 60 μH for 3.3 m^2 of area, which results in a low voltage stress of the power floor winding and the small size of the series resonant capacitors.

As a prototype, the whole power floor area was set as 1500 mm × 2100 mm. Considering the bottom size of the mobile robots (200 mm × 300 mm) and implementation complexity, the subwinding size l_{SW} is selected as 100 mm. As a consequence, there are 315 (=15 × 21) subwindings in the whole power floor.

Figure 27.3(a) shows the simulation result of the magnetic field intensity on the power floor whereas Figure 27.3(b) shows that of the vertical component when there is no pick-up. In Figure 27.3(b), the adjacent subwindings have the opposite magnetic polarity. The simulation was performed using the ANSOFT Maxwell ver. 14.0.

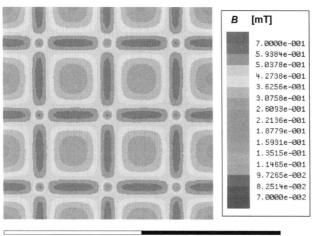

B [mT]

7.0000e-001
5.9384e-001
5.0378e-001
4.2738e-001
3.6256e-001
3.0758e-001
2.6093e-001
2.2136e-001
1.8779e-001
1.5931e-001
1.3515e-001
1.1465e-001
9.7265e-002
8.2514e-002
7.0000e-002

Figure 27.3 Simulation results of the magnetic field intensity on the power floor: (a) absolute value of magnetic field intensity and (b) vertical component of the magnetic field, where the blue area corresponds to a negative sign and the red area corresponds to a positive sign.

```
0          200          400(mm)
```
(a)

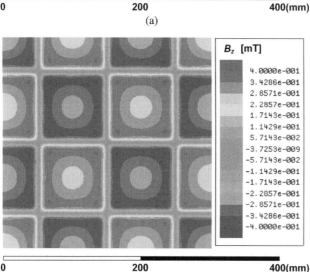

B$_z$ [mT]

4.0000e-001
3.4286e-001
2.8571e-001
2.2857e-001
1.7143e-001
1.1429e-001
5.7143e-002
-3.7253e-009
-5.7143e-002
-1.1429e-001
-1.7143e-001
-2.2857e-001
-2.8571e-001
-3.4286e-001
-4.0000e-001

```
0          200          400(mm)
```
(b)

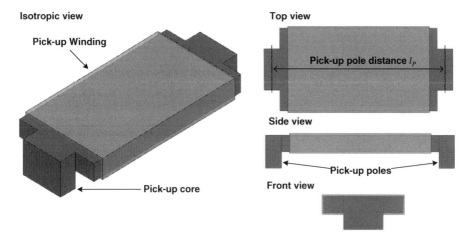

Figure 27.4 Detail of the structure of a pick-up, where the number of turns is 40 and the self-inductance is 280 μH.

27.3.2 Evenly Displaced Multiple Pick-ups

Figure 27.4 shows the structure of a pick-up in detail, where an enameled copper wire is wound up around the ferrite pick-up core. The contact area of the pick-up poles is appropriately minimized to reduce the chance of overriding neighborhood subwindings by the poles. However, its size should not be too small to avoid saturation of the pick-up core. The distance between the pick-up poles l_P must be carefully chosen to maximize the chance of powering when considering the size of the subwinding l_{SW} of the power floor. As shown in Figure 27.5(a), when l_P is

Figure 27.5 Powered (upper) and unpowered (bottom) pick-up positions: (a) short pick-up pole distance and (b) long pick-up pole distance.

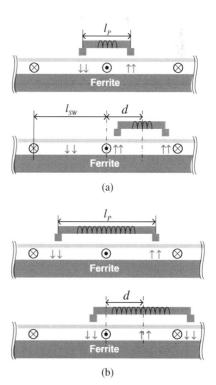

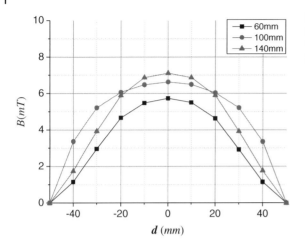

Figure 27.6 Simulation results of the magnetic field intensity in a pick-up core versus pick-up displacement *d* for different pick-up pole distances l_p.

shorter than l_{SW}, there is an unpowered section where the pick-up poles are on the subwinding of the same magnetic polarity as the pick-up moves from left to right on the power floor. When l_p is longer than l_{SW}, there is an unpowered section as well, forthe same reason, as shown in Figure 27.5(b). Therefore, the pick-up pole distance must not be too short or too long, that is, $l_p \sim l_{SW}$.

The simulation results of the magnetic field intensity in the pick-up core versus pick-up displacement *d* for three different pick-up pole distances (60 mm, 100 mm, 140 mm) is shown in Figure 27.6. The simulation was performed using the same ANSOFT Maxwell ver. 14.0. In the 60 mm case, the magnetic field intensity is weak and decreases steeply. In the 140 mm case, it is strong at the center but also decreases steeply. In the 100 mm case, however, it is strong enough and has the broadest field intensity profile; hence, this value is chosen for the optimum value of l_p.

The calculation of mutual inductance [18] between the power floor and pick-up must be useful for the systematic design of the IPTS; however, it is left for further work because of the irregularity of the two-dimensional position.

With only a pick-up under the robot, there are lots of unpowered pick-up positions on the power floor, as shown in Figure 27.7, where each square corresponds to the subwindings and the dark and bright areas show different magnetic polarities. When both pick-up poles are just on the floor cable, as shown in Figure 27.7(a) and (b), the magnetic flux does not circulate through the pick-up core but is canceled out at the tips of each pick-up pole. When the poles of

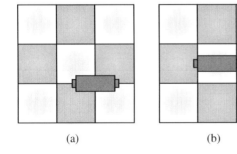
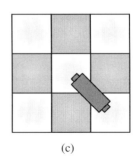

(a)   (b)   (c)

Figure 27.7 Unpowered pick-up positions when one pick-up only is used: (a) both pick-up poles are on the horizontal boarder, (b) both pick-up poles are on the vertical boarder, and (c) both pick-up poles are on the area with the same magnetic polarity.

each pick-up are on the areas with the same magnetic polarity, as shown in Figure 27.7(c), the magnetic flux does not circulate through the pick-up core as well.

To avoid the unpowered cases as much as possible, the multiple pick-up structure should be adopted. Considering the limited bottom area of the proposed mobile robot, three pick-ups are used. Figure 27.8 shows the possible arrangement of the proposed multiple pick-up structure of evenly displaced angles. The star marks indicate the powering pick-ups that can receive power from the power floor. Nine cases are selectively shown for Figure 27.8(a) and

Figure 27.8 Various positions of the multiple pick-up on the power floor, where the star marks indicate powering of pick-ups: (a) upright pick-up and (b) 45 degree rotated pick-up.

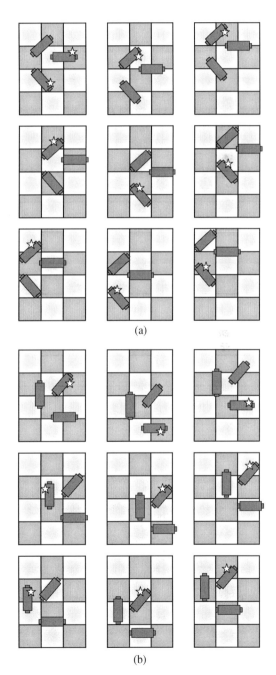

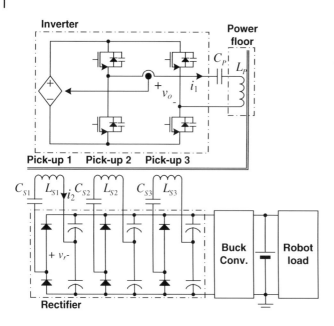

Figure 27.9 Schematics of the proposed IPTS including an inverter for power floor driving and a buck converter for the three pick-ups.

(b), which demonstate that at least one pick-up among three pick-ups may receive power in most cases. The power receiving probability and the uniformity of the output voltage can be substantially increased by the proposed multiple pick-up structure.

27.3.3 An Inverter, Rectifiers, and a Buck Converter

The inverter of the proposed IPTS for driving the power floor is controlled as an AC current source, which is very robust to abrupt load change, as shown in Figure 27.9. To make this current source, the output current of the inverter is sensed and given feedback to control the output voltage of the inverter. A series resonant capacitor C_P is inserted between the inverter and the power floor to nullify the reactance of the power floor inductance L_P. For guaranteeing the zero voltage switching (ZVS) operation of the inverter, the resonant frequency ω_{rp} of L_P and C_P is selected to be 10% lower than the inverter switching frequency ω_s as follows [5]:

$$\omega_s > \omega_{rp} = \frac{1}{\sqrt{L_P C_P}} \tag{27.1}$$

$$\omega_s = \omega_{rs} = \frac{1}{\sqrt{L_{S1} C_{S1}}} = \frac{1}{\sqrt{L_{S2} C_{S2}}} = \frac{1}{\sqrt{L_{S2} C_{S2}}} \tag{27.2}$$

The collected powers from the three pick-ups are provided to their corresponding input rectifiers, which are connected in parallel to add to the power. Since the inner space of the mobile robot is small, half-bridge rectifiers for voltage doubling are used to halve the numbers of diode heat sinks and coil turns compared to full-bridge rectifiers, as shown in Figure 27.9. For the resonance of pick-up inductances L_{S1}, L_{S2}, and L_{S3}, series resonant capacitors C_{S1}, C_{S2}, and C_{S3} are inserted, respectively. The resonance frequency ω_{rs} of each inductance and capacitance is the same as the inverter switching frequency ω_s of (27.2) to maximize the power transfer capability.

The output voltage of each rectifier is not uniform but varies for different pick-up positions over the power floor, whereas the on-board battery of the robot needs constant input voltage. Therefore a DC–DC converter is used for the voltage regulation; a buck converter driven by the LM2576 chip is selected due to its large input voltage range and moderate conversion efficiency.

27.4   Example Design and Experimental Verifications of the Design

The output power and system power efficiency were experimentally verified. The output power of the IPTS was measured at the input port of the battery for various positions and angular displacements of the pick-up set on the power floor, and the input power of it was measured at the DC input port of the inverter by a power meter, Yokogawa, WT210. A 100 kHz switching frequency was selected after considering the pick-up induced voltage, ferrite material characteristics [19], and component availability. Two-dimensional contour plots of the measured output power are shown in Figure 27.10. First, the characteristic of a pick-up on a subwinding of the power floor was explored, as shown in Figure 27.10(a), where the large black area corresponds to the output power less than 1.5 W and the small red area for 13.0 W. From (27.3) below, its spatial average output power P_{avg} for 100 ($N_X = 10$, $N_Y = 10$) points is found to be 4.60 W. Then the characteristics of three pick-ups were explored. Figure 27.10(b) is an example of the upright positioned multiple pick-ups and Figure 27.10(c) is an example of the 45 degree rotated multiple pick-ups; their spatial average output powers are 10.45 W and 12.30 W, respectively, which are a little less than 13.80 W, which is three times that of a pick-up power. Hence, the time average output power for a random walk robot can be higher than the power consumption of the robot of 10 W. For the uniformity comparison of output power, a standard deviation of it is defined as shown in (27.4) below, which was measured as low as 1.5~4.9 W, as depicted in Figure 27.10:

$$P_{avg} = \frac{1}{N_X N_Y} \sum_{i=1}^{N_X} \sum_{j=1}^{N_Y} P_{ij} \tag{27.3}$$

$$\sigma_P = \sqrt{\frac{1}{N_X N_Y} \sum_{i=1}^{N_X} \sum_{j=1}^{N_Y} (P_{ij} - P_{avg})^2} \tag{27.4}$$

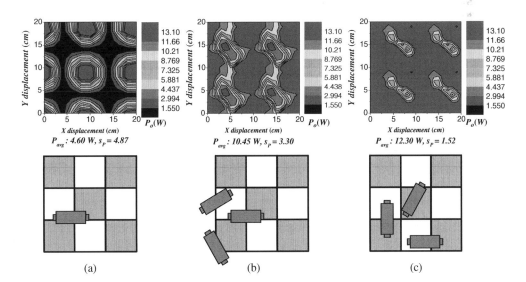

Figure 27.10 Measured output power of the DC-DC converter for various pick-up positions. The lower pictures show the relative locations of the pick-ups at the reference positions ($X = 0$, $Y = 0$) of the upper diagrams. (a) a pick-up only at upright direction, (b) three active pick-ups at upright direction, (c) three active pick-ups at 45° rotated direction.

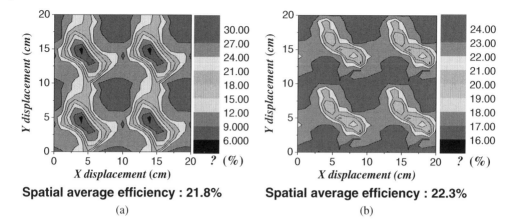

Spatial average efficiency : 21.8%   **Spatial average efficiency : 22.3%**

(a)   (b)

Figure 27.11 Measured system power efficiency for various pick-up positions over the power floor: (a) three active pick-ups at the upright direction and (b) three active pick-ups at the 45 degree rotated direction.

The power efficiency measurement results for various pick-up positions and angular displacements are shown in Figure 27.11, where the spatial average efficiency, which is defined below, is about 22%:

$$\eta_{avg} = \frac{1}{N_X N_Y} \sum_{i=1}^{N_X} \sum_{j=1}^{N_Y} \eta_{ij} \tag{27.5}$$

$$\sigma_\eta = \sqrt{\frac{1}{N_X N_Y} \sum_{i=1}^{N_X} \sum_{j=1}^{N_Y} (\eta_{ij} - \eta_{avg})^2} \tag{27.6}$$

The power efficiency is very low due to the large area of the power floor, 1500 mm × 2100 mm, where the idle power of it without pick-ups is 30 W and the low efficiency of the buck-converter is 80%. Major loss comes from the eddy current loss and hysteresis loss of the ferrite cores. The size of a ferrite block is 100 mm × 100 mm × 10 mm, and the ferrite material is an Mn–Zn type with conductivity of 5 Ω m. For economic reasons, a low-quality ferrite material was used; however, the power efficiency can be enhanced if a high-quality ferrite core is used. Excluding these losses, the power efficiency becomes as high as 93%. The power loss distributions of the proposed system using a low-quality ferrite core and a high-quality ferrite core [19] are shown in Figure 27.12. The floor core loss can be almost halved by using the high-quality ferrite core, which corresponds to a 9.4% efficiency enhancement.

The experimental waveforms of the proposed IPTS are shown in Figure 27.13, where a ZVS operation of the inverter is verified, as shown in Figure 27.13(b).

It was demonstrated that the mobile robots with the two-hour on-board battery can operate for more than 6 hours without recharging, which was ceased to prevent mechanical wearing out. The major parameters of the proposed IPTS are summarized in Table 27.2. The nominal primary current is 7 A_{rms}, the calculated mutual inductance between the power floor and one pick-up coil is 5 μH, the pick-up side quality factor is $Q = \omega_o L_s / R = 4.4$, and the airgap between the floor core and pick-up core is about 10 mm.

The implemented IPTS and mobile robots for demonstration are shown in Figure 27.14. For the much wider area adoption of the proposed IPTS, the idle power loss due to a large core loss and long wire length needs to be improved. The segmentation of the power floor and location

Figure 27.12 The power loss distributions of the proposed low-quality ferrite case and a high-quality ferrite case.

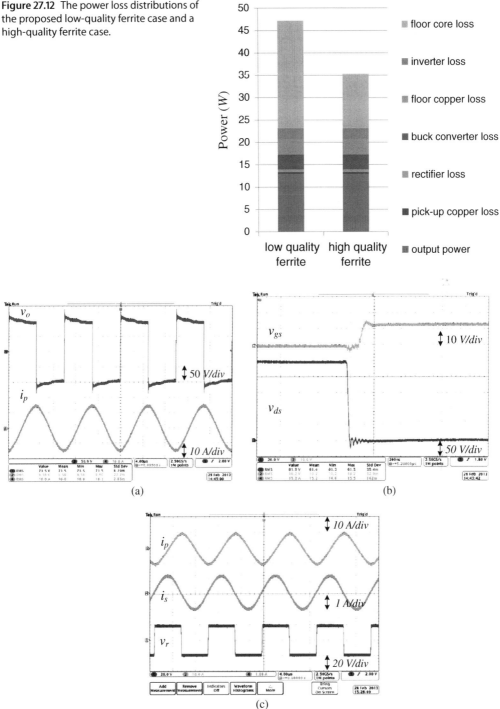

- floor core loss
- inverter loss
- floor copper loss
- buck converter loss
- rectifier loss
- pick-up copper loss
- output power

(a)

(b)

(c)

Figure 27.13 Experimental waveforms of the proposed IPTS: (a) inverter output voltage v_o and primary current i_p, (b) v_{gs} and v_{ds} of the FET of the inverter to verify ZVS operation, and (c) secondary current i_s and rectifier input voltage v_r.

Table 27.2 Parameters of the proposed IPTS

Operating frequency	100 kHz
Nominal primary current I_p	7 A$_{\text{rms}}$
L_p	60 μH (3.3 m^2)
C_p	47 nF
L_{s1}, L_{s2}, L_{s3}	300 μH
C_{s1}, C_{s2}, C_{s3}	8.4 nF
Calculated mutual inductance	5 μH
Q-factor of the pick-up	4.4
Core-to-core airgap	10 mm

sensing technique can also be adopted for the improvement of power efficiency. The position sensing is not necessarily accurate and the position can be sensed by image processing using a camera or Wi-Fi based technique, even GPS at outdoor environments.

27.5 Conclusion

The proposed IPTS was verified to provide persistent power for two-dimensional mobile robots. The proposed one-layer winding method eliminates crossing-over points such that fabrication of the power floor can be simple and cheap, and the airgap between the power floor and pick-ups can be minimized for a higher power transfer. The evenly displaced multiple pick-up structure makes it possible to get continuous output power and to reduce the fluctuation of it substantially. The spatial average output power of the IPTS was demonstrated to be larger than the requirement of mobile robot power of 10 W for more than 6 hours.

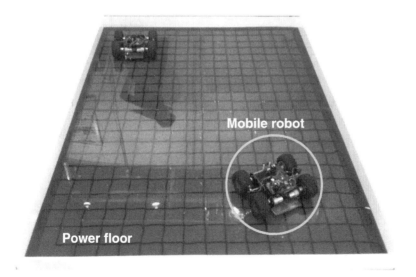

Figure 27.14 Implemented mobile robots and power floor.

Problems

27.1 In this chapter, the positions of three pick-up coils are determined by the heuristic method. Can you optimally design the positions of the coils by a systematic approach such as the probability function for random direction and position of a robot?

27.2 Discuss the merits and demerits of the proposed IPTS for robots, comparing other IPT systems such as resonant shoes and resonant mat suggested here.

References

1 R.C. Luo and K.L. Su, "Multilevel multi sensor-based intelligent recharging system for mobile robot," *IEEE Trans. on Ind. Electron.*, vol. 55, no. 1, pp. 270–279, January 2008.

2 A. Kottas, A. Drenner, and N. Papanikolopoulos, "Intelligent power management: promoting power-consciousness in teams of Mobile," in *ICRA*, Kobe, 2009, pp. 1140–1145.

3 M.C. Silverman, D. Nies, B. Jung, and G.S. Sukhatme, "Staying alive: a docking station for autonomous robot recharging," in *ICRA*, Washington, 2002, pp. 1050–1055.

4 C. Cai, D. Du, and Z. Liu, "Advanced traction rechargeable battery system for cableless mobile robot," in *AIM*, Kobe, 2003, pp. 235–239.

5 K.W. Klontz, D.M. Divan, and D.W. Novotny, "An actively cooled 120 kW coaxial winding transformer for fast charging electric vehicles," *IEEE Trans. on Ind. Applic.*, vol. 31, no. 6, pp. 1257–1263, Nov. 1995.

6 J.G. Hayes, G. Egan, J.D. Murphy, S.E. Schulz, and J.T. Hall, "Wide load range resonant converter supplying the SAE J-1773 electric vehicle inductive charging interface," *IEEE Trans. on Ind. Applic.*, vol. 35, no. 4, pp. 884–895, July 1999.

7 A. Kawamura, K. Ishioka, and J. Hirai, "Wireless transmission of power and information through one high frequency resonant AC link inverter for robot manipulator applications," *IEEE Trans. on Ind. Applic.*, vol. 32, no. 3, pp. 503–508, May 1996.

8 H. Abe, H. Sakamoto, and K. Harada, "A noncontact charger using a resonant converter with parallel capacitor of the secondary coil," *IEEE Trans. on Ind. Applic.*, vol. 36, no. 2, pp. 444–451, March 2000.

9 S. Lee and R.D. Lorenz, "Development and validation of model for 95% efficiency 220 W wireless power transfer over a 30 cm air gap," *IEEE Trans. on Ind. Applic.*, vol. 47, no. 6, pp. 2495–2504, November 2011.

10 K.W. Klontz, D.M. Divan, D.W. Novotny, and R.D. Lorenz, "Contactless power delivery system for mining application," *IEEE Trans. on Ind. Applic.*, vol. 31, no. 16, pp. 27–35, January 1995.

11 J. Huh, S.W. Lee, W.Y. Lee, G.H. Cho, and C.T. Rim, "Narrow width inductive power transfer system for online electrical vehicles," *IEEE Trans. on Power Electron.*, vol. 26, no. 12, pp. 3666–3679, December 2011.

12 S. Lee, J. Huh, C. Park, G.-H. Cho, and C.T. Rim, "On-line electric vehicle using inductive power transfer system," in *ECCE*, Atlanta, GA, 2010, pp. 1598–1601.

13 C. Liu, A.P. Hu, and X. Dai, "A contactless power transfer system with capacitively coupled matrix pad," in *ECCE*, Phoenix, AZ, 2011, pp. 3488–3494.

14 A.P. Hu, C. Liu, and H.L. Li, "A novel contactless battery charging system for soccer playing robot," in *I2MTC*, Victoria, BC, 2008, pp. 985–990.

15 W.X. Zhong, X. Liu, and S.Y.R. Hui, "A novel single-layer winding array and receiver coil structure for contactless battery charging systems with free-positioning and localized charging features," *IEEE Trans. on Ind. Electron.*, vol. 58, no. 9, pp. 4136–4144, January 2011.

16 B. Lee, H. Kim, C.-T. Rim, S. Lee, and C. Park, "Resonant power shoes for humanoid robots," in *ECCE*, Phoenix, AZ, 2011, pp. 1791–1794.

17 P. Meyer, P. Germano, M. Markovic, and Y. Perriard, "Design of a contactless energy transfer system for desktop peripherals," *IEEE Trans. on Ind. Applic.*, vol. 47, no. 4, pp. 1643–1651, July 2011.

18 Y.P. Su *et al.*, "Mutual inductance calculation of movable planar coils on parallel surfaces," *IEEE Trans. on Power Electron.*, vol. 24, no. 4, pp. 1115–1124, April 2009.

19 Samwha Electronics, "Ferrite meterial characteristics" [online]. Available at: http://www.samwha.com/electronics/product/product_ferrite_mat.html.

Part VI

Special Applications of Wireless Power

In this part, special topics in wireless power transfer are explained. Magnetic field focusing is one of the areas that is an emerging and quite unique topic. Wireless nuclear instrumentation is also an emerging area, which has become important after the Hukushima nuclear accident. Finally, the future of wireless power is addressed.

Wireless Power Transfer for Electric Vehicles and Mobile Devices, First Edition. Chun T. Rim and Chris Mi.
© 2017 John Wiley & Sons Ltd. Published 2017 by John Wiley & Sons Ltd.

28

Magnetic Field Focusing

28.1 Introduction

Beam-forming or spatial signal processing has been adopted in numerous studies of technologies such as radar, sonar, and wireless communications in an effort to obtain a high signal-to-noise ratio (SNR) [1–4]. Unlike conventional beam-forming technologies, a focusing magnetic field is not well recognized in wireless power transfer (WPT), magnetic field communication (MC), and magnetic induction tomography (MIT). Researchers have attempted to generate a flat magnetic field density over a large transmitting coil area to realize the free positioning of a receiving coil in WPT applications; however, this is not magnetic field focusing [5]. References [6] to [8] tried magnetic field redistribution with fixed metal loop coils, which were excited by an incoming magnetic field; however, this allows only a certain output flux distribution. The chronic problems of conventional WPT, MC, and MIT, which mostly involve electromagnetic interference [9–14], the omnidirectional field distribution [15–17], and the poor resolution [18–25], can be resolved or mitigated by adopting magnetic field focusing.

A novel synthesized magnetic field focusing (SMF) technology, which appropriately controls the vector of the magnetic field by means of a few current-controlled transmitting (Tx) coils, is explained in this chapter. The proposed SMF technology, with its basic operation, is verified by both simulations and experiments. A one-dimensional (1-D) experimental kit consisting of ten evenly distributed Tx coils spaced 5 cm apart with a 90 cm-long ferrite core was fabricated. Magnetic flux of 1.5 μT was well centralized within a focusing resolution of 12.5 cm at a distance of 10 cm from the Tx coil, which is four times that of a non-focusing coil and which may increase with a larger number of coils. It was verified that the side-lobe can be improved by 9.5 dB by reducing the resolution by 28%.

28.2 Overview of 1-D SMF Technology

Arrays of Tx coils and receiving (Rx) points, represented as discrete current sources and dots, respectively, are modeled for an SMF analysis, as shown in Figure 28.1(a). To obtain the focused magnetic field distribution over the Rx plane, the magnitude of each current source should be determined. This differs from the well-known inverse problem [26], in which the current source distribution is determined from the measured magnetic field. Therefore, a new algorithm is developed for the SMF in this chapter.

Wireless Power Transfer for Electric Vehicles and Mobile Devices, First Edition. Chun T. Rim and Chris Mi.
© 2017 John Wiley & Sons Ltd. Published 2017 by John Wiley & Sons Ltd.

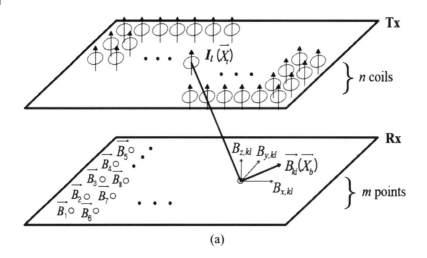

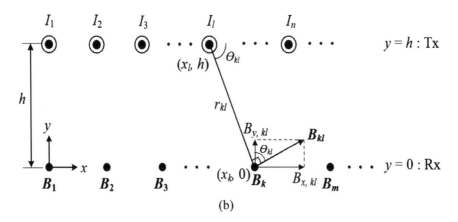

Figure 28.1 Schematics of the proposed SMF technology: (a) 2-D model and (b) 1-D model.

As shown in Figure 28.1(b), the magnetic field density vector $\overrightarrow{B_{kl}}$ generated from the current source I_l in free space can be derived for the 1-D model according to the Biot–Savart law, as follows [27]:

$$\overrightarrow{B_{kl}} = B_{x,kl}\overrightarrow{x_0} + B_{y,kl}\overrightarrow{y_0} = \frac{\mu_0 I_l}{2\pi r_{kl}}(\sin\theta_{kl}\overrightarrow{x_0} + \cos\theta_{kl}\overrightarrow{y_0}), \quad k = 1, 2, \ldots, m, \quad l = 1, 2, \ldots, n$$

$$\because r_{kl} = \sqrt{(x_k - x_l)^2 + h^2}, \quad \theta_{kl} = \tan^{-1}\left(\frac{y}{x_k - x_l}\right) \tag{28.1}$$

The total magnetic field density, which is the sum of all contributions made by all of the current sources, is then derived as follows:

$$\overrightarrow{B_k} = \sum_{l=1}^{n}\overrightarrow{B_{kl}} = \overrightarrow{x_0}\sum_{l=1}^{n}B_{x,kl} + \overrightarrow{y_0}\sum_{l=1}^{n}B_{y,kl} \equiv B_{x,k}\overrightarrow{x_0} + B_{y,k}\overrightarrow{y_0} \tag{28.2}$$

A matrix form of (28.2) can be separately represented for x and y components, as follows:

$$
\begin{bmatrix} B_{x,1} \\ B_{x,2} \\ \vdots \\ B_{x,m} \end{bmatrix} = \begin{bmatrix} a_{11x} & a_{12x} & \cdots & a_{1nx} \\ a_{21x} & a_{22x} & \cdots & a_{2nx} \\ \vdots & \vdots & & \vdots \\ a_{m1x} & a_{m2x} & \cdots & a_{mnx} \end{bmatrix} \begin{bmatrix} I_1 \\ I_2 \\ \vdots \\ I_n \end{bmatrix} \quad \Leftrightarrow \mathbf{B_x} = \mathbf{A_x I} \quad \because a_{klx} = \frac{\mu_0}{2\pi r_{kl}} \sin \theta_{kl} \qquad (28.3a)
$$

$$
\begin{bmatrix} B_{y,1} \\ B_{y,2} \\ \vdots \\ B_{y,m} \end{bmatrix} = \begin{bmatrix} a_{11y} & a_{12y} & \cdots & a_{1ny} \\ a_{21y} & a_{22y} & \cdots & a_{2ny} \\ \vdots & \vdots & & \vdots \\ a_{m1y} & a_{m2y} & \cdots & a_{mny} \end{bmatrix} \begin{bmatrix} I_1 \\ I_2 \\ \vdots \\ I_n \end{bmatrix} \quad \Leftrightarrow \mathbf{B_y} = \mathbf{A_y I} \quad \because a_{kly} = \frac{\mu_0}{2\pi r_{kl}} \cos \theta_{kl} \qquad (28.3b)
$$

The current source distribution matrix $\mathbf{I}$ for the designed magnetic field density distribution $\mathbf{B}$ can then be determined by an inverse matrix operation from (28.3a) and (28.3b), as follows:

$$
\mathbf{B} = \mathbf{AI} \quad \rightarrow \quad \mathbf{I} = \mathbf{A}^T(\mathbf{AA}^T)^{-1}\mathbf{B} \tag{28.4}
$$

where

$$
\mathbf{A} \equiv \begin{bmatrix} \mathbf{A_x} \\ \mathbf{A_y} \end{bmatrix}, \mathbf{B} \equiv \begin{bmatrix} \mathbf{B_x} \\ \mathbf{B_y} \end{bmatrix}
$$

and $n \geq 2$ m.

Note that procedures (28.1) to (28.4) are generally applicable not only to a 1-D case but also to higher dimensional cases (i.e., 2-D and 3-D, which are not covered in this chapter). Also note that the focusing is independent of the frequency and is valid for both the DC as well as the RF frequencies.

To verify the proposed algorithm for SMF technology, a finite-elementmethod (FEM) simulation was conducted with DC current sources in free space, as shown in Figure 28.2(a). Ten Tx coils as infinitely long current sources were used to centralize the Rx magnetic fields 10 cm away from the Tx plane. The current value of each Tx coil was determined by (28.4), as tabulated in Table 28.1, where the x-component of the magnetic field density for the center Rx point was set to 1.5 μT and all of the other points were set to zero. The simulation result is in very good agreement with the theoretical magnetic field density at every pre-set Rx point, as shown in Figure 28.2(b). The magnetic field density between two neighboring Rx points, known as the side-lobe, cannot be directly controlled by (28.4), as (28.4) offers a guarantee only for pre-set Rx points. Thus, an appropriate method should be devised to mitigate the side-lobe as well. Figure 28.2(c) depicts an example of focusing a magnetic field at different Rx points. Applying (28.3) and (28.4) again, the x-component of the magnetic field density at Rx-2, which is 11.25 cm apart from the center, was set to be 1.5 μT, where the other points were set to zero.

28.3 Example Design and Experimental Verifications

Ten Tx coils with a narrow-and-long ferrite core were fabricated for the verification of simulation, as shown in Figure 28.3(a). The ferrite core has a relative permeability of 2000, a length of 90 cm, a height of 10 cm, and a width of 1 cm. There were 20 turns of the winding for each Tx coil and the induced magnetic field is roughly doubled due to the magnetic mirror effect [28].

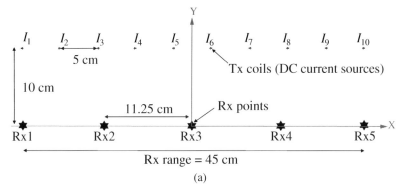

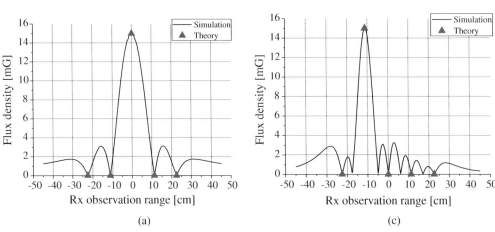

Figure 28.2 2-D FEM simulation results: (a) simulation model with ideal current sources, (b) comparison of the theoretical and simulated results for the centralized flux density of 1.5 µT, and (c) magnetic field focusing at Rx-2, when $I_1 = 0.9536$ A, $I_2 = -2.734$ A, $I_3 = 2.633$ A, $I_4 = 1.067$ A, $I_5 = -3.830$ A, $I_6 = 4.204$ A, $I_7 = -3.446$ A, $I_8 = 2.299$ A, $I_9 = -1.235$ A, $I_{10} = 0.3973$ A.

Table 28.1 Tx coil current distributions for a centralized flux density of 1.5 µT at an Rx center

	No side-lobe suppression (MSR = 10.4 dB)	With side-lobe suppression (MSR = 20 dB)
Current	Value (A-turn)	Value (A-turn)
$I_1 = I_{10}$	0.4877	0.0364
$I_2 = I_9$	−1.564	−0.2406
$I_3 = I_8$	2.917	1.273
$I_4 = I_7$	−3.824	−3.031
$I_5 = I_6$	2.113	1.989

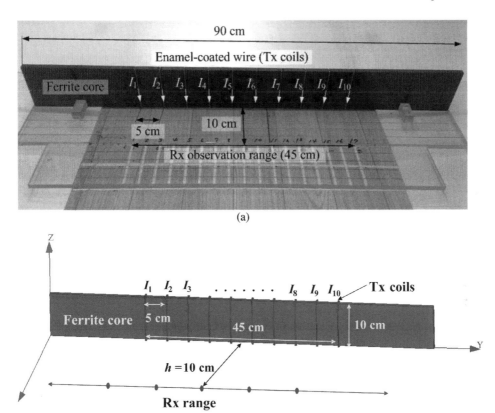

Figure 28.3 A prototype SMF: (a) experimental kit and (b) 3-D FEM simulation condition.

In order to avoid the influence of the Earth's magnetic field, a 60 Hz AC current for each Tx coil was provided with precise current control by a variable resistor. A Gauss meter (SPECTRAN NF-5035) was used to measure the magnetic field density distribution at the Rx points. Another 3-D FEM simulation for 1-D Tx coils with a core was performed with the physical parameters shown in Figure 28.3(b).

The experiment results for the focused and non-focused cases over an Rx range of 56 cm were compared with the simulation results, as shown in Figure 28.4(a), where only the right-hand side is presented, considering central symmetry. A resolution of 12.5 cm, as determined from the −3 dB point, was achieved for the focused case, whereas that for the non-focused case was 50 cm, which is a fourfold worse resolution. As predicted from (28.4), the resolution can be greatly enhanced by increasing the number of Tx coils. The measured main-lobe to side-lobe ratio (MSR) for the focused case was 10.4 dB. In order to increase the MSR, an appropriate current distribution of Tx coils was empirically determined for the suppression of the side-lobe, as shown in Table 28.1, where the magnetic flux density at two Rx points adjacent to the center were set to be 30% of flux density at the center (1.5 μT in this chapter) and having a 180 degree phase difference. In this way, the MSR was improved to 20 dB at a resolution expense of 28% but with a lower maximum current by 21%, as shown in Figure 28.4(b). The slight discrepancy between the simulation and experimental results comes mostly from the finite core length, the unequal core adhesion, and the inaccuracy of the simulation.

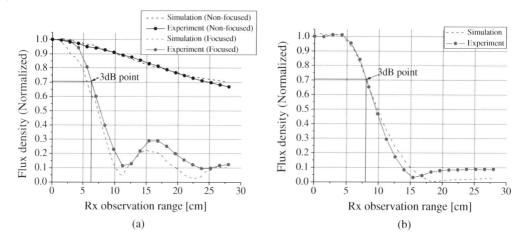

Figure 28.4 Comparison of the simulation and the experiment: (a) focused and non-focused cases and (b) case involving side-lobe suppression (high MSR).

28.4 Conclusion

SMF technology was theoretically proposed and verified by both a simulation and a 1-D experiment, with the results showing good agreement. Magnetic fields are conventionally known to be widespread in nature and they are not generally considered to be focusable by any means; however, the results here show that this is no longer the case, as a magnetic field can now be focused as desired. In this chapter, it was demonstrated that the focus of a magnetic field can be improved by fourfold when using ten Tx coils with controlled currents. The spatial resolution of the magnetic field can be arbitrarily enhanced by increasing the number of Tx coils, independent of the frequency, which is quite different from beam forming. Furthermore, the side-lobe can be mitigated significantly by appropriate selection of the current distribution. The proposed theory shows that 2-D or 3-D magnetic field focusing is also possible, which will lead to innovations in relation to cases involving high resolutions and low levels of electromagnetic interference in the areas of wireless power transfer, 3-D RF tags, magnetic induction tomography, magnetic therapies, and magnetic field communications.

Problems

28.1 Can you constitute any theory that can predict the resolution of the proposed SMF?

28.2 Can you find any theory to predict the side-lobes of the proposed SMF?

References

1 Fakharzadeh, S.H. Jamali, P. Mousavi, and S.N. Safieddin, "Fast beamforming for mobile satellite receiver phased arrays: theory and experiment," *IEEE Trans. on Antennas Propagation*, vol. 57, no. 6, pp. 1645–1654, 2009.

2 M.V. Ivashina, O. Iupikov, R. Maaskant, W.A. Cappellen, and T. Oosterloo, "An optimal beamforming strategy for wide-field surveys with phased-array-fed reflector antennas," *IEEE Trans. on Antennas Propagation*, vol. 59, no. 6, pp. 1864–1875, 2011.

3 W. Yao and Y.E. Wang, "Beamforming for phased arrays on vibrating apertures," *IEEE Trans. on Antennas Propagation*, vol. 54, no. 10, pp. 2820–2826, 2006.

4 R. Bernini, A. Minardo, and L. Zeni, "Distributed sensing at centimeter-scale spatial resolution by BOFDA: measurements and signal processing," *IEEE Photon. J.*, vol. 4, no. 1, pp. 48–56, 2012.

5 E. Waffenschmidt, "Free positioning for inductive wireless power system," in *2011 IEEE ECCE Conference*, pp. 3480–3487.

6 H. Tanaka and H. Iizuka "Kilohertz magnetic field focusing behavior of a single-defect loop array characterized by curl of the current distribution with delta function," *IEEE Antennas Wireless Propagation Letters*, vol. 11, pp. 1088–1091, 2012.

7 D. Banerjee, J. Lee, E.M. Dede, and H. Iizuka, "Kilohertz magnetic field focusing in a pair of metallic periodic-ladder structures," *Appl. Phys. Lett.*, vol. 99, pp. 093501/1–093501/3, 2011.

8 H. Tanaka and H. Iizuka, "Significant improvement of magnetic field focusing ability in actively-tuned resonant loop array," *IEEE Antennas Wireless Propagation Letters*, vol. PP, no. 99, pp. 1–4, 2015.

9 A. Kurs, A. Karalis, R. Moffatt, J.D. Joannopoulos, P. Fisher, and M. Soljacic, "Wireless power transfer via strongly coupled magnetic resonances," *Science*, vol. 317, no. 5834, pp. 83–86, 2007.

10 Z. Yan, Y. Li, C. Zhang, and Q. Yang, "Influence factors analysis and improvement method on efficiency of wireless power transfer via coupled magnetic resonance," *IEEE Trans. on Magnetics*, vol. 50, no. 4, pp. 1–4, 2014.

11 R. Hui, W. Zhong, and C.K. Lee, "A critical review of recent progress in mid-range wireless power transfer," *IEEE Trans. on Power Electron.*, vol. 29, no. 9, pp. 4500–4511, 2014.

12 H. Hwang, J. Moon, B. Lee, C. Jeong, and S. Kim, "An analysis of magnetic resonance coupling effects on wireless power transfer by coil inductance and placement," *IEEE Trans. on Consum. Electron.*, vol. 60, no. 2, pp. 203–209, 2014.

13 Z.N. Low, R.A. Chinga, R. Tseng, and J. Lin, "Design and test of a high-power high-efficiency loosely coupled planar wireless power transfer system," *IEEE Trans. on Ind. Electron.*, vol. 56, no. 5, pp. 1801–1812, 2009.

14 K. Lee and D.H. Cho, "Diversity analysis of multiple transmitters in wireless power transfer system," *IEEE Trans. on Magnetics*, vol. 49, no. 6, pp. 2946–2952, 2013.

15 Y. Won, S. Kang, K. Hwang, S. Kim, and S. Lim, "Research for wireless energy transmission in a magnetic field communication system," in *2010 IEEE ISWPC Conference*, pp. 256–260.

16 M. Masihpour and J.I. Agbinya, "Cooperative relay in near field magnetic induction: a new technology for embedded medical communication systems," in *2010 IEEE IB2Com Conference*, pp. 1–6.

17 M. Dionigi and M. Mongiardo, "Multi band resonators for wireless power transfer and near field magnetic communications," in *2012 IEEE IMWS Conference*, pp. 61–64.

18 A.J. Peyton, Z. Yu, G. Lyon, S. Al-Zeibak, J. Ferreira, J. Velez, F. Linhares, A.R. Borges, H.L. Xiong, N.H. Saunders and M.S. Beck, "An overview of electromagnetic inductance tomography: description of three different systems," *Measurement Science and Technology*, vol. 7, no. 3, pp. 261–271, 1996.

19 Z. Zakaria, R. Rahim, M. Mansor, S. Yaacob, N. Ayub, S. Muji, M. Rahiman, and S, Aman, "Advancements in transmitters and sensors for biological tissue imaging in magnetic induction tomography," *Sensors*, vol. 12, no. 6, pp. 7126–7156, 2012.

20 S. Watson, R.J. Williams, H. Griffiths, W. Gough, and A. Morris, "Magnetic induction tomography: phase versus vector-voltmeter measurement techniques," *Physiol. Measurements*, vol. 24, no. 2, pp. 555–564, 2003.

21 R. Merwa and H. Scharfetter, "Magnetic induction tomography: evaluation of the point spread function and analysis of resolution and image distortion," *Physiol. Measurements*, vol. 28, no. 7, pp. 313–324, 2007.

22 D.N. Dyck, D.A. Lowther, and E.M. Freeman, "A method of computing the sensitivity of electromagnetic quantities to changes in materials and sources," *IEEE Trans. on Magnetics*, vol. 30, no. 5, pp. 3415–3418, 1994.

23 H. Krause, I.G. Panaitov, and Y. Zhang, "Conductivity tomography for non-destructive evaluation using pulsed eddy current with HTS SQUID magnetometer," *IEEE Trans. on Appl. Supercond.*, vol. 13, no. 2, pp. 215–218, 2003.

24 H. Wei L. Ma, and M. Soleimani, "Volumetric magnetic induction tomography," *Measurement Science and Technology*, vol. 23, no. 5, pp. 1–9, 2012.

25 M. Yan, C. Jiang, C. Yao, and C. Li, "Development of a focusing pulsed magnetic field system for *in vivo* experiments," *IEEE Trans. on Dielectr. Electr. Insul.*, vol. 20, no. 4, pp. 1327–1333, 2013.

26 D. Gursoy and H. Scharfetter, "Optimum receiver array design for magnetic induction tomography," *IEEE Trans. on Biomed. Engng*, vol. 56, no. 5, pp. 1435–1441, 2009.

27 D.J. Griffiths, "Magnetostatics," in *Introduction to Electrodynamics*, 3rd edn, Prentice Hall, New Jersey, Ch. 5, Sec. 2, pp. 223–228, 1999.

28 W. Lee, J. Huh, S. Choi, X.V. Thai, J. Kim, E.A. Al-Ammar, M.A. El-Kady, and C.T. Rim, "Finite-width magnetic mirror models of mono and dual coils for wireless electric vehicles," *IEEE Trans. on Power Electron.*, vol. 28, no. 3, pp. 1413–1428, 2013.

29

Wireless Nuclear Instrumentation

29.1 Introduction

The availability of emergency countermeasures after a severe accident is the most critical issue in nuclear power plant (NPP) safety [1–8]. Since the Fukushima accident, reliable and continuous measurements of the NPP during a severe accident are critical to support decision making that is adaptable to rapidly varying accident environments. From several previous severe accidents, it was found that the loss of measurements is the major cause of delays in crucial decisions such as a seawater injection and public evacuation, and these delays consequently lead to uncontrollable public fears of NPPs.

A sequence of measurement loss after a severe accident can be determined as follows:

1. A beyond design basis accident involving significant core degradations occurs [1].
2. Major instruments and power/communication lines are exposed to extremely high temperature, pressure, and moisture, which are mainly due to both reactor failures and poor accident management.
3. The instruments and connected cables are damaged and permanent instrument failure happens, when repairs of the instruments are not available due to a high radiation environment during the severe accident [3, 4].

To overcome the physical failure of currently installed instruments and cables, which are designed by following the equipment qualification (EQ) based on design-based accidents, several methodological approaches for enhancing the reliability of equipment have been researched [5–8]. The previous methodologies can be classified into the following three categories, while a configuration of the collective results is depicted in Figure 29.1:

1. "Problem definition": intensified temperature and pressure profiles were suggested for reviewing and/or designing the protection of the NPP equipment against severe accident environments.
2. "Increase in redundancy": additional instrument channels were applied as extra redundancies to deal with the malfunctions and degradations of the existing channels.
3. "Physical reinforcement": instead of replacing every damageable equipment, physical protective remedies against extreme temperature and pressure conditions were introduced for existing instruments and cables.

To identify the design requirements of 2 and 3, the peak temperature and pressure during a severe accident can be determined by [8] and [9]. As shown in Figure 29.2, a temperature profile during 72 hours after an accident having a peak temperature of 627 °C and a long-term ambient temperature of 187 °C was evaluated based on simulations of various points in the containment

Wireless Power Transfer for Electric Vehicles and Mobile Devices, First Edition. Chun T. Rim and Chris Mi.
© 2017 John Wiley & Sons Ltd. Published 2017 by John Wiley & Sons Ltd.

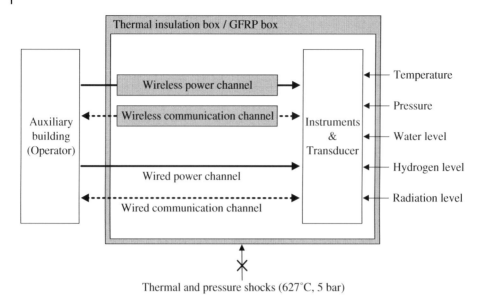

Figure 29.1 Configuration of the proposed highly reliable power and communication system for essential measurements in the NPP.

building. A hydrogen explosion can be considered as the worst condition that puts maximum stress on the equipment. From equivalent experiments in [12], the peak pressure value can be determined as 5 bar (or 72.5 psig).

To deal with the loss of conventional wired power and communication cables, reserved wireless channels were suggested for both the power channel and the communication channels by

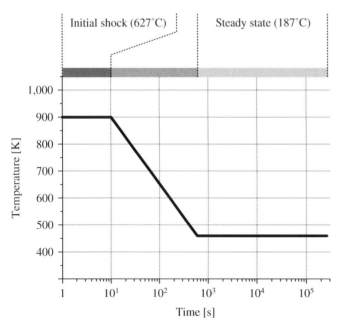

Figure 29.2 Dynamic temperature profile of the containment building during a severe accident lasting 72 hours.

adopting an inductive power transfer system (IPTS) and radio frequency (RF) communication, respectively [7, 8, 10, 11].

Direct use of the temperature and pressure conditions in 1 as design requirements of the NPP equipment is impractical due to the extremely high cost. Therefore, a thermal insulation box and a glass-fiber reinforced plastic (GFRP) box were conceptually proposed to protect only some of the equipment, which is essential for evaluating the NPP integrity, from extremely high temperature and pressure [7, 8].

In this chapter, the design principles of "increasing redundancy" and "physical reinforcement" are proposed and experimentally verified with relevant prototypes. As the wireless power channel, the 10 W-level IPTS using a dipole coil resonance system (DCRS) is designed over a 7 m distance, where the target distance matches the length of the main route of the conventional power/communication cables from the inner wall to the outer wall of the containment building shown in Figure 29.1. The wireless communication channel is composed of two Zigbee modules covering a 10 m to 20 m range without any data loss for 72 hours. The design of the proposed thermal insulation box consisting of a water-layer and a microporous insulator is provided based on a simplified model in [13] to isolate the equipment from the heat shock. A GFRP box having a thickness of 10 mm was designed to protect the proposed IPTS coils from the heat and pressure. Both the insulation box and the GFRP box were experimentally verified using a fabricated high-temperature chamber that can mimic the temperature profile in Figure 29.2.

The rest of this chapter is organized as follows. Section 29.2 presents design principles of the proposed highly reliable power and communication system consisting of the wireless power/communication channels, the thermal insulation box, and the GFRP box. Section 29.3 shows experimental verifications of the performance of the proposed redundant wireless channels and the physical protection boxes with consideration of high temperature and radiative environments. Conclusions are provided in Section 29.4.

29.2 Design of the proposed highly reliable power and communication system

In this section, the proposed highly reliable power and communication system is categorized into four subsystems of the 10 W-level wireless power channel, RF wireless communication channel, thermal insulation box, and GFRP box. Both the wireless power and communication channels are applied with currently installed wired channels to increase reliability. Both the thermal insulation box and the GFRP box work with any NPP equipment that is likely to be exposed to extreme temperature and pressure. In the following subsections, the design requirements of each subsystem are thoroughly investigated and fulfilled by prototype design, where the design requirements of each subsystem are identified in Table 29.1.

29.2.1 10 W-Level Wireless Power Channel

As shown in Figure 29.3, a target distance of 7 m was selected for a wireless power transfer to power sensors and transducers in NPP, when considering typical routes of the power cables from an inner wall to an outer wall of the containment building. Due to the extreme environments, which dynamically vary during a severe accident, most of the wireless power transfer methods in NPPs are highly restricted, as tabulated in Table 29.2. As a result of comparative evaluation of the robust powering characteristic, the IPTS method having a dipole coil at each

Table 29.1 Summarized design requirements of each subsystem.

Increase in redundancy with wireless channels		Physical reinforcement	
Inductive power transfer system (IPTS)	RF communication	Thermal insulation box	GFRP box
• 5 W powering over 7 m • Limited dipole length of 2 m • No EMI issue with other equipment used together • Operating with ferrite cores having a low Curie temperature of ~300 °C • Operating with high-frequency applicable wire having a low insulator melting temperature of ~200 °C • Requiring additional physical reinforcements under the given ambient temperature and pressure (187 °C ~ 627 °C, 5 bar)	• Data transfer over 7 m • No cross interference with the IPTS • No error or data loss for 72 hours • Satisfying low operating temperature of Zigbee modules/microprocessors (~85 °C) • Requiring additional physical reinforcements under the given ambient temperature and pressure (187 °C ~ 627 °C, 5 bar)	• Keeping the cavity temperature lower than 85 °C for 72 hours with the given temperature outside the cavity (187 °C ~ 627 °C) • Imperfect thermally isolated system due to conductive incoming cables from the outside into the cavity • Maintaining structural integrity for the given ambient temperature and pressure (187°C ~ 627°C, 5 bar)	• Maintaining structural integrity for the given ambient temperature and pressure (187 °C ~ 627 °C, 5 bar) • No disturbance to the magnetic field or electric field for use together with IPTS and RF communication

transmitting (Tx) and receiving (Rx) side is selected. The output power of the proposed ITPS is set as 10 W according to the load estimation in [8], which can cover typical power consumptions of a microprocessor, an RF communication module, and an instrument including its transducer.

The Curie temperature of typical Mn–Zn ferrite cores, which comprise the dipole coils, is approximately 300 °C and the typical operating temperature of the cable is also limited to as low as 200 °C, assuming that a Teflon coated cable is used. Moreover, the ferrite core is fragile and susceptible to corrosion. Therefore, additional physical reinforcement is required when considering vulnerable coil characteristics under extreme environments.

As shown in Figure 29.4, an EMI issue is reviewed by using finite element method (FEM) simulations, where the proposed 10 W-level IPTS is designed as follows [11]. An aluminum shielding box is introduced to satisfy the magnetic field emission restriction in the containment building, which is depicted in Figure 29.5. The magnetic field density inside a 1 mm-thick aluminum box satisfies the magnetic field emission restriction required by the Nuclear Regulatory Commission (NRC) of 105 BpT (= 178 nT) at 20 kHz [14]. Note that the proposed wireless channel operates only for the loss of wired channels during a severe accident; hence, there is no issue of EMI with the currently installed equipment, which would be no longer available during a severe accident.

Figure 29.3 Configuration of the proposed wireless power channel installed in the containment building with the conventional wired channel from the auxiliary building, where a quarter of the containment building is illustrated.

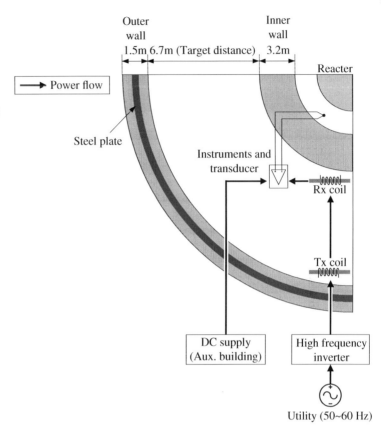

Table 29.2 Comparison among possible wireless power channels.

Methods	Advantages	Disadvantages	Remarks
RF	• Relatively good in environment containing steam, vapor, and/or debris (as compared to the laser type)	• Impossible to penetrate water layers and/or metal objects • High EMF/EMI	Unacceptable
Laser	• No EMI/EMF due to the highest straightness	• Impossible to penetrate obstacles • Difficult to overcome environments containing steam, vapor, and/or debris	Unacceptable
CMRS	• Relatively good for long-distance wireless power transfers (due to the high-quality factor)	• Large diameter of coils • Complex structure • Hypersensitive performance	Unacceptable
IPTS	• Simple structure • High output power • Long-distance power transfer with a low quality factor	• High EMF/EMI • An EMI/EMF shield can be achieved by using a metal plate	Acceptable

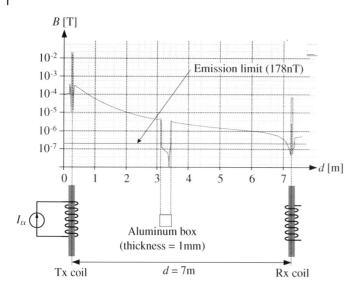

Figure 29.4 IPTS simulation result with an aluminum shielding box at the operating frequency of 20 kHz when $I_{tx} = 10$ A and 10 turns for each coil.

29.2.2 RF Wireless Communication Channel

The design of the proposed wireless communication channel is quite similar to the wireless power channel design. Considering the required communication distance of 7 m, which is same as the required wireless power transfer distance, RF communication is selected, where the magnetic field communication (MFC) technique is not preferred due to the technical immatureness and the cross-interference with IPTS. Among the available RF communication candidates, Zigbee is selected due to its low power consumption and relatively good data transfer ability.

Just as with the vulnerable issues in the previous section, applicable Zigbee modules and auxiliary electronic circuits have their maximum operating temperature of from 85 °C to 120 °C, which is mainly due to the performance of state-of-the-art semiconductor devices inside the module. Hence, an additional reinforcement should be installed together with the proposed wireless communication channel to enable it to endure the extreme ambient conditions.

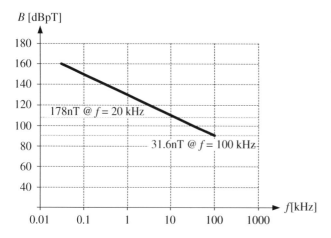

Figure 29.5 Magnetic field emission restriction from the Nuclear Regulatory Commission (NRC), RE101: 178 nT and 31.6 nT at the operating frequencies of 20 kHz and 100 kHz, respectively [14].

Figure 29.6 Block diagram of the proposed wireless communication channel with a Zigbee-type communication.

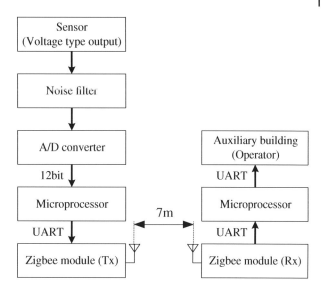

Figure 29.6 shows a block diagram of the proposed wireless communication channel, which can substitute for a conventional current transducer converting a voltage signal to a current signal ranging from 4 mA to 20 mA. Commercial Zigbee modules are used with microprocessors, which convert analog input to digital output, and single or cascaded noise filters can be used together depending on the input noise level.

29.2.3 Thermal Insulation Box

As identified from the temperature profile in Figure 29.2, the initial thermal shock of 627 °C and relatively high long-term ambient temperature of 187 °C are critical to the NPP equipment. The main roles of the proposed thermal insulation box are protecting the selective equipment from the thermal shock and providing operable temperature to the equipment.

The design of the proposed thermal insulation box is conducted based on the simplified heat transfer model in [13] and FEM simulation. As shown in Figure 29.7(a), the thermal insulation box consists of two layers with high thermal coefficients. The external layer is a microporous insulator, which is Super-G from Microtherm in this chapter, having a thickness of 75 mm. The inner water layer having various thicknesses in a range of 100~300 mm also delays heat transfer from ambient to the cavity, which is prepared for the equipment installation. As shown in Figure 29.7(b), the cavity temperature does not exceed 80 °C for 72 hours of simulation, where the ambient condition follows the temperature profile in Figure 29.2. Stainless steel is used to build the frame of the thermal insulation box due to its robustness against high temperature and excess moisture.

29.2.4 GFRP Box

The GFRP box is designed to provide the equipment with a protective case from high pressure including the mechanical shock of scattered objects due to hydrogen explosions or severe vibration caused by natural disasters. The material of GFRP is selected by considering that GFRP has no degrading effect on the proposed wireless power and communication channels. Figure 29.8(a) shows a design example of the proposed GFRP box for the 2 m-long dipole coil

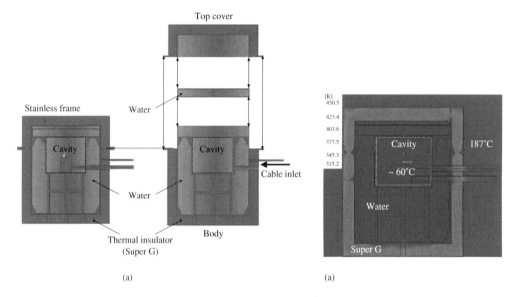

Figure 29.7 Overall configuration and an FEM simulation result of the proposed thermal insulation box: (a) assembled thermal insulation box (left) and assembly parts (right) and (b) heat transfer simulation result with the temperature profile in Figure 29.2 for 72 hours.

used in the proposed wireless power channel. Satisfying both design equations (29.1) and (28.2) below [8], a thickness of 10 mm was selected for the proposed GFRP box to achieve both structural integrity against a pressure shock of 5 bar and heat transfer delay against the initial heat shock of 627 °C.

Figure 29.8(b) and (c) show cross-sectional diagrams of the proposed GFRP box with the dipole coil inside. The heat transfer from ambient to the dipole coil inside the GFRP box can be determined w.r.t. time t as follows:

$$k\frac{A}{d}(T_o - T_i) + Q_g = mc\frac{dT_i}{dt} \tag{29.1}$$

where k, A, d, T_o, T_i, m, c, and Q_g are the thermal conductivity of GFRP, heat conduction cross-section area, thickness of GFRP, ambient temperature, dipole coil temperature, mass, specific heat of dipole coil material (ferrite), and heat generation of the dipole coil, respectively.

Considering the external pressure shock, the thickness of the CFRP box can be determined as follows:

$$d > \frac{w}{\sqrt{8\sigma/3P}} \tag{29.2}$$

where d, w, P, and σ are GFRP box thickness, GFRP box width, forced pressure on the GFRP box, and ultimate strength of GFRP, respectively.

When the GFRP box is 10 mm thick, the maximum bending stress of 5 bar is lower than the tensile strength of the material. The cavity temperature of the proposed GFRP box slowly increases without any overshoot up to 187 °C, which is the long-term ambient temperature during a severe accident. For example, the proposed GFRP box having 10 mm of thickness can

Figure 29.8 The proposed GFRP box for a dipole Tx coil of the proposed wireless power channel: (a) overall configuration, (b) side view, and (c) plan view.

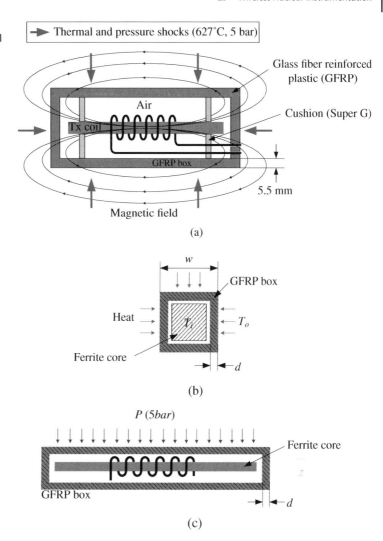

delay the cavity temperature increase up to 187 °C for 10 hours, where it is assumed that the proposed dipole coil with its power loss of 100 W is located in the cavity.

29.3 Example Design and Experimental verifications

The design results of the proposed highly reliable power and communication system in the previous section are verified with each prototype of four subsystems. To mimic the extreme temperature condition, a temperature controllable chamber is fabricated to verify the applicability of the thermal insulation box and the GFRP box. Considering the lack of previous studies on the defectiveness of power electronics devices under high radiation, an experimental study was also conducted for various high-power-applicable circuit components, which are essential to wireless power channels, under a radiative environment having a maximum accumulated dose of 27 Mrad.

29.3.1 10 W IPTS over a 7 m Distance: Performance Test with High-Temperature and Conductive Obstacles

The 10 W powering performance of a DCRS with 2 m-long dipole coils over a 7 m distance was verified in [11]. Changes in several key operating conditions, such as the coil inductance and the compensation capacitances, were experimentally evaluated with respect to a high ambient temperature and conductive obstacles between the Tx coil and the Rx coil.

For a high-temperature operating test, a quarter-scale dipole coil was used with two types of high-frequency applicable polymer film capacitors when considering the commercial temperature test chamber TD500, which has a maximum control range from −20 °C to 150 °C.

Figure 29.9 shows variations in the inductance and capacitance of the IPTS according to a temperature sweep ranging from 30 °C to 150 °C. As identified from Figure 29.9(a), the inductance variation did not exceed 1.6% of its nominal value, where the IPTS having a quality factor of 100 allows for an inductance variation of 2%. The series equivalent resistance of the dipole coil was constant during the test, as shown in Figure 29.9(a).

On the other hand, capacitance variations of two polymer film capacitors, where one is a plastic-cased type and the other one is an epoxy-lacquer-coated type, were more prominent than that of the coil inductance variation. The capacitance of the plastic-cased type increased by 2% and the capacitance of the epoxy-lacquer-coated type decreased by 1%, as shown in

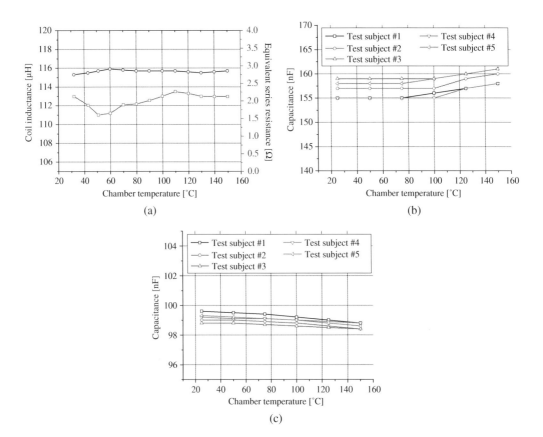

Figure 29.9 High-temperature operating test results: (a) quarter-scale dipole coil, (b) plastic-cased film capacitor, and (c) epoxy-lacquer-coated film capacitor.

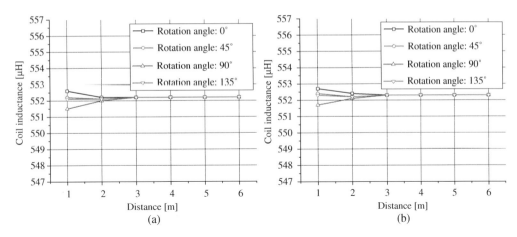

Figure 29.10 Coil inductance variation measurement test with 1 m²-sized steel plate according to distance and angles for two different operating frequencies: (a) maximum variation was 0.2% at 20 kHz operation, and (b) maximum variation was 0.18% at 100 kHz operation.

Figure 29.9(b) and (c), respectively. The capacitance variation can be neglected when both capacitors are used together when considering their complementary characteristics against temperature change.

To identify the effect of conductive obstacles on the powering characteristic, the dipole coil inductance was measured, where a 1 m²-sized stainless steel plate was in a different position within a 1 m to 6 m distance range and a 0° to 135° angle range between the aluminum plate and the coil. As shown in Figure 29.10, the coil inductance changed less than 0.2% from its nominal value for each operating frequency case.

From the performance verifications in this section with previous works in [11], it is proven that the IPTS having a relatively low quality factor of 100 is applicable as the emergency back-up power source under a severe accident environment.

29.3.2 Zigbee Wireless Communication: Data Loss Measurements with Conductive Surroundings

As shown in Figure 29.11, the proposed wireless communication channel was fabricated with a commercial Zigbee module (XB24CZ7PIS-004) and a microprocessor (Dspic30f6012A), where

Figure 29.11 Fabricated Tx module for the proposed wireless communication channel, which is identical to the Rx module.

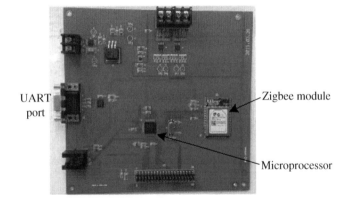

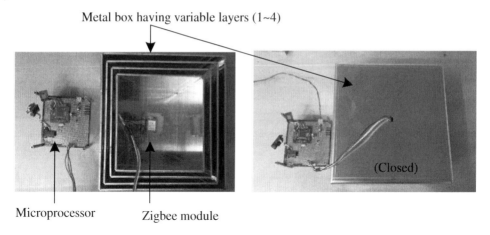

Metal box having variable layers (1~4)

(Closed)

Microprocessor   Zigbee module

Figure 29.12 Data loss measurement condition with metal boxes.

the measured total power consumption of devices above was less than 0.5 W. For evaluating the applicability of the commercial Zigbee modules during a severe accident, two conductive surroundings were applied during the test. Data losses between the Tx module and the Rx module were measured when one of them was located inside a metal box with variable layers, as shown in Figure 29.12. Two different metals were used: one was stainless steel and the other was aluminum. There was no data loss in the 7 m-distance Zigbee communication with both the stainless steel box and the aluminum box, even though the number of layers increased by up to four in both cases.

To mimic ionized spray under hydrogen and steam emissions in the containment building, a conductive saline solution having variable concentrations from 0% to 10% was used, as shown in Figure 29.13. Fabric soaked with saline solution, which mimics the excess of moisture around the Zigbee module, as shown in Figure 29.13(a), did not cause any data loss during the communication. On the other hand, data loss increased as the depth of the saline solution increased, as one of the Zigbee modules sank into the solution, as shown in Figure 29.13(b) and (c).

29.3.3 Dynamic Temperature Test Ranging from 627 °C to 187 °C: Thermal Insulation Box and GFRP Box

A 3.4 m^3-sized high-temperature test chamber, which is controlled by following the temperature profile in Figure 29.1, was fabricated for performance tests of both the thermal insulation box and the GFRP box, as shown in Figure 29.14.

The fabricated thermal isolation box, which has a Super G layer and a water layer for a cavity of 0.01 m^3, is depicted in Figure 29.15(a). As shown in Figure 29.15(b), the Super G layer was intensified with the design margin of 20% in thickness to compensate for a structural defect in the stainless frame, which is mainly due to imperfect welding conditions. The comparison between the experimental result and the simulation results for 72 hours is shown in Figure 29.15(b). After the dynamic temperature test, the cavity temperature was 62 °C, which is slightly lower than the simulation result due to the design margin of 20% in the thickness of the microporous insulator layer.

A half-scale GFRP box with a 1 m dipole coil was fabricated for the dynamic temperature test due to the limited size of the fabricated test chamber, as shown in Figure 29.16(a) and (b). The cavity temperature of the fabricated GFRP box reached 187 °C 16 hours after initiating the

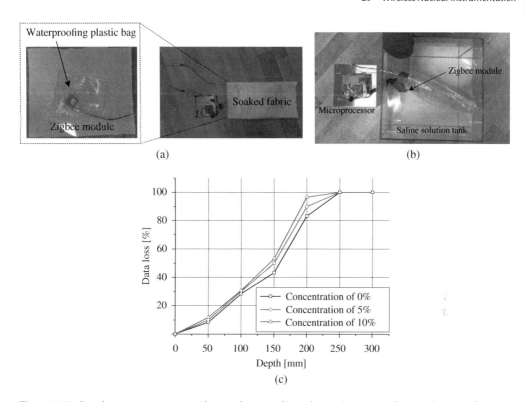

Figure 29.13 Data loss measurements with a conductive saline solution: (a) test condition with soaked fabric, (b) test condition with a water tank filled with the saline solution, and (c) test results according to different depths and concentrations.

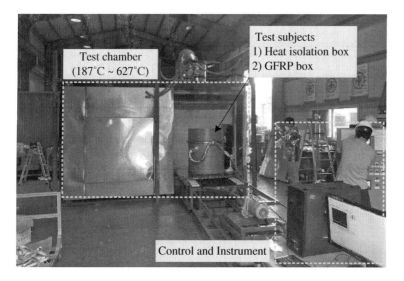

Figure 29.14 Fabricated dynamic temperature test chamber.

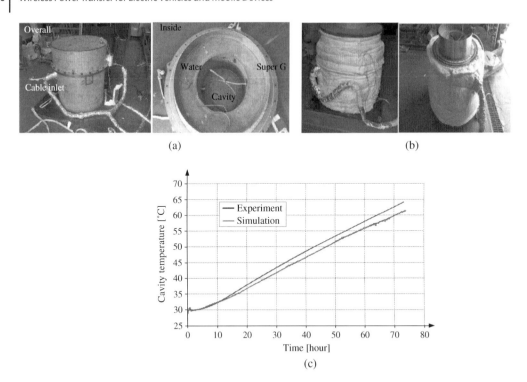

Figure 29.15 Fabricated thermal insulation box and its experimental result for 72 hours: (a) thermal insulation box, (b) additional Super G layer around the thermal insulation box, and (c) comparison between the experimental result and the simulation result.

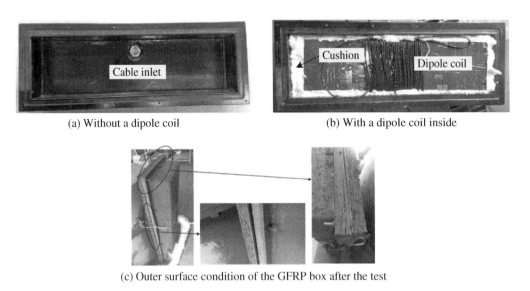

(a) Without a dipole coil

(b) With a dipole coil inside

(c) Outer surface condition of the GFRP box after the test

Figure 29.16 Fabricated half-scale GFRP box with the dipole coil.

test. Comparing the designed heat transfer delay of 10 hours, a slightly longer delay time was achieved with the help of the Super G cushion, which was mainly used to fix the dipole coil location against vibrations or shocks. Due to the initial heat shock of 627 °C, the outer surface of the GFRP box was partially damaged, as shown in Figure 29.16(c); however, every important joint part was still robust after the test.

29.3.4 High Radiation Test of 27 Mrad: Power Electronics Device Defects

The use of semiconductor-based components is inevitable in order to build the proposed wireless power and communication channels. Although low-power applicable electronic devices have been reviewed under a γ-ray environment for space craft applications [15], most medium-to-high power applicable devices are not tested for their application under a high γ-ray irradiation. In this section, 10 circuit components, which were carefully selected considering their frequent use in various power electronics applications, were tested under a high-radiation environment to evaluate their application in a severe accident, as tabulated in Table 29.3. Considering the accumulated dose of LBLOCA (large break loss of coolant accident), the devices were tested under the maximum accumulated dose of 27 Mrad at the Korea Atomic Energy Research Institute, where the dose rate of the test environment was 1.2 Mrad/h. Experimental data acquisition was not continuously conducted due to the limitation in controlling the γ-ray experiment environment; however, the experimental results are meaningful as an initial survey to figure out the applicability of power conversion circuits to NPPs.

As shown in Figure 29.17(a) and (b), there were no considerable changes in capacitance of the film capacitors, where the breakdown voltages were decreased by 32% for the epoxy-lacquer-coated types and 3% for the plastic-cased types after the irradiation. The breakdown voltage was measured with METREL MI 3201, which judges the breakdown when there is a leakage current of 1 mA between two test nodes. Considering constant capacitance regarding the accumulated dose, the test on the film capacitors was closed at 1.6 Mrad.

The breakdown voltage of SCS120AG, having an original breakdown voltage of 930 V, which is a silicon carbide (SiC) diode, was measured, as shown in Figure 29.17(c). The γ-ray irradiation was intermittently stopped to check for recovery characteristics during the radiation test. The breakdown voltage drastically decreased by 21% at the initial irradiation compared to the

Table 29.3 Summary of the irradiation test.

Test items	Survivability	Remarks
Film capacitor	O	3 ~ 30% decrease in the breakdown voltage
SiC diode	O	2.5% decrease in the breakdown voltage
Zener diode	O	–
Op-amp	X	Failure after 0.6 Mrad
SiC JFET	Δ	6% increase in D-S resistance
BJT	O	–
MCU	X	Failure after 20 krad
IGBT	O	–
SMPS	X	Failure in internal IC devices
Zigbee module	X	Failure in internal IC devices

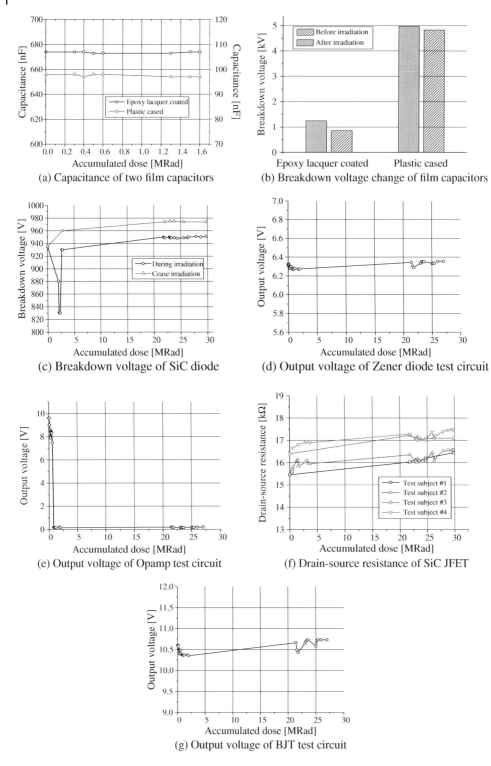

(a) Capacitance of two film capacitors

(b) Breakdown voltage change of film capacitors

(c) Breakdown voltage of SiC diode

(d) Output voltage of Zener diode test circuit

(e) Output voltage of Opamp test circuit

(f) Drain-source resistance of SiC JFET

(g) Output voltage of BJT test circuit

Figure 29.17 Measurement results of the irradiation test for various power electronics devices.

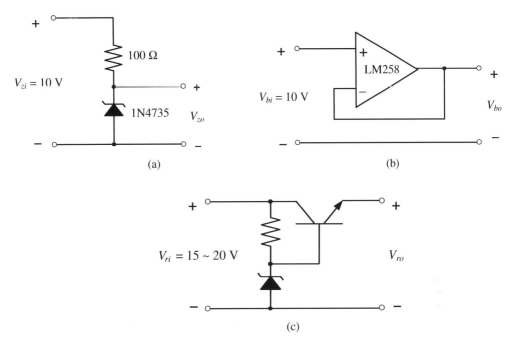

Figure 29.18 Irradiation test circuits: (a) Zener diode, (b) op-amp, and (c) BJT.

original value. During interruptions of the irradiation, the breakdown voltage temporally restored a little; however, it decreased again when the irradiation was restarted.

The Zener breakdown voltage was measured with a test circuit including a Zener diode (1N4735A), shown in Figure 29.18(a), where the theoretical output voltage V_{zo} is 6.2 V with an input voltage V_{zi} of 10 V. As identified from Figure 29.17(d), there was no significant defect in Zener breakdown voltage during the test.

The operation of operational amplifiers (op-amps) were tested with a buffer circuit configuration in Figure 29.18(b). As shown in Figure 29.17(e), the output voltage V_{bo} of the buffer, which is designed to follow the input voltage V_{bi}, gradually decreased after the initial irradiation; then V_{bo} dropped to zero when the accumulated dose was 0.6 Mrad.

A slightly increased drain-to-source (D-S) resistance of the junction gate field effect transistor (JFET) was measured, where its gate and source had the same potential, as shown in Figure 29.17(f). The 6% increased D-S resistance indicates higher conduction losses; hence, thermal management systems, such as a heat sink and a fan, should be designed with sufficient margins when they are installed with the JFET in the high radiation condition. However, the increased D-S resistance was decreased to the initial value two days after the experiment was terminated.

Using the normal operation of the Zener diodes, a linear regulator circuit was fabricated to evaluate the performance of the bipolar junction transistor (BJT), as shown in Figure 29.18(c). As shown in Figure 29.17(g), the regulator output voltage V_{ro} was maintained as 10.7 V, which matched well with the theoretical value, with the variable input voltage V_{ri} ranging from 15 V to 20 V.

The microprocessor (MCU), which was used for prototyping the proposed wireless communication channel, malfunctioned and was never restored at the initial stage of the test when the accumulated dose was only 80 krad.

Due to the difficulties of on-line experiments, the following components were tested after the irradiation of 27 Mad. (1) The insulated gate bipolar transistor (IGBT) had no operational defects. (2) A commercial switched mode power supply (SMPS), which is UHE-15/2000-Q12-C from MURATA Co.,Ltd in this chapter, did not survive due to a control circuit failure. (3) A commercial Zigbee module, which was used for the proposed wireless communication, had a permanent defect after the test.

The test results are summarized in Table 29.3. It is noteworthy that the passive components and the semiconductor devices, having relatively large doping areas, tend to be robust under the high radiation environment, where every integrated-circuit-based device was disabled.

29.4 Conclusion

A highly reliable power and communication system for the essential equipment in NPPs has been designed with relevant physical reinforcements and has been experimentally verified in this chapter. Both the temperature and pressure profiles of the containment building for 72 hours after a severe accident were determined and firstly applied in NPP equipment designs. Wireless power and communication channels were proposed for increasing redundancy in conventionally installed wired channels, where both IPTS and Zigbee communications were adopted under extreme environments. Two major difficulties in the equipment design, high temperature of 627 °C, and high pressure of 5 bar were solved by adopting a thermal insulation box and a GFRP box. The performances of prototypes of each subsystem were fully verified using the dynamic temperature test with the fabricated high-temperature chamber. A γ-ray irradiation test of 30 Mrad was conducted for various power electronics devices to evaluate their use in a high radiation environment, which causes critical defects in most semiconductor devices. Due to its versatile applicability to currently installed equipment without any interference, the proposed wireless channels and protective boxes can be widely applied to both currently operating NPPs and future NPPs as practical responses to the Fukushima accident.

Problems

29.1 Estimate the impact of iron debris between the Tx and Rx coils under a severe accident in NPPs. Does it severely deteriorate IPT?

29.2 If the proposed IPTS is operating in the high temperature of 180 °C, what happens to the coil and capacitor?

References

 1 International Atomic Energy Agency (IAEA), "Severe accident management programmes for nuclear power plants," IAEA Safety Standards, No. NS-G-2.15, Austria, 2009.
 2 International Atomic Energy Agency (IAEA), "IAEA international fact findings expert mission of the Fukushima Daiichi NPP accident following the Great East Japan earthquake and tsunami," IAEA Mission Report, Austria, 2011.
 3 Electric Power Research Institute (EPRI), "EPRI Fukushima Daini independent review and walkdown," EPRI 2011 Technical Report, USA, 2011.
 4 Institute of Nuclear Power Operations (INPO), "Special report on the nuclear accident at the Fukushima Daiichi nuclear power station." INPO 11-005, USA, 2011.

5 T. Takeuchi *et al.*, "Development of instruments for improved safety measure for LWRs," in *5th International Symposium on Material Testing Reactors*, Bariloche, Argentina, October 28–31, 2012.

6 Government of Japan, Nuclear Emergency Response Headquarters, "Report of Japanese government to the IAEA ministerial conference on nuclear safety," Japan, 2011.

7 S.J. Yoo, B.H. Choi, S.Y. Jung, and Chun T. Rim, "Highly reliable power and communication system for essential instruments under a severe accident of NPPs," in *Transactions of the Korean Nuclear Society Autumn Meeting*, Gyeongju, South Korea, October 23–25, 2013.

8 S.J. Yoo, B.W. Gu, B.H. Choi, S.I. Lee, and Chun T. Rim, "Development of highly survivable power and communication system for NPP instruments under severe accident," in *Transactions of the Korean Nuclear Society Autumn Meeting*, Pyeongchang, South Korea, October 30–31, 2014.

9 S.I. Lee, H.K. Jung, "Development of the NPP instruments for highly survivability under severe accidents," in *Transactions of the Korea Society for Energy Engineering Autumn Meeting*, Jeju, South Korea, November 21–22, 2013.

10 S.Y. Jung, B.H. Choi, S.J. Yoo, B.W. Gu, and Chun T. Rim, "A study on the application of wireless power transfer technologies in metal shielding spaces," in *Transactions of the Korean Institute of Power Electronics Annual Meeting*, Gyeongju, South Korea, July 2–5, 2013.

11 B.H. Choi, E.S. Lee, J.H. Kim, and Chun T. Rim, "7 m-off-long-distance extremely loosely coupled inductive power transfer systems using dipole coils," in *IEEE Energy Conversion Congress and Expo*, Pittsburgh, PA, September 14–18, 2014.

12 Electric Power Research Institute (EPRI), "Large-scale hydrogen burn equipment experiments," EPRI NP-4354s, USA, 1985.

13 M. Yoo, S.M. Shin, and H.G. Kang, "Development of instrument transmitter protecting device against high-temperature condition during severe accidents," *Science Technol. Nucl. Inst.*, pp. 1–8, 2014.

14 Office of Nuclear Regulatory Research, Regulatory Guide 1.180, US Nuclear Regulatory Commission, USA, 2003.

15 M.V. O'Bryan *et al.*, "Compendium of recent single event effects for candidate spacecraft electronics for NASA," in *IEEE Nuclear and Space Radiation Effects Conference*, San Francisco, CA, July 8–12, 2013.

30

The Future of Wireless Power

30.1 Future Areas of WPT

It is always difficult to predict the future with confidence because of the unpredictable nature of future itself. However, this is one of the purposes of prediction, that is, to prepare for the unpredictable future by surveying all possibilities that we can imagine. Keeping in mind this limitation, we can talk about the future of WPT.

In the near future, we will see many diversified applications of WPT in many new fields. These new areas include various mobile devices, home appliances, electrified new transportations (electric trains, electric ships, electric airplanes), drones, robots, industry automation, military equipment, security/monitor sensors, logistics, electric shelf labels (ESLs), medical devices, toys, tools, and Internet of things (IoT), which will embrace most moving things. If I pick up a few fascinating examples, WPT table platform for office devices, WPT TVs, and WPT ESLs have promising markets in the near future.

Of course smart phones and electric vehicles will remain the greatest areas for a few years, which became most popular WPT areas since 2015. More and more smart phones are adopting WPT and the power level of WPT is ever increasing from 5 W to 10 W and even 20 W for quick charging. For a higher power level, the thermal problem is of great concern because of the increased power loss in WPT. Electric vehicle (EV) charging will also be very important in the immediate future. Stationary charging will be much more widely used, where cheap, low weight, compact, quick charging, wide tolerances in airgap and lateral displacement will be major technical issues. Dynamic charging will also be more widely adopted, but interoperability may be of great concern in the future.

30.2 Competing Technologies in Future WPT

A few cases of competing technologies are comparatively assessed for futuristic trend estimation.

30.2.1 IPT vesus CPT

Though inductive power transfer (IPT) is overwhelmingly widely used now, capacitive power transfer (CPT) is finding its own application areas. Fundamentally, CPT is not good for long distance or large airgap applications because of drastically reduced capacitive current, which results in lower power. Switching frequency of MHz is quite often used in CPT to increase the capacitive current, which usually results in lower efficiency.

Wireless Power Transfer for Electric Vehicles and Mobile Devices, First Edition. Chun T. Rim and Chris Mi.
© 2017 John Wiley & Sons Ltd. Published 2017 by John Wiley & Sons Ltd.

Despite these limitations, CPT needs no core and makes it possible to fabricate compactly in size. If properly designed, CPT may have over 80% efficiency for a few W to kW power level; this is because there is no core loss and the virtue of a higher switching power device such as SiC and GaN.

IPT will play a major role in WPT; however, in the future CPT will draw more attention because of its own merits.

30.2.2 Coreless versus With Core

It is often postulated that to use a core is requisite for IPT because it significantly helps to focus on the magnetic field and to shield the unwanted electromagnetic field (EMF). This postulation, however, ought to have been questioned a long time ago. It is true that the core is a good material for reducing magnetic resistance like the conductor for reducing electric resistance. Thus, the core can change magnetic path as we want, which results in higher power transfer with high efficiency if properly designed.

I had been curious about the role of the core in IPT. For example, let us think of a transmitting (Tx) coil as composed of winding cables placed on a core plate. The magnetic field generated from the Tx coil is at most twice the Tx coils without a core according to magnetic mirror theory. In practice the magnetic flux density increase due to the core is typically only 50%~60%, which is too small considering the penalty of core loss, increased volume, and cost due to core use. Of course, the use of a core is essential when a plate of receiving (Rx) coil is displaced close to the Tx coil, where the cores of Tx and Rx coils constitute a closed magnetic path. Under this condition, we can get high power transfer as well as lower EMF, which is the typical application case of IPT.

Recall the coupled magnetic resonance system (CMRS), where usually no core can be used because of MHz operation. With a virtue of an air coil, which means no core is used, high-frequency operation is possible and a relatively higher induced voltage is achieved, which results in higher power transfer with reduced output current. CMRS was once believed to be only a viable solution for a long-distance power transfer, which has turned out to be not true by myself. The merits of CMRS such as high-frequency operation and long-distance power transfer can be explained by the air coil. The use of four coils and an extremely high quality factor (Q) are demerits of CMRS. Moreover, CMRS is nothing more than a special form of an IPT system, where resonant coils, that is, Tesla coils, may be successively used. The key characteristics of CMRS come from the air coil characteristics.

Therefore, it can be said that a coreless coil, that is, an air coil, can be used for a high-frequency long-distance WPT where no appropriate core is available now and the role of the core is not significant. Moreover, the use of a coreless coil has a unique characteristic that there is no resonance frequency change due to Tx and Rx magnetic coupling, which is completely different from conventional IPT with a core. It is a headache when designing IPT to deal with inductance change due to airgap and misalignment changes, but the air coil has no such problem. This good characteristic of the air coil is misleadingly known to be that of CMRS.

To summarize, more study must be paid to coreless applications in IPT, which is not necessarily concerned with CMRS.

30.2.3 kHz versus MHz

So far a few kHz to a few hundred kHz operating frequency, which is usually the same as the switching frequency of converters in WPT, has been widely used. However, MHz frequency is preferably used in some special applications. MHz operation is not often used because of the many restrictions such as limited allowed frequencies, lower power efficiency, and increased electromagnetic emission. Because unwanted RF radiation from an IPT system increases at the MHz range and increased use of that frequency for radio communications, only 6.78 MHz and 13.56 MHz are widely used now for WPT. Moreover, even modern high-frequency power switches cannot effectively cope with MHz converters with high efficiency. Nevertheless, despite the drawbacks of MHz operation, a higher induced voltage or capacitive current is achieved in MHz range for WPT.

If an appropriate core material is available in the future, MHz operated WPT will be more common. Especially the following three conditions for a core are crucial for IPT in the future:

1) High-frequency characteristics (> 1 MHz)
2) High permeability (>1000)
3) High magnetic flux density (>1 T)

For an advance core to be available now, one or two of the conditions above should be met. A breakthrough must be made in the future for widespread use of IPT. The second and third conditions can be mitigated to 100 and 0.5 T, if necessary. It would deserve a Nobel Prize if any engineer or scientist finds such an innovative magnetic material, the "magic core."

30.2.4   μW to MW

Usually, the power level of WPT ranges at present from sub W to several hundred kW. However, this will be extended much more in both directions.

First, much lower power will be needed due to increased distributed network sensors, that is, IoT sensors. Very low power of μW to mW applications will be abundant in the near future. Actually, energy harvesting is one such area that could be used to get a very low energy or power source; however, it is quite unreliable and too small to get enough power from the environment. For example, sunlight or artificial light could be a power source of such low power applications, but the light source is not available when it becomes dark or shaded. Thermoelectric power generation is also unreliable when the temperature difference between hot and cold points eventually disappears. In the future when IoT may dominate worldwide use in many areas, we need a reliable power source that does not rely on fortune or probability. WPT can be a good candidate for the low power of μW to mW applications, where a long-range power transfer even up to several tens of meters is sometimes required.

Second, much higher power will also be needed due to increased power level applications. One of them is train transportation, where 1 MW to 20 MW of power is needed. Several MW of power have been used in induction heating and high-frequency welding, which can technically be easily extended to WPT.

30.2.5   Burst Static Charging versus Continuous Dynamic Charging

For smartphone or EV charging, we use a static charger, where quick-time charging is preferred, but the road-powered electric vehicle (RPEV) gets power from roadway power rails, where continuous charging or powering is preferred. Each WPT method has merits and demerits; the former has freedom of range but the latter has freedom of time, and vice versa. The former relies on an energy storage element (battery) whereas the latter relies on roadway infrastructure, but both of them are expensive and often bulky. The detailed cost and convenience will determine specific application areas or will compete with each other.

30.2.6 Stationary WPT versus Ubiquitous WPT

Like airplanes and robots, degrees-of-freedom (DoFs) becomes important in WPT. The omni-directional WPT concept has been clearly proposed by Ron Hui [1] in his paper. This concept is quite useful in the future application of WPT where customers do not want to be bound to time and range anxieties. This omnidirectional WPT concept can be more generalized to 6 DoF WPT, where the 3-D position and 3-D direction freedom of wireless power transfer is guaranteed. Think about the recently commercialized Galaxy smartphone, where stationary IPT is adopted, which is only 1 DoF; thus a 6 DoF WPT is truly a ubiquitous WPT, where wireless power is available at any time at any place.

30.2.7 Loop Coil versus Dipole Coil

Like the loop antenna and dipole antenna, the loop coil and dipole coil are now used in WPT. What are fundamental differences between the loop coil and dipole coil? The dipole antenna is predominantly used now because of its compactness compared to the loop antenna. In other words, the dipole antenna is of a line whereas the loop antenna is of a circle, so the dimension of the dipole antenna is less than that of the loop antenna. This dimensionless feature is preserved for the dipole coil.

For a longer distance power delivery, either a large diameter of loop coil or a long length of dipole coil is needed even though the dipole coil has a little better magnetic coupling compared to the loop coil. Especially for the 6 DoF applications, the two coils should orthogonally displace each other, where only the dipole coil can be a flat-type coil structure because of its dimensionless property.

One demerit of the dipole coil is that the core must be used for the dipole whereas the loop coil may not need a core. Therefore, there will be no single answer remaining in the future, fortunately.

30.2.8 Active Tuning versus Detuning

A very cumbersome problem in WPT is the resonant frequency change due to airgap and displacement changes, as discussed above. A remedy for it is to use an active tuning method by adopting variable capacitance or variable inductance.

Another method is to change the operating frequency to maximize output voltage, power, or efficiency. Without tuning, wireless power delivery usually decreases so drastically that in practice only a few % of power can be obtained.

On the other hand, a resonant circuit is designed to be detuned so that power delivery is insensitive to the frequency change; of course, this merit is the result of technical penalties such as a lower power level and lower efficiency.

30.2.9 WPT versus Battery

What is the relationship between WPT and a battery? Are they a good alliance or enemy? As discussed above, they are competitors in dynamic charging; however, they are a good alliance in static charging. More EVs will be used if convenient WPT becomes available, and vice versa. Even in dynamic charging, a good battery of less price and weight is welcome for RPEVs because they also need batteries for emergency energy use and a high power source.

In the future, the price, capacity, weight, size, lifetime, charging time, efficiency, and reliability of a battery will definitely be improved. Roughly speaking, price and other major performances improve twice every 5–10 years. Therefore, a real breakthrough for the battery is expected within 10–20 years, and engineers and scientists should be prepared for this change in the same way as the "1$ CPU," which opened the digital era.

What will be the corresponding breakthrough in WPT?

30.2.10   EVs versus ICEs

What will be between EV versus the internal combustion engine (ICE)? If an innovative battery and WPT were to be available in the future, will ICE cars disappear? Probably yes, if this trend goes for a few decades. However, it seems not likely to happen in the near future. As discussed above, the limitations of EVs are in its operation time and range. Compared to ICEs, there are much less charging stations or charging infrastructures needed for EVs. Refuel time for ICEs is about 2 minutes, but recharging time for EVs is at most 20 minutes. Even though the recharging time has been drastically reduced to 2 minutes, now a 10 to 60 times larger power capacity of WPT and reduced lifetime of the battery with poor efficiency are problems. The interoperable roadway infrastructure of WPT can mitigate these problems and may accelerate the time to reach the breakeven point of EVs and ICEs.

30.2.11   Optical PT versus RF PT

Even though they have not been much addressed in this book, optical power transfer (PT) and RF PT will find more applications than ever because they are also viable solutions to WPT. Optical PT, especially infrared PT, may be useful for indoor applications where even weak EMF is not allowed, such as sensitive medical environments. RF PT may be very useful for outdoor applications where continuous and long-distance power is required. As discussed in the first part of the book, stratosphere drones and electric airplanes for short distance carrier are good candidates for RF PT.

30.2.12   Linked versus Linkless Control

In WPT, the information of output voltage and load status of Rx is usually transmitted to Tx so that the frequency or current level is appropriately controlled, which can be called "linked control." Although the information may not be needed for Tx, the information can instead be estimated from the currents or voltages on the Tx side, which can be called "linkless control." If working, linkless control has many merits such as simple and low cost; however, this method inherently lacks robustness against abnormal situations such as metal objects between Tx and Rx and malfunctions of Rx.

30.3   Conclusions

We cannot tell which technology will dominate in the future, but we can see a variety of possibilities and therefore can prepare for the worst and the best future scenarios. Definitely, the application of WPT will increase in the future. Especially, the speed of applications of WPT will be accelerated in the near future.

Of course, new obstacles and challenging problems will be in future WPT applications as "the devil is in the detail" throughout human history. This is why you are here; those who understand and overcome "the devil" will open the future. Many innovations are expected to be made in the next decade and I hope that the readers of this book become the leaders of WPT.

Reference

1 W.M. Ng, C. Zhang, D. Lin, and S.Y. Ron Hui, Two- and three-dimensional omnidirectional wireless power transfer, *IEEE Trans. on Power Electronics*, vol. 29, no. 9, pp. 4470–4474, September 2014.

Index

Note: Page numbers followed by "F" indicate figures.

Wireless Power Transfer for Electric Vehicles and Mobile Devices, First Edition. Chun T. Rim and Chris Mi.
© 2017 John Wiley & Sons Ltd. Published 2017 by John Wiley & Sons Ltd.

Printed and bound by CPI Group (UK) Ltd, Croydon, CR0 4YY